Nanoengineered Nanofibrous Materials

NATO Science Series

A Series presenting the results of scientific meetings supported under the NATO Science Programme.

The Series is published by IOS Press, Amsterdam, and Kluwer Academic Publishers in conjunction with the NATO Scientific Affairs Division

Sub-Series

I. Life and Behavioural Sciences	IOS Press
II. Mathematics, Physics and Chemistry	Kluwer Academic Publishers
III. Computer and Systems Science	IOS Press
IV. Earth and Environmental Sciences	Kluwer Academic Publishers
V. Science and Technology Policy	IOS Press

The NATO Science Series continues the series of books published formerly as the NATO ASI Series.

The NATO Science Programme offers support for collaboration in civil science between scientists (countries of the Euro-Atlantic Partnership Council. The types of scientific meeting generally supporte are "Advanced Study Institutes" and "Advanced Research Workshops", although other types (meeting are supported from time to time. The NATO Science Series collects together the results (these meetings. The meetings are co-organized bij scientists from NATO countries and scientists fror NATO's Partner countries – countries of the CIS and Central and Eastern Europe.

Advanced Study Institutes are high-level tutorial courses offering in-depth study of latest advance in a field.
Advanced Research Workshops are expert meetings aimed at critical assessment of a field, an identification of directions for future action.

As a consequence of the restructuring of the NATO Science Programme in 1999, the NATO Scienc Series has been re-organised and there are currently Five Sub-series as noted above. Please consu the following web sites for information on previous volumes published in the Series, as well as details (earlier Sub-series.

http://www.nato.int/science
http://www.wkap.nl
http://www.iospress.nl
http://www.wtv-books.de/nato-pco.htm

Series II: Mathematics, Physics and Chemistry – Vol. 169

Nanoengineered Nanofibrous Materials

edited by

Selcuk Guceri

Drexel University,
Philadelphia, PA, U.S.A.

Yuri G. Gogotsi

Drexel University,
Philadelphia, PA, U.S.A.

and

Vladimir Kuznetsov

Boreskov Institute of Catalysis,
Novosibirsk, Russia

Assistant Editor:

Jennifer Wright

Springer-Science+Business Media, B.V.

Proceedings of the NATO Advanced Study Institute on
Nanoengineered Nanofibrous Materials
Belek-Antalya, Turkey
1–12 September 2003

A C.I.P. Catalogue record for this book is available from the Library of Congress.

ISBN 978-94-017-5120-9 ISBN 978-1-4020-2550-1 (eBook)
DOI 10.1007/978-1-4020-2550-1

Printed on acid-free paper

Contents

Chapter 2. Physics and Chemistry of Nanofibers

Chapter 3. Simulation and Modeling

Chapter 4. Applications

4.1 Biomedical Applications

Chapter 5. Nanomaterials, Nanoparticles, and Nanostructures

Preface

The recent intensity in research world-wide on nanostructured materials has evidenced their potential use and impact in a variety of fields, such as electronics, sensors, biological sciences, computer and information technology. Polymeric fibrous materials at the nanoscale are the fundamental building blocks of living systems. From the 1.5 nm double helix strand of DNA molecules, including cytoskeleton filaments with diameters around 30 nm, to sensory cells such as hair cells and rod cells of the eyes, nanoscale fibers form the extracellular matrices for tissues and various organs. Specific junctions between these cells conduct electrical and chemical signals that result from various kinds of stimulation. The signals direct normal functions of the cells such as energy storage, information storage, retrieval and exchange, tissue regeneration, and sensing. Based upon these "blueprints" laid out by nature, it is reasonable to infer that the availability of nanoscale (less than 100 nm diameter) fibers made of carbon (nanotubes) or polymers having adjustable electrical conductivity will open new opportunities in science and technology. Nanofibers of conducting polymers and their composites, including those containing nanotubes, are the fundamental building blocks for the construction of devices and structures that perform unique new functions that can lead to new "enabling" technologies. The combination of conductive polymer technology with the ability to produce nanofibers puts us in a position to introduce important new capabilities in the rapidly growing field of biotechnology and information technology. Areas that benefit include: scaffolds for tissue engineering and drug delivery systems based on nanofibers; wires, capacitors, transistors and diodes for information technology; sensor technology; biohazard protection and systems for energy transport, conversion and storage, such as batteries and fuel cells.

Recognizing these developments, a group of scientists ranging from undergraduate students to senior researchers gathered in Antalya, Turkey from September 1-12, 2003 under the sponsorship of the NATO Advanced Study Institute on Nanoengineered Nanofibrous Materials to provide an advanced teaching/learning platform as well as to further the discussions and development of research in this emerging field. The ASI served to disseminate state of the art knowledge related to fundamentals and recent advances in nanofibrous materials for biomedical, electronic, power and air filtration applications. In particular, the characterization and fabrication of fibrous composite materials and nanotubular materials were discussed. Current research, covering a wide range of nanosized materials, their physical and chemical properties, as well as recent achievements in this field, were discussed and outlines for future directions in terms of technological developments and product commercialization in such fields as electronics, personal protection, biomedicine and sensors were given. Participants became familiar with the most recent developments in nanostructured fibrous materials and their potential use. The ASI brought together scientists from basic and applied research areas from both NATO and partner countries to initiate further interactions aimed at translating basic research achievements into engineering applications.

A total of 87 scientists from 14 different countries participated in our ASI, making it a truly international event. In all, 22 tutorial lectures, 30 short talks and over 40 posters were presented. A broad range of speakers from universities and industrial and government research laboratories from around the globe participated in this meeting. These proceedings reflect their insights in the area of nanoengineered nanofibrous materials.

This volume is complementary to various specialized books or more generalized books on nanomaterials and/or nanotechnology. It aims to present an overview of research activities in nanofibrous materials. The volume has been organized into five chapters corresponding to the following objectives: the **first chapter** is designed to instruct young scientists in the most advances methods on the *Formation of Nanofibers and Nanotubes Production*; the **second chapter** presents information on the P*hysics and Chemistry of Nanofibers*; the **third chapter**, due to perspectives of computation approaches, concerns the *Simulation and Modeling* of nanosystems and nanoobjects; the **fourth chapter** guides young researchers in applications of materials in areas of *Biomedical Applications*, *Nanotube-Based Devices, Electronic Applications of Nanotubes and Nanofibers*, *Nanofluidics and Composites*; and the fifth chapter deals with recent developments in *Nanomaterials, Nanoparticles, and Nanostructures*. All papers in this book have been peer-reviewed prior to publication. We believe this volume will be of major interest to researchers and students working in the area of materials science and engineering, nanotechnology, biomaterials, and sensors.

The contribution by Dr. Salim Çıracı of the organizing committee in making the site arrangements is gratefully acknowledged. We would like to recognize the dedicated, hard work of Jennifer Wright. She single-handedly managed the intense pre-conference activities, helped to ensure this ASI was a successful event and served as the assistant editor of this volume. Without her contributions, both the ASI and this book would not have been possible. We would like to recognize Nicole Porreca, who developed and fine tuned our successful proposal to NATO, Katrin Cowan, who helped in making with ASI a successful, smooth running event and YAPSIAL Corporation and the efforts of Selen Önal, who printed and paid for our conference program. Finally, we express our sincere gratitude to the NATO Science Committee for granting us the award that enabled us to both arrange this meeting and to publish the proceedings. Additional financial support for the meeting was provided by A.J. Drexel Nanotechnology Institute at Drexel University and by the U.S. National Science Foundation.

Selçuk Güçeri | Vladimir Kuznetsov | Yury Gogotsi

NANOFIBER TECHNOLOGY:
Bridging the Gap between Nano and Macro World

Frank K. Ko
Fibrous Materials Research Laboratory, Department of Materials Science and Engineering, Drexel University, Philadelphia, Pa. 19104, U.S.A

1. INTRODUCTION

Nanofibers are solid state linear nanomaterials characterized by flexibility and an aspect ratio greater than 1000:1. According to the National Science Foundation (NSF), nanomaterials are matters that have at least one dimension equal to or less than 100 nanometers[1]. Therefore, nanofibers are fibers that have diameter equal to or less than 100 nm. Materials in fiber form are of great practical and fundamental importance. The combination of high specific surface area, flexibility and superior directional strength makes fiber a preferred material form for many applications ranging from clothing to reinforcements for aerospace structures. Fibrous materials in nanometer scale are the fundamental building blocks of living systems. From the 1.5 nm double helix strand of DNA molecules, including cytoskeleton filaments with diameters around 30 nm, to sensory cells such as hair cells and rod cells of the eyes, nanoscale fibers form the extra-cellular matrices or the multifunctional structural backbone for tissues and organs. Specific junctions between these cells conduct electrical and chemical signals that result from various kinds of stimulation. The signals direct normal functions of the cells such as energy storage, information storage and retrieval, tissue regeneration, and sensing.

Analogous to nature's design, nanofibers of electronic polymers and their composites can provide fundamental building blocks for the construction of devices and structures that perform unique new functions that serve the needs of mankind. Other areas expected to be impacted by the nanofiber based technology include drug delivery systems and scaffolds for tissue engineering, wires, capacitors, transistors and diodes for information technology, systems for energy transport, conversion and storage, such as batteries and fuel cells, and structural composites for aerospace structures.

Considering the potential opportunities provided by nanofibers there is an increasing interest in nanofiber technology. Amongst the technologies

S. Guceri et al. (eds.), Nanoengineered Nanofibrous Materials, 1-18.

including the template method [2], vapor grown [3], phase separation [4] and electrospinning [5-22], electrospinning has attracted the most recent interest. Using the electrospinning process, Reneker and co-workers [5] demonstrated the ability to fabricate nanofibers of organic polymers with diameters as small as 3 nm. These molecular bundles, self-assembled by electrospinning, have only 6 or 7 molecules across the diameter of the fiber! Half of the 40 or so parallel molecules in the fiber are on the surface. Collaborative research in MacDiarmid and Ko's laboratory [6,9] demonstrated that blends of nonconductive polymers with conductive polyaniline polymers and nanofibers of pure conductive polymers can be electrospun. Additionally, in situ methods can be used to deposit 25 nm thick films of other conducting polymers, such as polypyrrole or polyaniline, on preformed insulating nanofibers. Carbon nanotubes, nanoplatelets and ceramic nanoparticles may also be dispersed in polymer solutions, which are then electrospun to form composites in the form of continuous nanofibers and nanofibrous assemblies. [7] Specifically, the role of fiber size has been recognized in significant increase in surface area; in bio-reactivity; electronic properties; and in mechanical properties:.

1.1 Effect of Fiber Size on Surface Area

One of most significant characteristic of nanofibers is the enormous availability of surface area per unit mass. For fibers having diameters from 5 to 500 nanometers, as shown in Figure 1, the surface area per unit mass is around 10, 000 to 1,000,000 square meters per kilogram. In nanofibers that are three nanometers in diameter, and which contain about 40 molecules; about half of the molecules are on the surface. As seen in Figure 1, the high surface area of nanofibers provides a remarkable capacity for the attachment or release of functional groups, absorbed molecules, ions, catalytic moieties and nanometer scale particles of many kinds.

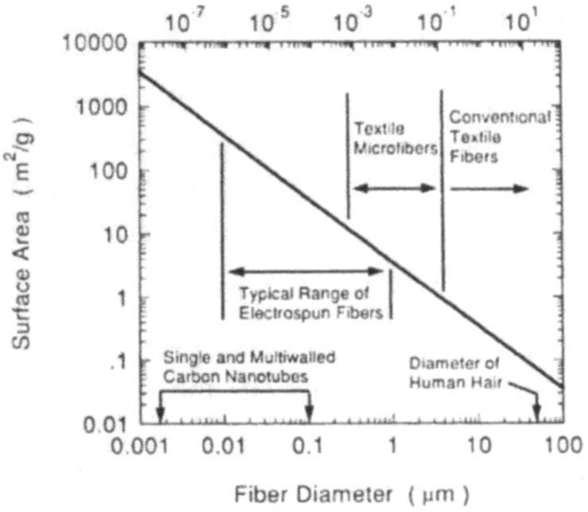

Figure 1: Effect of fiber diameter on surface area

1.2 Effect of Fiber Size on Bioactivity

Considering the importance of surfaces for cell adhesion and migration experiments were carried out in the Fibrous Materials Laboratory at Drexel University using osteoblasts isolated from neonatal rat calvarias and grown to confluence in Ham's F-12 medium (GIBCO), supplemented with 12% Sigma fetal bovine on PLAGA sintered spheres, 3-D braided filament bundles and nanofibrils. [8]. Four matrices were fabricated for the cell culture experiments. These matrices include 1) 150 - 300 μm PLAGA sintered spheres 2) Unidirectional bundles of 20 μm filaments 3) 3-D braided structure consisting of 20 bundles of 20 μm filaments 4) Nonwoven consisting of nanofibrils.. The most proliferate cell growth was observed for the nanofibrils scaffold as shown in the Thymadine-time relationship illustrated in Figure 2. This can be attributed to the greater available surfaces for cell adhesion as a result of the small fiber diameter which facilitates cell attachment.

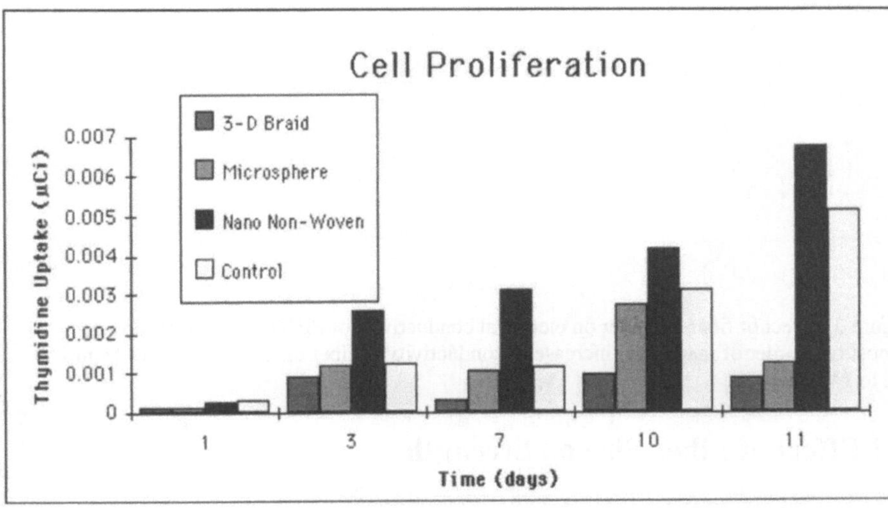

Figure 2. Fibroblast cell proliferation as indicated by the Thymidine uptake of cell as a function of time showing that Polylactic-glycolic acid nanofiber scaffold is most favorable for cell growth.

1.3 Effect of Fiber Size on Electroactivity

The size of conductive fiber has an important effect on system response time to electronic stimuli and the current carrying capability of the fiber over metal contacts. In a doping-de-doping experiment, Norris et al [9] found that polyaniline/PEO sub-micron fibrils had response time an order of magnitude faster than that of bulk polyaniline/PEO. There are three types of contact to a nano polymeric wire exist: Ohmic, Rectifying, and Tunneling. Each is modified due to nano effects. There exist critical diameters for wires below which metal contact produces much higher barrier heights thus showing much better rectification properties. According to Nabet [23], by reducing the size of a wire we can expect to simultaneously achieve better rectification properties as well as better transport in a nano wire. In a preliminary study [24], as shown in Figure 3 it was demonstrated, using sub-micron PEDT conductive fiber mat, that significant increase in conductivity was observed as the fiber diameter decreases. This could be attributed to intrinsic fiber conductivity effect or the geometric surface and packing density effect or both as a result of the reduction in fiber diameter.

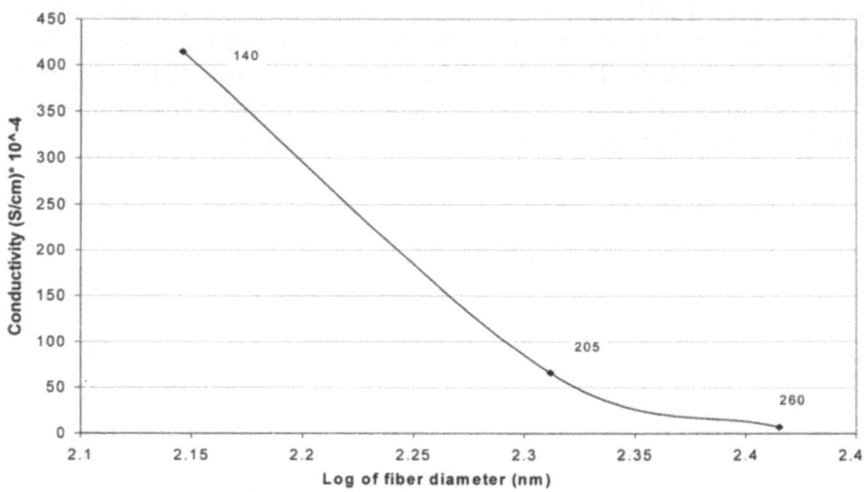

Figure 3. Effect of fiber diameter on electrical conductivity of PEDT nanofibers showing almost two order of magnitude increase in conductivity as fiber diameter decreases from 260 nm to 140 nm.

1.4 Effect of Fiber Size on Strength

Materials in fiber form are unique in that they are stronger than bulk materials. As fiber diameter decreases, it has been well established in glass fiber science that the strength of the fiber increases exponentially, as shown in Figure 4a., due to the reduction of the probability of including flaws. As the diameter of matter gets even smaller as in the case of nanotubes, Figure 4b., the strain energy per atom increases exponentially, contributing to the enormous strength of over 30 GPa for carbon nanotube.

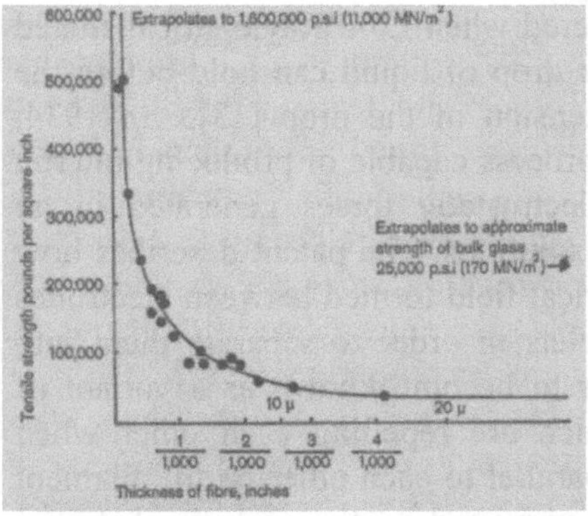

Figure 4a. Dependence of glass fiber strength on fiber diameter [25]

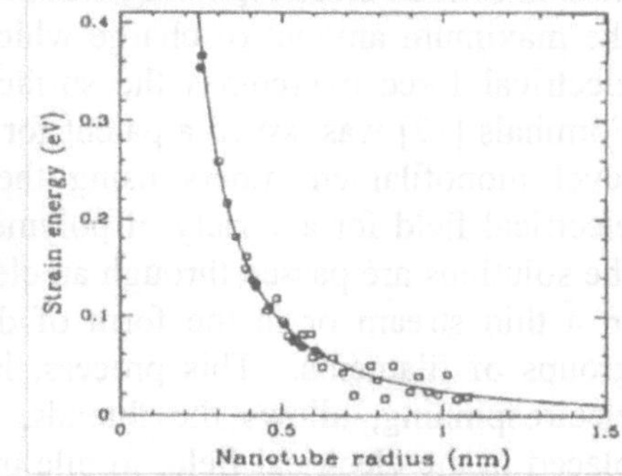

Figure 4b. Strain energy as a function of nanotube diameter [26]

Although the effect of fiber diameter on the performance and processibility of fibrous structures have long been recognized the practical generation of fibers down to the nanometer scale was not realized until the rediscovery and popularization of the electrospinning technology by Professor Darrell Reneker almost a decade ago [10]. The ability to create nanoscale fibers from a broad range of polymeric materials in a relatively simple manner using the electrospinning process coupled with the rapid growth of nanotechnology in the recent years have greatly accelerated the growth of nanofiber technology. Although there are several alternate methods for generating fibers in a nanometer scale, none matches the popularity of the electrospinning technology due largely to the great simplicity of the electrospinning process. In this paper we will be focused on the electrospinning technology. The relative importance of the various processing parameters in solution electrospinning is discussed. The structure and properties of the fibers produced by the electrospinning process are then examined with particular attention to the mechanical and chemical properties. There is a gradual recognition that the deceivingly simple process of electrospinning requires a deeper scientific understanding and engineering development in order to capitalize on the benefits promised by the attainment of nanoscale and translate the technology from a laboratory curiosity to a robust manufacturing process. To illustrate the method to connect properties of materials in the nano-scale to macro-structures, the approach of multi-scale modeling and a concept for the translation of carbon nanotube to composite fibrous assemblies is presented.

2. ELECTROSPINNING OF NANOFIBERS

The technology of electrostatic spinning or electrospinning may be traced back to 1745 when Bose created an aerosol spray by applying a high potential to a liquid at the end of a glass capillary tube. The principle behind what is now known as electrospinning was furthered when Lord Rayleigh calculated the maximum amount of charge which a drop of liquid can hold before the electrical force overcomes the surface tension of the drop [13]. In 1934, Formhals [12] was issued a patent for a process capable of producing micron level monofilament fibers using the electrostatic forces generated in an electrical field for a variety of polymer solutions. The patent describes how the solutions are passed through an electrical field formed between electrodes in a thin stream or in the form of droplets in order to separate them into groups of filaments. This process, later to become known as a variant of electrospinning, allows the threads, which are repelling each other when placed in the electrical field, to pile up parallel to each other on the filament collector in such a way that they can be unwound continuously in skeins or ropes of any desired length.

The operational principle of electrospinning is quite simple. In this non-mechanical, electrostatic technique, a high electric field is generated between a polymer fluid contained in a spinning dope reservoir with a capillary tip or a spinneret and a metallic fiber collection ground surface. When the voltage reaches a critical value, the charge overcomes the surface tension of the deformed drop of the suspended polymer solution formed on the tip of the spinneret, and a jet is produced. The electrically charged jet undergoes a series of electrically induced bending instabilities during its passage to the collection surface that results in the hyper-stretching of the jet. This stretching process is accompanied by the rapid evaporation of the solvent molecules that reduces the diameter of the jet, in a cone-shaped volume called the ''envelope cone''. The dry fibers are accumulated on the surface of the collection plate resulting in a non-woven mesh of nano to micron diameter fibers. The process can be adjusted to control the fiber diameter by varying the electric field strength and polymer solution concentration, whereas the duration of electrospinning controls the thickness of fiber deposition [12]. A schematic drawing of the electrospinning process is shown in Figure 5.

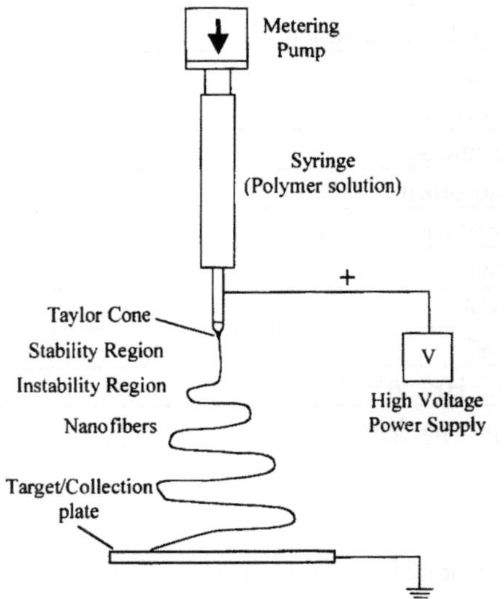

Figure 5. Schematic drawing of the electrospinning process showing the formation of nanofibers under the influence of an electronic field.

Numerous polymers have been electrospun by an increasing number of researchers around the world. Examples of some of the polymers that have been successfully spun are shown in Table 1. Solvents of varying pH, polymers with molecular weights ranging from 10,000 to 300,000 and higher have been electrospun. In our laboratory, we have employed a dimensionless parameter, the Berry Number (Be), to guide our electrospinning process by relating fiber diameter to the Be, which is an indication of molecular conformation.[27-30] The Be is the product of the intrinsic viscosity and polymer concentration. Be has been found to play a major role in spinnability and fiber diameter control. For example, using polylactic acid (PLA), it was found that if the Be is lower then 1, bead formation will occur due to the low level of viscosity does not favor fiber formation during the electrospinning process. On the other hand, for Be greater than 1, fiber diameters steadily increase as shown in Table 2.

Table 1: Examples of polymers that have been electrospun

Polymer	Solvent
Rayon	Caustic soda
acrylic resin	dimethyl formamide
PE, PPE	melted
PEO	water
PPTA (Kevlar)	98% sulfuric acid
polyester	1:1 dichloromethane:trifluoroacetic acid
DNA	70:30 water:ethanol
PAN and Pitch	dimethyl formamide (DMF)
styrene-butadiene-styrene triblock	75:25 tetrahydrofuran/dimethylformamide

PAN and Pitch	dimethyl formamide
PVDF	DMF
PLA	chloroform
Bombyx mori silk, spider silk	Formic acid
PANi/ PEO	chloroform
HEMA	50:50 formic acid/ethanol
CNT/PAN	DMF
GNP/PAN	DMF
Polyurathane	DMF
Nylon-6	95:5 HFIP/DMF
Poly(ethylene-co-vinyl alcohol)	2-propanol/water

Table 2. Polymer chain conformation and fiber morphology corresponding to four regions of Berry number

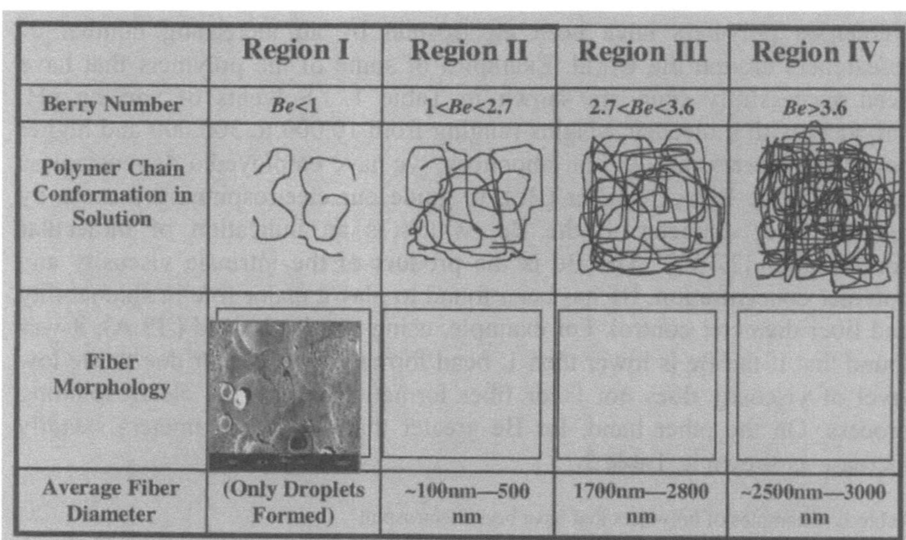

	Region I	Region II	Region III	Region IV
Berry Number	$Be<1$	$1<Be<2.7$	$2.7<Be<3.6$	$Be>3.6$
Polymer Chain Conformation in Solution				
Fiber Morphology				
Average Fiber Diameter	(Only Droplets Formed)	~100nm—500 nm	1700nm—2800 nm	~2500nm—3000 nm

3. STRUCTURE AND PROPERTIES

The characterization of nanofiber requires a multitude of tools. Figure 6 shows the spectrum of methodologies and instruments suitable for the characterization of nanofibers Scanning electron microscope (SEM) is one of the most popular that has been used as reported widely in the literature [5,10,16,31] to measure the diameter of electrospun fibers as well as the general morphology. Kim and Reneker [11] showed that SEM could also be used to examine the fibers cross-section.

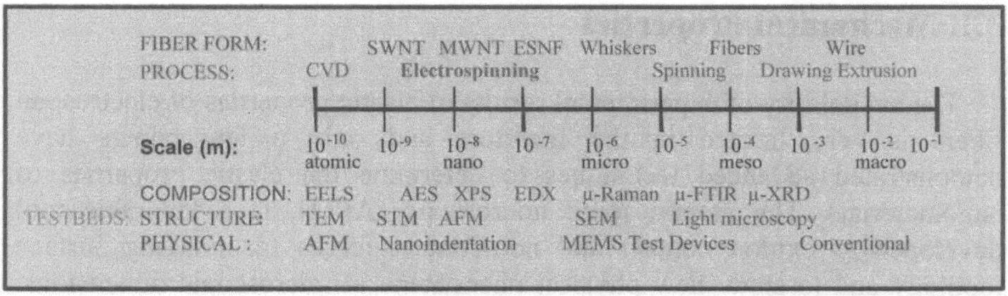

Figure 6. Corresponding scale of processing, testing and analysis techniques for electrospun nanofibers

They stated that the electrospun fibers are mostly circular in cross-section but other shapes such as ribbons and coils could be produced. Another methodology that has been utilized to measure the diameter of smaller diameter fibers is transmission electron microscopy (TEM). Fang and Reneker [31] analyzed the electrospun DNA by using TEM and reported diameters as low as 62 nm. Ko et al. [32] used the TEM to study the morphology and structural order for the electrospun PAN fiber, PAN/CNT and PAN/GNP nanocomposites. Ye et al. [33] used TEM technique to study the alignment of SWNT in two different electrospun polymers (PLA and PAN), as well as study the effect of the heat treatment of electro-spun PAN/CNT composites on the alignment of the CNT. Other researchers have utilized equipment such as x-ray diffraction (XRD) and differential scanning Calorimetry (DSC) to observe whether there is any degree of crystallinity in the electrospun samples. When spinning from solution, there have been some reports by Deitzel [22] that limited crystallization occurs but not to a high degree using PEO.

Although most of the fibers that are produced though electrospinning are circular solid filaments ; there are occasions when tubes, ribbons, coils, and beaded structures can be produced. Tubes have been discovered by several groups and it was shown in our laboratory that there was a high likelihood of forming tubes when electrospinning into water. In a similar fashion, ribbons are more than likely the result of collapsed tubes.

The most common form of collecting the nanofibers is in the form of 2-D fibrous nonwoven mats. Mats such as these are well suited for filtration and tissue culturing. Alternatively, single fibers or linear fiber assemblies (yarns) can be produced. It has been shown by Ko [34] that nanofibrous yarns, can be directly produced from the electrospinning process under well controlled processing conditions for some polymers. These yarns tended to be on the micron scale but were composed of numerous fibers that were on the nano scale that were naturally twisted together throughout the process to form the yarn. This will pave the way to large scale production of nanofibers based planar and 3-D fibrous structures such as woven, knitted and braided fabrics, thus bridging the gap between nano-structures and macrostructures.

3.1. Mechanical Properties

The availability of experimental results of elastic properties of electrospun fibers is very limited in the literature and only a few papers have demonstrated advanced techniques to determine the elastic properties of nanomaterials. The atomic force microscope (AFM), is a favorable tool developed to exploit contact and non-contact forces for imaging surface topology and to study new physical phenomena at microscopic dimensions. The heart of the AFM is the cantilever and tip assembly, which is scanned with respect to the surface. For contact imaging the sample is scanned beneath the atomically sharp tip mounted on the cantilever providing a repulsive force. Atomic resolution can be obtained by very light contact and measuring the deflection of the cantilever due to the repulsion of contacting atomic shells of the tip and the sample [33]. Silicon device technology has introduced micro-fabrication techniques to produce silicon, silicon oxide, or silicon nitride micro-cantilevers of extremely small dimensions, on the scale of 100 μm x 100 μm x 1 μm thick. Micro-cantilevers exhibit force constants around 0.1 N/m. In contact imaging, the measurement of cantilever deflection is performed directly (quasi-statically) as the tip is being scanned over the surface [29]. Ko et al. [34] used this AFM technique to measure the elastic modulus of a single fiber produced from an electrospun PAN polymer solution and carbon nanotube reinforced PAN.

Kracke and Damaschke [35] used the AFM to measure the nanoelasticity of thin gold films. They based the calculation of the modulus on the following relationships:

$$\frac{dF}{d\delta h} = \left(\frac{2}{\pi^{0.5}}\right) E^{*} A^{0.5} \qquad (1)$$

$$\frac{1}{E^{*}} = \frac{\left(1-\nu_1^2\right)}{E_1} + \frac{\left(1-\nu_2^2\right)}{E_2} \qquad (2)$$

where: E1, E2, υ_1& υ_2 are the elastic moduli and Poisson ratios of the sample and the tip respectively.

$$A = \pi r^2 \qquad (3)$$

$$F = 2rE^{*}\delta h \qquad (4)$$

where: r is the tip radius and δh, is the deformation between tip and sample

Sundararajan et al. [36] used the AFM technique to evaluate the elastic modulus and their bending strength of nanobeams. By using this technique, the measured values were comparable to those measured in bulk. Mechanical properties and the deformation behavior of materials having ultra-fine microstructures have also been illustrated by using a nanoindentation machine following the same guidelines. In addition, Micro-Raman Spectroscopy (MRS) has been used to measure the load transfer in short-fiber, high-modulus/epoxy composites as a function of fiber orientationto loading direction [37] as well as monitoring the strain in short fiber composites caused by fiber to fiber interactions.

3.2 Chemical Properties

Raman spectroscope is a powerful techniques for detecting the presented elements on the nanoscale. It is also useful for investigating the state of carbon in great detail due to its sensitivity to changes in translational symmetry. Duchet et al. [38] used Raman spectroscopy to correlate the molecular structure of the nanoscopic tubules with their conducting properties. Raman spectroscopy has been used for the characterization of carbon nanotubes (CNT) by many researchers [39].
Rao et al. [40] reported that, there are two frequencies or bands of the single walled carbon nanotubes (SWCNT) that can be detected by Raman. One is known as the radial band at $160 cm^{-1}$ and the other is called the tangential band at $1590\ cm^{-1}$. They showed that the first band could be detected at higher Raman shift values ($10\ cm^{-1}$ higher) when it was dispersed in a solution. The ratio (I (D-band)/I (G-band)) has been cross-correlated to crystallite size using x-ray diffraction data this correlation is given by the following equation:

$$L_a = 4.4\left(\frac{I_D}{I_G}\right)^{-1} \qquad (5)$$

where: L_a is the crystallite size in (nm)

Huang et al. [41] showed a linear relationship between the I_D/I_G band and the $1/L_a$ for commercialized carbon fibers produced from both PAN and PITCH. In the same paper they showed that there is a variation of the Raman spectra between the skin and core regions. These spectra show that I_D/I_G decreases from core to skin, which indicate smaller crystallite values in the fiber core more so than on the skin.

Laser Raman spectra have been used by Y. Matsumoto et al. [42] to examine the carbon nanofibers and films produced by hot filament-assisted sputtering and it was found that the films showed three Raman bands at 1581, 1368 and 1979 cm^{-1} with the first two peaks belong to polycrystalline graphite and the last peak corresponds to carbine, one of the allotropes of carbon.

4. MULTISCALE, HIERARCHICAL MODELING

In order to establish a framework to connect the properties of materials in the nanoscale to macrostructures, a fiber architecture based model is introduced. We will illustrate the methodology using nanotube reinforced nanofibrils and their assemblies. Specifically the nanocomposite fibrils produced by the co-electrospinning process is demonstrated in this study by spinning mixtures of CNT and polymer solution. A schematic illustration of the co-electrospinning process is shown in Figure 7a. The concept of CNT nanocomposites (CNTNC) has been reported elsewhere showing the orientation of the CNT in a polymer matrix through the electrospinning process by flow and charge induced orientation as well as confinement of the CNT in a nanocomposite filament [43].

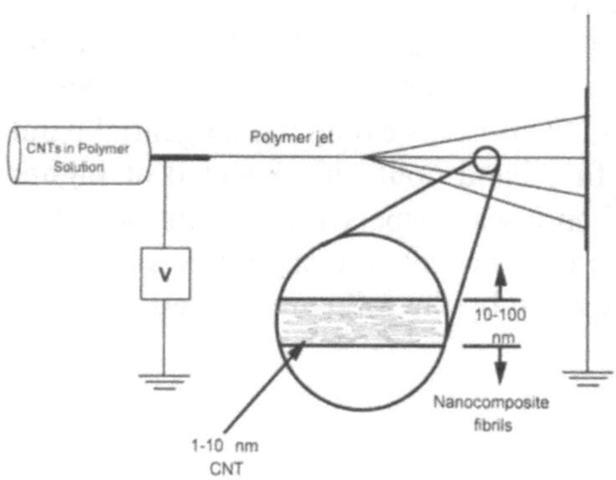

Figure 7a. Co-electrospinning of CNT

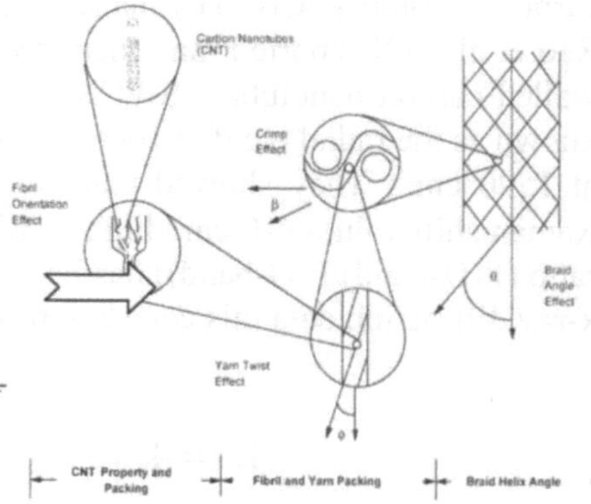

Figure 7b. Concept of CNTC

The nanofibril composite can also be subsequently deposited as a spunbonded nanofibril mat for subsequent processing into composites or for use as an un-impregnated nonwoven mat. Alternately, by proper manipulation, the CNTNC filaments can be aligned as a flat composite filament bundle or twisted to further enhance handling and/or tailoring of properties in higher order textile preforms for structural composites. As shown in Figure 7b, by twisting the nanocomposite fibrils, off-axis angular orientation may be introduced to the nanocomposite filament in order to tailor the composite filament modulus.

4.1. Modeling of Geometric Properties of CNTNC

The translation of properties to macrostructures begins with a consideration of the packing of the CNT. Depending on the nature of the CNT packing, the volume fraction of CNT can range from .75 to .90 for open, square and close packing respectively, as shown in Figure 8.

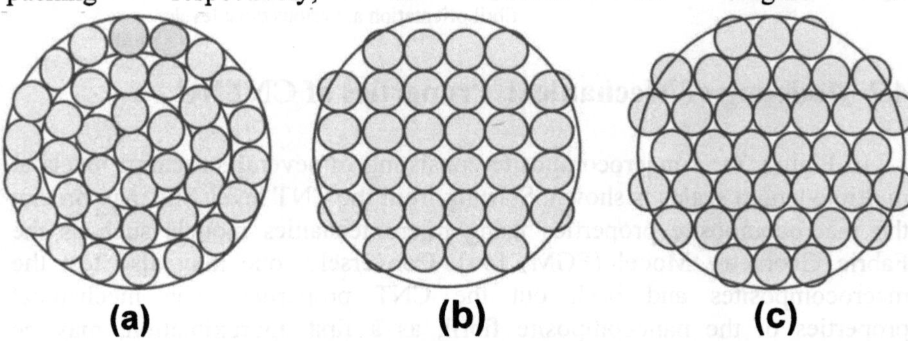

Figure 8 Idealized fibrillar packing: (a) open packing; (b) square packing; (c) close packing.

A twisted bundle of fibrils is called a yarn wherein the fibrils are no longer aligned to the yarn axis. Instead, the fibrils assume a helical geometry in the yarn, as shown in Figure 9a. For fibril yarn composites, fiber volume fraction is equivalent to its fiber packing fraction, and related to fibril helix angle, θ, fibril diameter, d, and number of twist per unit length, T, by equation (6) :

$$V_f = d + \frac{\tan\theta}{\pi d \mathrm{T}}^{\,-2} \qquad (6)$$

For a 12K yarn with 100 nm diameter fibrils, the relationship between fibril volume fraction, fibril orientation (surface helix angle), and twist level can be established as shown in Figure 9b. Clearly, for a given twist inserted to the fibril bundle, fibril volume fraction decreases as fibril orientation angle increases.

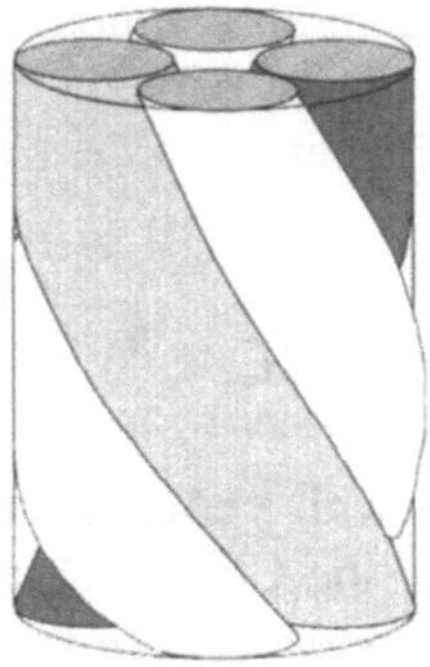

Figure 9a. Geometry of twisted yarn

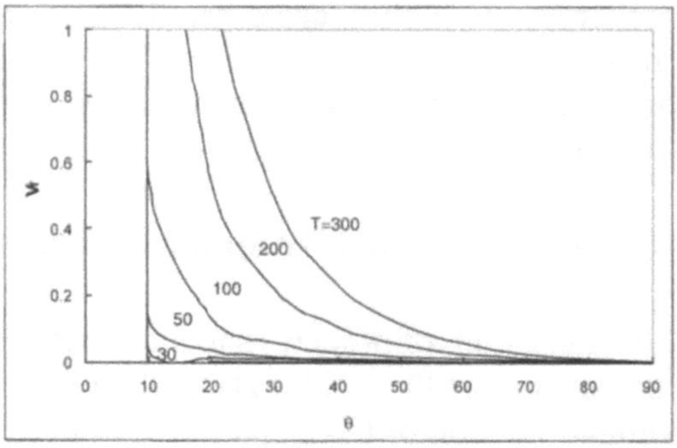

Figure 9b. Relationship of fibril volume fraction to fibril orientation at various twist levels.

4.2 Modeling of Mechanical Properties of CNTNC

In Figure 7b, a macrocomposite consisting of several structural levels at multiple length scales is shown. Starting from the CNT level, one may predict the macrocomposite properties using micromechanics models such as the Fabric Geometry Model (FGM) [44]. Conversely, one may also test the macrocomposites and back out the CNT properties. The mechanical properties of the nanocomposite fibril, as a first approximation, may be modeled as a short fiber composite with CNT playing the role of a short fiber that has a high aspect ratio. For an aligned CNT fibril assembly of CNT of length l, the longitudinal modulus is given by: [45]

$$E_{fibril} = \eta_l E_{cnt} \nu_{cnt} + E_m (1 - \nu_{cnt}) \tag{7}$$

where, ν_{cnt} is the volume fraction of the CNT, E_{cnt} is the longitudinal modulus of the CNT and E_m is the modulus of the matrix material. η_l is an efficiency parameter and is given by:

$$\eta_l = 1 - \frac{\tanh(\frac{1}{2}\beta l)}{\frac{1}{2}\beta l} \tag{8}$$

l is the length of the CNT and β is given by:

$$\beta = \sqrt{\frac{8G_m}{E_{cnt} d^2 \ln \frac{2R}{d}}} \tag{9}$$

where G_m is the shear modulus of the matrix, d is the fibril diameter and 2R is the inter- fibril spacing.

For a twisted assembly of the nanocomposite fibrils embedded in a resin matrix, it is referred to as a yarn composite, with fibril orientation, ϕ, The modulus of the yarn composite can be related to the modulus of CNT by equation (10) :

$$E_{yarn} = \eta_l E_{cnt} V_{cnt} Cos^4 \phi + E_m (1 - V_{cnt}) \tag{10}$$

By taking the packing density of CNT and fibrils into consideration, using the CNT properties shown in Table 3, the tensile modulus of the composite fibril and yarn can be plotted as a function of CNT and fibril orientation in Figures 10a and 10b respectively:

Table 3. CNT and Polymer Matrix Properties

CNT	
Modulus of Elasticity, E_{cnt}	1 TPa
Length, l	1000 nm
Diameter, d	1 nm
Matrix	
Modulus of Elasticity, E_m	3.5 GPa

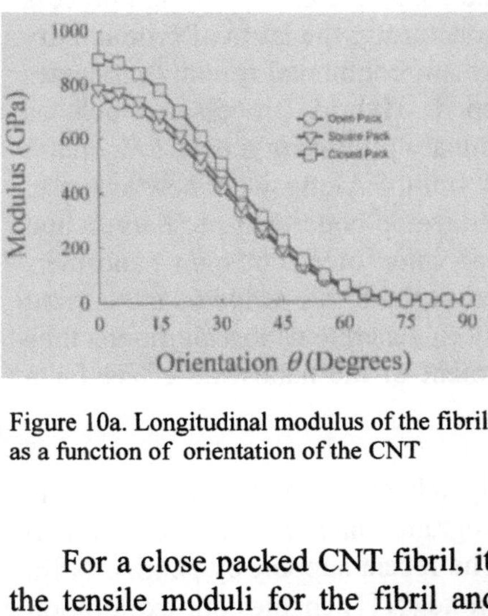

Figure 10a. Longitudinal modulus of the fibril as a function of orientation of the CNT

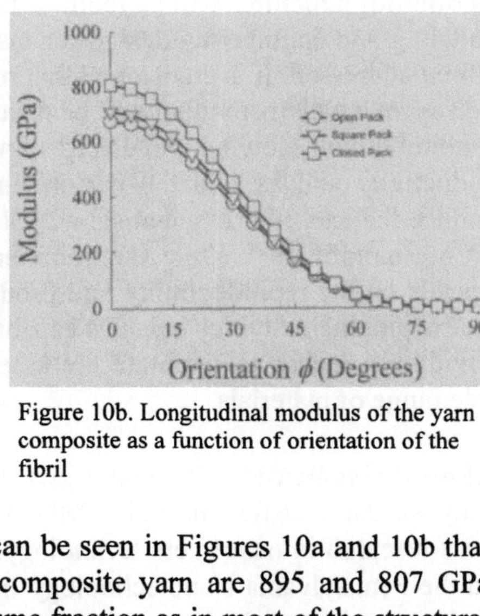

Figure 10b. Longitudinal modulus of the yarn composite as a function of orientation of the fibril

For a close packed CNT fibril, it can be seen in Figures 10a and 10b that the tensile moduli for the fibril and composite yarn are 895 and 807 GPa respectively. Assume a 60% fibril volume fraction as in most of the structural composites the moduli of the unidirectional composite and quasi isotropic composite consisting of the CNT fibrils are 538 and 203 GPa respectively. This agrees well with the prediction by Harris et.al. [46] showing a more than three fold increase in modulus compared to aluminum and high modulus carbon fiber reinforced composites.

5. CONCLUSIONS

Nanofiber technology is an important branch of the growing discipline of nanotechnology. Materials in nanofiber form not only lead to superior functions due to the nano-effect, but also provide a means to deliver functions to higher order structures. Electrospinning is an attractive process capable of producing polymeric fibers having diameters ranging over several orders of magnitude, from the micrometer range to the nanometer range. Under certain spinning conditions for some polymer solutions, continuous yarn containing nanofibers can be produced. This phenomenon opens the door to a practical means of connecting nanostructured materials to macroscopic structures. To illustrate the concept of hierarchical translation of the properties of nanomaterials to higher order structures, a mulltiscale modeling approach is introduced using carbon nanotube reinforced composite fibrillar assemblies as an example.

Encouraged by the simplicity of the electrospinning process and early demonstration of unique properties of nanofibers, there is an explosion of research activities worldwide. It is envisioned that higher level of scientific understanding of nanofibrous materials and creative applications of nanofibrous structures will be realized in the coming decade. Through process modeling and engineering design for manufacturing, the issue of productivity will be addressed. It is envisioned that more environmental friendly processes such as melt electrospinning will be developed. Hybrid processes, such as combining melt blowing and electrospinning are promising means to increase productivity and expand the performance limit. Along with new ways to produce the nanofibers, scientists will also develop better ways to harvest and test the nanofibers. While the commercial value of electrospun nanofibers depends on the reproducibility and productivity, the true value of the current electrospinning activities is in it's capability to generate nanoscale fibers, thus providing a means to facilitate the assessment of the nanoscale effect for a wide range of materials.

Acknowledgements: The author gratefully acknowledges the support of the Army Research Office through a MURI program which enables the initiation of his research in nanofibers technology. The recent support by NASA, NSF, and the Pennsylvania Nanotechnology Institute (NTI) enables our exploration of the co-electrospinning technology in multifunctional nanofibers and nanocomposites. This work was made possible by a NATO travel grant to Turkey to attend the ASI.

References

1. M.C.Rocco, R.S. William, and P. Alivisiatos, editors, Nanotechnology Research Directions: IWGN Workshop Report, National Science and technology Council, September 1999
2. H. Allcock and F. Lampe, Contemporary Polymer Chemistry, (1990)

3. Y.Y. Fan, H.M. Cheng, Y.L. Wei, G. Su, and Z.H. Shen, Carbon, 38, 6, 789, (2000)
4. T. Hongu and G.O. Phillips, New Fibers (1997)
5. D.H. Reneker and I. Chun, Nanotechnology, 7, 216, (1996)
6. A.G. MacDiarmid, W.E. Jones, Jr., I.D. Norris, J. Gao, A.T. Johnson, Jr., N.J. Pinto, J. Hone, B. Han, F.K.Ko, H. Okuzaki, M. Llaguno, Synthetic Metals, 119, 27, (2001)
7. F.K. Ko, W.B. Han, A. Rahman, H. Shimoda, and O. Zhou, Proceeding, Am. Soc. Composites, (2001)
8. D.H. Reneker, Report for the University of Akron, (1997)
9. I.D. Norris, M. Shaker, F.K. Ko and A.G. MacDiarmid, Synthetic Materials, 114, 109, (2000)
10. J. Doshi and D.H. Reneker, Journal of Electronics, 35, 151, (1995)
11. J-S. Kim and D.H. Reneker Polymer Engineering and Science, Vol. 39, No. 5, 849, (1999)
12. A. Formhals, US Patent 1,975,504, Oct. 2, 1934
13. G. Taylor, Proc Roy Soc London A, 313, 453, (1969)
14. C.J. Buchko, L.C. Chen, Y. Shen, and D.C. Martin, Polymer, 40, 7397, (1999)
15. P.K. Baumgarten , Journal of Colloid and Interface Science, Vol. 36, No. 1, 71, (1971)
16. L. Larrondo and R. St. John Manley, Journal of Polymer Science, 19, 909, (1981)
17. S. Hayati, A.I. Bailey and T. F. Tadros, Nature, 319.2, 41 (1986)
18. S. Hayati, A.I. Bailey and T. F. Tadros, Journal of Colloid and Interface Science, Academic Press, 117, No. 1, 205, (1987)
19. D.F.H. Smith, IEEE Transactions on Industry Applications, 1A-22, No. 3, 527, (1986)
20. Y.Y. Fan, H.M. Cheng, Y.L. Wei, G. Su, and Z.H. Shen, Carbon, 38, 921, (2000)
21. Website, http://www.hillsinc.net/nanofiber.shtml
22. J. Deitzel, N.C.B. Tan, J. Kleinmeyer, J. Rehrmann, D. Tevault, D. Reneker, I. Sendijarevic, A. McHugh, ARL-TR-1989, (1999)
23. B. Nabet, Priviate communication
24. A. El-Aufy, B. Nabet, F. Ko, Carbon Nanotube Reinforced nanocomposites for Wearable Electronics, Polymer Preprints, Vol.44, No. 2, ACS, September,2003
25. J.E. Gordon, The New Science of strong materials, Princeton,1984
26. M.S. Dresslhaus, G. Dresslhaus, and Ph. Avouris editors, Carbon Nanotubes, Springer, 2000
27. B.L. Hager and G.C. Berry, Journal of Polymer Science, 20, 911, (1985)
28. G.C. Berry, H. Nakayasu, and T.G. Fox, Poly Phy Edn 17, 1825, (1979)
29. G.C. Berry, The Journal of Chemical Physics, 46, 4, (1966)
30. Ali, Ashraf. Carbon Nanotube Reinforced Carbon Nano Composite Fibrils By Electrospinning. PhD thesis, Drexel University, October 2002
31. X. Fang, and D.H. Reneker, Journal of Macromolecular Science-Physics, B36, 169, (1996)
32. F. K. Ko, S. Khan, A. Ali, Y. Gogotsi, N. Naguib, G. Yang, C. Li, H. Shimoda, O. Zhou,, M. J. Bronikowski. R. E. Smalley and P. A. Willis. Proceedings, AIAA Conference, Denver, Colorado, April 23rd, 2002
33. H. Ye, N. Naguib, S. Khan, A. Ali, Y. Gogotsi, and F.K. Ko, Texcomp-6 Conf., (2002)
34. F.K. Ko, A. Ali, Y. Gogotsi, G. Yang, C. Li, and P. Willis, Fiber Society Conf. Fall, (2002)
35. B. Kracke and Damaschke, Applied Physics Letters, 77, 3, 361, (2000)
36. S. Sundararajan, B. Bhushan, T. Namazu, and Y. Isono, Ultramicroscopy, 0, in press, (2002)
37. M.L. Mehan, L.S. Schadler, Composites Science and Technology, 60, 1013, (2000)
38. J. Duchet, R. Legras, and S.D. Champagne, Synthetic Metals, 98, 2, 113, (1998)
39. P. Corio, P.S. Santos, V.W. Brar, Ge.G. Samsonidze, S.G. Chou, and M.S. Dresselhaus, Chemical Physics Letters, 370, 675, (2003)

40. A.M. Rao, J. Chen, E. Richter, U. Schlecht, P. C. Eklund, R. C. Haddon, U. D. Venkateswaran, Y.K. Kwon, and D. Tománek Physical Review Letters, 86, 17, 3895, (2001)
41. Y. Huang and R.J. Young, Carbon, 33, 2, 97, (1995)
42. Y. Matsumoto, M.T. Oo, M. Nakao, K. Kamimura, Y. Onuma, and H. Matsushima, Materials Science and Engineering, B74, 218, (2000)
43. Ko, F., Gogotsi, Y., Ali, A., Naguib, N., Ye, H., yang, G., Li, C., Willis, P., Electrospinning of Continuous carbon Nanotube-Filled Nanofiber Yarns, Adv. Mater 2003, 15, No.14. July 17
44. F. Ko, Processing of Preforms for Composites, in T.G. Gutowski, ed., Advanced Composites Manufacturing, Wiley Interscience, 1997
45. H.L. Cox, The elasticity and strength of paper and other fibrous materials, Brit. J. Appl. Physics, 3: 72-9, 1952.
46. C.E. Harris, J.H. Starnes, Jr., and M.J. Shuart, An Assessment of the State of the Art in the Design and Manufacturing of Large Composite Structures for Aerospace vehicles, NASA/TM-2001-210844, April 2001

MECHANISM OF CARBON FILAMENTS AND NANOTUBES FORMATION ON METAL CATALYSTS

V.L. KUZNETSOV
Boreskov Institute of Catalysis, Lavrentieva 5, Novosibirsk, 630090 Russia

Abstract: The consideration of the formation mechanisms of carbon deposits on surface metal catalysts led us to conclusion that the majority of these mechanisms include some common steps. The most important of which are the formation of metal or metal carbide particle oversaturated with carbon atoms and the nucleation of carbon deposits on the metal surface. A thermodynamic analysis of the carbon nucleation on the metal surface was performed. The master equation for the dependence of critical radius of the carbon nucleus on the reaction parameters, such as the reaction temperature, the catalyst nature, the supersaturation degree of catalyst particle by carbon was obtained. This equation and the phase diagram approach were used for a discussion of the different scenarios of carbon deposit formation, namely, encapsulated metal particles, carbon fibers and filaments, bamboo-like nanotubes, multi-wall and single-wall carbon nanotubes.

1. INTRODUCTION

Different types of carbon fibers (CF), filaments and nanotubes (CNT) are known today. Their family includes filaments and fibers, which can be produced via carbonization of natural or synthetic organic polymers or mesophase pitch in inert atmosphere [1]. CF can be also produced in a carbon arc [2]. Other important representatives of this family of carbon materials are produced by the decomposition of hydrocarbons or carbon monoxide on the surface of metallic catalysts. These groups of processes are often termed catalytic chemical vapour depositions (CCVDs). The first time carbon fibers and tubes were observed was by Radushkevich and Lukyanovich in 1952. The electron microscopy pictures presented in their paper [3] clearly demonstrate the formation of multi-wall nanotubes (MWNTs). However, the importance of this observation was realised completely only after discovery of single-wall nanotubes (SWNTs) in arc discharge products by S.Iijima in 1991 [4]. A huge number of papers (several thousands), many reviews and several books concerning CNT and CF appeared after this date [5-8].

S. Guceri et al. (eds.), Nanoengineered Nanofibrous Materials, 19-34.

At the present, nanofilaments and nanotubes have attracted great attention not only because of their fascinating structural features and properties, but also because of their potential technological applications. It is possible to separate the CNT and CF applications into two different groups. The first traditionally uses ensembles (great number) NT or CF while the second is based on the properties of a single (individual) NT. Thus, the **first group** includes the composite materials (where NT and CF are used as reinforcing components), materials for chemical current sources, adsorbents, catalysts supports etc. Carbon fibers are of technological importance because their composites have strength-to-weight properties superior to those of any other materials. The crucial factor to fiber strength is the alignment of graphitic hexagons along the fiber axis, with structurally coherent fibrils extended over long distances. At the present, the technologies of production of ex-polymer or mesophase pitch based fibers are well developed and these fibers are commercially available [9]. **The second group** deals with components for electronic components (field emitting devices, diodes, transistors, high frequency resonators, micro-wave generators, scanning microscope probe tips, nanopipettes etc.). The CF and NT applications are considered in a recently published book and reviews [10-12].

Macroscopic amounts of relatively good quality NT are produced by several groups around the world, and the theoretical understanding of the electronic properties of CNT has reached a significant level. At the same time, the available techniques allow only poor control over the growth of the nanotubes and yield numerous byproducts such as amorphous carbon, polyhedral carbon nanoparticles, and carbon encapsulated catalyst particles. The separation of nanotubes from impurities is a difficult task, mainly because the reactivity of the nanotubes is similar to that of the byproducts, and nanotubes may thus be damaged or destroyed during purification procedures. These problems create one of the most critical factors for the wide NT applications related to the cost of their production. Accordingly, research is being carried out on new ways of producing CNT and CF with cost as low as a few percent of current values. Another problem concerning the controlled manipulation of nanoscale objects can be the limited development of electronic devices. In this respect it seems the generation of self-assembled and/or self-aligned structures is the most promising path to explore further. The development of more effective methods of CNT and CF can be based only on an understanding of the formation mechanism of nanotubes, which is crucial when designing techniques for the controlled mass production of pure nanotubes material.

Here we examine various important aspects related to the catalytic synthesis and mechanism of formation of carbon filaments and nanotubes. Thus, we will not consider production of CF via noncatalytic CVD and formation of CF produced via the thermal treatment of carbon polymers and/or graphitic mesophases, which have been widely used in industry for many years. We continue to estimate the applicability of thermodynamic analysis and phase diagram approach for the design of new catalytic systems required for the development of effective production of CF and CNT. We compare the numerical results for the critical radius of carbon nucleus calculated for different metals (Fe, Ni, Pt, Pd, Rh, Co) with experimental data on the formation of single-wall carbon nanotubes. A new formation mechanism of bamboo-like carbon nanotubes is also proposed.

2. SHORT REVIEW OF PROPOSED FORMATION MECHANISM OF CF AND CNT ON METAL CATALYSTS

The formation of metal particles encapsulated in carbon shells, CF and CNT are usually considered as products of different processes. However, it is important to mention that all these products can be produced using the same metal catalysts. Often these catalysts have the same composition, but when used in different experimental conditions they lead to the formation of different products. There are several mechanisms that have been proposed to explain the formation of these rather different products, but these mechanisms do not explain the formation of all varieties of products using unified theory.

2.1. Formation Mechanism of Carbon Filaments and Nanotubes

Let us consider the formation mechanisms of CF and CNT to elucidate the common steps and features. These mechanisms were carefully considered in a number books and reviews, for example see ref. [5-9, 13, 14]. Due to space constraints, we will provide only their general description. These mechanisms can be subdivided into several groups:

1. The first group can be considered mechanisms, which include the stage of carbon-metal particle formation independently of carbon source: decomposition of hydrocarbon or CO on solid metal catalyst or condensation of preliminary evaporated metal and carbon atoms from gas phase. In 1972, Baker et. al. were the first [15] to suggest such a mechanism for the formation of filamentous carbon deposits on isolated metallic particles by the catalytic decomposition of hydrocarbons. They proposed that temperature and concentration gradients were the main driving forces for this process. According to them, suggested mechanism hydrocarbons are decomposed on the front exposed surfaces of metal particles and release hydrogen as a gas and carbon, which dissolves within the particle volume. A temperature gradient results from overheating part of the catalyst particle during the exothermal hydrocarbon decomposition reaction and gives rise to a gradient of the carbon concentration in the volume of the catalyst particle. The dissolved carbon diffuses through the particle and precipitates at the cooler part of the metal particle. The diffusion of carbon through the catalyst particle is believed to be the rate-limiting step in the growth of filaments.

There are many conflicting reports concerning the detailed composition of the catalyst particle under reaction conditions of CNF formation. Successive formation of metal carbides and their decomposition with carbon deposition was proposed [14, 16-18]. At high temperatures formation of carbides were not registered.
Related mechanism of SWNTs formation ("root-growth mechanism") from evaporated carbon and metal atoms, which is generally accepted now [19-21] proposes: the initial production of liquid metal catalyst particles saturated with carbon; the nucleation of nanotubes via C segregation at the particle surface and the growth of the nuclei by incorporation of C at the interface nanotube/particle.
Note that under the arc discharge and laser vaporization conditions, a temperature gradient occurs following the diffusion of the particles away from the hot zone of the interelectrode plasma or of the vaporization plume. In that case, the carbon precipitation is caused by the decrease in temperature, and hence in carbon solubility.

2. The second group consists of mechanisms, which are based on surface diffusion of carbon around the metal particle, rather than the bulk diffusion of carbon through the catalyst particle [22]. However, the surface diffusion of carbon structures, proceeding without any carbon dissolution in metal particle, seems to be improbable. These mechanisms can be realized at relatively low temperatures (700 -800 K), when the diffusion of carbon into the metal particle is limited by temperature, or when hydrocarbon decomposition product can be stabilized at the metal surface by hydrogen atoms.
3. Some authors suggest that catalyst particles take part only in the tube initiation process. According to this model, the metal particle initiates the formation of a primary tube, which consists of prefect graphitic layers cylindrically rolled around the metal particle [23]. The thickening of the primary tube by the carbon deposition on the outer surface of tube proceeds without participation of the metal particle. A similar mechanism was proposed proceeding via initial fullerene molecule adsorption on the metal surface with formation of a pre-formed carbon nanotube, which can serve as a nucleus for the growth of another carbon nanotube [24]. Carbon atoms can then be added and the elongated fullerene molecule can grow into a nanotube with an open end to sustain continued growth. The growth of a single nanotube stops when the ends are closed without dangling bonds. Note that this mechanism can only be realized in the presence of free carbon atoms in a gas phase.
4. For the growth of SWNTs it was proposed that metal atom or small metal clusters provide tube growth by moving around the open tube tip and bonding carbon atoms from the gas phase [25]. This mechanism fails to answer the question surrounding nanotubes rope formation, in which SWNTs are usually observed. Although it is possible that this growth mode really takes place during the synthesis, its contribution to the SWNT formation is rather negligible.

A model for the formation of cylindrical layers around the hollow core with use of a **thermodynamic approach** was suggested by Tibbetts [26].

In this model the graphitic planes are elastically strained in forming a tree-ring structure. This gives rise to an extra elastic term in the considered free energy equation. However according to Tibbetts model the free energy change precludes formation of carbon filaments with inner radius smaller then 5 nm and synthesis of carbon nanotubes of much smaller diameter can't be explained. In our opinion such thermodynamic approach is most perspective in description of filamentous carbon deposit growth but the model suggested by Tibbetts must be extended and stage of carbon nucleus formation must be considered also.

2.2. Phase Diagram Analysis of Carbon Deposition on Metal

To understand the role of the catalyst in CCVD, let us consider the carbon deposition on the metal surface in terms of a metal-carbon phase diagram. CCVD can proceed in two temperature regions (at high and moderate temperatures, respectively). Processes that occur at high temperatures (arc-discharge or laser ablation SWNT growth) correspond to the line ***a′-b′-c′-d′*** in *Figure 1.* The process starts after the initial vaporization of a target consisting of a metal catalyst mixed with graphite powder at a temperature higher than 4000 K. A condensation of the metal-carbon vapor proceeds under a temperature gradient in an inert gas (He, Ar) and leads to the formation of liquid metal-carbon alloy (point ***a′***). Carbon nucleation on the liquid catalyst particle begins when the reaction temperature decreases below the temperature corresponding to the point ***b′***. At this point, the liquid alloy becomes supersaturated with carbon to a value critical for deposition of carbon at the solid-liquid interface (carbon-metal interface). The equilibrium concentration of carbon in a metal particle below that temperature is determined by the liquidus line. As the temperature drops below the value corresponding to the solidus line (point ***c′***) the metal particle becomes hard (point ***d′***) and the carbon deposition rate drops to relatively low values.

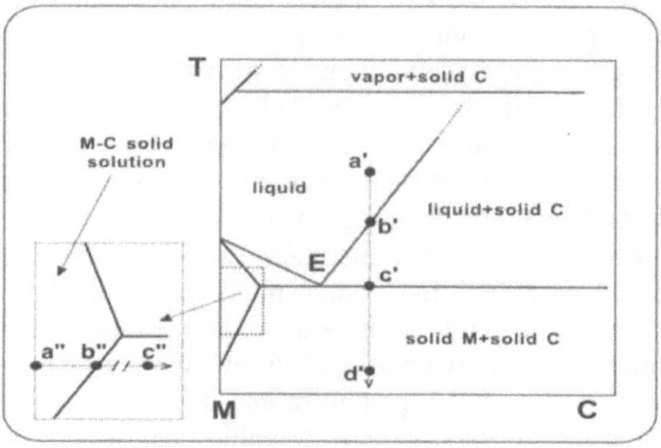

Figure 1. Schematic metal-carbon phase diagram consideration. Lines (***a′-b′-c′-d′***) and (***a″-b″-c″***) correspond high temperature and moderate temperature carbon deposition on metal surface.

The decomposition of hydrocarbon or CO on the surface of metal particles, which leads to the formation of SWNT or CF, proceeds in isothermal conditions at moderate temperature (see ***line a″-b″-c″*** in the region below eutectic temperature). A high degree of carbon supersaturation is required for the SWNT growth, as will be demonstrated below. These requirements can be relatively easily achieved using methods based on preliminary metal-carbon mixture evaporation. At the same time, for the region below eutectic temperature, high carbon supersaturation (point ***c″***) can be reached using a highly dispersed metal particle. In some cases metal particles can melt even at the temperature below eutectic temperature due to their small size or addition of specific promoters. The most difficult problems of this are related with sintering of disperse metal particles and carbon diffusion limitations. To overcome these problems, chemists use two approaches: (1) the decomposition of volatile organometallic compounds in a reaction gas flow containing hydrocarbon or CO [27-30] and (2) the reduction of metal oxide solid solutions characterized by high reduction temperature [31-33]. Both approaches allow *in situ* prepared dispersed metal particles with a high carbon supersaturation to provide the optimal nucleation conditions (see analysis below). Thus, metal catalysts take part in initial reagents activation and serve as media for carbon dissolution providing the dramatic decrease of temperature solidification of carbon to form the metal-carbon interfaces responsible for the formation different carbon deposits.

2.3 Carbon Nucleation on Metal as a Key Step Of Carbon Deposits Formation

The phase diagram approach and analysis of reaction mechanisms allows the proposition that carbon nucleation is a common and key step for the formation of all type of filamentous carbon deposits. It determines the critical size of the carbon nucleus and finally the type of carbon deposit obtained, because the size of tubes or graphite plates (in the case of filamentous carbon) cannot be smaller than the critical size of carbon nucleus. In the case when several nuclei form at the same metal particle, they can grow independently. It can be understood in terms of relatively high stability of nanocarbon species even at high temperatures up to 1800 K. *Figure 2.* summaries the literature data on the stability carbon species to sintering (graphitization). So, single wall carbon nanotubes (SWNT) double their diameter only after thermal treatment at the temperature higher then 1700 – 1800 K [34], while multi-wall nanotube formation from SWNT ropes occurs only at temperatures higher than 2300 K [35].

Other evidence of high thermal stability of nanosize graphite species came from the study of the graphitization of carbon materials, the formation of stable mosaic closed graphitic nanostructures on the annealed diamond surface at the temperature higher than 1800 K was observed [36]. Thus, once formed the nanosize carbon nuclei are stable enough and the nucleation step can determine the carbon deposit type. Multiple nucleations of single wall nanotubes leading to the formation of nanotube ropes are the most remarkable example of the last statement. *Figure 3.* demonstrates the formation of the SWNT rope at different stages of its growth. The presence of such particles was demonstrated earlier [37-40] and we also observed these particles in typical soot sample obtained by the arc discharge method [41] (Fig. 3). Note that in our TEM images the length of the nanotubes

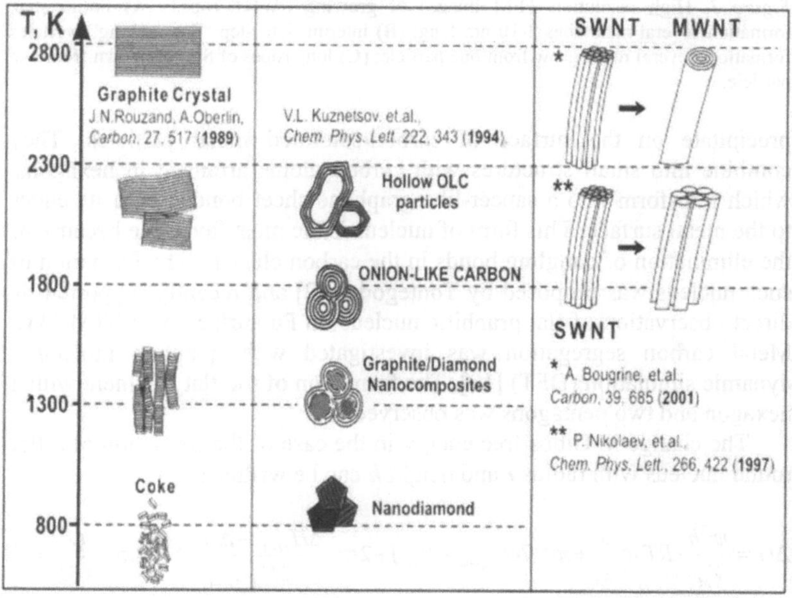

Figure 2. Thermal stability of different carbon species to sintering and graphitization.

extending from the particle is variable and depends on the considered particle. This strongly suggests that they represent different stages of SWNT growth, such as in Fig. 3(a) for the initial step (nucleation and growth of short tubes), Fig. 3(b) for an intermediate step (radial extension of the tubes), and Fig. 1(c) for the final state (formation of ropes of more than 1 μm length).

3. CARBON NUCLEATION ON METAL SURFACES

3.1 Choice of Carbon Nucleus Model and Thermodynamic Analysis of Carbon Nucleation

We have developed a model of graphite-like species nucleation on the surface of metal particles, which is based on classic theory of nucleation.

We proposed that at the beginning of the nucleation a few carbon atoms

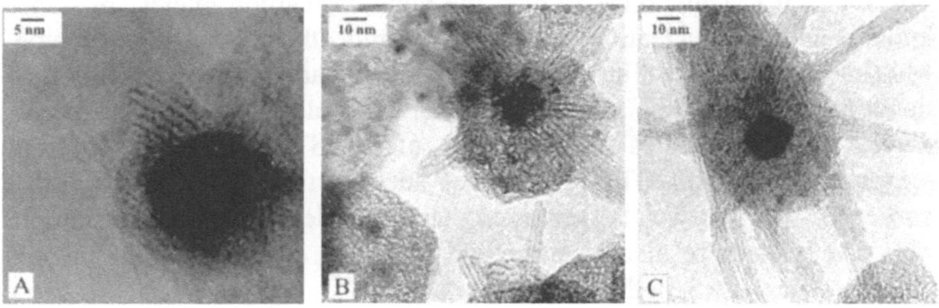

Figure 3. High resolution TEM images of growing SWNT ropes: **(A)** initial step, formation several nanotubes 3-10 nm long; **(B)** intermediate step "Sea urching" particles formation, several ropes grow from one particle; **(C)** long ropes of SWNT grown from one particle.

precipitate on the surface of carbon-saturated metal particles. They combine into small structures with carbon atoms arranged in hexagons, which transform into a saucer-like graphene sheet bonded with its edges to the metal surface. This form of nucleus is the most favorable because of the elimination of dangling bonds in the carbon cluster. The formation of such nucleus was proposed by Tontegode [42] and recently supported by direct observation of flat graphitic nucleus on Pd surface with STM [43]. Metal carbon segregation was investigated with quantum molecular dynamic simulation (DFT) [19]. The formation of the flat fragment with a hexagon and two pentagons was observed.

The change in Gibbs free energy in the case of the formation of a flat round nucleus with radius $\boldsymbol{r}$ and height $\boldsymbol{h}$ can be written as:

$$\Delta G = \frac{\pi r^2 h}{V_M} \cdot RT \ln\frac{x}{x_0} + \pi r^2\left(2\sigma_{graphite} - W_{ad}\right) + 2\pi r \cdot \frac{\Delta H_{M-C} - \Delta H_{C-C}}{2N_A \cdot r_{C-C}} + 2\pi r \cdot \frac{Q}{4.5h} \quad (1)$$

(for details see [21]). In this equation the first term presents the free energy change when graphitic carbon precipitates from the metal-carbon solution and $\boldsymbol{V_M}$ is the molar volume of graphite, $\boldsymbol{T}$ is the reaction temperature, $\boldsymbol{x/x_0}$ is the carbon saturation coefficient of the metal-carbon solution.. The second term is the free energy change caused by the interaction of the carbon nucleus with the catalyst surface, where $\sigma_{graphite}$ is the specific surface energies of graphite and $\mathbf{W_{ad}}$ is a work of adhesion of metal graphite system. The third term estimates the energy of chemical bonding between a border atom of the nucleus and the metal surface, and $\boldsymbol{\Delta H_i}$ are the enthalpies of formation of M-C and C-C bonds, $\boldsymbol{r_{C\text{-}C}}$ is the distance between two neighbouring carbon atoms in the zigzag edge of a graphite sheet, $\boldsymbol{N_A}$ is the Avogadro constant. The fourth term reflects the strain energy due to the bending of the graphitic layer of the nucleus during its bonding with the metal surface, where $\boldsymbol{Q}$ is a constant of continuum elasticity formalism ($\boldsymbol{Q}$ is equal to 4.4 eV for an uncapped cylinder, see Ref. [21,44]).

Taking $d(\Delta G)/dr=0$ for critical nucleus one can easily deduce from equation (1) the master equation (2) for the radii of critical nucleus

$$r_{crit}=\left(\frac{\Delta H_{M-C}-\Delta H_{C-C}}{2N_A\cdot r_{C-C}}+\frac{Q}{4.5h}\right)\cdot\left[\frac{RT\cdot h}{V_M}\ln\frac{x}{x_0}+\left(W_{ad}-2\sigma_{graphite}\right)\right]^{-1}. \quad (2)$$

3.2. The Comparison of Experimental Data with Results of Thermodynamic Analysis of the Size of Critical RADII of Carbon the Carbon Nucleus

Equation (2) expresses the dependence of the critical radius of the carbon nucleus on the different reaction parameters, which can be measured experimentally. We have used this equation to estimate numerically the critical radius for pure Fe, Co, Ni, Pd and Pt catalysts. The values of thermodynamic functions and constants that we used in our estimations are presented in Table I. The simultaneous analysis of all factors, which influence $\mathbf{r}_{crit}$, is a rather difficult task. So we have considered the influence of each factor separately with other fixed parameters. Note that in our estimations we did not take into account the dependence of some thermodynamic functions, such as ΔH_{C-C}, ΔH_{M-C}, $W_{ad}(M-C)_{melt}$, or σgraphite on the temperature. We suggest that, in our case, this influence is negligible. The results of our estimates are presented in *Figure 4*. Note that this estimate gives us a minimal value for the radius of the carbon deposit. Nuclei with smaller radii are not stable, but nuclei with larger sizes can continue to grow. So, for comparison in *Figure 4.*, we have used the experimental values corresponding to the smallest observed size of carbon deposits.

There is a large spread for the values of metal-carbon enthalpies reported in the literature depending on the calculation or experimental methods; hence, the choice of the parameters is very important. We have tried to take into the consideration data from the same source or measured by the comparable methods. We have used data of Miyazaki [45], where M-C bond enthalpy were empirically calculated using a bond energy bond order (BEBO) approach.

We have estimated numerically the dependence given by equation (2) between rcrit and the reaction temperature in presence of pure metal catalysts both for solid and melted state of the metal. Note that we have used a medium value of Wad - 0.1 J/m^2 for the estimation of rcrit for solid metal

TABLE 1. The values of thermodynamic functions and parameters used for the calculation of critical radiuses [21].

Symbol	Used values
ΔH_{C-C}	473.2 kJ/mol
ΔH_{Fe-C}	245.2 kJ/mol
ΔH_{Co-C}	221.3 kJ/mol

ΔH_{Ni-C}	191.2 kJ/mol
ΔH_{Pd-C}	250.2 kJ/mol
ΔH_{Pt-C}	200.4 kJ/mol
ΔW_{ad}(Fe-C)solid	0.1 J/m^2
W_{ad}(Fe-C)melt	2.2 J/m^2 for Fe-C eutectics (17 at. % C)
W_{ad}(Ni-C)solid	0.1 J/m^2
W_{ad}(Ni-C)melt	1.8 J/m^2 for Ni-C eutectics (10 at. % C)
W_{ad}(Co-C)melt	2.1 J/m^2 for Co-C eutectics (12 at.% C)
W_{ad}(Pd-C)melt	1.9 J/m^2 for Pd-C eutectics (10 at.% C)
W_{ad}(Pt-C)melt	0.7 J/m^2 for Pt-C eutectics (17 at.% C)
$\sigma_{graphite}$	0.08 J/m^2
x/x_0	2

particles. The critical radius for a liquid metal particle is much smaller than that for a solid particle due to much higher W_{ad} for metal melts (W_{ad} - 1.8-2.2 J/m^2). We have marked the change of the state of the catalyst particles, which can occur at temperatures close to the corresponding metal-carbon eutectic temperatures in *Figure 4 A*. However, some factors can decrease the solidification temperature of metal particles and thus expand the temperature range of formation of smaller nuclei. We suggest that the production yield of SWNTs in the presence of promoters such as S, Bi, Pb increases due to the decrease of the melting point of the multicomponet system (catalyst-promoter-carbon) as compared to the pure metal-carbon mixture. Our estimates of the relationship between rcrit and the reaction temperature presented in *Figure 4A* are in good agreement with the experimental data extracted from the literature and confirm the idea that single-wall carbon nanotubes grow on the liquid metal catalysts.

Additional improvement of the yield of SWNT can be reached using the promoters, which not only decrease the melting point of catalytic particles but provide the high values of W_{ad} (characterized with higher energy metal-carbon bond (or ΔH_{MC})). Also, one can not exclude that highly dispersed metal particles could be in a liquid-like state during nucleation even at relatively low temperatures because their formation occurs *in situ* via the condensation of metal atoms or small clusters distributed on the supports or prepared via decomposition of organometallic compounds in gas phase [46]. The last approach appears to be the most promising for the development of large scale low temperature SWNT synthesis.

The sensitivity of the calculations to the variation of used parameters was estimated in [47]. In these estimations we used fixed temperatures equal to 973 and 1473 K for the solid and melted metal-particle state, respectively. It was shown that the influence of the nature of the catalyst, which is responsible for the values ΔH_{M-C}, W_{ad} (M-C) and x/x_0, is the most significant when the metal catalyst is in a solid state. For the liquid catalyst, the influence of these parameters is not so pronounced. As an example, the nanotube formation on Mo catalysts can be considered.

These catalysts provide the formation of SWNTs at 850-1200°C. Due to the high melting temperature of Mo, we suggest that in the reaction conditions Mo is in a solid state. However, due to higher values of $\Delta H_{M\text{-}C}$ and W_{ad} (M-C) for Mo in comparison with Fe and Ni, it provides the formation of nanotubes with diameters 10-20 Å even in solid state [48]. After analysis of equation (2), it is obvious why small metal particles produce nanotubes while larger particles are encapsulated in graphitic shells. The difference in behavior can be explained in terms of differences of saturation coefficients, which can attain much higher values for small particles. Thus, small particles can be easily oversaturated with carbon and are characterized by smaller rcrit values, which are preferable for the formation of nanotubes.

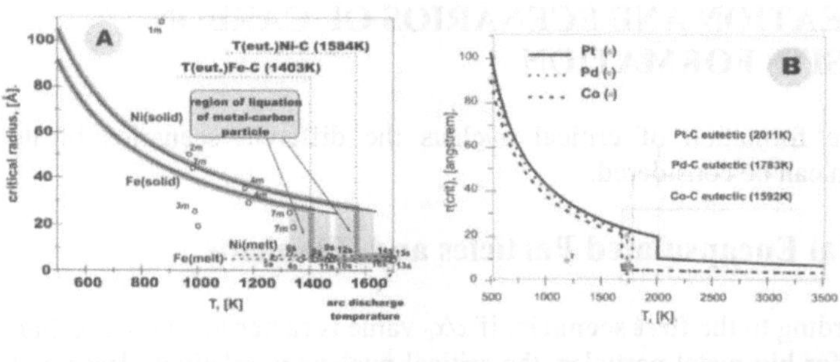

Figure 4. Dependence of r_{crit} on the reaction temperature for (**A**) Fe,Ni and (**B**) Pt, Pd, Co catalysts. The points marked by circles and triangles for Fe and NI in fragment (**A**) and marked by circles, squares and triangles, respectively for Pt, Co and Pd catalysts in fragment (**B**), correspond to the smallest radii observing in experimental papers.

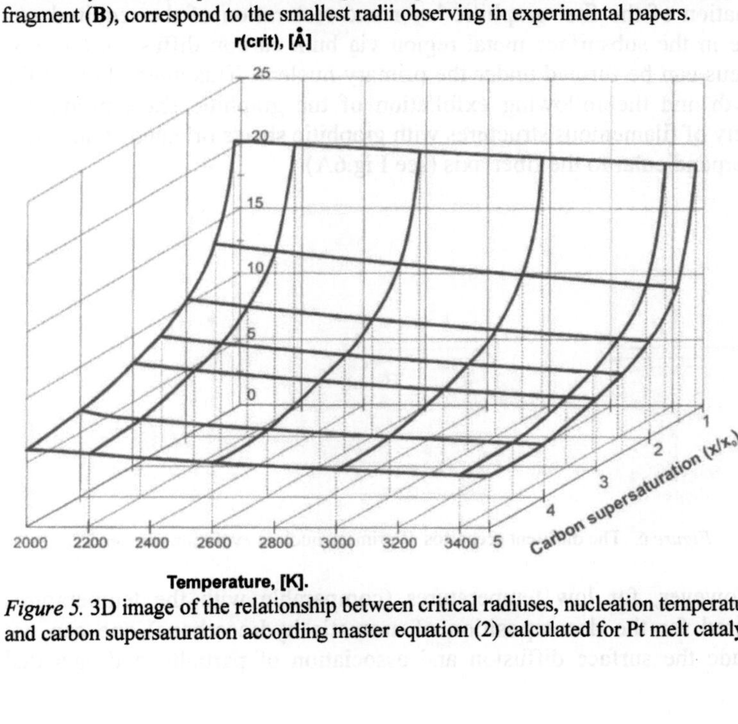

Figure 5. 3D image of the relationship between critical radiuses, nucleation temperature and carbon supersaturation according master equation (2) calculated for Pt melt catalyst.

3.2.1 Analysis of experimental and calculated data for SWNT produced with Pt catalysts

Due to relatively low W_{ad} for Pt in comparison with other metals we obtained relatively higher values of r_{crit} (see Fig.4B). *Figure 5.* shows three dimensional diagram for the dependence of nucleus critical radius on nucleation temperature and Pt particle carbon supersaturation (W_{ad} was fixed). Comparison of this data with the observed size of SWNT (13-30 Å) allows us to propose that carbon nucleation on Pt catalysts proceeds at a relatively high temperature and there is a high supersaturation of catalyst with carbon.

4. THE RELATIONSHIP BETWEEN CARBON NUCLEATION AND SCENARIOS OF CARBON DEPOSITS FORMATION

After the formation of critical nucleus the different scenarios of its evolution can be considered.

4.1.Metal Encapsulated Particles and Filaments

According to the **first scenario**, if x/x_0 value is rather low ($x \approx x_0$), than is usual for big metal particles, the critical nucleus is relatively large and continues to grow, which results in the formation of graphitic sheets covering a significant part of the surface of the metal particle. After the formation of the first graphitic layer and restoration of the required x/x_0 value in the subsurface metal region via bulk carbon diffusion, the next nucleus can be formed under the primary nucleus. This alternation of the growth and the following exfoliation of the graphitic sheet produce a variety of filamentous structures with graphitic sheets oriented at an angle, or perpendicular to the fiber axis (see Fig.6A).

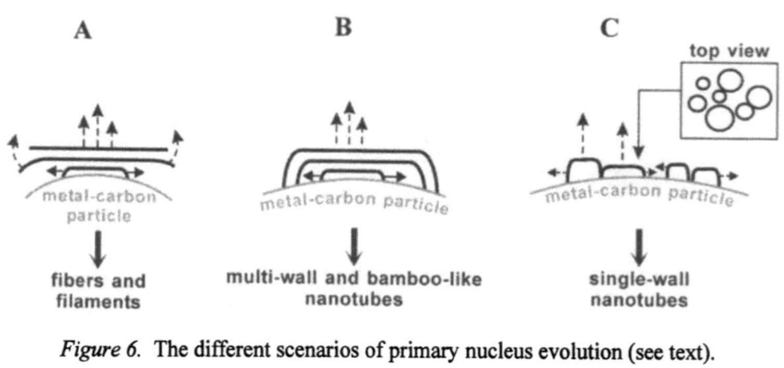

Figure 6. The different scenarios of primary nucleus evolution (see text).

However, for low temperatures (comparable with the temperatures required for the decomposition of aromatic hydrocarbons) one can not exclude the surface diffusion and association of partially hydrogenated carbon species leading to the formation of a large graphite shell on the metal particles. Further decomposition of hydrocarbons on the surface of the primary shell leads to the encapsulation of catalyst particles inside graphitic shells.

4.2. Multi Wall and Bamboo-like Nanotubes

In the **second scenario** *(Fig.6B),* a new nuclei form under the primary one, with edges bonded to the metal particle surface. It can be achieved using disperse metal particles, where the nucleation can proceed only with higher x/x_0 values than in the first case. After the production of a primary nuclei in the case of diffusion limitations the x/x_0 value drops to one insufficient for the further formation of the nucleus and all carbon atoms diffusing through the metal particle are consumed for the nanotube growth. It should be noted that the internal tube radius cannot be smaller than the critical radius of the nucleus, while the external walls have a diameter that is comparable to the size of the metal particle.

The developed thermodynamic formalism was used to describe the formation of **bamboo-like carbon nanotubes** for the case when carbon diffusion through the catalyst particle is at the rate determining step of the reaction. *Fig. 7.* presents the variation of the value of supersaturation of catalyst particles with carbon versus time. Initially, metal particles must be saturated with carbon. The value of supersaturation required for the initial carbon deposition can be estimated from equation (2). According to equation (2), a smaller metal particle diameter of higher supersaturation is required. After the primary nuclei formation the next nuclei cannot form because the drop of carbon supersaturation values lower the level sufficient for the formation of another critical nucleus, whose radii are comparable to the size of the catalyst particle. Before the supersaturation value reaches the level sufficient for the formation of the next nucleus and the partition formation diffusion flow of carbon atoms will be partially consumed for the tube growth. Thus, the cyclic changes of carbon supersaturation value (near the level sufficient for the formation of critical nucleus) are responsible for the formation of bamboo-like tubes. The periodical partition formation can also lead to the periodic heat release, which can result in the periodical melting of the metal particle. Such periodical melting of the catalyst particle corresponds to the pulsating regime of bamboo-like tube growth, which was proposed in some papers [49-51].

4.3. Single Wall Carbon Nanotubes

In the **third scenario** (Fig.6C) very high x/x_0 values are required. In this case several nuclei precipitate on the surface of the same metal particle and interact to form the tightly packed ropes. If the nucleation frequency is high enough, then a size of the single nanotube tends to be a minimal value that corresponds to the critical nucleus radius.

When the carbon nucleus critical radius is considerably smaller than that of the metal particle and the saturation of the metal-carbon particles by carbon is sufficiently high, formation of a multitude of nuclei can proceed. In this case, multiple nuclei grow independently until their diameters reach certain sizes, which allow them to interact with each other. This interaction results in the formation of mosaic structures, consisting of growing nuclei. "Sea-urchin" particles, which consist of SWNT ropes growing radially from metal particle, represent an example of such structures.

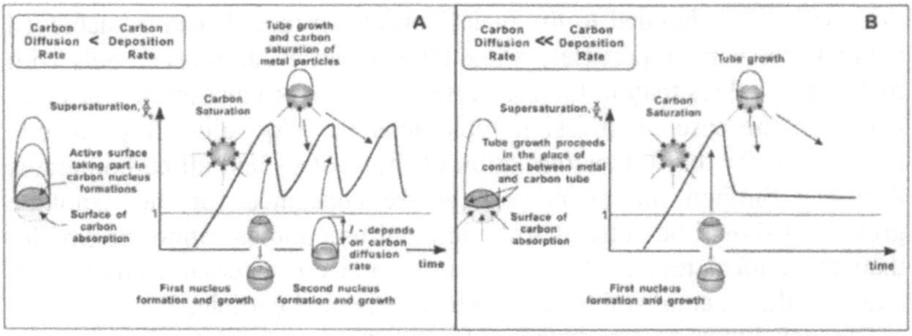

Figure 7. The variation of carbon supersaturation value (x/x_0) of catalyst particle versus time causing the formation bamboo-like nanotubes.

5. CONCLUSION

The consideration of the formation mechanisms of carbon deposits on surface metal catalysts leads us to conclusion that the majority of these mechanisms include some common steps. The most important of these are the formation of metal or metal carbide particles oversaturated with carbon atoms and the nucleation of carbon deposits on the metal surface. This allows considering the formation of different carbon deposits within unified (theory) approach. Thus, we propose that the nucleation is a crucial step (stage) for the formation of most carbon deposits and that a nucleus structure can finally determine the type of carbon deposits. Thermodynamic analysis of the carbon nucleation on the metal surfaces to estimate the influence of reaction parameters on the carbon nucleation step was performed. As a result, a functional dependence between the critical radius of the carbon nucleus (r_{crit}) and reaction parameters such as reaction temperature, carbon supersaturation degree, and parameters which characterize the metal catalyst (metal-carbon bond strength, work of adhesion) was obtained. The analysis of this dependence in combination with metal-carbon phase diagram analysis allows: to formulate the relationship between the size of carbon nucleus and type of carbon deposit; to lighten the optimization of single-wall nanotubes growth conditions; to estimate the real carbon supersaturation of metal-carbon particles and to propose the mechanism of bamboo-like carbon nanotubes formation.

ACKNOWLEDGEMENTS

This work was partially supported by INTAS grants Nos. 00-237, 01-254, and Award No. NO-008-X1 of US CRDF.

References

1. Dresselhaus, M.S., Dresselhaus,.G. Sugihara, K., Spain, I.L., and Goldberg, H.A. (1988) *Graphite Fibers and Filaments,* Springer Verlag, Berlin, Heidelberg, New York.
2. Bacon, R., (1960) *J.Appl.Phys.* **31**, 283.
3. Radushkevich, L.V. and Lukyanovich, V.M. (1952), Zh. Fiz. Khim. **26**, 88.
4. Iijima, S. (1991), *Nature*, London **354**, 56.
5. Ebbesen T. (ed) (1997), *Carbon Nanotube: Preparation and Properties*, CRC Press, Boca Raton, FL.
6. Saito, R., Dresslhaus, G., and Dresselhaus, M. S. (1998*) Physical Properties of Carbon Nanotubes*, Imperial College Press, London.
7. Harris, P.J F (1999) *Carbon Nanotubes and Related Structures. New Materials for the Twenty-First Century,* Cambridge University Press, Cambridge.
8. Dresselhaus, M. S., Dresselhaus, G., and Avouris, P. (Editors) (2001) *Carbon Nanotubes: Synthesis, Structure, Properties, and Applications*, Springer-Verlag.
9. Burchell, T.D. (editor) (1999) *Carbon Materials for Advanced Technologies*, Pergamon an Imprint of Elsevier Science, Elsivier Science Ltd., Amsterdam, Lausanne, New York.
10. Bonard, J. M., Kind, H., Stockli, T., and Nilsson, L. A. (2001), *Solid-State Electronics*, **45**, (6), 893.
11. Rakov, E.G. (2001) *Russian Chemical Reviews (English),* 70(10), 827-863.
12. Eletskii, A.V. (2002) *Physics-Uspekhi (Rus),* **45**(4) 401.
13. Rakov, E.G.(2000), *Russian Chemical Reviews,* **69**, 41.
14. Chesnokov, V.V, Buyanov, R.A. (2000) *Russian Chemical Reviews,* 69,675.
15. Baker, R.T.K., Barber, M.A., Harris, P.S., Feates, F.S., Waite, R.J., *J. of Catal.* 26 (1972) 51.
16. Oberlin, A., Endo, M., Koyama, T. (1976) *J. Cryst. Growth*, 32, 335.
17. Buyanov, R.A., Chesnokov, V.V., Afanasiev, Babenko, V.S. (1977), *Kinetika i Kataliz (Rus.),* **18**, 1021.
18. De Bokx, P.K., Kock, A.J.H.M., Boellard, E., Klop, W., Geus, J.W. (1985), *J. Catal.* **96**, 454. Kock, A.J.H.M., de Bokx, P.K., Boellard, E., Klop, W., Geus, J.W. (1985), *J. Catal.* **96**, 468. Boellard, E., de Bokx, P.K., Kock, A.J.H.M., Geus, J.W. (1985), *J. Catal.* **96**, 481.
19. Gavillet, J.; Loiseau, A.; Journet, C.; Willaime, F.; Ducastelle, F.;Charlier, J.-C. (2001), *Physical Review Letters*, **87**(27, Pt. 1), 275504.
20. Loiseau, A.; Willaime, F. (2001) in *NATO Science Series, Series E: Applied Sciences,* Kluwer Academic Publishers, **372**, 133.
21. Kuznetsov, V.L., Usoltseva, A.N., Chuvilin, A.L., Obraztsova, E.D., Bonard, J.-M. (2001), *Phys. Rev.B.* **64** (23), 5401.
22. Oberlin, A., Endo, M., Koyama, T., *Carbon* 14 (1976) 133.
23. Gamaly, E.G. (1997) Ebbesen T. (ed) (1997), *Carbon Nanotube: Preparation and Properties*, CRC Press, Boca Raton, FL, p.163.
24. Qin, Lu-C., Iijima, S. (1997) *Chemical Physics Letters* **269**, 65.
25. Kiang, C.-H. W. A. Goddard III, *Phys. Rev. Lett.* 76 (1996) 2515.
26. Tibbetts, G. G., *J. Cryst. Growth*, 66 (1984) 632.
27. Satishkumar, B. C., Govindaraj, A., Sen, Rahul, Rao, C. N. R. (1998), *Chem. Phys. Lett.* **293**, 47.
28. Cheng, H. M., Li, F., Su, G., Pan, H. Y., He, L. L., Sun, X., Dresselhaus, M. S. (1998), *Appl. Phys. Lett.* **72**, 3282.
29. Nikolaev, P., Bronikowski, M., Bradley, R., Rohmund, F., Colbert, D., Smith, K., Smalley, R. E. (1999), *Chem. Phys. Lett.* **313**, 91.
30. Zhou, W. Ooi, Y., Russo, R., Papanck, P., Luzzi, D., Fischer, J., Bronikowski, M., Willis, P., Smalley, R. E. (2001), *Chem. Phys. Lett.* **350**, 6.
31. Bacsa, R. R., Laurent, Ch., Peigney, A., Vaugien, T., Flahaut, E., Bacsa, W. S., Rousset A. (2002), *J. Am. Ceram. Soc.*, **85**, 11, 2666.

32. Govindaraj, A., Flahaut, E., Laurent, Ch., Peigney, A., Rousset, A., Rao C. N. R. (1999), *J. Mater. Res.*, **14**, 6, 2567.
33. Flahaut, E., Govindaraj, A., Peigney, A., Laurent, Ch., Rousset, A., Rao, C. N. R. (1999), *Chem. Phys. Lett.*, **300**, 1-2, 236.
34. Nikolaev, P.N., Thess, A., Rinzler, A.G., et al., (1997), *Chem. Phys. Lett.*, **266** (5-6), 422.
35. Bougrine, A., Dupont-Pavlovsky, N., Naji, A., et al. (2001), *Carbon* **39,** 685.
36. Kuznetsov, V. L., Chuvilin, A. L., Butenko, Y. V., et al. (1998), *Chem. Phys. Lett.* **289,** 353.
37. Y. Saito, M. Okuda, N. Fujimoto, T. Yoshikawa, M. Tomita, T. Hayashi, *Jpn. J. Appl. Phys.* 33 (1994) L526.
38. H. Kataura, A. Kimura, Y. Ohtsuka, S. Suzuki, Y. Maniwa, T. Hanyu, Y. Achiba, *Jpn. J. Appl. Phys.* 37 (1998) L616.
39. S. Subramoney, R.S. Ruoff, D.C. Lorents, R. Malhotra, Nature 366 (1993) 637.
40. Y. Satio, M. Okuda, M. Tomita, T. Hayashi, Chem. Phys. Lett. 236 (1995) 419.
41. E.D. Obraztsova, J.-M. Bonard, V.L. Kuznetsov, V.I. Zaikovskii, S.M. Pimenov, A.S. Pozarov, S.V. Terekhov, V.I. Konov, A.N. Obraztsov, A.P. Volkov, *Nanostructured Mater.,* 12 (1999) 567.
42. Tontegode A. Ya.(1988), *Poverhnost* (Rus) 8, 13.
43. Dulot F., Eugene J., Kierren B., Malterre D., Eltsov K.N. (1999) *Phys. Low-Dim. Struct.* **1**/2. P. 217.
44. Tomanek, D., Zhong, W., Krastev, E. (1993), *Physical Review B,* **48,** 15461.
45. E. Miyazaki, *J. Catal.* 65 (1980) 84-94.
46. Bronikowski, M. J.; Willis, P. A.; Colbert, D. T.; Smith, K. A.; Smalley, R. E. (2001), *J. Vac. Sci. Technol.* **A19**, 1800.
47. Kuznetsov, V. L., Usoltseva, A. N., Chuvilin, A. L., Obraztsova, E. D., Bonard, J-M. (2002), *2001 MRS Fall Meeting,* v. 707, Z 6.22.
48. Hahner, J. H., Bronicovski, M. J., Azamian, B. R. et al. (1998), *Chem. Phys. Lett.* **296,** 195.
49. De Jong K. P., Geus J. W. (2000), *Catal. Rev.-Sci. Eng.* **42,** 481.
50. Wang, X., Hu, W., Liu, Y., Long, C., Xu Y., Zhou, S., Zhu, D., Dai, L. (2001), *Carbon* **39**, 1533.
51. Blank, V.D., Gorlova, I.G., Hutchison, J.L., Kiselev N.A., Ormont A.B., Polyakov E.V., Sloan J., Zakharov D.N., Zybtsev S.G. (2000) *Carbon* **38,** 1217.

CCVD SYNTHESIS OF SINGLE- AND DOUBLE-WALLED CARBON NANOTUBES

E. FLAHAUT, A. PEIGNEY and Ch. LAURENT
CIRIMAT, UMR CNRS 5085, Université Paul Sabatier
31062 Toulouse Cedex 04, FRANCE
flahaut@chimie.ups-tlse.fr

Abstract: We describe the synthesis of single and double-walled carbon nanotubes by an original catalytic chemical vapor deposition method based on the selective reduction of oxide solid solutions in H_2-CH_4 atmosphere. For example, the reduction of an α–$Al_{2-2x}Fe_{2x}O_3$ solid solution produces pristine Fe nanoparticles at a temperature high enough to catalyze the decomposition of CH_4 and the *in situ* formation of carbon nanotubes (CNTs). The so-obtained products are CNTs-metal-oxide composite powders with a very homogeneous dispersion of the CNTs. This method was extended to $Mg_{1-y}M_yAl_2O_4$ and $Mg_{1-z}M_zO$ solid solutions (with M = Fe, Co, Ni or a binary alloy), to obtain CNT-M-$MgAl_2O_4$ and CNT-M-MgO composites powders, respectively. The use of MgO-based catalysts is of special interest because of their easy removal by dissolution in an HCl aqueous solution, allowing the preparation of isolated CNTs. Some control of the number of walls and diameter distribution can be achieved through the CCVD parameters, such as the composition of the H_2-CH_4 atmosphere, the reduction temperature and the time spent at that temperature. A global overview of the CCVD synthesis of single- and double-walled carbon nanotubes is proposed through a review of some of the most striking results in the field.

Keywords: Double-walled carbon nanotube, CCVD, Synthesis, Methane, Oxide solid solution

1. INTRODUCTION

This paper will review some of the most striking results obtained in the field of the synthesis of single-walled and double-walled carbon nanotubes (SWNTs and DWNTs, respectively) by catalytic chemical vapour deposition (CCVD). After briefly introducing the main synthesis methods that are used to produce SWNTs (arc-discharge, laser ablation and CCVD), we will get an overview of the different technical choices that have been made to achieve the CCVD synthesis of SWNTs. This part of the paper will also be illustrated by experimental results obtained by our group. The emphasis will be put on the understanding of the

S. Guceri et al. (eds.), Nanoengineered Nanofibrous Materials, 35-45.

phenomena that play a key-role for the specific synthesis of carbon nanotubes (CNTs) with a small number of walls (SWNTs, DWNTs).

CNTs were first brought to light in 1991 by Iijima. These 1D hollow objects can be schematically described by the rolling over itself of a fragment of a graphene sheet. There are numerous different possible structures, described by two indices (n, m). CNTs are typically only a few nanometers wide, and their length can range between a few hundreds of nanometers and up to 100 μm. For this reason, they have a very high aspect ratio (typically between 5000 and 50000). They also have a very high specific area (the theoretical value for a closed SWNT is 1300 m^2/g), which decreases quickly as the number of layers increases[1]. Their mechanical properties are exceptional: they have a very important stiffness, the Young modulus is typically close to 1 TPa for SWNTs, but they also exhibit a very high flexibility. Their electrical properties are remarkable because they can behave either as metallic conductors or as semiconductors, depending only on their structure.

Three main synthesis methods have been developed for the synthesis of SWNTs. Historically, the first method used is the arc-discharge between two graphite electrodes. For the synthesis of SWNTs, one of the two electrodes (the anode) contains a catalyst (often present as a powder placed in a hole bored in the electrode). This method has been used since 1992 [2] and generally requires deep purification steps to remove fullerene material, amorphous carbon, graphitic particles and soot, which always form together with the CNTs, although important modifications and improvements of the process have now been proposed. Laser ablation was first reported in 1995 by Guo *et al.*[3] and allows the synthesis of good quality ropes of SWNTs (70-90%). The main drawback of this method is its cost but it could be envisaged as a continuous synthesis process. The last method is CVD, which was first restricted to carbon nanofibres and then multi-walled CNTs, until the first syntheses of SWNTs by CCVD were reported in 1997. CCVD is especially interesting because it is based on the catalytic decomposition of hydrocarbons or carbon monoxide over catalytic particles, and this is a rather low-cost and possibly large-scale process. To transpose the knowledge accumulated for years during the synthesis of carbon fibres, researchers first had to understand the role played by the catalytic particles: they have to be kept at a nanometric size for the synthesis of SWNTs and DWNTs. The most important point is that these particles have to keep nanometric dimensions until the catalytic decomposition of the carbon source actually starts. Different ways have been proposed to keep the metal particles as small as possible: working at the lowest possible temperature (which will determine the carbon source to be used), using refractory metals, opting for fast thermal treatments or generating the nanoparticles in situ at high temperature. This last choice will be described in details later on.

What are the typical CCVD conditions? the carbon supply is generally CO or an hydrocarbon (CH_4, C_2H_2, C_2H_4, C_6H_6, etc.), often mixed together with an inert gas (Ar, N_2, He) or with H_2. The working temperature depends strongly on the carbon source and must fulfil two *a priori* contradictory conditions: being *high enough* to allow the catalytic decomposition of the carbon source and the diffusion of carbon species, but at the same time be *low enough* to avoid pyrolysis, that is to say non-catalytic self-decomposition of the carbon source. The pressure is generally close to one atmosphere (but not systematically), and it can have important effects on the quality of the CCVD products. The flow rate has to be adapted both to the amount of catalyst present in the CCVD reactor, the kind of CNTs (SWNTs, DWNTs or MWNTs) and, of course, the nature of the carbon source. Thermodynamics and kinetics must be taken into account. The reactors used are typically quartz tubes fitted inside one-stage tubular furnaces. Two-stage furnaces are sometimes used: the catalyst or its precursor is placed in the first stage, and the catalytic reaction takes place in the second stage, at a higher temperature. Vertical furnaces are used when fluidised beds are required.

Different strategies have been developed over the years to achieve the CCVD synthesis of SWNTs and DWNTs. We will, for first, review the synthesis of SWNTs on catalytic nanoparticles produced by decomposition of organometallic precursors. Next, we will focus on the use of supported catalytic nanoparticles, prepared (i) in situ by reduction of oxide solid solutions, (ii) in situ after impregnation of a support by catalytic precursors or (iii) pre-formed before the introduction of the carbon source.

2. SYNTHESIS OF SWNTs and DWNTs BY THE CCVD PROCESS

2.1 Catalytic nanoparticles formed in situ by decomposition of organometallic precursors

Sublimated metallocenes have often been used as a precursor for the synthesis of SWNTs. For example, Satishkumar *et al.*[4] have compared the efficiency of different metallocenes ($M(C_5H_5)_2$ with M = Fe, Co or Ni) and that of $Fe(CO)_5$, at 1100°C in both cases. The carbon source was C_2H_2 (diluted in Ar). The SWNTs prepared by this method had a very narrow diameter distribution, between 1 and 1.5 nm. Fe and Co gave better results than Ni, and amorphous deposit was present in all cases. However, the use of a mixture of metallocenes gave cleaner tubes, with less amorphous deposit. Ci *et al.*[5] used a similar process (ferrocene and C_2H_2 diluted in Ar) and reported that the optimal temperature range was between 900 and 1000°C. They obtained SWNTs (diameter ranging from 0.7 to 1.6 nm), individual or in small bundles, with less amorphous deposit than previously reported. However, their sample contained numerous Fe nanoparticles, probably encapsulated into graphitic shells. They also studied the influence of the partial pressure of C_2H_2 and found an optimum for 5 Torr. Cheng *et al.*[6,7] have replaced C_2H_2 by C_6H_6 (together with H_2) and added a small amount of thiophene (temperature between 1100 and 1200°C, reaction time between 1 and 30 minutes, or longer). They reported the synthesis of long and wide ropes (or ribbons) of SWNT bundles, with rope diameters of 100 μm and length up to 3 cm. The ribbons consisted in roughly-aligned bundles of aligned SWNTs (diameter ranging from 1 to 3 nm). Thiophene (source of S) was claimed to increase the yield of SWNTs. The SWNTs presented some amorphous carbon deposit on the surface of the bundles, together with catalyst particles and carbon blacks, suggesting that the product would require some purification before use. Nikolaev *et al.*[8] have reported a very promising process (HipCO), based on the catalytic decomposition of CO over Fe nanoparticles formed by thermal decomposition of $Fe(CO)_5$. The pressure of CO was varied between 1 and 10 atm. has an important effect on the selectivity of the process: increasing the pressure led to a narrower diameter distribution of the SWNTs, together with a decrease of the mean diameter. The optimal conditions are a temperature of 1200°C and a pressure of 10 atm. In these conditions, the process (which could be a continuous one) yielded around 15 mg of product per hour. The SWNTs are individual or gathered into small bundles, and are decorated with numerous metal particles. These latter can be removed by air oxidation (200°C, 24h) followed by HCl washing[9]. Ajayan's group[10] has recently reported the synthesis of very long SWNTs ropes, up to 20 cm. The process involved ferrocene as the catalyst, and n-hexane as the carbon source, with additions of thiophene. The large ropes had a diameter close to 15 μm but were made of SWNTs ranging from 1.1 to 1.7 nm. As in the previous case, sulphur seemed to increase the selectivity towards SWNTs. The process used a vertical furnace and the use in continuous was proposed.

2.2 Supported catalytic nanoparticles formed in situ by reduction of oxide solid solutions

This process was first proposed by our group in 1997 [11] for the synthesis of SWNTs together with MWNTs. It is based on the selective reduction of $Al_{2-2x}Fe_{2x}O_3$ solid solutions by H_2/CH_4 mixtures at a temperature between 900 and 1070°C. Heating (from room temperature), dwelling (if any dwell) and cooling down are done in the H_2/CH_4 environment. The characterisation of the samples is achieved both by macroscopic (Carbon elemental analysis, BET surface area measurements, X-Rays diffraction (XRD), thermal analysis) and microscopic (SEM, TEM, Raman spectroscopy) techniques. We have proposed a macroscopic characterization method based on the determination of the carbon quantity in the powder (C_n) and on specific surface area measurements[1, 12]. This allows us to calculate a parameter (ΔS) representing the quantity of CNTs and a parameter ($\Delta S/C_n$) representing the quality of the carbon, a higher figure denoting the presence of tubes with a smaller diameter and/or CNTs with less walls and/or more carbon in a tubular form[1]. We have shown that the selective reduction in H_2 of oxide solid solutions such as α-$Al_{2-2x}Fe_{2x}O_3$ leads to Fe-Al_2O_3 nanocomposite powders containing Fe nanoparticles (< 10 nm in diameter) both inside and at the surface of the Al_2O_3 grains. The particles start to form at a relatively high temperature (in the range 600-800°C). It was further shown[11] that the reduction in a

H_2-CH_4 atmosphere allows to use the freshly formed surface Fe nanoparticles to form the CNTs. CH_4 decomposes on the nanoparticles and supplies them with carbon which accumulates until the saturation in carbon is reached. The nucleation and growth of CNTs then takes place, stopping the particle growth. The so-obtained powders contain an enormous amount of SWNTs and thin MWNTs, with a very homogenous dispersion of the CNTs which probably could not be matched by methods involving a mechanical mixing. Incidentally, to the best of our knowledge, this was the first time that the formation of SWNTs by CCVD from an hydrocarbon was reported. Several parameters related to the starting oxide material have been investigated in order to increase the proportion of CNTs with respect to the other carbon species (carbon nanofibres, graphitic shells, disordered carbon) in the first place and then to improve the selectivity of the process towards SWNTs or DWNTs.

These parameters include the iron content[12], the crystallographic form[13] and the specific surface area[14] of Al_2O_3-based solid solutions. The effects upon the CNT formation of some parameters related to the experimental conditions of the reduction, such as the composition of the H_2-CH_4 atmosphere, the reduction temperature and the time spent at that temperature have also been examined[15-18].

The nature of the metallic phase (Fe, Co, Ni and their binary alloys) was also investigated[18, 19] with $MgAl_2O_4$-based solid solutions (This was not possible using Al_2O_3-based oxides because only ions at the oxidation state +III could be stabilized in the oxide matrix). Preliminary work on the $MgAl_2O_4$ system[18] (Fe, Co or Ni in different concentrations) indicated that an increase in the transition metal content of the catalyst yielded more CNTs (up to a metal content of 10.0 wt. %), but caused a decrease in the carbon quality. The best compromise was to use 6.7 wt. % of metal in the catalyst. Co gave superior results with respect to both the quantity and quality of the CNTs. In the case of Fe, the quality was notably hampered by the formation of Fe_3C particles. Working with a binary alloy between Co and Mo was then found to be the best compromise between an important quantity of CNTs while preserving a good quality[19]. The control of the experimental CCVD parameters using MgO-based[20] solid solutions allowed us to obtain a rather good selectivity towards the number of walls of the CNTs, 80% of which are SWNTs and DWNTs. The CNTs are, again, very well dispersed within the oxide matrix grains and their specific surface area can be as high as 980 m^2/g [21] (Co with an addition of Mo). Furthermore, it was also shown that when using MgO-based solid solutions, a simple soaking in HCl allows separating the CNTs from the metal and MgO matrix without damaging them[20]. All these experiments on different systems revealed the necessity of reducing *true solid solutions*. For example in the case of $Mg_{1-x}Co_xO$, we have investigated the solubility limit of CoO within MgO, in our synthesis conditions[20]. We have thus varied x from 0 to 0.4 and characterised by XRD the corresponding oxides (Fig. 1). It is clear from the XRD data that from $x = 0.15$, the oxide is no more a solid solution and the peaks corresponding to Co_3O_4 are clearly visible, their intensity increasing with the Co loading. In these conditions, the oxide is rather a mixture of Mg_{1-}

$_{y}Co_{y}0$ (with $y < x$) and Co_3O_4. This latter oxide, more easily reducible than Co^{2+} in the solid solution, will thus be reduced first, at lower temperature. As a consequence, it will give larger Co particles (because it is also very likely that at such a low temperature CH_4 could not yet be catalytically decomposed on these particles), which, in turn, will later lead to the growth of carbon nanofibres.

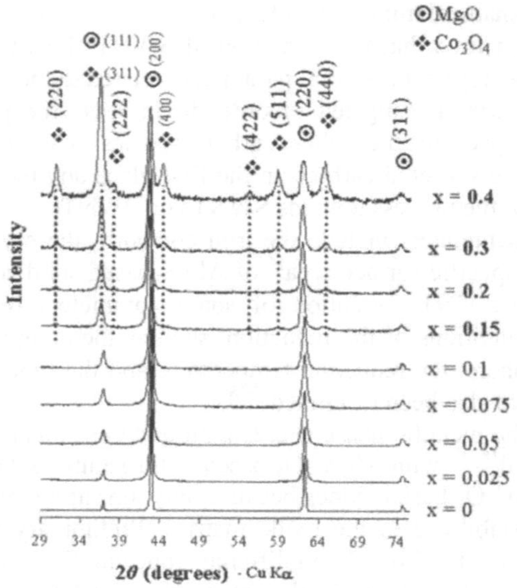

Figure 1: X-Ray diffractograms of $Mg_{1-x}Co_xO$ oxides ($0 \leq x \leq 0.4$).

This is verified rather clearly on the SEM images of the CNTs-Co-MgO composite powders prepared by CCVD on these oxides (Fig. 2): carbon nanofibres appear very clearly as soon as $x \geq 0.2$. The ratio of carbon nanofibres then increases regularly with the Co loading, as well as their thickness. The observation of such fibres, related to the presence of the oxide Co_3O_4 in the starting catalyst, is similar to what happens in the case of the $Al_{2-2x}Fe_{2x}O_3$ system when $x \geq 0.1$. In that case, the catalyst is in fact a mixture of two separated solid solutions, one being hematite-rich. Similarly, the early reduction of the latter is responsible for the formation of large Fe particle which later catalyse the formation of very similar carbon nanofibres.

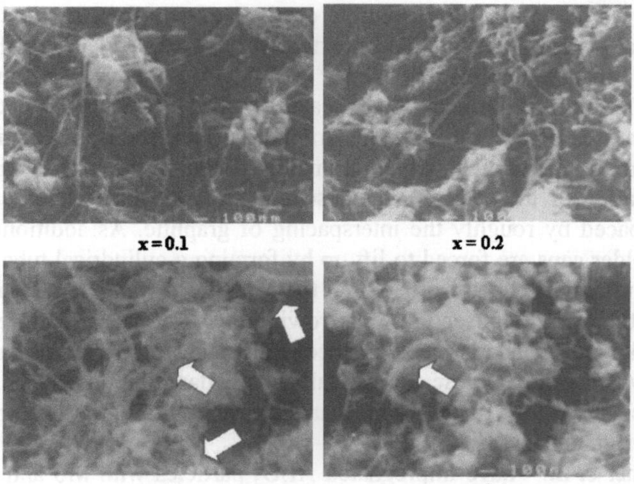

Figure 2: SEM images of the CNT-Co-MgO composite powders obtained from $Mg_{1-x}Co_xO$ oxides (with x = 0.1, 0.2, 0.3 and 0.4).

Our latest results obtained with the MgO system[21] are very promising for the large-scale production of DWNTs because we managed to prepare in one unique batch 1.3 g of CNTs (77% of which are DWNTs, only less than 5% triple-walled CNTs; purity: around 98 at.% of carbon) from 10 g of $Mg_{0.99}(Co_{0.0075}Mo_{0.0025})O$ [5]. These DWNTs have been fully characterised[21, 22] and Raman spectroscopy even suggests that the inner and outer wall of some of the DWNTs could be observed separately (RBM peaks).

In order to improve both the accessibility of the metal and the diffusion of the gases by increasing the open porosity, oxides foams[23] have also been prepared by the gelcasting-foam method. Processing the solid solution in the form of a highly (98%) porous foam allowed to multiply by 4 the amount of CNTs, because the surface accessible for CH_4 decomposition during the CCVD was greatly enhanced.

2.3 Supported catalytic nanoparticles formed after impregnation of a support by salts in solution

The first example in this category is also the first example of synthesis of SWNTs by CCVD. Reported by Dai *et al.*[24], it was based on the catalytic decomposition of CO on Mo nanoparticles supported over Al_2O_3 particles. The Al_2O_3 particles were first impregnated by a solution of a Mo salt in methanol. The catalyst was calcined and ground before the CCVD step. This led to the growth of mainly individual SWNTs, with diameters ranging from 1 to 5 nm, in close correlation with the size of the catalytic particle found attached to the tube end. The growth mechanism proposed by the authors (*Yarmulke* mechanism) was original and proposed the formation in the early stage of a graphene cap (the so-called yarmulke) with its edges strongly chemisorbed onto the metal. Newly arriving carbon would continue to assemble on the surface of the catalyst. Since the basal plane of graphite has an extremely low surface energy, the total surface energy diminishes. Newly arriving carbon will continue to assemble on the surface of the catalyst. There are three places for additional carbon to go. (1) The original surface shell can continue to grow around the particle, which ultimately results in the over-coating and deactivation of the catalyst. (2) A second cap can form underneath the first, spaced by roughly the interspacing of graphite. As additional caps form, older caps are forced to lift up by forming a cylindrical tube whose open end remains chemisorbed onto the catalytic particle. (3) Carbon can add to the cylindrical section of a growing layer. A crucial feature of the yarmulke mechanism is the avoidance at all stages of growth of any open edge, which would expose energetically costly dangling bonds (see ref. 21 for a scheme of the *Yarmulke* mechanism, from the descriptions provided in the original paper by Dai *et al.*[24]).

Hafner *et al.*[25] have impregnated Al_2O_3 particles with Mo and Fe/Mo mixtures (dissolution in ethanol) and used either CO or C_2H_4 as the carbon source. In the case of C_2H_4, a pre-reduction of the oxide was realised by heating the catalyst in a mixture of Ar and H_2, before introducing the hydrocarbon. The authors reported that only pure Mo led to the formation of only SWNTs, with a very narrow diameter distribution (0.8-0.9 nm). In the case of the Fe/Mo mixture and whatever the carbon source, mixtures of SWNTs and DWNTs were obtained in various ratios. In the case of C_2H_4, increasing the temperature from 700°C to 850°C led to an increase of the ratio of DWNTs from 30% to around 70%, respectively. However, the samples contained important amounts of graphitic shells. A model was proposed to describe the variation of the energy of formation of SWNTs and carbon shells versus their diameter. The comparison of the data indicated that nanoparticles having a diameter less than 3 nm would favour the formation of SWNTs. On the contrary, the formation of carbon shells would be favoured if the diameter was larger than 3 nm. Harutyunyan *et al.*[26] have studied more in details the influence of the pre-reduction of the nanoparticles before the introduction of the carbon source. They used also Fe or Fe/Mo mixtures as the catalyst, supported over Al_20_3 (impregnation from solutions in methanol). They reported that it was possible to grow SWNTs at a temperature as low as 680°C if the catalytic nanoparticles were previously reduced (H_2:He 1:9 at 500°C between 10 and 20 hours), before the introduction of the carbon source (CH_4). The resulting SWNTs were individual or gathered into small bundles and contained some amorphous deposits, as well as catalyst particles. As many other authors, they note that Fe:Mo yields more SWNTs than Fe alone. Our last example of catalytic nanoparticles prepared by impregnation of a support was reported by Maruyama *et al.*[27] and deals with the impregnation of a zeolite with Fe and Co salts. The carbon source, alcohol (methanol or ethanol), is the original part of the work. The pressure of alcohol is rather low (5 Torr) and the optimal temperature comprised between 800 and 900°C. They reported that the diameter of the SWNTs increases with the CCVD temperature.

2.4 Pre-reduced supported catalytic nanoparticles

In the following examples, the catalytic nanoparticles were deliberately formed before the introduction of the carbon source. The first example comes from Resasco's group and involves Co and Mo as the catalytic species, and SiO_2 as the support. The carbon source is CO [28, 29].The characterisation of the CCVD samples by Temperature Programmed Oxidation (TPO) analysis allows an easy comparison of the samples (TPO measurements are conducted by passing a continuous flow of 5% O_2 in He, while the temperature is linearly increased at a rate of 12°C min). The corresponding salts were dissolved in water and used to impregnate the SiO_2 support, before air calcination followed by reduction in H_2 at 500°C. CO together with He was then introduced at 700°C. Pure Mo did not allow the production of SWNTs, and pure Co led to a mixture of SWNTs and MWNTs. However, a clear synergistic effect of the combination of the two metals was brought to light, leading to almost only SWNTs (no MWNT, no graphitic carbon, but however presence of metal nanoparticles). This was seen clearly by TPO and confirmed by TEM observations. The last example concerns the use of calibrated Fe nanoparticles prepared by decomposition of $Fe(CO)_5$ in organic acids (oleic, lauric, octanoic)[30]. The stabilised metal clusters were then reduced in H_2/Ar at 800°C before introducing a hydrocarbon (C_2H_4 at 800°C or CH_4 at 900°C). A clear correlation was found between the original size of the Fe clusters and the diameter of the corresponding SWNTs. However, the samples did not contain only SWNTs and the proportion of CNTs with two or more walls increased with the Fe particle diameter. This is in good agreement with the model proposed by Hafner *et al.*[25], described previously. This is also in good agreement with our experimental results and confirms once more the importance of controlling the size of the catalytic nanoparticles. Replacing C_2H_4 by CH_4 resulted in cleaner CNTs (less amorphous deposit). It was also reported that the hydrocarbon partial pressure had to be increased when large CNTs were to be formed (large Fe clusters).

3. CONCLUSION

CCVD is a very efficient method to grow SWNTs and DWNTs. Many different processes are available and the common feature for all of them is the use of catalytic nanoparticles, which have to be as small as possible. There is a general agreement that Fe and Co give better results than Ni, as far as SWNTs and DWNTs are concerned (this is not true anymore for MWNTs). The nanoparticles can be produced by different methods. When they are formed by decomposition of an organometallic precursor, the best results are obtained with iron (ferrocene or $Fe(CO)_5$). The addition of small amounts of sulphur (thiophene) has a positive effect and increases the selectivity towards SWNTs. This could be attributed to the lowering of the melting point of the catalytic metal. When the catalytic nanoparticles are supported over oxides (typically Al_2O_3, $MgAl_2O_4$, SiO_2 or MgO), Co seems to be the more efficient metal whatever the way the nanoparticles are produced (selective reduction of an oxide solid solution, impregnation of the support by a precursor, generally a salt such as a nitrate or chloride, which is then converted to an oxide by thermal decomposition). Mo plays an important role, when it is mixed with another metal (Fe, Co), and always increases the yield; however, it does not systematically increase the selectivity towards SWNTs (this seems to depend to some extent on the carbon source used). Some results indicate that the higher the proportion of Mo the higher the selectivity (with CO), but some others find an opposite behaviour (using CH_4). Amongst carbon sources, CH_4 gives the cleaner CNTs (with less or no amorphous carbon deposit at all). Unfortunately, the use of CH_4 requires higher temperatures than C_2H_2 and C_2H_4 and thus is not adapted to all the substrates. The use of alcohols as the carbon source is promising (ease of use, low cost). There is no "magic" temperature for the CCVD process and it has to be optimised for each catalyst, and for each carbon source. CCVD is a very flexible method, allowing some degree of control of the diameter of the CNTs (control of the diameter distribution) and the number of walls (distribution of the number of walls). This is mainly obtained by the control of the diameter of the starting catalytic nanoparticles. However, some important characteristics, such as the length or the chirality of the CNTs, are still out of control (whatever the synthesis method).

References

1. Peigney, A., Laurent, Ch., Flahaut, E., Bacsa, R., Rousset, A. (2000) Specific surface area of carbon nanotubes and bundles of carbon nanotubes, *Carbon* 39, 507-514.
2. Cho, W., Hamada, E., Kondo, Y., Takayanagani, K. (1996) Synthesis of carbon nanotubes from bulk polymer, *Appl. Phys. Lett.* 69, 278-279.
3. Guo, T. Nikolaev, P., Thess, A., Colbert, D., Smalley, R. E. (1995) Catalytic growth of single-walled nanotubes by laser vaporization, *Chem. Phys. Lett.* 243, 49-54.
4. Satishkumar, B. C., Govindaraj, A., Sen, Rahul, Rao, C. N. R. (1998) Single-walled nanotubes by the pyrolysis of acetylene-organometallic mixtures, *Chem. Phys. Lett.* 293, 47-52.
5. Ci, L., Xie, S., Tang, D., Yan, X., Li, Y., Liu, Z., Zou, X., Zhou, W., Wang, G. (2001) Controlable growth of single-walled carbon nanotubes by pyrolizing acetylene on the floating Fe catalyst, *Chem. Phys. Lett.* 349, 191-195.
6. Cheng, H. M., Li, F., Su, G., Pan, H. Y., He, L. L., Sun, X., Dresselhaus, M. S. (1998) Large-scale and low-cost synthesis of single-walled carbon nanotubes by the catalytic pyrolysis of hydrocarbons, *Appl. Phys. Lett.* 72, 3282-3284.
7. Cheng, H., Li, F., Sun, X., Brown, S., Pimenta, M., Marucci, A., Dresselhaus, G., Dresselhaus, M. (1998) Bulk morphology and diameter distribution of single-walled carbon nanotubes synthesized by catalytic decomposition of hydrocarbons, *Chem. Phys. Lett.* 289, 602-610.
8. Nikolaev, P., Bronikowski, M., Bradley, R., Rohmund, F., Colbert, D., Smith, K., Smalley, R. E. (1999) Gas-phase catalytic growth of single-walled carbon nanotubes from carbon monoxide, *Chem. Phys. Lett.* 313, 91-97.
9. Zhou, W. Ooi, Y., Russo, R., Papanck, P., Luzzi, D., Fischer, J., Bronikowski, M., Willis, P., Smalley, R. E. (2001) Structural characterisation and diameter-dependent oxidative stability of single wall carbon nanotubes synthesized by the catalytic decomposition of CO, *Chem. Phys. Lett.* 350, 6-14.
10. Zhu, H., Xu, C., Wu, D., Wei, B., Vajtai, R., Ajayan, P. (2002) Direct synthesis of long single-walled carbon nanotube strands, *Science* 296, 884-886.

11. Peigney, A., Laurent, Ch., Dobigeon F., Rousset, A. (1997) Carbon nanotubes grown in situ by a novel catalytic method *J. Mater. Res.* 12, 613-615.
12. Peigney, A., Laurent, Ch., Dumortier, O., Rousset, A. (1998) Carbon nanotubes-Fe-Alumina nanocomposites. Part I: influence of the Fe content on the synthesis of powders, *J. Eur. Ceram. Soc.* 18 1995-2004.
13. Laurent, Ch., Peigney, A., Rousset, A. (1998) Synthesis of carbon nanotube-Fe-Al_2O_3 nanocomposite powders by selective reduction of different $Al_{1.8}Fe_{0.2}O_3$ solid solutions, *J. Mater. Chem.* 8, 1263-1272.
14. Laurent, Ch., Peigney, A., Flahaut, E., Rousset, A. (2000) Synthesis of carbon nanotubes-Fe-Al2O3 powders : influence of the characteristics of the starting $Al_{1.8}Fe_{0.2}O_3$ oxide solid solution, *Mat. Res. Bull.* 35, 661-673.
15. Peigney, A., Coquay, P., Flahaut, E., De Grave, E., Vandenberghe, R. E., Laurent, Ch. (2001) A study of the formation of single- and double-walled carbon nanotubes by a CVD method, *J. Phys. Chem. B* 105, 9699-9710.
16. Peigney, A., Laurent, Ch., Rousset, A. (1999) Influence of the composition of a H_2-CH_4 gas mixture on the catalytic synthesis of carbon nanotubes-Fe/Fe_3C-Al_2O_3 nanocomposite powders, *J. Mater. Chem.* 9, 1167-1177.
17. Flahaut, E., Peigney, A., Laurent, Ch. (2003) Double-walled carbon nanotubes in composite powders, *J. Nanosci. Nanotech.* 3, 1-8
18. Govindaraj, A., Flahaut, E., Laurent, Ch., Peigney, A., Rousset, A., Rao, C. N. R. (1999) An investigation of carbon nanotubes obtained from the decomposition of methane over reduced $Mg_{1-x}M_xAl_2O_4$ (M = Fe, Co, Ni) spinel catalysts, *J. Mater. Res.* 14, 2567-2576.
19. Flahaut, E., Govindaraj, A., Peigney, A., Laurent, Ch., Rousset, A., Rao, C. N. R. (1999) Synthesis of single-walled carbon nanotubes using binary (Fe, Co, Ni) alloy nanoparticles prepared in situ by the reduction of oxide solid solutions, *Chem. Phys. Lett.* 300, 236-242.
20. Flahaut, E., Peigney, A., Laurent, Ch., Rousset, A. (2000) Synthesis of single-walled carbon nanotube-Co-MgO composite powders and extraction of the nanotubes, *J. Mater. Chem.* 10, 249-252.
21. Flahaut, E., Bacsa, R., Peigney, A., Laurent, Ch. (2003) Gram-Scale CCVD Synthesis of Double-Walled Carbon Nanotubes, *Chem. Commun.* 1442-1443
22. Colomer, J. F., Henrard, L., Flahaut, E., Van Tendeloo, G., Lucas, A. A., Lambin, Ph. (2003) Rings of Double-Walled Carbon Nanotube Bundles, *Nano Lett.* 3, 685-689.
23. Rul, S., Laurent, Ch., Peigney, A., Rousset, A. (2003) Carbon nanotubes prepared in situ in a cellular ceramic by the gelcasting-foam method, *J. Eur. Ceram. Soc.* 23, 1233-1241.
24. Dai, H., Rinzler, A., Nikolaev, P., Thess, A., Colbert, D., Smalley, R. E. (1996) Single-wall nanotubes produced by metal-catalyzed disproportionation of carbon monoxide, *Chem. Phys. Lett.* 260, 471-475.
25. Hafner, J., Bronikowski, M., Azamian, B., Nikolaev, P., Rinzler, A., Colbert, D., Smith, K., Smalley, R. E. (1998) Catalytic growth of single-wall carbon nanotubes from metal particles, *Chem. Phys. Lett.* 296, 195-202.
26. Harutyunyan, A., Pradhan, B., Kim, U., Chen, G., Eklund, P. C (2002) *CVD* Synthesis of Single Wall Carbon Nanotubes under "Soft" Conditions, *Nano Lett.* 2, 525-530.
27. Maruyama, S., Kojima, R., Miyauchi, Y., Chiashi, S., Kohno, M. (2002) Low-temperature synthesis of high-purity single-walled carbon nanotubes from alcohol, *Chem. Phys. Lett.* 360, 229-234.
28. Kitiyanan, B., Alvarez, W., Harwell, J., Resasco, D. E. (2000) Controlled production of single-wall carbon nanotubes by catalytic decomposition of CO on bimetallic Co-Mo catalysts, *Chem. phys. Lett.* 317, 497-503.
29. Alvarez, W., Kitiyanan, B., Borgna, A., Resasco, D. E. (2001) Synergism of Co and Mo in the catalytic production of single-wall carbon nanotubes by decomposition of CO, *Carbon* 39, 547-558.
30. Cheung, C., Kurtz, A., Park, H., Lieber, C. M. (2002) Diameter-Controlled Synthesis of Carbon Nanotubes, *J. Phys. Chem. B* 106, 2429-2433.

PRECISE SEMICONDUCTOR, METAL AND HYBRID NANOTUBES AND NANOFIBERS

V.Ya. PRINZ
Institute of Semiconductor Physics,
Siberian Division of Russion Academy of Sciences,
ave.Lavrent'eva,13, Novosibirsk, 630090, Russia

Abstract: This overview focuses on the fabrication technology and on the properties of precise nanotubes and nanofibers formed from monocrystalline A3B5 or Si/GeSi strained heterofilms, or from hybrid films. Experimental results on the batch formation of solid-state micro- and nanotubes, and helical fibers are described. The nanotubes and nanofibers fabricated offer much promise as building blocks for nanoelectronic and nanomechanical devices, their fabrication technology being quite compatible with the well-established integrated-circuit technology.

1. INTRODUCTION

Nanotechnologies aimed at fabricating novel nanoelectronic and nanomechanical devices and creating novel materials, including nanofiberous ones, are rapidly progressing [1].

Many problems remain to be solved in this field of research until these applications become practical [1, 2]. The most important issue is to control the size and position of self-formed tubes, shells and fibers more precisely with atomic-scale accuracy. In nanostructuring, Nature gives unprecedented examples which Mankind still fails to surpass. Nature constructs natural objects from perfectly precise building blocks, individual molecules and atoms, assembling them in more complex formations using self-formation and self-organization processes. An example of natural nanofibers is given by desoxyribonucleic acid (DNA), whose diameter is only 2 nm, with a length of up to several meters. Most importantly, DNA can be controllably transported through nanotubes, for instance, through bacteriophage proboscises 5 nm in diameter. Mankind is now coping with the problem of creating precise fibers with ultimately small, atomic-scale diameters, and assembling them in desirable architectures.

Some of the first steps towards this goal were made in our works [3-10] describing a new method that makes it possible to reach molecular precision in nanostructuring. This method employs a self-rolling

S. Guceri et al. (eds.), Nanoengineered Nanofibrous Materials, 47-63.

procedure intended for rolling cylindrical nano-objects from strained semiconductor heterofilms detached from substrates. With this method, we succeeded in obtaining a new class of nanostructures (tubes, scrolls, spirals, rings, etc.). We fabricated free-standing solid nanotubes with diameters down to 2 nm [3] and lengths up to 1 cm, whose properties make it possible to consider them as fibers. We proposed methods for the directional rolling of films and formed ordered arrays of nanotubes and other nano-objects that can be used as templates for producing polymer nanofibers [5, 6, 10].

The present review focuses on new methods for fabricating precise nanotubes, nanoshells and nanofibers. The emphasis on the new method is intended for batch production of semiconductor single-crystal and metal nanofibers.

2. METHODS FOR FABRICATING PRECISE MICRO- AND NANOTUBES

Apparently, to produce high-precision nanoobjects, one needs to use precise building blocks – atoms, molecules, or macromolecules. Analyzing the present-day solid-state technology, we see only one type of artificial precise building blocks – molecular-beam epitaxy (MBE)-grown ultra-thin epitaxial layers. The thickness of such layers can be controlled *in situ* with atomic precision and can be made as small as one monolayer (ML). Such ultimately thin layers constitute planar macromolecules. Molecular-beam epitaxy permits a monolayer-by-monolayer growth of multi-layered strained uniform heterostructures. Yet, the individual layers as heterostructures are immovable. This circumstance seriously restricts the possibilities of their practical application.

Recently, we proved it possible to detach ultra-thin (down to 2 ML) strained bilayer films from substrates on which these films were preliminarily grown on, by selective etching of underlying sacrificial layers provided between the films and the substrates [3]. Using this possibility, we proposed a new method permitting precise fabrication of three-dimensional (3D) micro- and nanoshells of cylindrical geometry (tubes, scrolls, rings, spirals, etc.) by the self-rolling of strained solid heterofilms [3-23]. Like carbon nanotubes [1], the precise nanotubes and nanoshells formed in this manner from semiconductors, dielectrics, or metals, can be used as building blocks for nanoelectronic and nanomechanical devices and as templates for nanofibers.

It is of utmost importance that, since the starting multilayered films have precise *in situ* controlled thickness and composition, the new method enables obtaining precise nano-objects. Below, we give a brief overview of an extended version of the fabrication methods of such objects, and report some new results on nanostructuring in the fields where substantial progress has been currently made.

The method for fabricating nanotubes from GaAs/InAs strained heterostructures [3-4] is schematically illustrated by *Fig. 1.* The lattice constants of free GaAs and InAs layers differ considerably ($\Delta a/a = 7.2\%$). Epitaxial growth of these layers on, say, an InP substrate makes it possible to obtain a biaxially strained bifilm whose InAs layer is in-plane compressed to lattice-match the InP substrate, and the GaAs layer is tensile-stressed *(see Fig. 1, a, b)*. The GaAs/InAs bifilm can be detached from the substrate by selective etching of the underlying AlAs sacrificial layer preliminarily grown on the substrate. As the bifilm detaches from the substrate, the interatomic forces in the compressed and stretched layers tend to accommodate the interatomic distances. These forces in the InAs and GaAs layers *(F_1 and F_2, respectively)* are oppositely directed and produce a moment of force that acts to bend the bifilm *(Fig. 1, c)*. As a result, the planar bifilm rolls up in a tube-scroll.

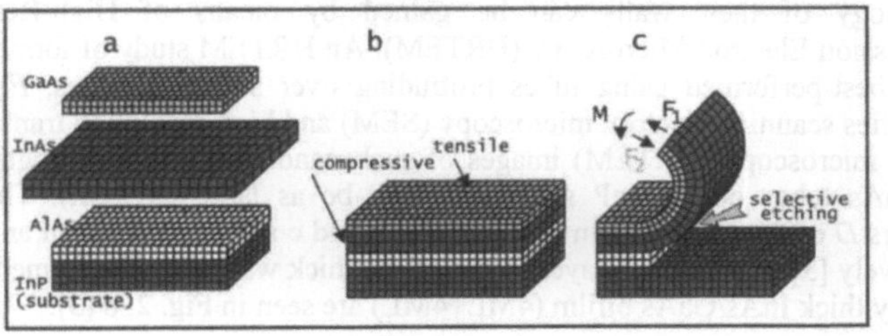

Figure 1..Schematic representation of the method for forming free-standing bent or rolled-up several monolayers thick films. a - free 2ML-thick InAs and GaAs layers with naturally mismatched lattice constants ($\Delta a/a = 7.2\%$); b - matching of the layers at the interface between them during their pseudomorfic epitaxial growth; c - bending of the GaAs/InAs two-monolayer thick film after its partial detachment from the substrate during selective etching of the underlying AlAs sacrificial layer.

The AlAs sacrificial layer can be selectively dissolved in an HF-based solution that does not etch the GaAs and InAs top layers.

The diameter D of self-formed tubes depends on the thickness d of the starting heterofilm and on its elastic stress field. This diameter can therefore be precisely predefined in the MBE process in the range between several hundreds of micrometers and several nanometers. The diameter D of the rolled-up film can be predicted by the formula [24]

$$D = \frac{1}{3}\frac{1}{\Delta a/a}\frac{(d_1+d_2)^3}{d_1 d_2}, \qquad (1)$$

where d_1 and d_2 are the thicknesses of the two layers and $\Delta a/a$ is the lattice mismatch. In derivation of the above formula, the equality between Young's module of the two materials was assumed. Our experiments show [3] that this formula is valid for rolled-up bifilms with d>4ML. It should be noted that, conceptually, we have a molecular technology that deals with inorganic 2D-macromolecules (molecular MLs). Since the thickness and composition of molecular MLs can be precisely controlled, the nanostructuring method outlined above enables obtaining precise nanoobjects.

The key point in detaching an ultra-thin film from the substrate is the free access of etchant to the sacrificial layer that becomes possible as the strained film bends away from the substrate. It is the factor that allows using ultra-thin sacrificial layers that provide for the even etch front and high quality in nanotube fabrication.

3. EXPERIMENTAL RESULTS

High quality of MBE-grown heterostructures makes it possible to obtain tubes with diameters down to 2 nm and atomically smooth and uniform walls. Moreover, not only nanotubes, but also 3D nanostructures of many other types can be obtained from strained bilayers. To fabricate them, the self-rolling and bending processes were applied to planar lithographically patterned mesa-structures of various (e.g., triangular, square, circular, etc.) geometries. In this manner, spirals, rings, and other complex shells have been fabricated [3-9].

Reliable information on the dimensions of rolled nanotubes and on the morphology of their walls can be gained by means of High-Resolution Transmission Electron Microscopy (HRTEM). An HRTEM study of formed tubes can be best-performed using tubes protruding over substrate edges. *Figure 2.* exemplifies scanning electron microscopy (SEM) and high-resolution transmission electron microscopy (HRTEM) images of such nanotubes (the full length of the InAs/GaAs tubes on the InP substrate could be as large as 1 cm). The inner diameters *D* of the tubes shown in Figs. 2, a, b and c are 1 μm, 120 nm and 3 nm, respectively [3]. Individual bilayers forming the thick wall of a tube formed from a relatively thick InAs/GaAs bifilm (4ML+4ML) are seen in Fig. 2, b [3].

Scroll-shaped tubes prepared from thin initial bi-films roll in hollow cylinders with continuous single-crystal walls [3, 7]. Fig.3. The nanotubes with diameters less than 100nm proved to be real tubes, not scrolls [3, 7]. The presence of dangling bonds on the film surface facilitates low-temperature bonding: the layers are very flexible and readily stick to each other as the film rolls in the tube.

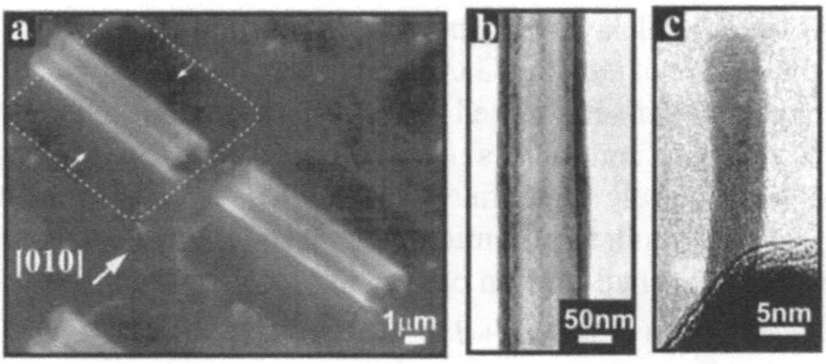

Figure 2. SEM and HRTEM images of InGaAs/GaAs nanotubes rolled up from bi-layered films. a - single-wall scrolls after their collision at the end of the formation process in which the scrolls were formed by rolling a lithographically predefined strip from its both ends, b - multiwall (six-turn) nanotube rolled from a 4ML GaAs +4ML In_xGa_{1-x} As bifilm (x=0.6); c) single-wall nanotube rolled-up from a 2ML GaAs + 1ML InAs bifilm.

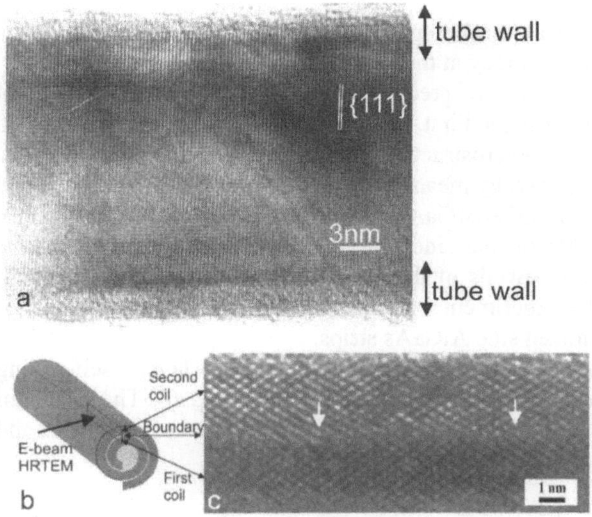

Figure 3. a) HRTEM image of a nanotube formed from 2MLGaAs+1 ML $In_xGa_{1-x}As$(x=0.8) film: atomic planes are clearly seen, b) A schematic diagram of the electron-beam trajectory in the HRTEM study of a two-coil nanotube, c) A HRTEM cross-sectional image of the wall formed by two coils in an In0,5Ga0,5As/GaAs (5ML/14ML) tube of 200 nm diameter. The white arrows show dislocations at the bonded interface; the observation of these dislocations proves that the wall as indeed a single-crystal one.

Not only the inner diameter of formed tubes, but also their other parameters can be precisely predefined prior to starting the film rolling procedure. Using this approach, it appeared possible to obtain tubes consisting of up to forty film turns *(Fig. 4a)*. It should be noted that tubes with different numbers of film turns (and, hence, with different wall thicknesses) could be controllably fabricated on the same substrate (see Fig. 4, b) [3].

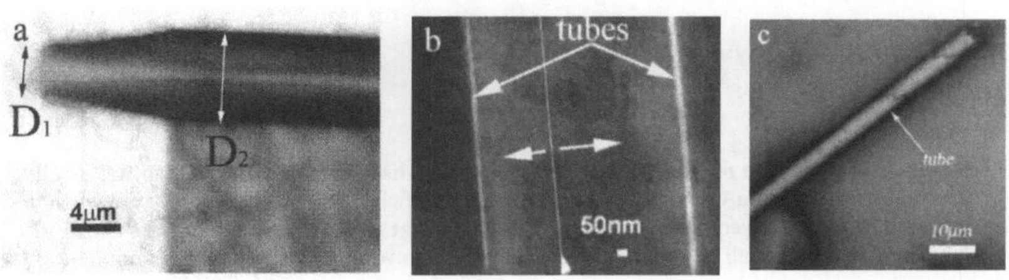

Figure 4. a) - photograph of an InGaAs/GaAs microtube fomed by 40 turns of the initial InGaAs/GaAs bifilm, D_1 and D_2 are the inner and outer diameters of the microtube. b - SEM image of nanotubes with different outer diameters formed from a 2ML GaAs+1ML InAs bifilm. The line in the figure marks the place from which the film rolling was initiated. The arrows show the rolling directions. The right and left tubes are made by 25 and 12 film turns, respectively. The typical tube length is 1-2 mm, and the inner tube diameter is smaller than 10 nm. c – three-turn microtube.

So far, we have dealt with nano-objects in precisely only two dimensions (precise *D*). The approach can be further extended to permit fabrication of objects perfectly controlled in three dimensions. Two examples concerning this point are given below [7, 14, 16]. For practical applications, it is of utmost importance not only to have precise-diameter tubes, but also tubes with perfectly controlled length and desired number of film turns as well as scroll separations between individual film turns controlled with high accuracy at the nanometer scale.

To fabricate nanotubes of precise length, we proposed a method that includes the following steps. First, an array of V-grooves was prepared on a starting multilayer heterostructure (substrate-AlGaAs-GaAs-AlGaAs-GaAs) (Fig. 5). Secondly, by means of selective MBE growth, a sacrificial layer and a strained InGaAs/GaAs bilayer were grown on the precise GaAs strips available on the sidewalls of V-grooves. The width of the GaAs strips on the sidewalls of the V-grooves was controlled with high accuracy. Under the experimental conditions adopted, no growth occurred on the adjacent oxidized side AlGaAs strips.

By the subsequent etching of the sacrificial AlAs layers, self-rolling of the InGaAs/GaAs strips in tubes was initiated (Fig. 6). The proposed technique ensures good reproducibility of all dimensions of nanotubes and their exact positioning on substrates [7].

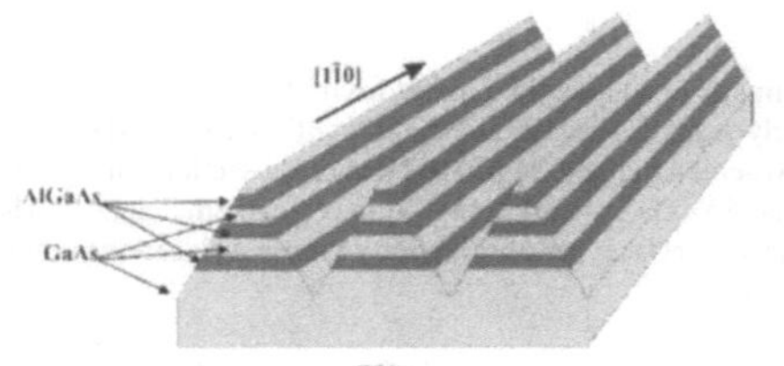

Figure 5. Schematic diagram of an array of V-grooves fabricated in the initial GaAs/AlGaAs/GaAs/AlGaAs heterostructure.

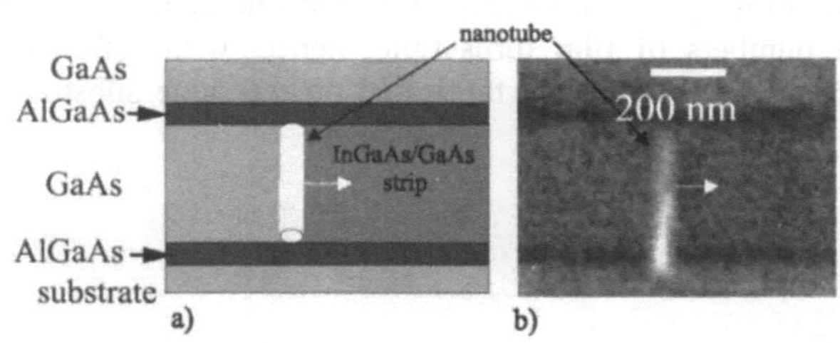

Figure 6. a) Schematic representation of the process in which an InGaAs/GaAs strip was scrolled in a nanotube. The InGaAs/GaAs strip and the AlAs sacrificial layer were selectively grown on the sidewall of a multilayer heterostructure; b) SEM image of the resultant InGaAs/GaAs tube fabricated on the sidewall of the GaAs layer. The white arrow shows the rolling direction.

Nanotubes described above possess a number of interesting mechanical, geometrical, electrical, magnetic, and quantum properties very important from a practical point of view. Among such properties of the tubes are their remarkable strength and elasticity. If compressed with a force normal to their axes, the tubes assume an elliptical shape. For example, our experiments [3] showed that multi wall, three-turn InGaAs/GaAs tubes of 1-μm diameter and 100-μm length fabricated from a 10-nm thick film can withstand a 10-g load. Our estimates show that dense arrays of tubes on the substrate surface prepared by the new technology can withstand a load amounting to several tons/cm 2. Tubes made by many film turns withstand the action of capillary forces. In the formation of single-turn tubes and nanofibers, which will be described in the next section, the most challenging problem is the deformation of these objects. Deformation of such delicate nanotubes and nanofibers, and their sticking to the substrate, occurs during the final drying procedure under the action of capillary forces produced by the evaporating liquid. A marvelous solution to this problem is offered by the supercritical drying technique, which uses a zero-surface-tension supercritical (SC) fluid and permits drying single-turn nanotubes without subjecting them to surface tension–induced mechanical actions [22,23]. Under supercritical conditions, there are no gas/liquid interfaces in the system, and the process guarantees a continuous density change from liquid to gas. This makes it possible to eliminate all surface-tension-induced problems that might result in mechanical failure of suspended parts of the system. We used SC carbon dioxide for successfull drying arrays of nanotubes.

The difference ΔP between the atmospheric pressure Pa and the water pressure Pw is given by the Laplace law:

$$\Delta P = P_a - P_w = 2\sigma/r \qquad (2)$$

where σ is the surface tension of water ($7.3 \cdot 10^{-2}$ N/m) and r is the radius of curvature of the water/air interface near the tube wall.

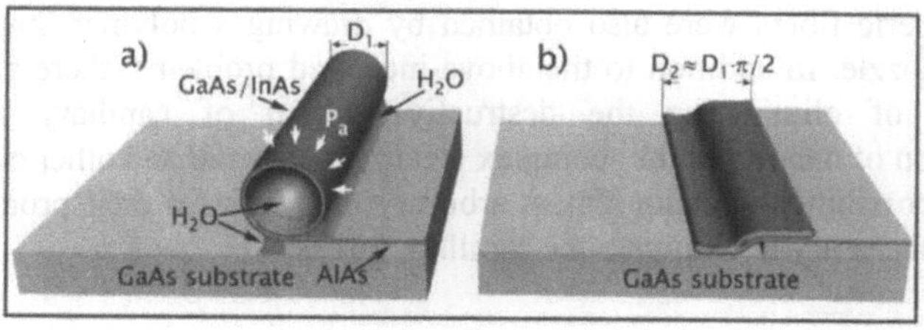

Figure 7.. Schematic illustration of a single-turn GaAs/InAs nanotube before and after collapsing. a- the tube and the narrow capillaries between the tube and the substrate are filled with water; b- collapsed tube after evaporation of the water.

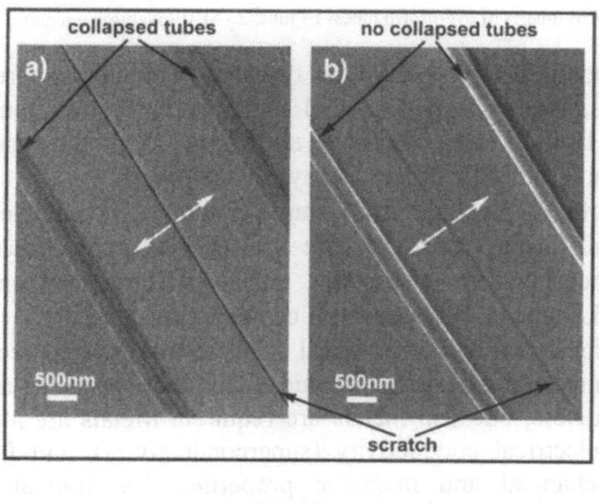

Figure 8. SEM images of collapsed and uncollapsed tubes rolled from a 12ML GaAs / 1ML InAs heterofilm. The white arrows show the direction of tube rolling from the scratches. The tube rolled to the left from the scratch a) is completely collapsed, stuck to the substrate and deformed such in case shown in Fig.7b

It is worth noting that precise nanostructuring requires the elimination of detrimental factors seriously affecting the final results of the fabrication technology. First, the etching procedure for sacrificial layers should retain other layers in the structure intact to the largest possible degree. The etching selectivity ratio of the two materials in the technology for fabricating precise tubes should be not less than 10^5 (for AlAs and GaAs layers, this ratio amounts to 10^9). Secondly, the drying procedure after the wet-etching of sacrificial layers should be performed under optimal conditions preventing the destructive action of capillary forces (here, for instance, we mean the supercritical CO_2 drying technique described in [23, 24]).

Polymeric fibers were produced employing the capillary filling of InGaAs/GaAs tubes with a polymer (polystyrene) with subsequent etching of the InGaAs/GaAs layers [8]. Another method was polymer drawing through a tube. To accomplish this goal, it is necessary to reinforce the tube walls. This was done by using multi-turn tubes (fig. 4) or by overgrowing tubes, which led to the formation of a system of precise hollow micro- or nanochannels in, say, a semiconductor material [3]. *Figure 9.* shows InGaAs/GaAs tubes epitaxially overgrown on the substrate.

Polymeric fibers were also obtained by drawing a polymer through a narrow nozzle. In addition to the above-indicated problems, there arises a problem of eliminating the destructive action of capillary forces. Integration of nanotubes into complex electrical, thermal and other circuits, and the possibility of using almost arbitrary materials for their production make it necessary to eliminate the capillary forces.

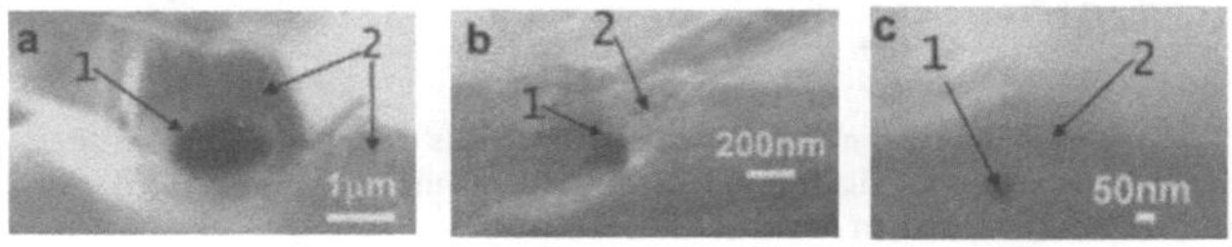

Figure 9. SEM images of overgrown tubes: 1 - tube, 2 - MBE-overgrown regions.

To make it possible to prepare other types of 3D nanostructures from strained bifilms, we developed several methods for unidirctional film rolling [5,6,10,18,19]. These methods allow one to produce not only single shells, but also highly ordered arrays of such shells.

Above, nanotubes prepared from single-crystal semiconductor films were considered. In our works [25], we showed how hybrid micro- and nanotubes (metal, semiconductor, metal-dielectric-semiconductor, semiconductor-Langmure-Blodgett film) can be fabricated.

Obviously, for a number of practical applications in nanomechanics, magnetoelectronics and vacuum electronics, nanofibers prepared not only from semiconductors, but also metals are required. Metals are attractive for their high electrical conductivity (superconductivity), and for their remarkable mechanical and magnetic properties. To fabricate metal nanotubes and nanofibers by the proposed method, it is required to create a metal bifilm composed by tensile-stressed and compressed layers, and also to provide for the possibility of detaching the bifilm from the substrate with the help of a sacrificial layer. In nanofiber manufacturing, we are oriented on the low-cost manufacturing process - sputtering of metals onto large-diameter substrates. It is well known that thin metal films can be prepared either tensile-stressed or compressed [26], with *in situ* controlled value and sign of the internal stress. The majority of refractory metals such as gold, chrome, nickel, tungsten or iron; when sputtered onto cool substrates under clean conditions; form tensile-stressed films [26]. Depositing films in the presence of impurities can change the situation. Then, one often finds the formation of compressive stresses. For instance, nickel or titan sputtered in a vacuum system in the presence of oxygen or hydrogen, form compressive stressed films[26].

The rich variety of metals with various chemical properties allows one to provide for the easy detachment of bifilms from substrates (the etchant for the sacrificial layer should not interact with the bi-film material) through an optimum choice of the sacrificial-layer material. It should be noted that, in the proposed technology, no special requirements are imposed on the substrate material, and one and the same substrate can be repeatedly used in the manufacturing process. Owing to the fact that metals usually display excellent plasticity, the Au/Ti nanotubes *(Fig. 10.)* prepared from them exhibit the following properties which are of primary significance for fabricating nanofibers: 1) the maximum tube length is restricted by the substrate diameter only; 2) after detachment of a tube from the substrate, the tube can be bent without any mechanical failure; 3) after plastic deformation of formed nanotubes, it is possible to obtain flat nanofibers with smooth edges, which can be used to weave, to knit, or to produce various architectures; 4) the tubes can be elongated by stretching.

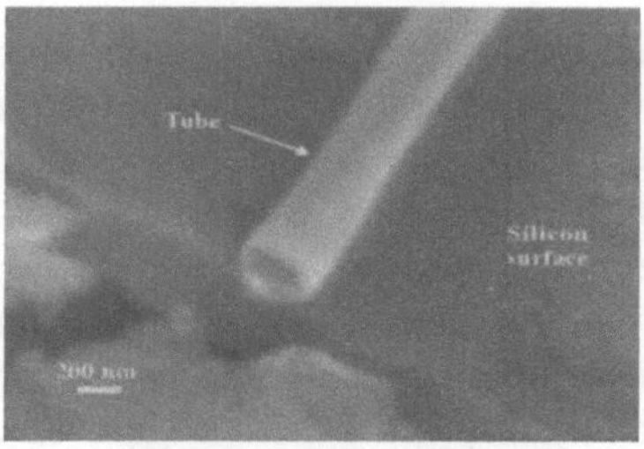

Figure 10. SEM image of an Au/Ti nanotube ruptured by cleaving the substrate. The monolithic cross-sectional structure of the nanotube can be easily seen.

The above-described stress-driven process makes it possible to produce not only free-standing and epitaxially overgrown nanotubes, but also a wide variety of other 3D nanostructures. To fabricate them, we applied the self-scrolling process to initial mesa-structures of various (e.g., rectangular, triangular, square, circular, etc.) geometries.

Using multilayer structures containing more than one bilayer we demonstrate the possibility of fabrication multiple tube configurations (tubes adjacent each other and Russian Dolls). More complex nanoobjects from bilayers with laterally modulated properties have been obtained [10].

4. SINGLE-CRYSTAL SEMICONDUCTOR AND METAL NANOFIBERS

For practical applications nanofibers prepared from semiconductors and metals are required. Solid-state nanofibers are attractive for quite a number of properties important for their applications, including high electrical conductivity (superconductivity), mechanical and magnetic properties. Unlike the traditional methods for creating (polymeric) nanofibers from polymer solutions and catalyst-assisted synthesis, we propose the preparation of nanofibers untra-thin films detached from substrates. Two fundamentally differing methods can be used to obtain nanofibers. The first method uses the nanostructuring of a grown thin film by means of electron-beam or mechanical lithography followed by its detachment from the substrate. The second is based on the selective epitaxial growth of an array of thin-film nanostrips followed by the bending procedure.

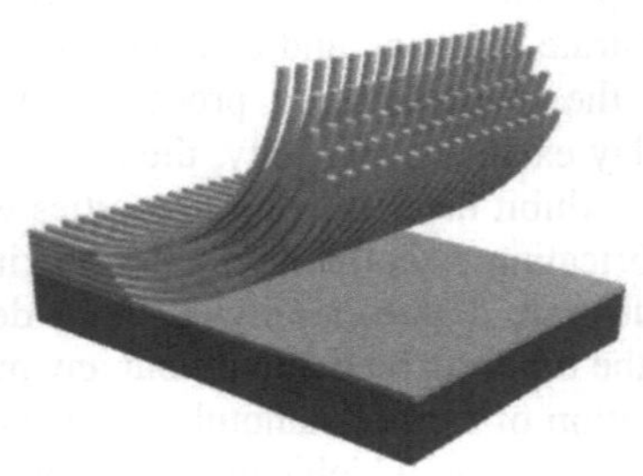

Figure 11. Schema of nanofibers dense array formation from multilayer thin film structure. The process includes lithographic patterning and stress-driven detachment of fiber from substrate.

Figure 12 illustrates the first experiments on forming semiconductor nanofibers fabricated from ultra-thin strips detached using a stress-driven process. The upper part of the figure schematically shows a lithographically patterned thin-film structure. Shown in the bottom is a SEM image of prepared nanofibers. Note that the nanofibers display excellent flexibility and mechanical strength. The final technological step was drying in air; the action of capillary forces seems to cause no damage to the fibers, just slightly displacing them. *Figure 13* exemplifies a single Si/SiGe (15 nm /5 nm) nanofiber prepared by means of mechanical lithography pattern. We failed to break the nanofiber by bending it. The nanofibers shown in *Fig. 12, 13* were prepared from a single bifilm; as a matter of fact, to prepare nanofibers, one can use multi-layered films consisting of up to 300 bilayers with compensated stresses, which will make it possible to obtain dense arrays of nanofibers. By placing a dense array of such tubes in a polymer, one can obtain a composite material.

The second method for fabricating precise fibers, with which the future undoubtedly lies, is based on selective epitaxy.

As starting precise building blocks here, precise thin-film strips grown on sidewalls of a multilayered structure are used (fig.12) [7]. Previously, we showed that the width of sidewall strips in MBE-grown heterostructures can be on the nanometer scale and perfectly controlled

[7]. It is important that this method can be used for batch production of precise nanofibers. One of possible variants is to use periodically profiled multi-layered structures (arrays of V grooves, for instance)

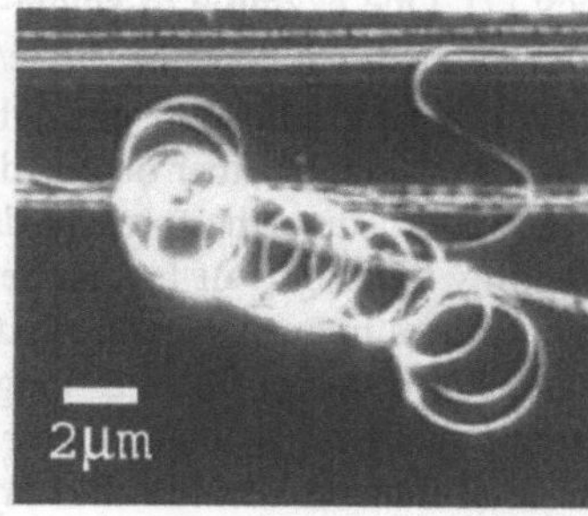

Figure. 13 Si/SiGe (15 nm /5 nm) nanofiber prepared by means of mechanical lithography.

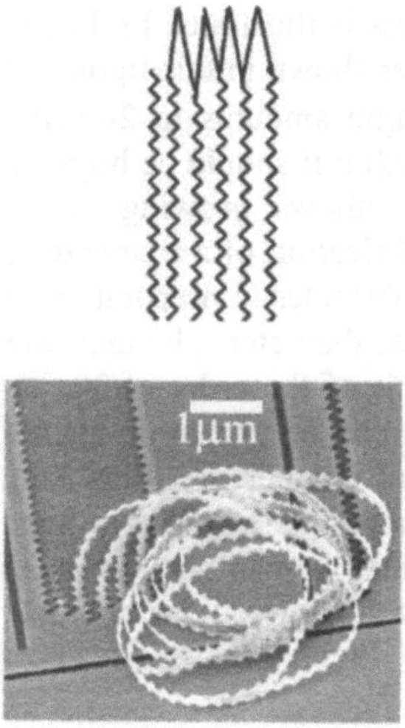

Figure. 12 SEM image of nanofibers formed by electron-beam lithography cutting of a thin Si/SiGe bifilm (15 nm /5 nm) followed by detachment from the substrate

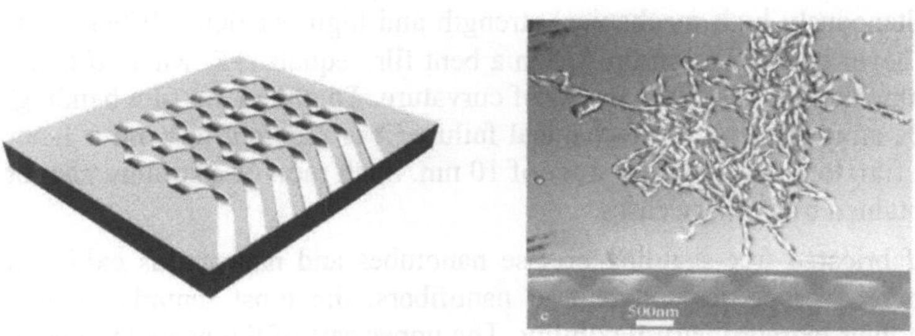

Figure 14. a) Scheme of formation of helical fibers. b) The helical fibers tangled by capillary forces during evaporation drying, but not broke. c) Segment of InGaAs/GaAs helical fiber dried in supercritical CO2.

In F*igure 14a* is shown a scheme of formation of helical fibers. Narrow strip of a (100) InGaAs/GaAs bilayer misoriented from the [010] direction, which were rolled up in a helical springs. The rolling of strips is clearly influenced by the anisotropy of Young's modul in A3B5 crystals [3]. The spacing between the spirals in a helical coil will be determined by the value of the misorientation angle.

Figure 14c shows a helical fiber obtained with the help of the supercritical CO_2 drying procedure [22, 23]. From this figure, perfect periodicity of the spiral is evident (the minimum diameter of the fabricated InGaAs/GaAs spirals was 7 nm).

As initial precise building blocks in the proposed approach, not only MBE-grown ultra-thin strained bifilms can be used, but also precise strips prepared from heterofilms. We showed previously that the strip width can be made as small as several nanometers, being predefined with high accuracy [7]. One example of single-crystal fibers is illustrated by Fig.13 The thickness and width of the Si/Ge/Si nanofiber shown in this figure are respectively 20 nm and 100 nm, while the length amounts to 200 μm. Here, we have a single fiber obtained from one film; it should be borne in mind, however, that the present-day epitaxy allows growing stacks composed by thousands of layers, permitting fabrication of a tremendous number of fibers in one process. Next, I see no obstacles to preparation of semiconductor or metal fibers as long as substrate diameter. The diameter of presently available semiconductor substrates is of the order of 30 cm, and the total number of ultra-thin layers which can be grown on such a substrate can be as high as 300; hence, the total number of 20-cm long fibers which can be obtained on one substrate amounts to 10^9. The total surface area of strained films grown on one substrate can amount to 30 m^2. In the immediate future, substrates of much larger diameters will become available.

5. PROPERTIES AND APPLICATIONS OF FORMED NANOSHELLS

The obtained solid-state fibers are very flexible, demonstrating simultaneously high mechanical strength and high elasticity. When bent, they never break. The strain $\Delta l/l$ in a bent film equals d/R, where d is the film thickness and R is the radius of curvature. To obtain, by film bending, a 10% strain leading to mechanical failure, it is required to bend a 1-nm thick film to a radius of the order of 10 nm. Such radii of curvature cannot be obtained in ordinary cases.

The fabricated free-standing precise nanotubes and nanospirals exhibit a number of properties inherent to nanofibers, the most remarkable one being their extraordinary flexibility. The upper part of the nanotube image in Fig. 2, c looks smeared. This smearing was caused by the oscillatory motion of the nanotube under the action of the electron beam. The amplitude of the oscillations increased with increasing the tube length protruding over the substrate edge.

Such a spiral nanofiber can easily be unbent by relaxing the internal stress in the structure using selective etching of one of the two layers. On the other hand, for practical applications in which fibers work in tension, it may be possible to produce arrays of tightened up fibers (helical springs).
Thus, the fabricated single-crystal (semiconductor or metal) nanofibers, possessing the main properties of common nanofibers, offer a number of additional merits. These include high precision, high stability to external factors (temperature, pressure, illumination, etc.). It is of importance that

such fibers can be assembled in well-ordered arrays and other desirable architectures.

The above-described nanotubes and other shells possess many extraordinary mechanical, electrical, optical properties very important from the practical point of view. Below, we focus on the quantum phenomena that can be observed in formed nanoshells and have considerable practical potential.

The theory of quantum phenomena in structures with curved surfaces is currently the subject of many studies and needs to be further developed. Our experimental results on fabrication of nanotubes and nanoshells initiated a series of theoretical works in the field of research dealing with quantum properties of such structures (e.g. [25, 26]). In [26], it was shown that any curved electron-gas-containing film placed in a magnetic field should generate spin-polarized currents. This effect opens up new possibilities in using spin currents in integrated solid-state spintronic devices.

First experiments on studying the magnetotransport properties of two-dimensional electron gas in rolled-up quantum wells were reported in [17]. The stress in a rolled film substantially affects the film-material properties [11, 27]. In [11], quantum properties of InAs/GaAs nanotubes were studied; the numerical results revealed spatial separation of charge carriers (electrons and holes) in nanotube walls (Fig. 15).

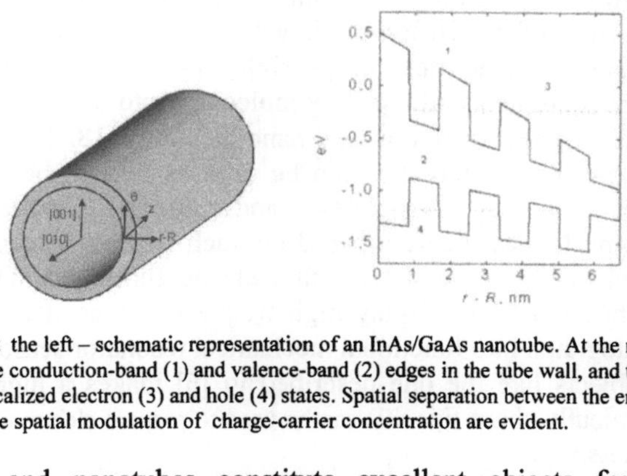

Figure 15. At the left – schematic representation of an InAs/GaAs nanotube. At the right – position of the conduction-band (1) and valence-band (2) edges in the tube wall, and the enrgy position of localized electron (3) and hole (4) states. Spatial separation between the energy levels and, hence, the spatial modulation of charge-carrier concentration are evident.

Micro- and nanotubes constitute excellent objects for studying properties of 2D electron gas on cylindrical surfaces, including charge-carrier ballistic transport and spin-orbital interaction; they also can be used for fabricating electronic interferometers (Aharonov-Bohm rings). Narrow-gap semiconductors with near-surface accumulation regions enriched with charge carriers show much promise as materials for producing narrow conducting channels and thin-walled shells. The possibility of creating conducting semiconductor tubes with 2D electron gas and metallic nanotubes was demonstrated in [20] and [15], respectively. Ultra-thin layers of wide-gap semiconductors can be used to prepare tunnel barriers.

A most intriguing and surprising property of InGaAs/GaAs nanotubes is their stability to oxidation at room temperature [3]. This finding can be tentatively attributed either to changes in the bandgap width of the nanostructured composite material due to quantum confinement or to surface passivation (during the selective removal of the sacrificial layer, the Ga dangling bonds on the GaAs surface have enough time to interact with F, yielding inert GaF).

These films open up fresh opportunities in studying properties of semiconductor interfaces. For instance, 2 ML-thick single-crystal InAs/GaAs films, in which all molecules can be considered as occupying simultaneously surface and hetero-interfacial sites, display such unusual chemical and mechanical properties as (i) stability against oxidation [3], (ii) bright manifestation of surface-tension forces in their elastic characteristics, and (iii) formation of a single-crystal monolithic wall structure by means of "flexible" room-temperature bonding [3,7].

From the viewpoint of using the above-described semiconductor and metal 3-D structures in electronics, the fact is of primary importance that these structures are arranged in out-of-plane configurations; for instance, the spiral and tube shown in [10] are attached to the substrate only at contact pads. This characteristic feature of the structures makes it possible to minimize the electrical interaction with the substrate and enables filling the tubes with desired substances intended for improving working device characteristics. The possibility of rolling films in tubes with more than 40 film turns was demonstrated. We believe that this possibility will open a new way towards fabricating very compact device elements and circuits.

Obviously, nanotubes custom-designed in their diameter and wall thickness have a broad field of possible applications as elements for producing nanodroplets and injecting molecules into living cells. By now, needles and syringes for micro- and nanoinjections [18, 19] and nanojets have been obtained. Nanotubes can be used as nanoprobes in scanning microscopes or as templates and precise masks. Surely, nanoelectromechanical devices based on such flexible systems will find wide application in sensors. Ultra-thin elastic films and 3D nanoshells formed from solid films display high frequency of oscillations and can easily be integrated into functional mechanical resonator sensors. An array of rolled objects like the one described in [6] makes a high density of almost atomically sharp tips that can be used to prepare electron emitters with enhanced emissivity.

6. CONCLUSIONS

Thus, the essence of the new nanofabrication methods can be summarized as follows. Ultra-thin strained layers released from the massive substrate tend to acquire a new equilibrium shape for which their elastic energy is minimal. Bending off from the substrate or rolling into cylindrical objects, the layers form precise shells. The final shape of the shells depends on the composition of the starting heterostructure and can be easily controlled since the whole technology is based on using standard technological procedures. The starting objects for forming the final three-dimensional semiconductor nanoobjects are uniform, single-crystal thin films, whose thickness can be preset with atomic precision in the MBE process. Another condition ensuring the high precision of sizes and shapes of the final objects is fixed lattice mismatch between dissimilar materials. This property of MBE-grown heterostructures in combination with the possibility of detaching strained heterolayers from substrates provides the possibility to obtain precise nanoobjects.

A rich variety of solid-state nanoshells of various shapes that can be obtained with the above-described technology is evident. The simplicity of the new method, its applicability to a broad class of materials and, finally, its compatibility with the mature integrated-circuit technology allows us to anticipate its wide practical applications in the future technology. Our methods allow fabrication of precise nanofibers. Unlike conventional methods for making nanofibers, our methods use film technologies. Only standard processes can be used to obtain thin uniform films and properly pattern them. Our technological step is the detachment of the ultra-thin strips obtained from patterned films off the substrate. At present, we know the only way to do this, the one employing the stress-driven process [3]. The most important feature of the technology is that it allows easy control of not only the shape of the fibers, but also their arrangement on the substrate by exerting control over the stress-driven assembly. By this technology, one can create architectures from various materials, - metals, semiconductors, superconductors, dielectrics, and even from unstrained materials attached to stained films. The above methods can be applied to all materials without exception. Using multi-layered structures, one can manufacture dense arrays of nanofibers. In the case of nanofibers prepared from sputtered films, the possibility of their batch production is quite clear.

The nanofibers obtained can be used in many fields where standard nanofibers are normally used, including filtration, protective clothing, biomedical applications such as wound dressing and drug delivery systems, mirrors for use in space, etc. Moreover, since our method allows manufacturing of nanofibers from various (magnetic, conducting, single-crystal) materials never previously used for this purpose, and also permits assembling nanofibers in various ordered architectures, this will undoubtedly initiate new fields of applications for nanofibers.

Our works initiated other publications in this field (see, for example, [27-31]). The forthcoming years will undoubtedly witness new, even more interesting findings and discoveries in this research area.

Acknowledgments: The present work was supported by the NEDO collaboration program "Nano-elasticity", by the Russian Foundation for Basic Research and by the Swiss National Science Foundation (through SCOPES contract)

References

[1] Moriarty, P. (2001) Nanostructured materials, *Rep. Prog. Phys.* **64,** 297-381.

[2] Sakaki, H. (1999) Fabrication of Atomically Controlled Nanostructures and Their Device Application, in: G. Timp, (eds.), *Nanotechnology*, Springer, New York, pp. 207-255.

[3] Prinz, V. Ya., Seleznev, V. A., Gutakovsky, A. K., Chehovskiy, A. V., Preobrazenskii, V. V., Putyato, M. A. and Gavrilova, T. A. (2000) Free-standing and overgrown InGaAs/GaAs nanotubes, nanohelices and their arrays, *Physica E,* **6,** N 1-4, 828-831.

[4] Prinz, V. Ya., Seleznev, V. A., Samoylov, V.A., and Gutakovsky, A. K. (1996) Nanoscale engineering using controllable formation of ultra-thin crack in heterostructures, *Microelectronic Engineering* **30,** 439-442.

[5] Prinz, A. V., Seleznev, V. A. and Prinz V. Ya. (2001) Integrated microneedle and method for its fabrication, Application № 99111533/14, 1999. Patent of 20. 09.2001.

[6] Prinz, V. Ya., Grutzmacher, D., Beyer, A., David, C., Ketterer, B. and Deccard, E. (2001) A new technique for fabricating three-dimensional micro- and nanostructures of various shapes superlattices, *Nanotechnology* **12,** 399-402.

[7] Prinz, V. Ya., Chehovskiy, A. V., Preobrazenskii, V. V., Semyagin, B. R. and Gutakovsky, A. K. (2002) A technique for fabricating InGaAs/GaAs nanostructures of precisely controlled lengths, *Nanotechnology* **13,** 231-233.

[8] Prinz, V. Ya., Golod, S. V., Mashanov, V. I. and Gutakovsky, A. K. (2000) Free-standing conductive GeSi/Si helical microcoils, micro- and nanotubes, *Inst. Phys. Conf. Ser.* **166,** 203-206.

[9] Golod, S. V., Prinz, V. Ya., Mashanov, V. I., Gutakovsky, A. K. (2001) Fabrication of conducting GeSi/Si micro-and nanotubes and helical microcoils, *Sem. Sci. Techn.* **16,** 181-185.

[10] Vorob'ev, A. B. and Prinz, V. Ya. (2002) Directional rolling of strained heterofilms. Semicond, *Sci. Technol.* **17,** 614-616.

[11] Osadchii, V.M. and Prinz, V.Ya. (2000) Scrolled heterostructures, *JETP Lett.* **72,** 312-317.

[12] Prinz, V. Ya., Seleznev, V. A., Sveshnikova, L. L. and Badmaeva, J. A. (2000) Precise micro- and nanotubes formed by scrolling Langmuir-Blodgett/GaAs/InGaAs films, Proc. 8th Int. Symp. "Nanostructures: Physics and Technology", St. Petersburg, pp. 92-94.

[13] Prinz, V.Ya., Vorob'ev, A.B. and Seleznev, V.A. (2002) Three-dimensional structuring using self-rolling of strained InGaAs/GaAs films, *Inst. Phys. Conf. Ser.* **170,** 319-323.

[14] Prinz, V. Ya., Vyun, V. A. (2003) to be published.

[15] Nastaushev, Yu. V., Prinz, V. Ya. and Svitasheva, S. N. (2003) A technique for fabricating Au/Ti micro- and nanotubes, *Nanotechnology* to be published.

[16] Prinz, V. Ya. and Prinz, A. V. (2003) to be published.

[17] Vorob'ev, A.B., Prinz, V. Ya. and Toropov A.I. (2002) Magnetotransport properties of two-dimensional electron gas in rolled-up quantum well. *International Conference on Superlattices, Nanostructures and Nanodevices* Toulouse, France.

[18] Prinz, A. V. and Prinz, V. Ya. (2003) Application of semiconductor micro- and nanotubes in biology, *Surface Science*, **532-535,** 911-915.

[19] Prinz, A.V., Prinz, V.Ya. and Seleznev V.A. (2003) Semiconductor micro- and nanoneedles for microinjections and ink-jet printing, *Microelectronic Engineering,* **67-68,** 782-788.

[20] Vorob'ev, A.B.. Prinz, V.Ya., Preobrazhenskii, V.V. and Semyagin, B.R. (2003) Free-standing InAs/InGaAs microtubes and microspirals on InAs100, *Jpn. J. Appl. Phys*, **42,**

[21] Prinz, V. Ya. (2003) A New Concept in Fabricating Building Blocks for Nanoelectronic and Nanomechanic Devices, *Microelectronic Engineering.*

[22] . Seleznev, V.A., Yamaguchi, H., Hirayama, Y. and Prinz V. (2003) Single-Turn GaAs/InAs Nanotubes Fabricated Using the Supercritical CO_2 Drying Technique, *Jpn. J. Appl. Phys.* **42,** No.7AL791 - L794.

[23] Chehovskiy, A.V. and Prinz, V.Ya. (2003) Application of supercritical fluids for fabrication of free-standing nanoobjects, in Nanostructures, *Physics and Technology.*

[24] Tsui, Y.C. and Clyne, T.W. (1997) An analitical model for predicting residual stresses in progressivily deposited coatings, *Thin Solid Films* **306,** 23.

[25] S.V. Golod, V.Ya. Prinz, P. Wägli, L. Zhang, O. Kirfel, E. Deckhardt, F. Glaus, C. David and D. Grützmacher (2003) Free-Standing SiGe/Si/Cr and SiGe/Si/SixNy/Cr Microtubes (to be published).

[26] R. Koch (1994) the intrinic stress of polycrystalline and epitaxial thin metal films, *J. Phys.: Condens. Matter* **6** 9519-9550.

[27] Entin, M.V. and Magarill, L.I. (2002) Electrons in a twisted quantum wire, *Phys.Rev. B.* ,**66,** 205308-1 – 205308-5.

[28] Kleiner, A. (2003) Chiral spin currents and quantum Hall effect in nanotubes, *Phys.Rev. B* .**67,** 155311.

[29] Jain, S.C., Willander, M. and Maes, H. (1996) Stress and strain in epilayers, stripes and quantum structures of 111-V compound semiconductors, *Semicond. Sci. Technol.* **11,** 641-671.

[30] Schmidt, O.G. and Eberl, K. (2001) Thin solid films roll up into nanotubes. Nature **410**, 168 + correction: *Nature* **412,** 42.

[31] P.O. Vaccaro, K. Kubota, and T. Aida, (2001) Strain-driven self-positioning of micromachined structures, *Appl. Phys. Lett.* **78** 2852-2854.

CARBON NANOPIPETTES: SYNTHESIS AND ELECTROCHEMICAL PROPERTIES

R. C. Mani, M. K. Sunkara, and R.P. Baldwin
Dept. of Chemical Engineering, University of Louisville, Louisville, KY 40292

Abstract: In this paper, we report the synthesis and electrochemical properties of a novel morphological manifestation of carbon nanostructures – "carbon nanopipette". Substrates in the form of wires or sheets were immersed vertically into a Microwave Plasma containing 1-2% CH_4/H_2. Short-term (1-2 hours) and long-term (24 hours) experiments were carried out. At the end of a long-term experiment, a dense carbon deposit was found at the tip of the Pt wire exposed to the plasma, while at a short distance away from the tip, there was a deposit of microcrystalline diamond film along with a dense growth of tapered carbon whiskers. These whiskers had a base of 1 µm, which tapered down to few nm. Energy Filtered Transmission Electron Microscopy (EFTEM) revealed that these nanostructures had a hollow core of a constant diameter of about 4-20 nm, while the outer shell is made up of helical sheets of graphite winding continuously around the hollow core. Hence these nanostructures were termed as "carbon nanopipettes". Short time-scale experiments performed on substrates electroplated with thin films of platinum resulted in arrays of nanopipettes. However, the aspect ratio of the nanopipettes increased from near tip region to regions downward on the substrate. The results using different time scales for growth showed that the length of these nanopipettes could be increased from a few microns to about 100 microns. The regions with carbon nanopipettes on platinum substrates were isolated by masking off the other parts with a non-conducting epoxy. Electrochemical studies were performed on these blanked substrates with dopamine in KCl solutions. Nanopipettes are open-ended on both sides, mechanically rigid, and have a tip as sharp as a conventional nanotube, whereas the body of the nanopipette is robust. Among other applications, the arrays of nanopipettes, as synthesized, could be encapsulated into polymer matrix for making a patch suitable for drug delivery applications.

S. Guceri et al. (eds.), Nanoengineered Nanofibrous Materials, 65-73.

1. INTRODUCTION

Research in nanoscale materials has been inspired by the scientific interest of discovering new phases of materials and novel properties they may possess due to their low-dimensionality. With the same motivation, a myriad of structural manifestations of carbon nanostructures have been reported. Conical graphite nanostructures have been discussed as early as 1960 [1], in which a model for graphite whiskers was proposed with the wrapping up of graphite sheets around the whisker axis. Later, conical graphite whiskers were synthesized by heating SiC crystals [2]. Since then, a variety of conical graphite whiskers have been reported such as graphite fullerene cones [3, 4], nanohorns [5], polyhedral crystals [6], conical crystals [7]. We have recently reported another novel morphology of carbon nanostructures called "Carbon Nanopipettes" [8], which is unlike any of the morphologies reported above for conical nanostructures. Carbon Nanopipettes are rigid, yet have a nano-sized tip. The base of this nanostructure is about 1 micron reducing to a very fine tip of about 4-20 nm. A constant uniform-diameter hollow core runs through its length and hence it is open-ended on both sides. Due to its very unique structure, we envision the potential applications of nanopipettes in fluid delivery (drug and ink delivery), nano-sized electrochemical probes, NSOM tips, etc.

Carbon in its several forms such as graphite, microcrystalline diamond, glassy carbon and nanocrystalline diamond has been used as electrodes for a long period of time. Microelectrodes are beneficial for their faster steady state time scales, higher currents for the same electrode area, less capacitance currents apart from their use in probing small areas [9]. With the same view in mind, carbon nanotubes have been recently studied for their electrochemical properties [10-12]. In the case of nanotubes, it has been shown that the pentagons at the tip of the nanotube, pentagon and heptagon defect pairs in the lattice of the nanotube are responsible for the oxygen reduction at their surface [10]. There is no reason to believe that the electrochemistry occurs in the interior parts of the nanotube. In fact, one would have to rely on the defect states within the graphene sheets for exterior parts of the nanotubes to be active toward significant electron transfer kinetics. On the other hand, the carbon nanopipettes could be considered as nanocrystals of graphite with edge planes showing up on the entire 3-dimensional surface including the tip. One would expect faster kinetics and longer stability with electrochemical response for reactions that rely on chemisorption for electron transfer. In fact, most of the reactions of interest in medical situations involve either amines or proteins that require chemisorption for electron transfer e.g. Dopamine (3-hydroxy tyramine hydrochloride) is a neurotransmitter, whose deficiency is responsible for Parkinson's disease.

2. EXPERIMENTAL

Substrates in the form of platinum wires, platinum-plated diamond substrates, Mo sheet substrates were used. These substrates were immersed in a Microwave Plasma with 1-2% CH_4/H_2 at 50 Torr pressure and 1100W power. Solid Boron was used to make sure that the diamond phase deposited was conducting. Experiments using blank Platinum and Mo were long term (24 hours), while platinum coated diamond and Mo substrates were short-term experiments (1-2 hours). Long-term experiments on Platinum had a dense deposit of carbon at the tip of the substrate. Further away from the tip were diamond crystals along with whisker-like structures emerging from them. Long-term experiments on Mo, gave a dense deposit over a large area from the top of the substrate immersed in to the plasma. Regions further away, microcrystals of diamond were observed. In this case, the pipettes were observed throughout the region of the dense deposit. Short-term experiments using electroplated platinum on Mo and diamond substrates resulted in arrays of nanopipettes with varying aspect ratios from the tip to the end of the substrate. The reader may refer to [8] for Transmission Electron Microscopy studies for detailed structural analysis of these carbon nanostructures.

Cyclic voltammetry (CV) experiments were carried out with a Bioanalytical Systems Model CV-50W Voltammetric Analyzer using a single-compartment, three-electrode cell with Ag/AgCl (3M NaCl) reference and Pt wire auxiliary electrodes. For electrochemistry, the samples were prepared by coating the regions not containing the dense array of nanopipettes with a non-conducting, chemically inert epoxy. Two samples with distinct types of overall morphology for nanopipette arrays were prepared: the first one had a dense array of nanopipettes pointing in all directions and the second sample had regular arrangement of nanopipette array with each one separated by a couple of μm distance. The underlying platinum wire is used for electrical contact. The simplicity of this procedure makes our synthesized structures (nanopipettes) extremely useful for practical applications.

3. RESULTS AND DISCUSSIONS

The synthesis procedure of immersing wires into dense part of plasma is not a typical procedure for diamond or carbon tube formation. However, in order to study the variation of carbon phases during deposition at different parts of the plasma, we have used direct immersion. The results of this method and corresponding results are first published in reference [13]. Figure 1 indicates the carbon nanopipettes grown on platinum substrates at regions away from the tip of the substrate. As seen, the nanopipettes were surrounded by microcrystals of diamond. These pipettes were grown over a period of 24 hours. The true length of the pipettes cannot really be determined here, as they have their base covered with diamond crystals. The tip of these substrates has a dense deposit containing nanocrystals of diamond and graphite.

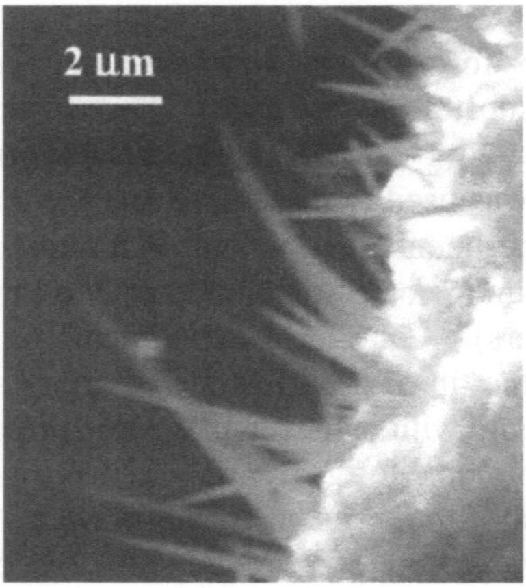

Figure 1. Carbon Nanopipettes grown along with microcrystals of diamond on a platinum substrate

We have demonstrated previously [8] that the growth of nanopipettes begins with a central nanotube (single or multiwalled) growth, followed by helical graphite sheets wrapping around it, falling just short of the pervious sheet. The resulting structure is a whisker whose outer shape is conical, but has an inner hollow core of a constant diameter. One can vary the thickness of the surrounding graphite sheets depending on the position in the plasma. The conditions within the dense part of the plasma vary quite widely in terms of temperature, radical and ion density, which further affect the etching and growth characteristics of these graphite sheets on the nanopipettes. The regions at the tip of the substrate immersed in the plasma are at higher temperatures. Here, there is a high tendency of deposition of competing carbon phases such as graphite in this region. Experiments on uncoated/coated Mo sheet substrates for 4 hours or more gave large nanopipettes, but along with other graphitic phase deposition. This could be explained by the tendency of Mo to promote carbon phase deposition. Similarly, short-term experiments with plain Mo substrates gave multiwalled carbon nanotubes. While using Pt-plated Mo sheets as substrates yielded carbon nanopipettes as shown in Figure 2(a). Even short-term experiments with platinum coated, diamond film covered Mo substrates yielded carbon nanopipettes (Figure 2(b)). Diamond forms a stable substrate that does not get etched in the plasma, hence gives a uniform array of carbon nanopipettes. The nanopipettes in an array form over a large area are necessary for several other applications involving drug delivery. In this regard, experiments are currently underway to increase the lengths of the pipettes to mm covering a large area.

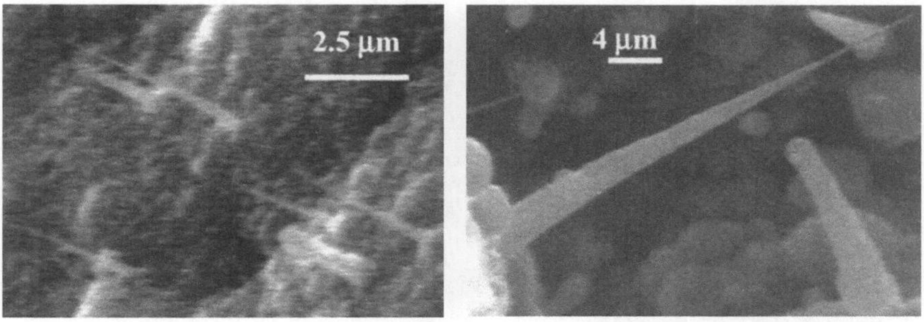

(a) *(b)*

Figure 2. Nanopipettes grown on Pt-plated (a) Mo substrate (b) Diamond on Mo substrate for 1-2 hours

The electrode samples containing carbon nanopipettes were tested for their electrochemical response in the following solutions: 1 mM $K_3Fe(CN)_6$ in 0.1M KCl, and 1 mM Dopamine in 0.1M KCl. $K_3Fe(CN)_6$ is typically used as a calibration analyte for electrodes, while dopamine was used, as it is one of the most well studied biologically important analyte by electrochemical detection. The first substrate tested for electrochemical response had a smaller region exposed to the solution. The approximate geometrical area of this region is 10^{-4} cm^2. This small region had approximately 10^8 pipettes/cm^2, which are not grown in any particular direction (Figure 3(a)). In accordance to this, the electrochemical response of dopamine as indicated in figure 3(b) was strongly controlled by diffusion. As shown in Figure 3(b), the anodic limiting current increased linearly with the scan rate (ν) (V/s) ^ 0.5. Increasing the scan rate beyond 1 V/s begins to show the peak shaped response, but this was accompanied by an increase in the background current, making it more difficult to detect the signal from the analyte.

The approximate electrochemical area calculated from the response to Dopamine by Randles-Selvick equation was very close to the geometrical area of the exposed surface (unmasked region containing nanopipettes). Similar behavior was reported for SWNT bundles [12] that were prepared in the form of thin papers. This is typical when the concentration boundary layers of individual nanostructures overlap. The advantage of using nanopipettes is in using them as a nanoelectrode. The small dimension of the electrode with fast kinetics helps in decreasing the time scales for detection. This can be used as an advantage in detecting intermediate analytes secreted during a complex reaction. In order to make use of the nano-scale of the pipette in a nanoelectrode, we would need to have them placed in the form of arrays, such that the distance between individual elements exceed the minimum needed for non-overlapping diffusion boundary layers. To understand this, we used a sample in which the nanopipettes were grown for short time on an uncoated platinum wire for 1 hour.

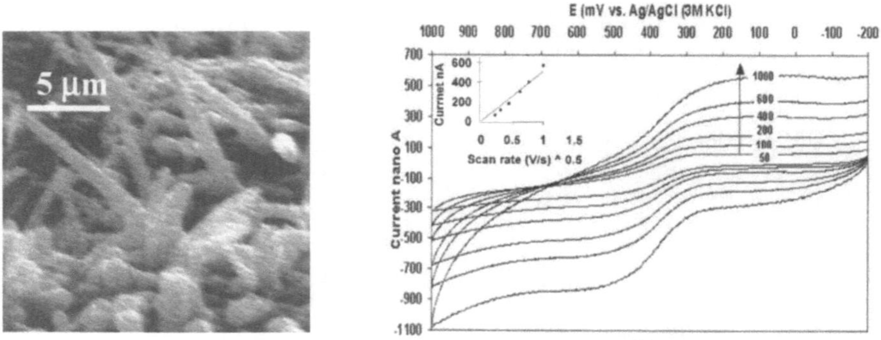

(a) (b)

Figure 3. (a) SEM image of the bunch of nanopipettes on an electrode (b) Cyclic Voltammogram of 1mM Dopamine in 0.1M KCl, at different scan rates indicated in mV/s. Inset shows the dependence of the anodic limiting current to the scan rate.

The sample contained nanostructures with varying aspect ratio as described in ref [8]. At a small distance below the tip (few mm), due to the competing growth and etching of the surrounding graphene sheets, the aspect ratios of these nanostructures change eventually giving fully developed, high aspect ratio nanopipettes. One such electrode was carefully coated with epoxy while keeping the region containing the nanopipette arrays exposed. The density of pipettes in this sample was lower, about $10^5/cm^2$. The SEM image in Figure 4 shows two kinds of nanopipette arrays with different aspect rations on this sample. All of the structures in this sample are vertical while keeping enough distance in between them.

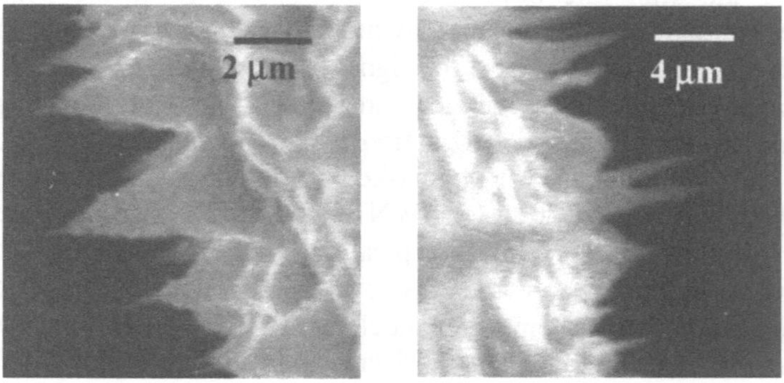

(a) (b)

Figure 4. Tip of the Pt substrate has (a) Cones with a central nanotube emerging and (b) Nanopipettes at a distance away

Figure 5(a) shows the response of such an electrode to 1mM Dopamine in 0.1M KCl solution before and after treatment in 0.5M H_2SO_4 at a scanning window of 2V to –2V. As seen in the Figure, the peak separation (ΔEp) before acid treatment was 300 mV and after acid treatment was 100 mV. This is similar to what we have observed for nanocrystals of graphite with edge planes exposed to the solution. The treated surfaces are extremely stable with time and repeated analysis over 1 year. The acid treatment for one cycle of potential sweep made a number of reactions including $K_3Fe(CN)_6$ reversible at the electrode surface. After initial treatment of the as-synthesized structures, no further treatment was necessary. The electrochemical response seems to be reproducible and stable over long periods of time.

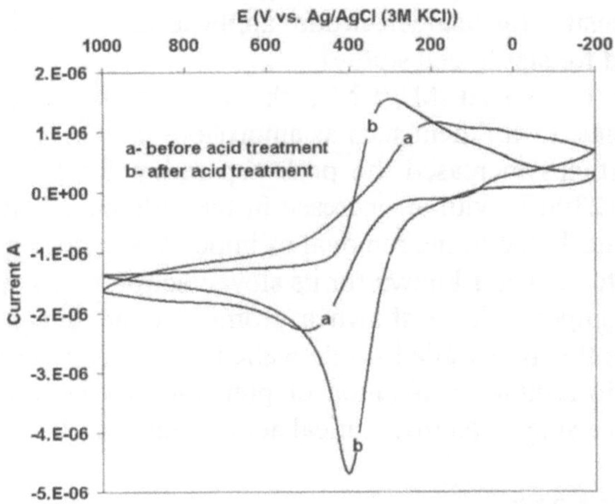

(a)

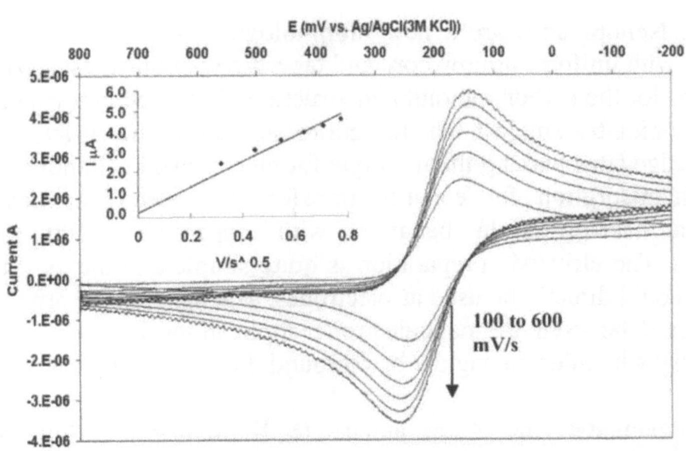

(b)

Figure 5. (a) CV of 1mM Dopamine in 0.1M KCl at 100mV/s before and after electrochemical treatment in 0.5M H_2SO_4 (b) CV of 1mM $K_3Fe(CN)_6$ in 0.1M KCl at 100mV/s to 600m/s with peak separation of about 100 mV. Inset shows the scan rate dependence of cathodic current with v (V/s) ^0.5.

Acid treatment was not always necessary for certain as-synthesized structures. Figure 5(b) shows the CV for 1mM $K_3Fe(CN)_6$ in 0.1M KCl solution. These pipettes, unlike the one seen in Figure 3(a) have a peak shaped behavior. The currents are also relatively higher, due to the increase in the exposed area. The inset in Figure 5(b) shows that the cathodic peak current dependence is linear with increase in the square root of the scan rate, indicating a diffusion-controlled behavior. The area calculated again from this experiment was approximately equal to the geometrical area. At the same time, the estimated surface area of all nanopipettes over this region is also similar to the geometrical area due to the density and length. It is highly possible to envision high aspect ratio, long (mm in length) nanopipette array with a low density for nanoelectrode applications involving fast cyclic voltammetry and for single-cell studies.

In the case of typical MWCNT, the heat treatment to remove the impurities associated with them such as amorphous carbon, graphitic particles and catalyst particles increased the peak separation for $K_3Fe(CN)_6$ analyte from 100mV to 230mV with an increase in the volume specific capacitance [12]. This was attributed to the removal of impurities exposing only the basal planes of graphite, which is known for its slow electron transfer kinetics. Thus our carbon nanopipettes have shown a promise as an interesting electrode material. Unlike the multiwalled/single walled carbon nanotubes, they do not need any specific sample purification or preparation. They can be used as synthesized with a single electrochemical acid-treatment.

4. CONCLUSIONS

Carbon Nanopipettes are a new morphological manifestation for carbon nanotubes with uniform hollow core and tapering wall thickness giving a high aspect ratio for the carbon nanotubular structure. The structural characteristics of the nanopipettes suggest that the entire surface of the structure contains graphene edge sites making them unique for electrochemical applications that require chemisorption for electron transfer. The electrochemical studies indicate a near-reversible behavior with dopamine (~100 mV peak separation). The electrode preparation is quite simple and the as-synthesized structures could directly be used as electrodes. The nanopipette array with low density could be used for nanoelectrode applications involving fast cyclic voltammetry while decreasing the background due to capacitance.

Acknowledgements: One of the authors (M.K. Sunkara) greatly appreciates partial financial support from NSF through CAREER grant (CTS 9876251). The authors also acknowledge Prof. K. Rajan and X. Li (Renesselaer Polytechnic Institute) for their help with the TEM Analysis.

References

1. Bacon, R. Growth, Structure, and Properties of Graphite Whiskers. J. Appl. Phys. 1960; 31: 283-290.
2. Haanstra, H.B., Verspui and G., Knippenberg, W.F. Columnar Growth of Carbon. J. Cryst. Growth 1972; 16: 71-79.
3. Ge, M. and Sattler, K. Observation of Fullerene cones. Chem. Phys. Lett. 1994; 220: 192-196.
4. Krishnan, A., Dujardin, E., Treacy, M.M.J., Hugdahl, J., Lynum, S and Ebbesen, T.W. Graphitic Cones and the Nucleation of Curved Carbon Surfaces. Nature 1997; 388: 451-454.

5. Iijima, S., Yudasaka, M., Yamada, R., Bandow, S., Suenaga, K., Kokai, F. and Takahashi, K. Nano-aggregates of single-walled graphitic carbon Nano-horns. Chem. Phys. Lett. 1999; 309: 165-170.
6. Gogotsi, Y., Libera, J.A., Kalashnikov, N. and Yoshimura, M. Graphite polyhedral crystals. Science 2000; 290: 317-320.
7. Gogotsi, Y., Dimovski, S. and Libera J.A. Conical crystals of graphite. Carbon 2002; 40: 2263-2267.
8. Mani, R. C., Li, X., Sunkara, M. K. and Rajan, K. Carbon Nanopipettes. Nano Letters 2003; 3: 671-673.
9. Michael, A. C. and Wightman, R. M. "Microelectrodes" In *Laboratory Techniques Electroanalytical Chemistry*, P. T. Kissinger and W. R. Heineman ed. NY: Marcel Dekker Inc, 1996.
10. Britto, P. J., Santhanam, K. V. S., Rubio, A., Alonso, J. A. and Ajayan, P. M. Improved Charge Transfer at Carbon Nanotube Electrodes. Adv. Mater. 1999; 11: 154-157.
11. Nugent, J. M., Santhanam, K. V. S., Rubio, A. and Ajayan, P. M. Fast Electron Transfer Kinetics on Multiwalled Carbon Nanotube Microbundle Electrodes. Nano Letters 2001; 1: 87-91.
12. Li, J., Cassell, A., Delzeit, L., Han, J. and Meyyappan, M. Novel Three-Dimensional Electrodes. Electrochemical Properties of Carbon Nanotube Ensembles. J. Phys. Chem. B 2002; 106: 9299-9305.
13. Mani, R.C., Sharma, S., Sunkara, M.K., Gullapalli, J., Baldwin, R.P., Rao, R., Rao, A.M. and Cowley, J.M. Synthesis and electrochemical characteristics of a nanocomposite diamond electrode. Electrochem. Solid St. Lett. 2002; 5: E32-E35.

INFLUENCE OF PLD AND CVD EXPERIMENTAL GROWTH CONDITIONS ON CARBON FILM NANO-STRUCTURE EVOLUTION

E. CAPPELLI[1], S. ORLANDO[2], G. MATTEI[1], C. SCILLETTA[1], F.CORTICELLI[3] and P. ASCARELLI[1]

[1] ***CNR-IMIP, P.O.B.10, 00016 Monterotondo, Rome, Italy.***
[2] ***CNR-IMIP, P.O.B.27, 85050,Tito Scalo, Potenza, Italy.***
[3] ***CNR-IMM, via P.Gobetti 101, 40129 Bologna, Italy.***

Abstract: Owing to their outstanding properties, many research efforts have been focused on the growth and application of carbon based materials, mainly single or multi-wall carbon nano-tubes, nano-horns and cones, fibres . Nano-structured carbon films have also been proved to be a very promising field emitting material, providing the nano-graphene structure have the constituting basal plane oriented perpendicular to the substrate. Pulsed Laser Deposition (PLD) and Chemical Vapour Deposition (CVD) methods have been used to prepare carbon films of different nano-structure ranging from amorphous carbon to nano-clustered graphite and nano-wires. The influence of experimental conditions (v.z. deposition system and parent species, substrate heating, working pressure, inert sustaining gases, RF or DC plasma assisted deposition) on the carbon atom clustering, particle aggregation and structure evolution has been thoroughly analysed. Nd:YAG pulsed laser ablation of a graphite target produced nano-structured films when assisted by plasma or at high substrate temperature. Hot Filament CVD was able to grow nano-graphite "petal"-like structures vertically oriented at moderate substrate temperature (650 °C) and sharp graphite tips and cones at high T (1650 °C). The presence of a DC plasma was able to produce big "urchin" structures characterised by long, entangled, tubular wires (stripes), without any catalyst addition. The film morphology have been characterised by SEM; film quality and nano-particle dimension have been estimated be Raman spectroscopy. Field emission properties have been measured by a planar / spherical configuration.

Keywords: carbon nano-structures, graphite nano-wires, RF plasma assisted PLD deposition, DC plasma assisted HF-CVD, Raman spectroscopy, SEM, Field Emission.

S. Guceri et al. (eds.), Nanoengineered Nanofibrous Materials, 75-81.

1. INTRODUCTION

The carbon atom has a very peculiar, outstanding property in nature, since, due to its ability to have many types of hybridisation (sp^3,sp^2 and sp) and, consequently, spatial orientation of chemical bonds, it is capable of constructing a wide variety of materials ranging from diamond graphite and polymers to new materials like fullerenes, nano-tubes, nano-fibres, whiskers, cones, horns and amorphous carbon. These structures show very different properties and have very interesting potential industrial applications including electronic, energetic, mechanics or even catalysis and biological applications [1-5]. Many research efforts have been concentrated in structure optimization and substrate patterning to obtain a controlled product for special electronic applications, mainly patterned and vertically oriented single or multi-wall carbon nanotubes (SW-CNT and MW-CNT). Preparation methods, ranging from DC arc-discharge, pulsed laser ablation and CVD, thermic or plasma assisted, have been used under special experimental conditions to obtain specific physical properties. Most of them made use of a specific catalyst (Co, Fe or Ni) of nano-sized dimension and regular distribution to obtain ordered tubular structure of definite diameter [6-11]. Other carbon based materials, like nano-structured particles, fibres, nano-horns and cones, but also "petal"-like graphite sheets have been produced and studied for potential applications both in electronic field and in catalysis [12-17].

In our work we explored the ability and limits of PVD and CVD deposition methods to deposit carbon with different nano-structures, as a function of substrate temperature and plasma activation of reactive species without using any catalyst or surface patterning.

2. EXPERIMENTAL

2.1 Pulsed Laser Deposition

The carbon films were deposited in a vacuum chamber (base level ~ 10^{-6} mbar) by pulsed laser ablation from a pyrolytic graphite target. A pulsed Nd:YAG laser (2nd harmonic λ=532 nm, hν=2.33 eV, repetition rate ν=10 Hz, pulse width τ=7 ns, fluence ϕ=5 J/cm^2) was used as the excitation source.

PLD depositions were performed at ~ 2x10-1 mbar Ar atmosphere, on Si<100> substrate at increasing temperature, ranging from RT to 900 °C. The distance between the substrate and the graphite rotating target was 5 cm.

A set of samples were deposited, in the same experimental conditions, in the presence of an Ar plasma, generated by a 13.56 MHz radio frequency source just over the <100> Si substrate, maintained at a negative bias potential. The working power was 60 W.

2.2 Hot Filament Chemical Vapor Deposition

The CVD depositions on mild HF/HNO_3 etched Si <100> substrates are performed at ~ 10÷15 mbar, 15% CH4, 15% H2 and 70% Ar with a total gas flow 100 sccm. The DC plasma assisted HF-CVD depositions are performed at 1.5÷2.0 x10^{-1} mbar, in the same reactive gases and flux. The Ta filament is heated up to 1800-2100 °C, while the substrate can be heated to 650-850°C. The distance between the substrate and the Ta filament is 10-13 mm. The plasma is generated by a DC source between the substrate and the filament just over the Si. The substrate is kept at -350, -400V negative bias potential. The working power was 60 W.

2.3 SEM and Raman Characterisation. Field emission

Micro-Raman measurements were performed in a back-scattering geometry, at RT, using a Dilor XY triple spectrometer equipped with a liquid nitrogen cooled charge coupled devise (CCD) detector and an Olympus microscope in a confocal mode. The spectra were excited with an Ar+ laser (514.5 nm, 3.8 mW) focused into a spot of 2 mm in diameter: This photon energy preferentially excites the p states associated to sp2 sites. The spectral resolution was 0.5 cm-1. Spectra were analysed by a peak fitting procedure to obtain frequencies, line-widths (FWHM) and intensities of band components. The film morphology have been studied by a Philips XL30 SEM apparatus.

Field emission measurements were obtained in a planar cathode and emispherical anode (1 mm diameter) variable distance configuration, at a pressure ~ 10^{-6} torr. The I/V characteristics were measured by a HP 4339-A high resistance meter.

3. RESULTS AND DISCUSSION

3.1 PLD and RF plasma assisted PLD

The effect of temperature on the morphology and structural properties of carbon films is shown in fig.1a-c. The SEM picture exhibits a macroscopic particle aggregation with increasing substrate temperature, ranging from RT to 900 °C. The corresponding Raman spectra show a considerable D (1350 cm^{-1}) and G (1580 cm^{-1}) peak evolution correlated to aromatic ring formation and cluster condensation. The D peak narrows and the ratio I_D/I_G increases, correlated to a temperature driven ordering phenomena of the aromatic rings into graphene layers and nano-graphite particles. The RF plasma activation of ablated carbon species causes a clustering effect starting at lower temperature; at T=900°C the Raman spectrum shows a peak shapes closely and resembles the slightly disordered HOPG. The I_D/I_G intensity ratio has been related by Tuinstra and Koenig [18] to XRD dimension of graphite nano-particle (L_a) along the basal plane, according to the equation:

$$\frac{I_D}{I_G} = \frac{44}{L_a}(A°)$$

Recently, for domains smaller than 20 Å, a direct correlation ($I_D/I_G = C' L_a^2$) has been suggested to better represent the system[19]. From this formula, at RT, we calculated an La ~7.3 Å size for PLD, compared to La ~13.5 Å size for RF-PLD. At T=900 °C we obtained closer values: 17.5 Å versus 20.3 Å. The nano-structured carbon showed a sensible enhancement of field emission properties, with a turn-on field of about 17 V/μm, compared to 23 V/μm for amorphous carbon.

3.2 CVD and DC plasma assisted CVD

The films deposited by HF-CVD show a new allotropic carbon form, nano-flakes or nano-ribbons, which consists of "petal"-like graphite layers with a thickness of about 20÷30 nm and an orientation perpendicular to the substrate (fig. 2). The Raman spectrum of film on Si <100> underlines the graphitic nature of the deposited nano-structures, compared to the slightly mechanically disordered HOPG2, with a sharp G peak (FWHM=28 cm^{-1}) and a high and sharp D peak (FWHM=30cm^{-1}). At higher T (~1650°C, on a Ta wire) the "graphitisation" process is more evident (fig. 3): very sharp graphite tips ($FWHM_{G\ peak}$=18 cm-1), grows on a very thin "petal"- like base-structure (fig.4), characterised by a few registered and interlaced graphene layers. The corresponding Raman spectra are characterised by the presence of a strong G peak and a contribution of bounded, "edge" type C-C bonds (weak D and D' peaks).
When HF-CVD operates in the presence of a high concentration of radicals and ions, generated by the DC plasma between the filament and the substrate, a strongly enhanced "out-of-equilibrium" growth of graphene structures was observed, with a local, diffused formation of big "urchin" objects (fig. 5) with very long, curled, entangled tubules made of small layered graphene planes, growing on a substrate structure, characterised by the near-equilibrium "petal"-like configuration.

The CVD system has been proven to be more suitable than PLD for the deposition of vertically nano-structured graphene ribbons, at relatively low T. These nanostructures, characterised by sharp edges, have been demonstrated to act as good field emitters [20].The fast, tubular growth of nano-fibers, obtained by plasma assisted CVD, can be explained by the very high active species concentration, which may give rise to a pure kinetic, out-of-control growth process.

4. CONCLUSIONS

Carbon films obtained by pulsed laser deposition are characterised by a structure evolution from smooth amorphous, at room temperature, to granular particles at 900°C. Increasing temperature, an aromatic ring formation and a progressive condensation is evident as measured by the Raman D and G mode evolution and an increase in the I_D/I_G ratio. The RF-plasma is able to induce this clustering effect at lower T.

The emission properties are directly related to the increase of aromatic ring clustering v.z. to the increase of the I_D/I_G ratio.

The CVD conditions (presence of radicals, quasi-equilibrium process, high CH_4 concentration and inert gas (Ar)) made easy to obtain nano-graphite structures, with ribbon or "petal"-like morphology, oriented perpendicularly to the substrate, along the growth direction. The Raman spectra are characterised by sharp G and D peaks, very similar to partially disordered HOPG. At higher T (~1650°C) the "graphitisation" process goes further: very sharp graphite tips, with a typical intense G peak, develop from a thin "petal"- like nest, characterised by some parallel, bounded and interlaced graphene layers.

The DC plasma is able to accelerate the graphene plane growth with a local formation of very long, entangled tubular structure even at relatively low temperature (~650°C). The lack of any catalyst or etching gas during the deposition hindered the nano-tube formation, favouring a layered-type fibre structure.

Acknowledgements: This work was supported by FIRB, a public funding for basic research from The Ministry for University and Research of Italy.

References

1. S. Iijima, *Nature* 354 (1991) 56.
2. H.W. Kroto, J.R. Heath, S.C. O'Brien, R.F. Curl, R.E. Smalley, *Nature* 318 (1985) 162.
3. J.Robertson, *Materials Sci. Engineering R37* (2002) 129-281.
4. R.J. Chen, S. Bangsaruntip, K. A. Drouvalakis, N. Wong Shi Kam, M. Shim, Y. Li, W. Kim, P. J. Utz, and Hongjie Dai, *Proc.Nat. Ac. Sci.* 100 (2003) 4984-4989.
5. Hongjie Dai, *Surf. Sci.*500 (2002) 218-241.
6. Z. F. Ren, Z. P. Huang, J. W. Xu, J. H. Wang, P. Bush, M. P. Siegal, and P. N. Provencio, *Science* 282 (1998) 1105-1107.
7. M. Chhowalla, K.B.K. Teo, C. Ducati, N.L. Rupesinghe, G.A.J. amaratunga, A.C. Ferrari, D. Roy, J. Robertson and W.I. Milne, *J. Appl. Phys.* 90 (2001) 5308.
8. D.J. Kim, K.S. Choi, Y.S. Cho, S.Y. Hong, J.B. Park,K.H. Son, *J. Appl. Phys.* 91 (2002)3847.
9. Cui, H.; Eres, G.; Howe, J.Y.; Puretkzy, A.; Varela, M.; Geohegan, D.B.; Lowndes, D.H. *Chem. Phys. Lett.* 374 (2003) 222-228.
10. Melechko, Anatoli V.; Merkulov, Vladimir I.; Lowndes, Douglas H.; Guillorn, Michael A.; Simpson, Michael L. *Chem. Phys. Lett.* 356 (2002) 527-533.
11. W. Z. Li, S. S. Xie, L. X. Qian, B. H. Chang, B. S. Zou, W. Y. Zhou, R. A. Zhao, and G. Wang, *Science* 274 (1996) 1701-1703.
12. M. Terrones, N. Grobert, J. Olivares, J. P. Zhang, H. Terrones, K. Kordatos, W. K. Hsu, J. P. Hare, P. D. Townsend, K. Prassides, A. K. Cheetam, H. W. Kroto, D. R. M. Walton, *Nature* 388, (1997) 52 – 55.
13. Dong, Jian; Shen, Wanci; Zhang, Baofa; Liu, Xuan; Kang, Feiyu; Gu, Jialin; Li, Dongsheng; Chen, NanPing, *Carbon* 39 (2001) 2325-2333.
14. J. Libera, Y. Gogotsi, *Carbon* 39 (2001) 1307-1318.
15. S. Iijima, T. Wakabayashi, Y. Achiba, *J.Phys.Chem.* 100 (1996) 5839-5843.
16. Y.Ando, X. Zhao, M. Ohkohchi, *Carbon* 35 (1997) 153-158.
17. N.G. Shang, F.C.K. Au, X.M. Meng, C.S. Lee, I. Bello, S.T. Lee, *Chem. Phys.Lett* .358 (2002) 187-191.
18. F. Tuinstra, J.L. Koening, *J.Chem.Phys.*53 (1970) 1126.
19. A.C. Ferrari, J. Robertson, *Phys. Rev.B* 61 (2000) 14095.
20. Obraztsov, A.N.; Volkov, A.P.; Zakhidov, Al.A.; Lyashenko, D.A.; Petrushenko, Yu.V.; Satanovskaya, O.P., *Appl. Surf. Sci.* 215 (2003) 214-221.

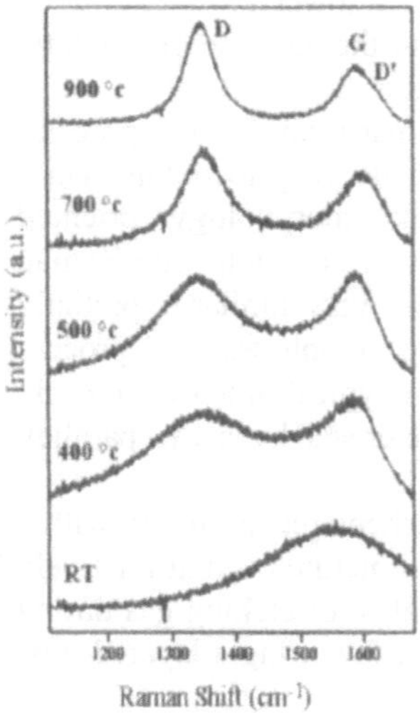

Fig.1a. Raman spectra of sample PLD deposited at increasing T

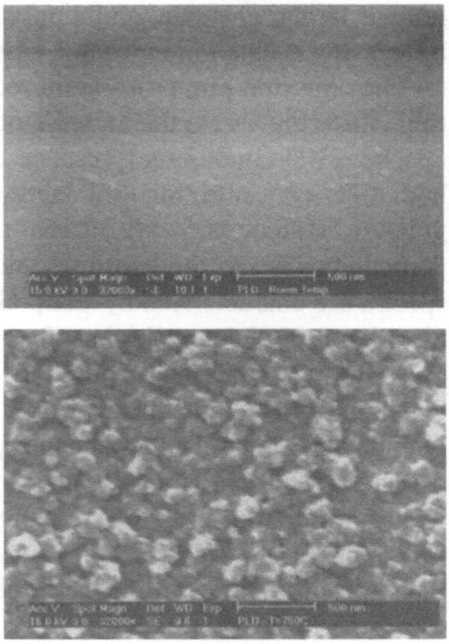

Fig. 1b,c. SEM pictures of film PLD deposited at RT (*upper*) and 900 °C.

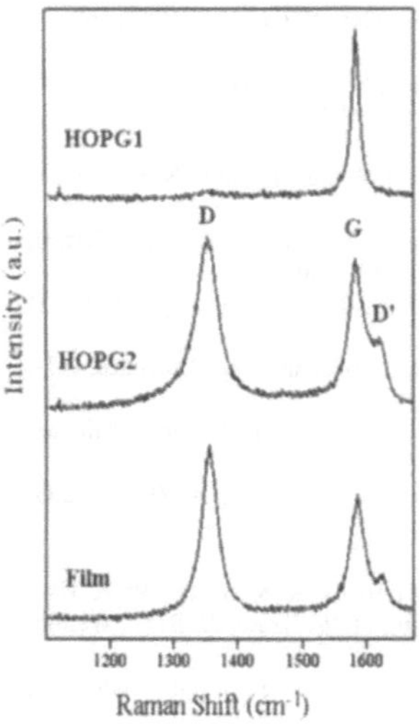

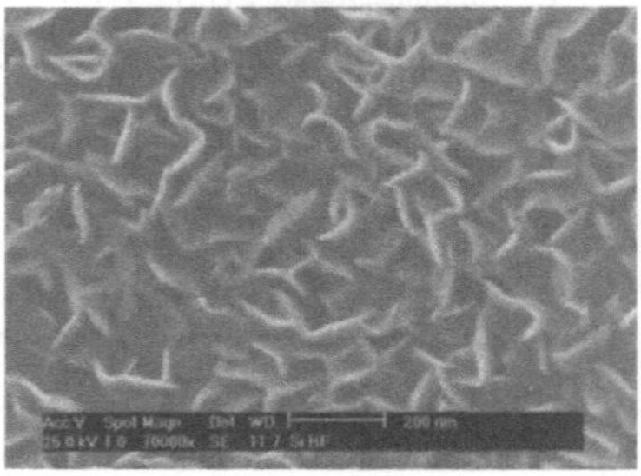

Fig.2. *left*. Raman spectra of " petal"-like graphene nano-structures, compared to HOPG graphite and mechanically slight disordered HOPG. *Right*. SEM picture of nano-sized "petal"-like graphene structures

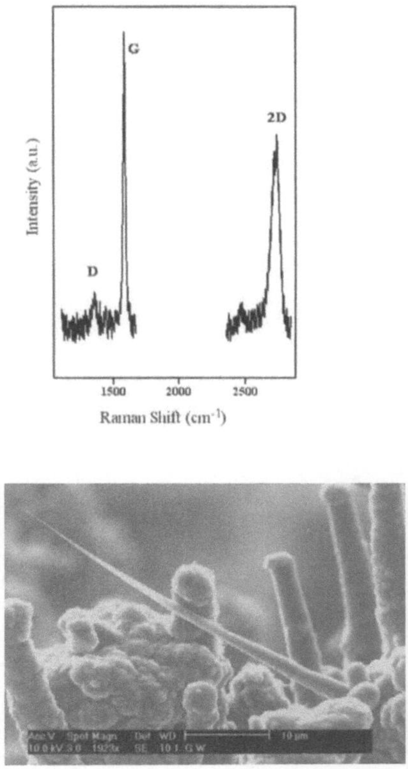

Fig.3. *lower.* Graphite whiskers and cones grown by HF-CVD on a Ta wire at 1650 °C. *Upper.* The graphite whiskers Raman spectra

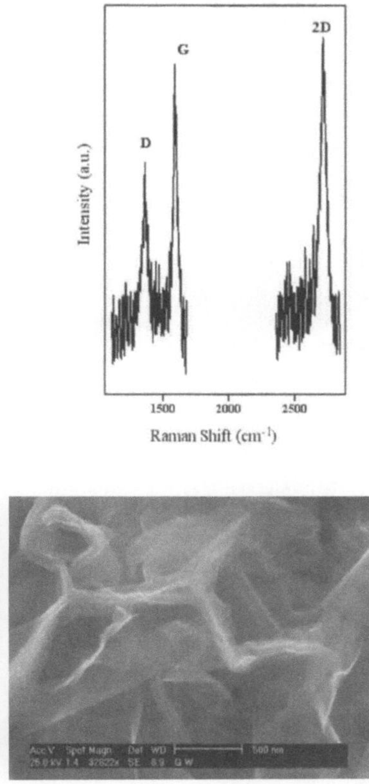

Fig.4. *lower.* SEM picture of ribbon or petal-like graphene layers grown by HF-CVD at at 1650 °C. *Upper.* The corresponding Raman spectra.

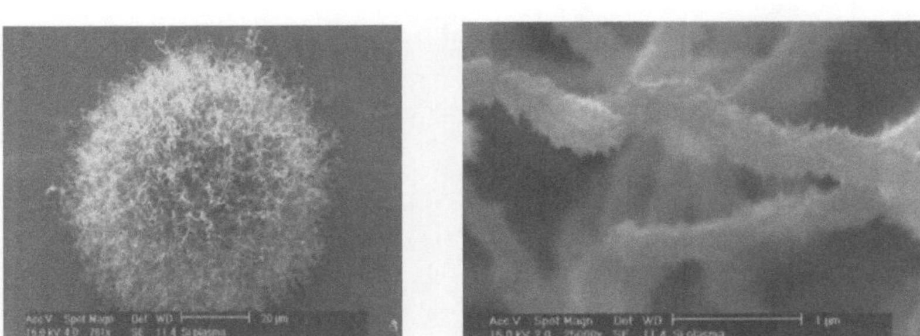

Fig.5. *Left.* The “urchin” morphology of long curled nano-graphene tubules obtained by DC plasma assisted HF-CVD at 650 °C. *Right.* An enlarged view of a tubulus.

CONTROLLED GROWTH OF NOVEL HOLLOW CARBON STRUCTURES WITH BUILT-IN JUNCTIONS

G. Bhimarasetti, M. K. Sunkara*, Uschi Graham**, C.Suh, K. Rajan***; *Department of Chemical Engineering, University of Louisville, Louisville, KY 40292*; University of Kentucky Center for Applied Energy Research, 2540 Research Park Drive, Lexington, Kentucky 40511**; Dept. of Material Science and Engineering, Rensselaer Polytechnic Institute, Troy, NY 12180****

Abstract: We describe a conceptually different technique to synthesize large inner diameter carbon tubular structures. This technique utilizes the wetting behavior of gallium with carbon to control the morphology *in-situ* during growth by controlled gas phase chemistry. The internal diameters can be varied from 100 nm to about 5 μm, with constant wall thickness of 10-20 nm. Using this technique we synthesized straight tubes, funnels, tube-on-cone, multi-junctioned tubules and Y-junctions with seamless joining. These carbon structures could find applications in nano/micro-fluidics and also as micro-reactors.

1. INTRODUCTION

Several different nano-structures of carbon have been synthesized since it was realized that carbon nano-structures having very interesting properties could find applications in numerous fields. Some of these structures include: nanotubes, helical nanotubes, cones, horns, conical crystals, micro-trees and nanopipettes (1-8). The diameters of these structures are in the nanometer scale. Larger inner diameter straight tubes have been synthesized using hydrothermal synthesis techniques with diameters ranging from 70-1300 nm (9). Carbon nanotubes filled with gallium, projected as nano-thermometers, have been synthesized by using evaporation of gallium oxide and gallium nitride powders (10-12). All the above-mentioned synthesis methods only yield straight tubes and cannot be used to synthesize structures with varying inner diameters. Merkulov et al., (13) used nickel catalyst and relative concentrations C_2H_2/NH_3 to synthesize "cylinder-on-cone" morphology. Again, the internal diameter of the structure is constant and determined by the nickel catalyst size as is the case with the traditional carbon nano-tube synthesis methods. Recently, we described a synthesis technique that can be

S. Guceri et al. (eds.), Nanoengineered Nanofibrous Materials, 83-90.

used to synthesize a variety of morphologies during growth by controlling the gas phase chemistry (14).

Our synthesis method is based on the variation in wetting behavior of gallium with carbon in the presence or absence of oxygen and nitrogen. The contact angle gallium creates with carbon can be varied either by changing the gas phase or by changing the temperature. Here we describe how one can utilize this behavior to control the morphology of the carbon structures.

2. EXPERIMENTAL

A microwave plasma CVD reactor is utilized in this work. A thin film of gallium is spread on various substrates: graphite, molybdenum and titanium. Micron-sized molybdenum powder is then dusted on to the gallium film (1-15% by weight of Gallium). This setup is exposed to 18% CH_4/H_2 plasma operated at 1100 watts microwave power and 40 torr reactor pressure. Typical growth times are one hour. The temperature of the gallium surface as measured using a pyrometer is about 700-750 ^{0}C. The synthesized structures are analyzed using a Scanning Electron Microscope (operated at 15, 20 and 25kV) and a Transmission Electron Microscope (200 kV).

3. RESULTS AND DISCUSSION

Conical and nozzle shaped structures are synthesized using only CH_4/H_2 as the feed gases. These structures are partially filled with gallium along the length as shown in Figure 1. Also, the inner diameters vary from hundreds of nano-meters to microns, with a constant wall thickness of 10-20 nm.

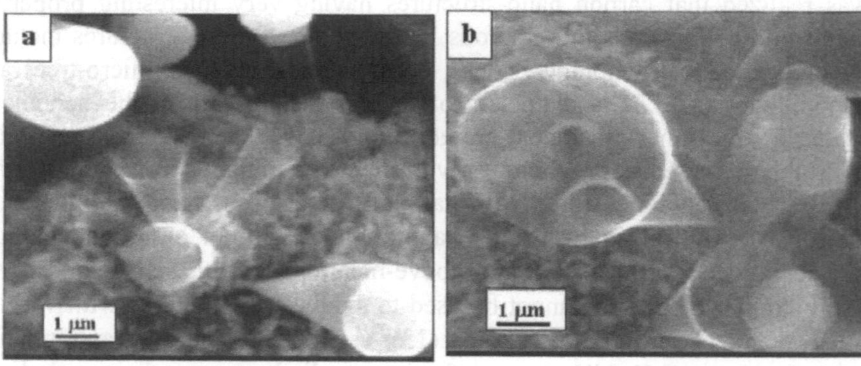

Figure 1. SEM micrographs of Conical structures in the form of nozzles synthesized using only CH_4/H_2 are shown in (a) and (b). The bright droplets are the gallium droplets on top of the conical structures.

3.1 Growth Mechanism

The contact angle gallium creates with carbon will be different in different gas phase environments. In the presence of oxygen or nitrogen, gallium wets carbon (15, 16), thus forming a flatter meniscus during growth (lower contact angle). The growth of the carbon tubular structures occurs by the following mechanism (*Figure 2*). Due to the presence of plasma during growth, gallium forms tiny droplets (few nanometers to microns). The presence of molybdenum promotes the precipitation of carbon at the Ga-Mo interface (*Figure 2a*). This forms a graphite sheet around the gallium droplet. Further addition of carbon then takes place at the Ga-C as shown in Figure 2b. The Ga-C interface 'pulls' the gallium in the direction of growth. The graphite wall that forms at the Ga-C is always tangential to the curvature of the gallium droplet. The angle the wall makes with the gallium droplet is the contact angle θ. As the growth proceeds, the structures grow in length. The conical angle ϕ of the evolving tubular morphology is determined by the contact angle θ by the following relationship:

$$\phi = 2\theta - 180^0 \qquad \text{Eq (1)}$$

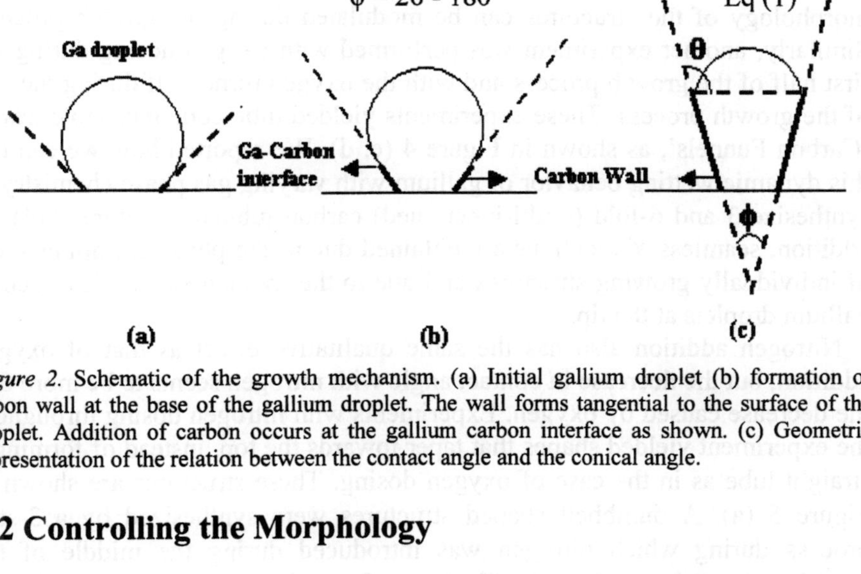

Figure 2. Schematic of the growth mechanism. (a) Initial gallium droplet (b) formation of cabon wall at the base of the gallium droplet. The wall forms tangential to the surface of the droplet. Addition of carbon occurs at the gallium carbon interface as shown. (c) Geometric representation of the relation between the contact angle and the conical angle.

3.2 Controlling the Morphology

As mentioned earlier the contact angle θ can be changed using different gas phase compositions. Based on this we can control the morphology on the carbon structures during growth. Experiments performed with oxygen dosing throughout the growth yielded straight tubes with constant diameters *(Figure 3)*. The presence of oxygen decreased the contact angles, thus yielding structures with smaller or no conical angles.

Other morphologies were synthesized by intermittent oxygen dosing experiments. In one experiment oxygen (5 sccm) was introduced during the second half of the growth process. During the first half of the growth with no oxygen added conical structures are formed. As oxygen is introduced in the growth process the structures become straighter. This procedure yields 'cone-tube' structures (Figure 4 a, b). Thus the

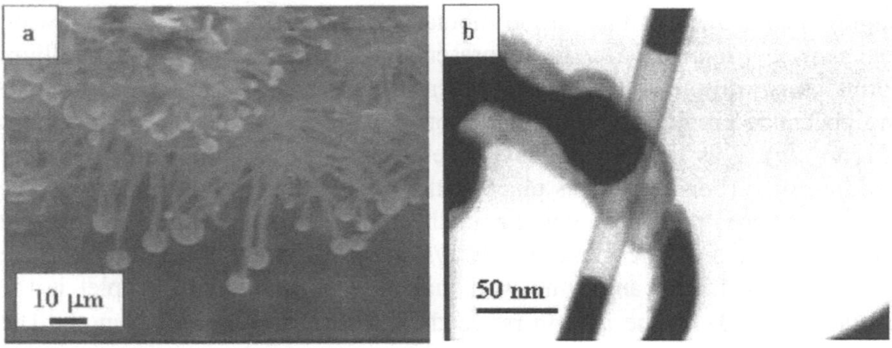

Figure 3. Straight tubes synthesized by oxygen dosing throughout the experiment.

morphology of the structures can be modulated during the growth process. Similarly, another experiment was performed with oxygen dosing during the first half of the growth process and with the oxygen turned off during the rest of the growth process. These experiments yielded tube-cone morphologies or 'Carbon Funnels', as shown in Figure 4 (c, d). We reported how we can use this dynamic wetting behavior of gallium with varying gas phase chemistry to synthesize 3 and 6-fold (multi-junctioned) carbon tubular structures (14). In addition, seamless Y-junctions are obtained due to the physical impingement of individually growing structures and due to the spontaneous coalescence of gallium droplets at the tip.

Nitrogen addition also has the same qualitative effect as that of oxygen addition, but the decrease in contact angle with nitrogen seems to be more that the decrease caused by oxygen. Experiments with nitrogen dosing throughout the experiment yielded shapes that taper towards the top, instead of forming a straight tube as in the case of oxygen dosing. These structures are shown in Figure 5 (a). A dumbbell shaped structures were synthesized by a 3-step process during which nitrogen was introduced during the middle of the experiments and was turned off after a few minutes and the growth was continued. This caused a 'pinch' in the structure at the center as shown in Figure 5 (b). The pinch was caused due to the drastic decrease in contact angle.

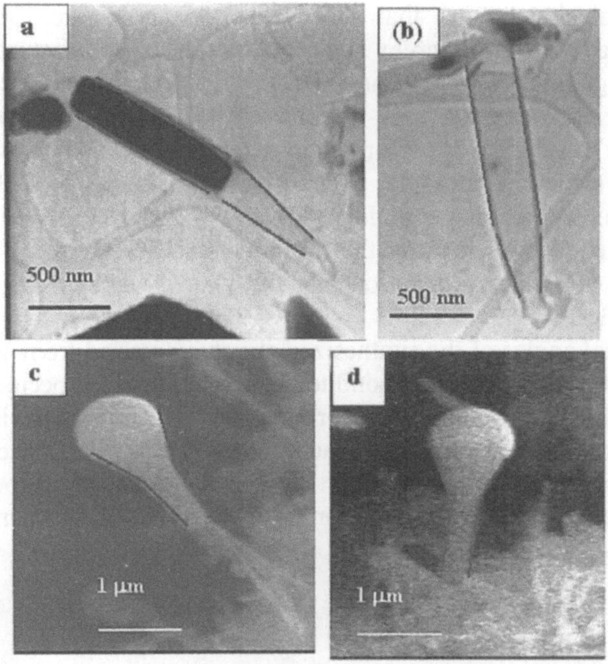

Figure 4. Two stage morphologies synthesized by oxygen dosing experiments. (a) No oxygen dosing followed by oxygen dosing, (b) oxygen dosing followed by no oxygen dosing.

Figure 5. Morphologies synthesized by nitrogen dosing experiments. (a) Nitrogen dosing throughout the experiment, (b) 'no nitrogen – nitrogen – no nitrogen' 3-step sequence experiment.

3.3 Wall Structure

TEM analysis showed that the walls are crystalline. Selective area diffraction gave a very complex spot pattern. The wall structure is currently being thoroughly investigated using HRTEM, RAMAN spectroscopy and EELS (electron energy loss spectroscopy). EDS (energy dispersive spectroscopy) analysis was performed to determine if molybdenum or gallium are present either in the wall structure or at the sides of the wall. No molybdenum or gallium was detected above the background signal. We showed that in all the structures the wall tapers at the end as shown in Figure 6 (14). This is further illustration that the wall is always tangential to the gallium droplet.

The conical angles of the synthesized structures varied from 10-50^0. These conical structures do not correspond to the conical angle expected if graphene sheets are rolled into cones. Also preliminary HRTEM analysis has shown that the layers are not parallel to each other. The wall is not comprised of parallel graphene sheets as seen in multi-walled carbon nanotubes. These factors indicate that the wall formation follows a different mechanism.

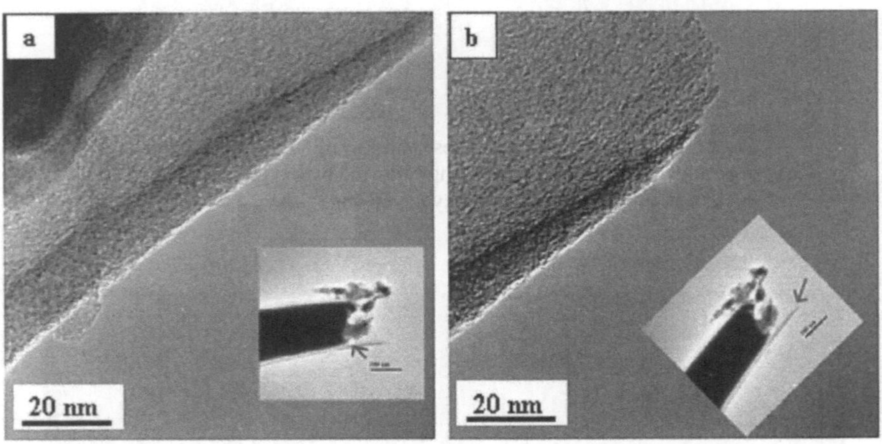

Figure 6. Thinning down of the wall towards the end of the tube.

We believe that the wall is comprised of a graphene sheet that is spirally rolled along the axis of the structure at the tip. During this process growth steps are formed, as the sheet is not fully rolled into a cylinder. Thus, these steps become the growth front for the further evolution of the wall. Carbon atoms add at the tip of the tube and then they are spiraled downwards. This maintains a constant wall thickness for most of the length of the tube except at the tip where the growth is initiated. This process causes the tapering at the tip of the growing structures. The tapering occurs only along the Ga-C interface.

This growth phenomenon is schematically shown in Figure 7.

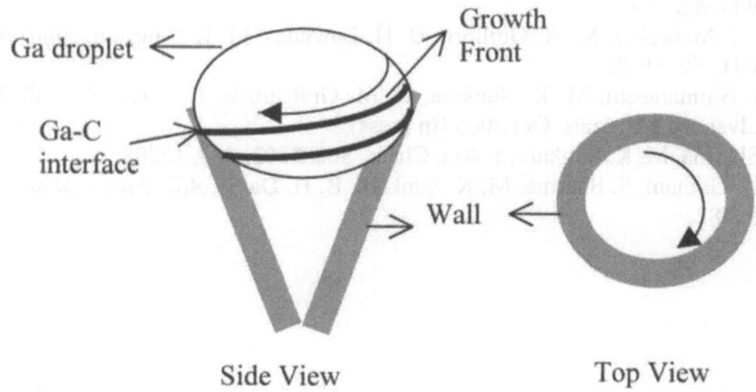

Figure 7. Schematic of the wall formation (a) Side on view and (b) top view of the spiral growth. The arrows indicate the direction of growth.

4. CONCLUSIONS

In summary, we described a synthesis strategy to control and modulate the morphology of carbon tubular structures in-situ during growth using gas phase chemistry. Although, we have shown only a few illustrations of the various morphologies that can be synthesized using this technique, a number of other morphologies can be synthesized by optimizing the process. For example, the tubes can be pinched to a certain diameter along the length at a certain point. These new class of carbon structures add a new dimension to controlled synthesis of several morphologies just by changing the gas phase chemistry. These larger inner diameter tubular structures can be very applicable in nano/micro-fluidics, micro reactor and also in electronic devices.

References

[1] S. Iijima, *Nature* **1991,** *354*, 56.

[2] S. Amelinckx, X. B. Zhang, D. Bernaerts, X. F. Zhang, V. Ivanov, J. B. Nagy, Science **1994,** *265*, 635.

[3] A. Krishnan, E. Dujardin, M. M. J. Treacy, J. Hugdahl, S. Lynum, T. W. Ebbesen, *Nature* **1997**, *388*, 451.

[4] S. Iijima, M. Yudasaka, R. Yamada, S. Bandow, K. Suenaga, F. Kokai, K. Takahashi, *Chem. Phys. Lett.* **1999,** *309*, 165.

[5] Y. Gogotsi, J. A. Libera, N. Kalashnikov, M. Yoshimura, *Science* **2000**, *290*, 317.

[6] Y. Gogotsi, S. Dimovski, J. A. Libera, *Carbon* **2002**, *40*, 2263.

[7] P. M. Ajayan, J. M. Nugent, R. W. Siegel, B. Wei, P. Kohler-Redlich, *Nature* **2000**, *404*, 243.

[8] R. C. Mani, X. Li, M. K. Sunkara, K. Rajan, *Nanoletters* (Web Pub Date – April 1) **(2003)**.
[9] J. Libera, Y. Gogotsi, *Carbon* **2001**, *39*, 1307.
[10] Y. Gao, Y. Bando, *Nature* **2002**, *415*, 599.
[11] Y. Gao, Y. Bando, *Appl. Phys. Lett.* **81**, 3966 (2002).
[12] Z. W. Pan, S. Dai, D. B. Beach, N. D. Evans, D. H. Lowndes, *Appl. Phys. Lett.* **2003**, *82*, 1947.
[13] V. I. Merkulov, M. A. Guillorn, D. H. Lowndes, M. L. Simpson, *Appl. Phys. Lett.* **2001**, *79*, 1178.
[14] G. Bhimarasetti, M. K. Sunkara, U. M. Graham, B. C. Davis, C. Suh, K. Rajan, Advanced Materials, Oct 2003 (In press).
[15] S. Sharma, M. K. Sunkara, J. Am. Chem. Soc. **2002**, *124*, 12288.
[16] U. H. Graham, S. Sharma, M. K. Sunkara, B. H. Davis, *Adv. Funct. Mater.* **2003**, 13, 582.

CARBON FILAMENT ROPE FORMATION

A. N. USOLTSEVA[1], V. L. KUZNETSOV[1], N. A. RUDINA[1],
M. YU. ALEKSEEV[2], L. V. LUTSEV[2]
[1]-Boreskov Institute of Catalysis, Pr. Akademika Lavrentieva 5, 630090 Novosibirsk, Russia, [2]-Research Institute "Ferrite-Domen", Chernigovskaya 8, St Petersburg, 196084, Russia

Abstract: Carbon filament growth via CO disproportionation on cobalt supported catalyst was investigated. For the first time the formation of the carbon filament ropes on Co catalysts was observed.

1. INTRODUCTION

The self-processes, which spontaneously assemble and organize various building blocks into hierarchical structures, have emerged as the most promising techniques [1] for the efficient production of carbon nanostructured materials. The formation of single-wall nanotubes (SWNT) ropes on metal catalysts is well known [2]. It can be considered as self-assembling process involving: multiple carbon nucleation (1) combined with Van der Waals interaction of individually growing tubes, resulting in nanotube rope formation (2). Here we demonstrate the rope formation from the individual carbon filaments produced on silica-modified Co-catalyst as self-assembling process.

2. EXPERIMENTAL

Reaction of catalytic CO decomposition in the presence of H_2 and Co-based catalyst supported onto Al_2O_3 was used for the synthesis of carbon nanodeposits. The reaction was carried out in a flowing quartz reactor at the temperature range 500-900°C and rates of CO and H_2 gas flow equal to 30 ml/min. Co-based catalyst was prepared via magnetron sputtering of Co/Si target in Ar atmosphere on Al_2O_3 plate. Initially the catalyst consisted of Co particles with 2 nm diameter in amorphous SiO_2 film on Al_2O_3 support. The typical thickness of Co-SiO_2 layer was about 0.5 μm.

S. Guceri et al. (eds.), Nanoengineered Nanofibrous Materials, 91-96.

The catalyst surface morphology before and after reaction was studied using a scanning electron microscope REM 100U (Russia). The genesis of Co-catalyst and the structure of formed carbon deposits were investigated with use of low and high resolution transmission electron microscopes: "JEOL" JEM-100CX and JEM-2010.

3. RESULTS AND DISCUSSION

3.1 Carbon Filament Rope Formation

Figure 1. presents the SEM image of the reaction products produced via catalytic CO decomposition (in presence of H_2) during three hours at 700°C on the Co-catalyst, which was pretreated at 900°C during six hours. One can see that the reaction mainly results in formation of coiled and curved carbon filaments. Detailed structure of coiled carbon fibers presents in figure 2. One can see that these fibers (with diameter ~100-500 nm) have a rope structure and consist of more thin primary filaments (d~ 10-50 nm).

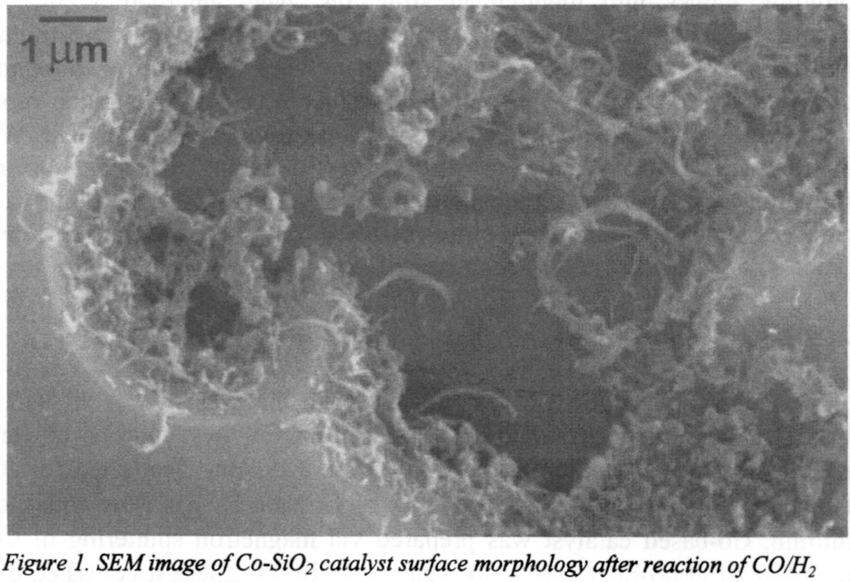

Figure 1. SEM image of Co-SiO_2 catalyst surface morphology after reaction of CO/H_2 decomposition at 700 °C.

Additional evidence of the rope structure was provided by the dark-field micrographs of the filament rope obtained with (0002) Bragg reflections of graphite lattice (figure 3). One can see bright stripes corresponding to the pile of graphite sheets of individual filament.

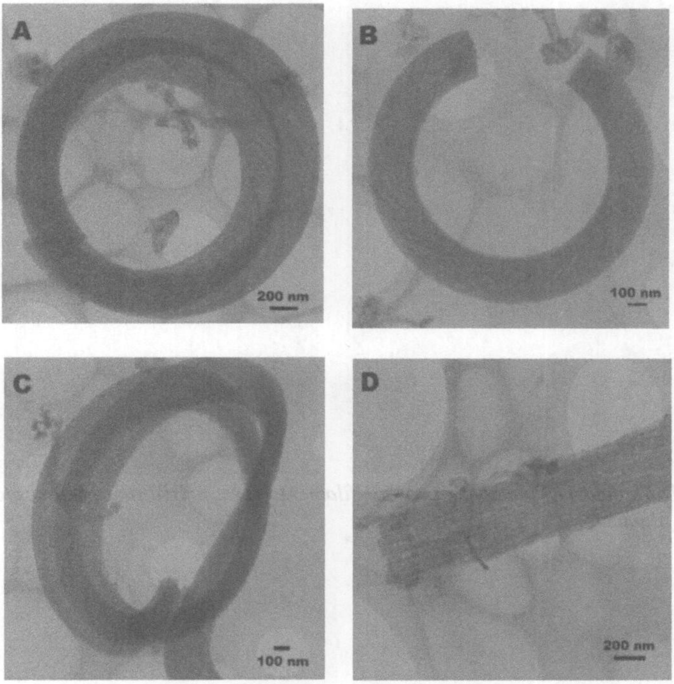

Figure 2. TEM images of carbon filament ropes.

A high resolution transmission electron microscopy (HR TEM) study reveals that the filaments in ropes consist of piles of graphite sheets, (fig. 4) which are oriented perpendicular to the filament axis ("pack of card" structure) and have a diameter about 30-50 nm. Ropes diameters are comparable with cross-section of catalysts particles (fig. 5). The formation of piles of graphitic sheets corresponds to the low values catalyst particle supersaturation with carbon. That is very reasonable for such big metal particles. Figure 4C. shows the contact region of two individual filaments in rope. It is rather difficult to estimate the distance between the filaments within contact regions because of the possibility of image overlays. However, from HR and dark field TEM images (fig. 4C and 3, correspondingly), we estimated that the uneven filament surfaces provide the formation of chinks between filaments up to 6-7 nm. In some cases the closure of edges of some part of graphitic sheets was observed (fig. 4B).

The formation of the ropes of carbon filaments was an unpredictable result, because usually one big catalyst particle provides the formation of one or

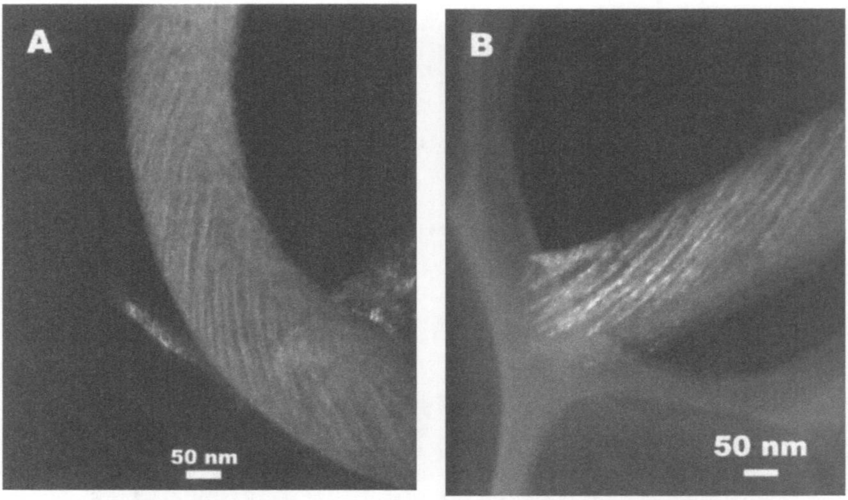

Figure 3. Dark field TEM images of carbon filament ropes. Bright stripes correspond to the individual filaments.

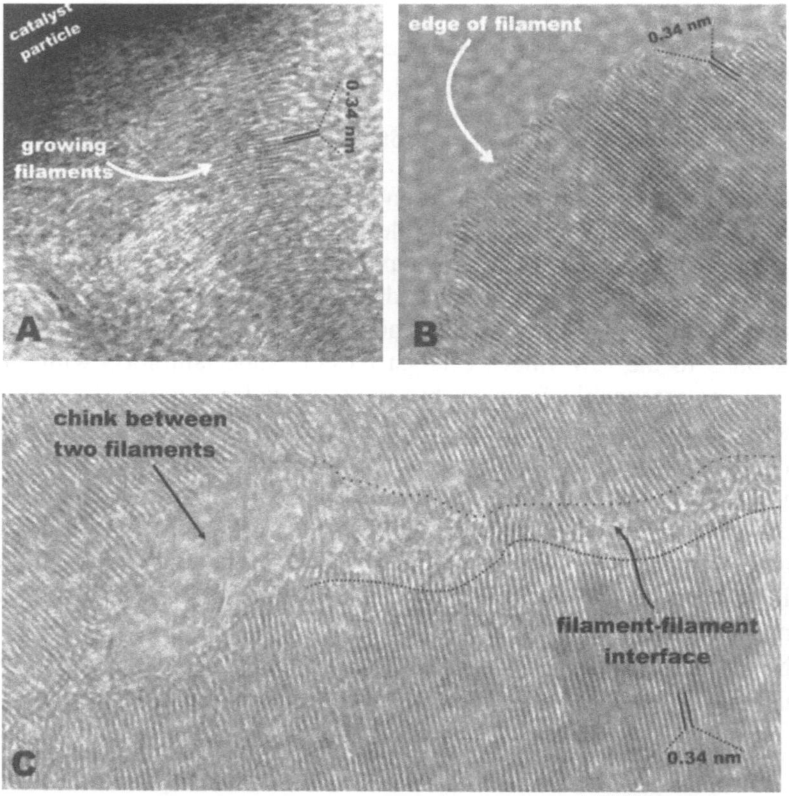

Figure 4. High resolution TEM images of carbon filament rope structure.

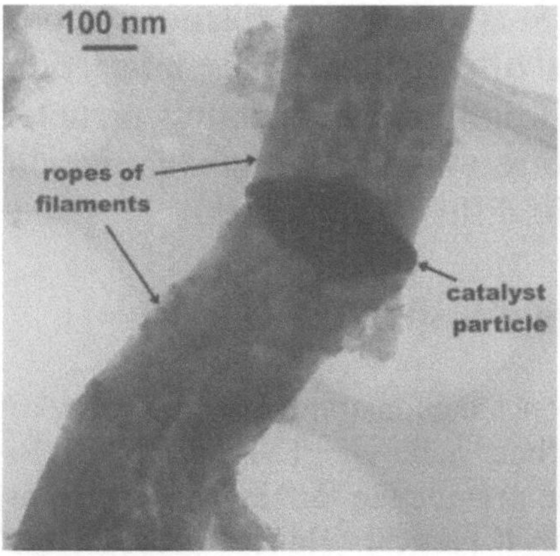

Figure 5. TEM image of catalyst particle with growing filament rope.

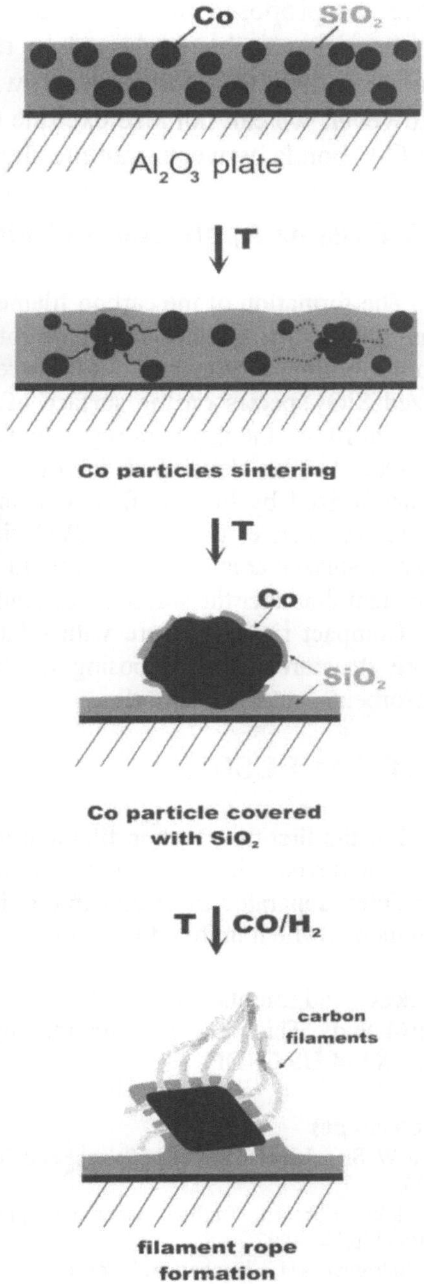

Figure 6. Proposed scheme of Co-Si catalyst genesis and carbon filament rope growth.

two carbon filaments [3] (growing from the opposite sides of particles) and stands as the reason the carbon filament rope formation can be described in the terms of the Co-Si catalyst genesis [4]. So, catalyst pretreatment at 900°C leads to the sintering of disperse Co particles with the formation of big metal particles, presumably, covered with a holed SiO_2 layer and/or SiO_2 islands. We suggest that the carbon filament rope formation is conditioned by specific structure of catalyst particles, which formed during sintering of initial Co particles separated by SiO_2 (figure 6). The carbon filament nucleation presumably occurs in many places of big metal particles surface free from SiO_2, which forms holed layer and/or islands. Thus filaments grow through the holes of SiO_2 shells of metal particle and then self-organized in the regular ropes. Note that the filaments have rough surface structure but they self-organize in ropes with a surprisingly well-ordered structure.

One can proposed that the reason for Self-assembling filaments ropes formation may be explained in terms of Van der Waals interaction between separated filaments growing from the same catalyst particle. However, we can not also exclude the direct binding with the formation of C-C bonds between graphite sheets of different filaments.

3.2 Possible Applications of Filament Ropes

The formation of the carbon filament ropes demonstrates the possibility of the production of the new type of carbon material. The control of Si concentration in primary Co-SiO_2 layer can provide the formation of different sized SiO_2 species on the surface of catalyst particle. Thus, it provides the possibility for the diameter regulation of primary filaments forming the rope.

Due to the developed internal structure, the filament ropes can be characterized by high surface area in comparison with traditional filaments materials; hence, typically CCVD filaments with diameters about 500 nm have a surface area of ~ 100 – 200 m^2/g and for filament ropes with the same filament diameter the surface area could increase up to 300 – 600 m^2/g.

Compact rigid structure with relatively high surface area and developed pore structure allow proposing the usage of carbon filament rope as new adsorbents, catalyst supports.

4. CONCLUSIONS

For the first time carbon filament rope formation was observed. This result was achieved via creation of multiple nucleation centres on big metal particles, separated by irreducible oxide species and independently produced filaments, which unite into the ropes.

Acknowledgments: We are thankful to Dr. A. L. Chuvilin for performing the TEM study. This work was supported by INTAS grants, 01-254, Award No. NO-008-X1 of US CRDF.

References

1. R.W. Siegel in Nanostructure Science and Technology, Kluwer Academic Publishers, 1999, p.25.
2. T.W. Ebbesen, Carbon nanotubes: preparation and properties, CRC Press, Boca Raton,Florida, 1997.
3. Rodriguez NM, Chambers A, Baker RTK. Catalytic engineering of carbon nanostructure Langmuir 1995; 11: 3862-3866.
4. A. N. Usoltseva, V. L. Kuznetsov, A. L. Chuvilin, N. A. Rudina, M. Yu. Alekseev, L. V. Lutsev, Carbon (2003) in press.

ELECTROSPINNING OF LOW SURFACE ENERGY QUATERNARY AMMONIUM SALT CONTAINING POLYMERS AND THEIR ANTIBACTERIAL ACTIVITY

Kazım Acatay, Eren Şimşek, Mert Akel, Yusuf Z. Menceloğlu

Sabancı University, Material Science and Engineering Program, Faculty of Engineering and Natural Sciences, Orhanli 34956 Tuzla/Istanbul, Turkey

Abstract: Antibacterial agent (quaternary ammonium salt) containing perfluorinated polymers had been synthesized at different agent concentrations. The polymers were dissolved in solvent mixture and electrospun, to increase effective surface area, at 12 kV which had resulted nanofibers with diameters as low as 40 nanometers and fluffy structures. The resultant electrospun webs' biocidal activities relative to the solution cast film samples and other biocides had been tested with *Escherichia Coli* bacteria containing aqueous medium. The potential application fields of the product may be antifouling applications at air filtering, marine industry and antibacterial applications in medicine.

Keywords: Electrospinning, Low-surface energy, Antibacterial, Antifungal.

1. INTRODUCTION

Quaternary ammonium compounds (QACs) have a large variety of usage areas from cosmetics to clothes softeners, but especially they are known to be good disinfectants. In proper concentrations, they are very effective against fungal attack [1]. QACs, having cationic nitrogen structure, have the general formula of $R_4N^+X^-$. In a quaternary structure, nitrogen atom is covalently bonded to four groups and the positive charge is balanced by a negative counterion.

QACs antibacterial ability is resulted from their amphiphilic structure and surfactant property which was first reported by Dogmak [2]. The antimicrobial action of the QACs is based on their damaging surfactant-like interaction with the membrane (cytoplasmic) of bacteria resulting the loss of the membrane permeability. At convenient concentrations, they can cause cell leakage and the death of the cell [1].

Quaternary structures are effective on both gram positive and gram negative bacteria, but they have a stronger antibacterial effect on gram

S. Guceri et al. (eds.), Nanoengineered Nanofibrous Materials, 97-106.

positive ones, since gram negatives have an extra protective membrane. Quaternary compounds are widely used because of their non-toxic and non-irritant property [3].

Quaternary ammonium compounds are not affected from the protein concentration of the environment and don't loose their affectivity on bacteria over the course of time [4]. The effectiveness of the agent against microorganisms is directly related to its area of contact with the microorganism's medium. Especially, if it has been attached into a non-dissolving phase, such as solid phase, and in contact with a non-solvent fluid, such as flowing air.

Electrospinning, or electrostatic fiber spinning, is a novel fiber manufacturing process to produce sub-micron, or nanometer, diameter polymeric fibers. Although its idea was first published by Formhals in 1934 [5] and some of preliminary studies were carried by Baumgarten [6], its reputation had risen in 90's due to increased interest to nanotechnology and relative ease of fiber fabrication by electrospinning [7, 8].

Electrospun fibers have several outstanding features such as very high surface to volume ratio, flexibility in surface functionality, and enhanced mechanical properties [9]. These impressive properties make the electrospun fiber mats recently emerging candidate for filtration, membrane, composite applications, tissue templating, biomedical applications such as protective clothing, medical prosthesis and wound dressing, electrical and optical applications and nanoscale tube fabrication [9].

The electrospun fibers are generated by subjecting polymeric solution (or melt) in a glass syringe to an electric field of several kilovolts to tens of kilovolts. The liquid droplet at the capillary tip of the syringe, which is kept by surface tension of the liquid, is distorted as the intensity of the electric field between the tip and ground increase. Previously hemispherical droplet is elongated, and turned to a conical shape known as the Taylor cone. Further increase of the attractive force of electric field results generation of a charged jet of fluid that is channeled to the ground. The discharged jet of fluid travels in a random pattern with continuous instable whipping motions. Meanwhile, the solvent in the traveling wet fiber evaporates and the charged fiber becomes thinner. The fibers are deposited on the grounded collector in a random, nonwoven manner.

The typical diameters of the collected dry fibers are ranged from several nanometers to a few micrometers. The thickness of the electrospun fibers differs depends on the fluid properties, such as the fluid's viscosity, conductivity, dielectric constant, surface tension, polymer's molecular weight and process operating parameters, such as flow rate of the solution, jet current, applied electrical potential and tip to ground distance [9, 10, 11]. Ambient parameters such as temperature, humidity and air velocity around the spinning chamber is also important [12]. Huang et al. stated that in open literature fifty different polymers, some in several different solvents, have been successfully electrospun to nanometer thickness nonwoven mats [12].

In this study, we synthesized a terpolymer consisting of a backbone with two kinds of functional side chains, one with quaternary structure and the other with perfluorinated structure for hydrophobic nature. Then, we investigated the effect of electrospinning on the antibacterial activity. Bulky structures could have difficulties to confront with microorganism membrane so the best way is to increase encountering chance of microorganism and quaternary structure, as every phenomenon occurs on the surface [13]. Therefore, we increased surface area by turning the quaternary structure containing polymers into nanofibers by electrospinning.

2. EXPERIMENTAL

2.1. Materials

The monomers; commercial grade methylmethacrylate (MMA) was purified by double passing through alumina powder packed column before use, and perfluoro alkyl ethyl acrylate (PFAEA, Fluowet AC 812), kindly supplied by Clariant, was used without purification. The quaternary ammonium salt containing monomer was synthesized with the reaction of dimethyl coconut amine and vinyl benzyl chloride, supplied from Fluka, in distilled water. Other commercial biocides, Nipaguard BPX and 2-bromo-2-nitropropane-1,3-diol was, also, supplied by Clariant and used as received. Azoisobutyronitrile (AIBN, Fluka) was used as polymerization reaction initiator and the solvents Tetrahydrofuran (THF) and Dimethylformamide (DMF) was used without purification.

2.2. Synthesis of Vinylbenzyl-Dimethylcocoammonium Chloride (VBDCC) Monomer

The reaction of quaternary ammonium salt containing monomer was carried out in distilled water environment at 50°C with vinyl benzyl chloride and dimethyl coconut amine as the reactants, in the presence of Na_2CO_3 as catalyst (Figure 1). The reaction was stopped when the opaque color of the mixture in the flask turned to clear, this took approximately 30 minutes. The water in the monomer reactor was removed under vacuum at room temperature.

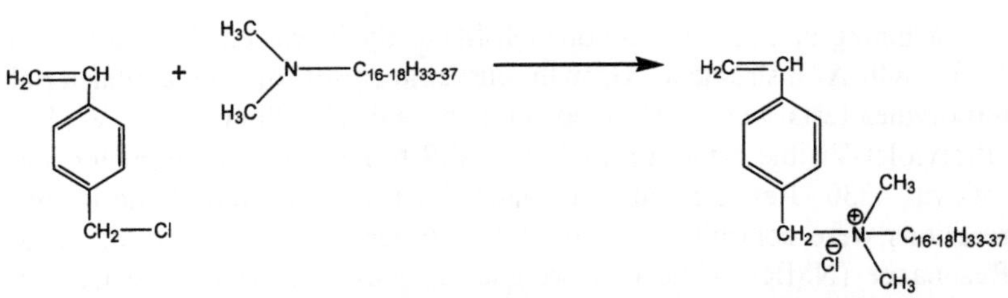

Figure 1. Reaction scheme for Vinylbenzyl-dimethylcocoammonium chloride synthesis.

2.3. Synthesis of Terpolymers

For the synthesis of vinylbenzyl-dimethylcocoammonium chloride - MMA - perfluoro alkyl ethyl acrylate terpolymer, polymerization reaction was carried in THF with AIBN as initiator at 70°C (Figure 2). There were five different sets, named as quat-free, 1%, 5%, 10% and 25%, according to the molar ratio of the quaternary monomer in the polymer. The compositions of the terpolymers are presented in Table 1.

Table 1. The molar compositions of the synthesized terpolymers.

	VBDCC (% molar)	MMA (% molar)	PFAEA (% molar)
Quat-free	0	90	10
1% polymer	1	89	10
5% polymer	5	85	10
10% polymer	10	80	10
25% polymer	25	65	10

The reaction conditions for the terpolymers, $[AIBN] / [M]_{total} = 1.02 \times 10^{-3}$, reaction temperature 70°C and reaction time was 24 hours. After the reactions were completed in the corresponding interval, the products were first dissolved in chloroform and then precipitated in *n*-hexane (solution/precipitator = 1/10). The terpolymers were obtained by filtration, and dried in vacuum for 12 hours at room temperature.

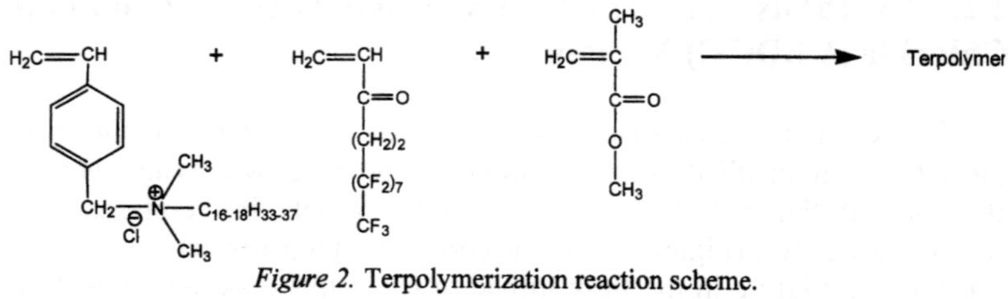

Figure 2. Terpolymerization reaction scheme.

2.4. Measurements

Scanning electron microscope (SEM) analysis was carried out with a JEOL 840-A instrument. Growth curve analyses of the antibacterial terpolymer tests were performed with the aid of a Shimadzu UV-3150 Ultraviolet-Visible-Near Infrared (UV-VIS-NIR) Spectrophotometer and Innova 4330 Refrigerated Incubator Shaker. The terpolymerization reactions were controlled by 500 MHz Varian Inova Nuclear Magnetic Resonance (NMR). Molecular weights of polymers were measured by polystyrene calibrated Waters Gel Permeation Chromatography (GPC) instrument.

2.5. Electrospinning

The electrospinning setup (Figure 3) employed in this study was consisted of a high voltage (HV) power supply (GPS HV power supply Model 2594), a collector screen connected to a grounded electrode, and a vertically located syringe controlled by a Univentor 801 Syringe Pump. The terpolymer solutions were prepared for 15 wt % polymer in THF-DMF (50/50 wt %) solvent mixture. In each of the four electrospinning processes, polymer solution was transferred into a syringe, with a 16 gauge stainless steel tip, and the flow rate of the syringe set to 5 µl/min. 12 kV voltages was applied to the polymer solution, while the distance between the syringe tip and the collector screen was kept at 10 cm.

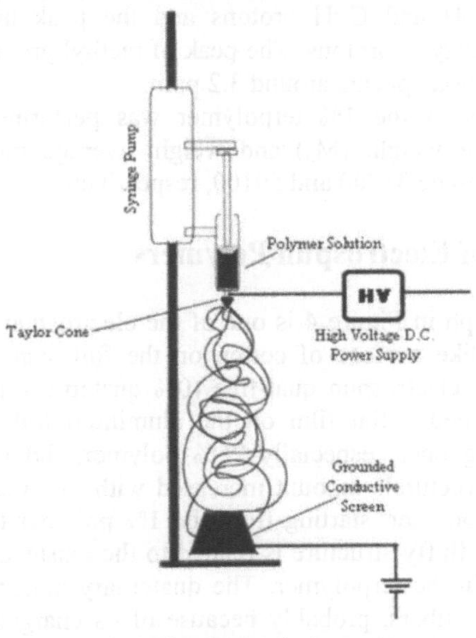

Figure 3. The electrospinning apparatus.

2.6. Biocide Activity Tests

A series of growth curve tests were carried out to investigate the antimicrobial activity of the polymers, with *Escherichia Coli* (XL1 blue), a gram positive bacteria. For this purpose, similar concentrations of bacteria solutions were prepared in falcon tubes by diluting the overnight grown bacteria, at 37°C in incubator shaker, with LB Broth growth media. Electrospun polymers were put into different tubes and metal nets were placed onto the electrospun polymers in the tubes, in order to make the swimming polymer sink completely into the bacteria solution. In each experiment, one more bacteria solution was used as control group. Growth curve analysis of the bacteria was performed by recording the bacteria population in each tube in certain time intervals, by measuring the absorption via UV/Vis/NIR at 600 nm, which is the specific wavelength for *E. Coli* dissolved in water. Antimicrobial activity investigation of the quaternary monomer and other biocides, such as industrial grade Nipaguard BPX and 2-bromo-2-nitropropane-1,3-diol, were also performed.

3. RESULTS AND DISCUSSION

3.1. Characterization

Chemical structure of the synthesized quaternary monomer was confirmed by ^{1}H NMR, where peaks around 7.4~7.6 ppm are assigned to the aromatic protons of VBC, the peaks between 5 and 7 ppm are assigned to the vinylic C-H and C=H protons and the peak around 4.9 ppm corresponds to benzylic protons. The peak of methyl groups were bonded to the nitrogen cation appears around 3.2 ppm.

GPC analysis of the 1% terpolymer was performed and number average molecular weight (M_n) and weight average molecular weight (M_w) were found to be 31700 and 59100, respectively.

3.2. Imaging of Electrospun Polymers

The photograph in Figure 4 is one of the electrospun terpolymers. A fluffy structure, like a piece of cotton on the foil, can be seen on the figure. While the electrospun quat-free (0% quaternary part containing) polymer had formed a flat film on the aluminum foil, the quaternary moiety containing ones, especially 25% polymer, did not stick to the surface. Fluffy structure's amount increased with the increasing ratio of the quaternary monomer, starting from the 1% polymer to the 25%. So, the emergence of fluffy structure is related to the existence of quaternary ammonium part in the terpolymer. The quaternary moiety has restricted uniform lie of the fibers, probably because of its charged structure, and fibers heaped. However, the quat-free sample has formed a flat surface on the grounded aluminum foil.

The Scanning Electron Microscope (SEM) images had displayed one more important distinction between QAC containing and not containing polymers (Figure 5): Fiber diameter distribution. The quat-free sample had uniform diameter distribution in the range of 150-250 nm. However, the 25% QAC containing sample showed a diameter distribution between 40-800 nm, where one group was between 400-800 nm and a second group was 40-120 nm thickness. Similar occurrence had observed by Bognitzki et al. [14], where they used tetrabenzylammonium chloride (TEBAC) salt. Due to fiber diameter decrease, average fiber diameter (AFD) significantly decreases. So, effective surface area has increased per volume. This is important for tuning of the material properties.

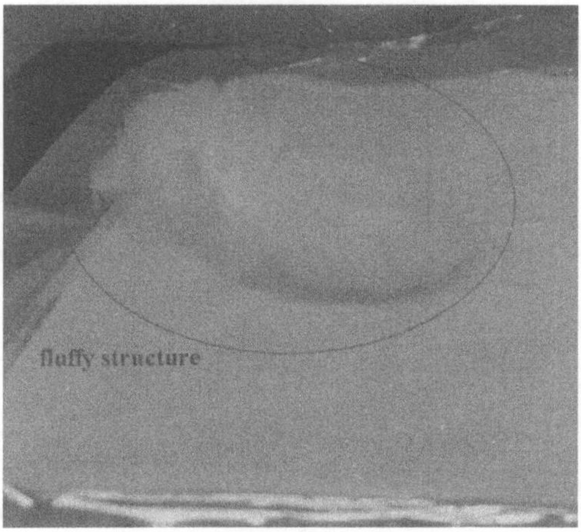

Figure 4. Photograph of an electrospun polymer in fluffy structure.

Figure 5. SEM images of (a) Quat-free polymer, (b) 25% terpolymer in fluffy structure, and (c) enlarged image of the same zone in b.

3.3. Testing the Biocidal Activity

The preliminary bacterial tests showed that 25% terpolymer (electrospun) has the most effective antibacterial property, which led to further biocide activity tests performed only with 25% terpolymer. Antimicrobial affectivity investigation of the quaternary monomer and other reference biocides Nipaguard and Bronopol was also performed (Figure 6).

Water soluble quaternary monomer was detected to be quite effective against bacterial attack. However, similar effect could not be observed with the electrospun, water non-soluble terpolymer (Figure 7). Obvious success of the fine powdered polymer when compared with the

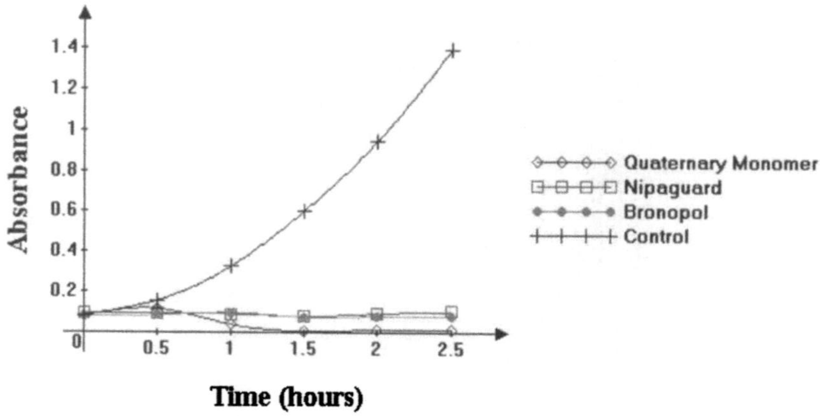

Figure 6. Tests with the quaternary monomer and other biocides.

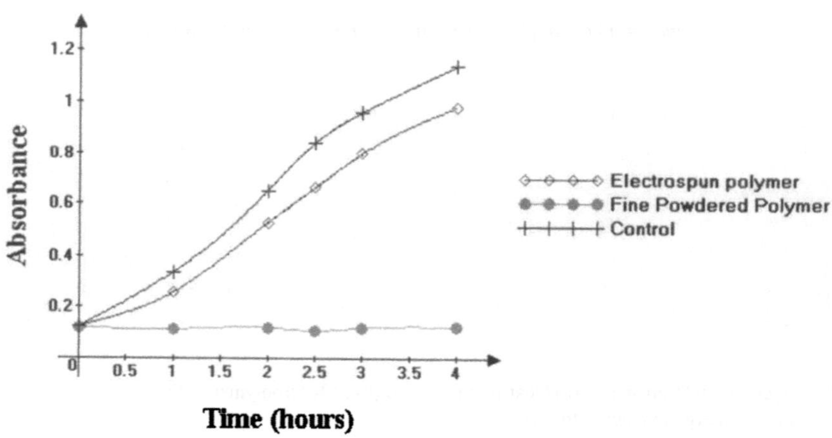

Figure 7. Comparison of the 25% electrospun polymer with the same composition but in the fine powdered form.

electrospun one led to further studies on hydrophobic property of the polymer. 25% QAC containing non-fluorinated polymer was synthesized and additional antibacterial tests were performed.

Figure 8 represents the surviving bacterial population versus time, while testing with non-fluorinated polymer. The graph shows that both electrospun and non-spun non-fluorinated polymers had a marked biocide efficiency. Consequently, it was determined that the fluorinated, electrospun polymer showed less biocide activity because the hydrophobic property prevented the bacteria solution from entering the inside of the fiber ball and interacting with the single fibers; so the electrospun fiber ball behaved like a big solid polymer having much smaller surface to volume ratio than expected.

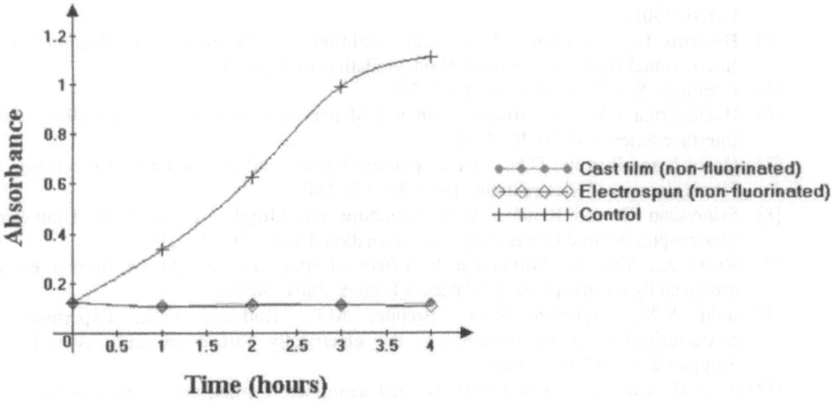

Figure 8. Tests with non-fluorinated polymer

4. CONCLUSIONS

In the present study, synthesis of quaternary ammonium salt containing polymers was performed and their antibacterial activity was observed. The electrospinning of QAC containing polymers resulted in fluffy structures, probably due to the presence of charge. The presence of salt (here QAC) resulted very thin fibers as well as normal fibers. The effective concentration of quaternary monomer in the terpolymer for biocide activity against *Escherichia Coli* bacteria was found to be 25% molar in the polymer. Antibacterial activity of fluorinated terpolymer electrospun was detected to be weak, due to its low surface tension. So, the bacteria containing liquid could not penetrate inside and contact with the interior electrospun fibers. This had decreased the effective surface for applications. Fluorinated polymer may not give effective results for liquid applications, but may show enhanced properties at air filter applications due to its non-wetting property and smaller average fiber diameter. Finally, present study proves adjustable characteristic of electrospun fibers and their ability to be easily tunable according to specific needs.

Acknowledgments: The authors would like to acknowledge Dr. Mehmet Ali Gülgün of Sabancı University for his help on SEM imaging, Brisa Inc. for their permission to use SEM, Clariant (Turkey) A.S. for their interest to supply chemicals, and Kıvanç Bilecen of Sabancı University for his help on antibacterial tests and his fruitful discussions on biological concepts.

References

[1] Massi L., Guittard F., Géribaldi S., Levy R., Duccini Y. Antimicrobial properties of highly fluorinated bis-ammonium salts. International Journal of Antimicrobial Agents 2003; 21, 20-26.
[2] Dogmak G. Eine neue Klasse von Disinfektionmitteln Dtsch. Med Wochenschr 1935; 61: 829-32.

[3] Yenen, Şadi, Anğ, Özdem, Anğ, Mine, *Tıbbi Mikrobiyoloji.* Istanbul: Nobel Tıp Kitap Evleri, 2001.
[4] Bessems E., The effect of practical conditions on the efficacy of disinfectants, International Biodeterioration & Biodegredation 1998; 41: 177-183.
[5] Formhals A., US Patent 1934; 1,975,504.
[6] Baumgarten P.K., Electrostatic spinning of acrylic microfibers, J. of Colloid and Interface Science 1971; 36: 71-9.
[7] Doshi J. and Reneker D.H., Electrospinning Process and Applications of Electrospun Fibers, Journal of Electrostatics 1995; 35: 151-160.
[8] Srinivasan G. and Reneker D.H., Structure and Morphology of Small Diameter Electrospun Aramid Fibers, Polymer International 1994; 36: 195-201.
[9] Koski A., Yim K., Shivkumar S. Effect of molecular weight on fibrous PVA produced by electrospinning. Materials Letters, 2004; 58:493-497.
[10] Shin Y.M., Hohman M.M., Brenner M.P., Rutledge G.C., Experimental characterization of electrospinning: the electrically forced jet and instabilities, Polymer 2001; 42:9955-9967.
[11] Fong H., Chun I., Reneker D.H. Beaded nanofibers formed during electrospinning, Polymer 1999; 49:4585-4592.
[12] Huang Z.M., Zhang Y.Z., Kotaki M., Ramakrishna S. A review on polymer nanofibers by electrospinning and their applications in nanofibers, Composites Science and Technology 2003; 63:2223-2253
[13] Chen Z. C. and Cooper S. L., Interactions between dendrimer biocides and bacterial membranes, Biomaterials 2002; 23: 3359-3368.
[14] Bognitzki M., Frese T., Wendorff J.H., Greiner A. Submicrometer shaped polylactide fibers by electrospinning, ACS-Polymeric Materials: Science and Engineering, 2000.

ON THE MECHANISM OF SINGLE-WALL CARBON NANOTUBE NUCLEATION IN THE ARC AND LASER PROCESSES: WHY BIMETALLIC CATALYSTS HAVE HIGH EFFICIENCY?

A.V. KRESTININ, M.B.KISLOV, A.G.RYABENKO
Institute of Problems of Chemical Physics, Russian Academy of Sciences, Chernogolovka, Moscow Region, 142432, Russia

Abstract: A new model of SWNT nucleation has been proposed in which an endohedral metallofullerene serves as the precursor of a nanotube nucleus, and the nucleus itself is formed as an adduct arising when metal atoms attach to the endofullerene shell. High efficiency of bimetallic catalysts is explained by double action of the metal catalyst in SWNT nucleation. First, one metal helps to create the formation of endometallofullerenes. Secondly, the other metal atoms attach to the endofullerene shell to transform an endofullerene into a nanotube nucleus. The relative efficiencies of La/Ni, Gd/Ni, Ce/Ni, Pr/Ni catalysts, in comparison to Y/Ni catalyst were measured and their high efficiency in SWNT formation has been demonstrated.

Key words: carbon/metal vapor condensation, single-wall carbon nanotube nucleation, bimetallic catalyst

1. INTRODUCTION

High efficiency of bimetallic catalysts such as Y/Ni or Co/Ni in single-wall nanotube (SWNT, buckytube) synthesis is the most intriguing feature of the nanotube formation mechanism. In addition, quite recently Takizawa and co-workers [1] discovered experimentally that an increase of yttrium concentration entailed a noticeable increase of the nanotube diameters. On the contrary, in the same experiments an increase of nickel concentration caused only a minor change in the nanotube diameters. These experiments showed

S. Guceri et al. (eds.), Nanoengineered Nanofibrous Materials, 107-114.

that two metals in the catalyst mixture take different parts in the process of SWNT formation.

The way of searching for an explanation for the mechanism of buckytube formation depends crucially on which of two existing approaches to its treatment is to be used. In the first approach, proposed by Smalley and co-workers [2,3], the process of SWNT growth is assumed to proceed along with condensation of carbon/metal vapor by absorption of small carbon molecules in the open end of the tube. In this case the temperature of buckytube growth should be as high as 2000 –2500 K. In the second approach [4,5], the SWNT formation process is supposed to be the condensed phase process of structural reconstruction and amorphous carbon serves as a source of carbon material for nanotube growth. In this case the growth process could proceed at a rather low temperature, which is evaluated as ≈ 1600 K [4]. Here we offer additional theoretical and experimental arguments supporting the first model of the SWNT formation mechanism.

2. THE SOURCE OF CARBON FOR SWNT GROWTH IN THE PROCESS OF CARBON/METAL VAPOR CONDENSATION

If one supposes buckytubes to grow from amorphous carbon, the change in Gibbs free energy ΔG in the following reaction should be less than zero:

$$\text{C (amorph.)} \rightarrow \text{C(SWNT)}, \Delta G = \Delta H - T\Delta S \qquad (1)$$

The entropy of SWNTs is much less than the entropy of amorphous carbon, because of high structural perfection of buckytubes. Thus, the enthalpy of formation of the amorphous carbon should be much higher than the value of the buckytubes to be sure the process (1) is running. According to the estimations made by De Bokx et al. [6] the energy contribution in enthalpy of the formation of amorphous carbon is because its disorder cannot not exceed 15 kJ/mol. Thus, the following inequality can be written:

$$\Delta H_f(\text{C, amorph.}) - \Delta H_f(\text{C, graphite}) < 15000 \text{ J/mol} \qquad (2)$$

The data presented in the literature [7] allows accurately estimating the enthalpy of buckytube formation as it follows [8]:

$$\Delta H_f(\text{C, SWNT}) - \Delta H_f(\text{C, graphite}) = 6290+7120/d^2 \qquad (3)$$

Thus, from (2) and (3) the following inequality results (J/C-mol):

$$\Delta H_f(\text{C, SWNT}) - \Delta H_f(\text{C, amorph.}) > 7120/d^2 - 8710 \qquad (4)$$

where d is the buckytube diameter (nm). From the equation (4) it follows that the enthalpy change together with the Gibbs energy change in the process (1) surpasses zero for nanotubes with the diameter of 0.9 nm and is smaller, that is, they can not grow using amorphous carbon as the source of material. In the arc discharge process, the nanotubes of different diameters form simultaneously. For example, by using Ce/Ni catalyst [9] the buckytubes of the diameters in the range of 0.83 nm - 1.6 nm grow. Thus, if not to reject the possibility of SWNT growth from amorphous carbon, one should admit that the SWNTs are growing in different ways depending on their diameters. Considering this option as doubtful, we assume that the source of carbon material for buckytube growth in the process of carbon/metal vapor condensation may be only small carbon clusters of gas phase or the same clusters in the adsorbed state on the buckytube surface.

3. KINETICS OF CARBON/METAL VAPOR CONDENSATION

Since the boiling temperatures of the metals commonly used as the catalyst lie well below the temperature of the graphite sublimation (which is near 4000 K), the process of vapor condensation always starts with the formation of pure carbon clusters. Analysis of the kinetics of pure carbon vapor condensation [10-12] showed the following.

1. Mainly because of poor thermodynamic stability of fullerene shells condensation of carbon vapor starts only as the gas temperature drops to ~ 2800 K. The whole process of condensation in the arc reactor under typical conditions for fullerene formation proceeds for a few milliseconds. The maximum rate of fullerene formation in the arc reactor is attained at 2200-2400 K.
2. A high yield of fullerenes, as in the experiments conducted under optimized conditions for fullerene synthesis, can be attained if only (i) the rate of cooling and mixing of carbon vapor with inert atmosphere is optimal and (ii) rather tough restrictions on the kinetics of carbon cluster growth are fulfilled.

What is happening in the presence of a mixture of carbon/metal vapors was discussed in detail elsewhere [8]. Typical parameters of the condensation process using Y/Ni catalyst in the arc reactor are listed in Table 1.

TABLE 1. Condensation of carbon/metal vapor in the arc discharge process of SWNT synthesis

Element (boiling temperature, T_{boil}, K)	Ni (3070)	Y (3610)	C
Metal concentration in the catalytic mixture, at.%	4 %	1 %	94 %
Metal vapor concentration at the stage of C-vapor condensation (vol. units)	~ 0.004	~0.001	~0.03 - 0.05
Temperature of metal vapor saturation, K	2270	2300	3300 - 3400
Temperature of metal vapor condensation starting, K	~ 2320 (est)	?	~ 2800*

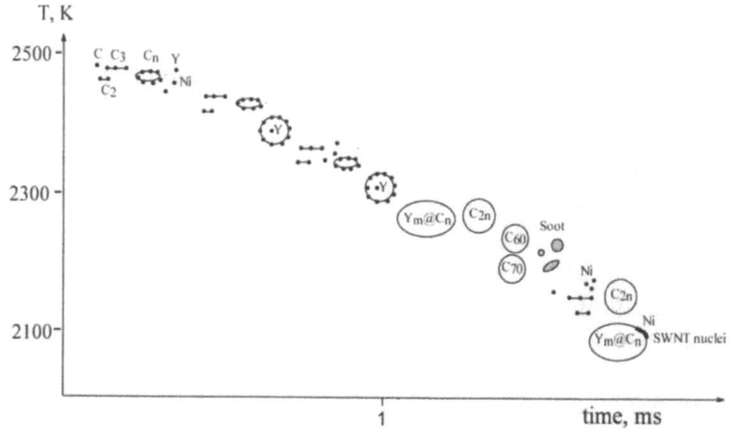

Fig.1. A picture of SWNT formation in the process of carbon/metal vapor condensation

Nickel and yttrium vapors become saturated after the gas temperature decreases to ≈ 2270 K and to ≈ 2300 K, respectively. The calculated temperatures give only the low limits for onset of the condensation of the metal vapors, since the estimation did not take into consideration that the process of condensation goes through the formation of metal/carbon compounds, which have a lower saturated pressure of metal vapor compared to the equilibrium vapor pressure for liquid metal. Unfortunately, the structure and thermodynamics of small carbon/metal clusters are entirely unknown at present, and this makes theoretical studies of condensation of carbon/metal vapor very difficult.

4. THE MODEL OF SWNT NUCLEATION IN THE PROCESS OF CARBON/METAL VAPOR CONDENSATION. WHY BIMETALLIC CATALYSTS ARE HIGHLY EFFICIENT?

Since metal vapors become supersaturated at temperatures, that are close to the temperature of the highest rate of fullerene formation (2200-2400 K), it is reasonable to suppose that metal atoms may enter a fullerene shell just as the latter is being formed. The possible picture is sketched in Fig.1 and assumes that small carbon clusters, for example, cycles of C_n can form adducts with a metal atom at a temperature of about 2300-2500 K, just when the rate of fullerene formation attains its maximum. In this case, coagulation of those cycles, followed by rearrangement of an unstable intermediate structure could yield a fullerene shell with one or more metal atoms inside. Maruyama and co-workers [13] performed molecular dynamic simulations of the clustering process in carbon vapor with metal atoms such as La, Y, Gd, Ce. They demonstrated that metal-carbon clusters MC_n ($n \approx 15 - 40$) with an open-cap appearance form easily at the temperatures of 2500-3000 K. These clusters grew larger with closing the open-cap structures at the n-value of about 40-50. Surprisingly, nickel atom at the final stage of metal/carbon cluster growth preferred to attach at the larger defect of the caged structure such as large rings of more than 7 or 8 members, and frequently moved in and out of the carbon cage. The especially high efficiency of bimetallic catalysts such as Y/Ni, La/Ni, Ce/Ni [1,9,14] is explained in the model by the double action of the metals in formation of a buckytube nucleus. First, metal atoms take part in formation of endometallo-fullerenes. For this purpose, condensation of metal vapor has to start on small carbon clusters before a fullerene shell begins to form. In this process the same metals suit very well which are very active in endofullerene formation, for example, yttrium, lanthanum, gadolinium. They possess high boiling temperatures, and their vapors being just in sufficiently low concentrations start precipitating at the temperatures of about 2400-2500 K. Second, metal atoms have to attach to the endofullerene shell for transformation of the latter into a buckytube nucleus. For this purpose such a metal suits well, which is only weakly active in endofullerene formation, and has a lower boiling temperature. The high performance of bimetallic catalysts lies in adjusting the values of two metal concentrations separately for each of these two purposes. Thus, experiments of Takizawa and co-workers [1] have their natural explanation, because only the metals forming endofullerene affect the nanotube diameters.

If the proposed model is true, different metals which are able to form endofullerenes should be equally efficient as a catalyst in mixtures with nickel. We measured efficiencies of a set of bimetallic catalysts in comparison to the Y/Ni one. The results are listed in Table 2. Details of the experiments are described elsewhere [8].

TABLE 2. Efficiency of bimetallic catalysts Me/Ni in the arc process of single wall carbon nanotube synthesis

Element	Y	La	Gd	Ce	Pr	Yb
T_{boil}, K	3610	3730	3553	3613	3633	1484
Optimized composition of the catalyst, at.%; He pressure	0.8% Y 3.2% Ni 550 Torr	0.6% La 3.2% Ni 550 Torr	0.7% Gd 3.3% Ni 550 Torr	0.8% Ce 3.2% Ni 450 Torr	0.8%Pr 3.2%Ni 300 Torr	varied
Relative efficiency of the catalyst in the arc reactor	1.0	0.5-0.7	~1.0	~1.0	0.5-0.7	Trace of SWNTs

The relative efficiency of a given catalyst was determined as the ratio of the SWNT quantity obtained in the optimized operational conditions for this catalyst to the quantity of SWNTs obtained in the optimized process with the Y/Ni catalyst. Measurement of SWNT content in as-produced material was performed by using absorption spectroscopy in near IR range of wavelengths.

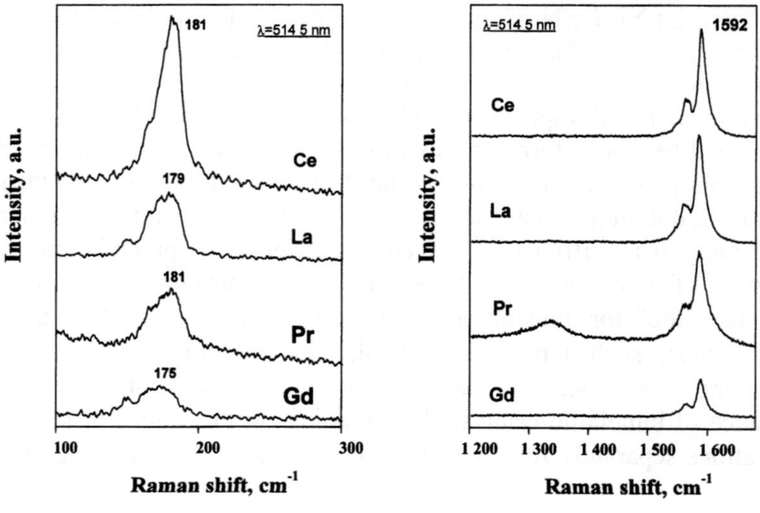

Figure 2. Raman spectra of SWNT samples in the region of the radial breathing modes (150-250 cm^{-1}) and of the G-band modes (near 1590 cm^{-1})

The relative efficiencies are sure to depend much less than the absolute efficiencies upon systematic errors, which are inevitable in a procedure of measuring of SWNT content. Replacement of yttrium in the catalyst for lanthanum, praseodymium and cerium affects crucially the evaporation of the carbon/metal mixture because of a large difference in the ionization energies of these metals. Nevertheless, by variation of the arc process parameters conditions have been revealed where catalytic efficiencies of these metals were nearly as high as the efficiency of the Y/Ni catalyst (Table 2). We did not manage to find any favorable conditions for SWNT synthesis using Yb/Ni catalyst, and we attribute this fact to the low boiling temperature of ytterbium. Raman spectra of SWNT samples produced with different catalysts are shown in Fig.2.

Acknowledgement: This work was supported in part by the Russian Foundation of Basic Research, grant # 02-03-33335 and by the International Science and Technology Center, grant # 1024.2. We are grateful to A.V.Raevskii for experimental assistance and helpful discussions and Obraztsova E.D. for Raman measurements.

References

1. Takizawa, M., Bandow, S., Yudasaka, M., Ando, Y., Shimoyama, H., and Iijima, S. (2000) Change of tube diameter distribution of single-wall carbon nanotubes induced by changing the bimetallic ratio of Ni and Y catalysts, *Chem. Phys. Lett.*, **326**, 351-357.
2. Thess, A., Lee, R., Nikolaev, P., Dai, H., Petit, P., Robert, J., Xu, C., Lee, Y.H., Kim, S.G., Rinzler, A.G., Colbert, D.T., Scuseria, G.E., Tomanek, D., Fischer, J.E., and Smalley, R.E. (1996) Crystalline ropes of metallic carbon nanotubes, *Science*, **273**, 483-487.
3. Scott, C.D., Arepalli, S., Nikolaev, P., and Smalley, R.E. (2001) Growth mechanism for single-wall carbon nanotubes in a laser-ablation process, *Apll. Phys. A*, **72**, 573-580.
4. Jost, O., Gorbunov, A.A., Möller, J., Pompe, W., Liu, X., Georgi, P., Dunsch, L., Golden, M.S., and Fink, J. (2002) Rate-limiting process in the formation of single-wall carbon nanotubes: pointing the way to the nanotube formation mechanism, *J. Phys. Chem. B*, **106**, 2875-2883.
5. Jost, O., Gorbunov, A.A., Pompe, W., Pichler, T., Friedlein, R., Knupfer, M., Reibold, M., Bauer, H.-D., Dunsch, L., Golden, M. S., and Fink, J. (1999) Diameter grouping in bulk samples of single-walled carbon nanotubes from optical absorption spectroscopy, *Appl. Phys. Lett.*, **75**, 2217-2219.
6. De Bokx, P.K., Kock, A.J.H.M., Boellaard, E., Klop, W., and Geus, J.W. (1985) The formation of filamentous carbon on iron and nickel catalysts. I. Thermodynamics, *J. Catal.*, .**96**, 454-467.
7. Lavin, J.G., Subramoney, S., Ruoff, R.S., Berber, S., and Tomanek, D. (2002) Scrolls and nested tubes in multiwall carbon nanotubes, *Carbon*, **40**, 1123-1130.
8. Krestinin, A.V., Kislov, M.B., Ryabenko, A.G. (2003) Endofullerenes with metal atoms inside as precursors of nuclei of single-wall carbon nanotubes, *J. Nanosci. Nanotech.*, in press

9. Liu, B., Wagberg, T., Olsson, E., Yang, R., Li, H., Zhang, S., Yang, H., Zou G., and Sundqvist, B. (2000) Synthesis and characterization of single-walled nanotubes produced with Ce/Ni catalysts, *Chem. Phys. Lett.*, **320**, 365-372.
10. Krestinin, A.V., and Moravsky, A.P. (1998) Mechanism of fullerene synthesis in the arc reactor, *Chem. Phys. Lett.*, **286**, 479-484.
11. Krestinin, A.V., and Moravsky, A.P. (1999) Kinetics of fullerene C_{60} and C_{70} Formation in a Reactor with graphite rods evaporated in electric arc, *Chem. Phys. Reports*, **18**, 515-532
12. Krestinin, A.V., Moravsky, A.P., and Tesner, P.A.(1998) Kinetic model of formation of fullerenes C_{60} and C_{70} in condensation of carbon vapor, *Chem. Phys. Reports*, **17**, 1687-1707.
13. Maruyama, S., Yamaguchi, Y., Kohno, M., and Yoshida, T. (1999) Formation process of empty and metal-containing fullerene – molecular dynamics and FT-ICR studies, *Fullerene Sci. Tech.*, **7**, 621-39.
14. Krestinin, A.V., Kiselev, N.A., Raevskii, A.V., Ryabenko, A.G., Zakharov, D.N., and Zvereva, G.I. (2003) Perspectives of single-wall carbon nanotube production in the arc discharge process, *Eurasian Chem. Tech. J.*, **5**, 7-18.

PRODUCTION OF BORON NITRIDE BY CARBOTHERMIC AND MECHANOCHEMICAL METHODS, AND NANOTUBE FORMATION

H. E. Çamurlu[1], A. Aydoğdu[2], Y. Topkaya[1] and N. Sevinç[1]
[1]Department of Metallurgical and Materials Engineering, Middle East Technical University, Ankara, Turkey; [2]General Directorate of Minerals Research and Exploration, Ankara, Turkey

Abstract: The formation of hexagonal boron nitride by carbothermic reduction of boron oxide and nitridation has been examined. Experiments were conducted in the temperature range of 1100-1500°C for durations between 15-240 minutes. Products were examined by X-ray, SEM and chemical analysis. The results showed that the reaction proceeds through a gaseous boron containing species, which is most probably $B_2O_3(g)$. It was found that all of the carbon was consumed and formation of boron nitride was complete in 2 hours at 1500°C.

Keywords: boron nitride, carbothermic, mechanochemical, nanotube

1. INTRODUCTION

Hexagonal boron nitride (BN) has a crystal structure similar to graphite; hence it is called 'white graphite'. However, it is an electrical insulator and intensively used as a carbon substitute for its much higher chemical inertness, especially at high temperature. BN is not wetted by many metallic melts (Fe, Cu, Zn). Thus, some important applications are boats and crucibles for evaporation of metals and high temperature lubrication [1].

BN does not occur in nature and was first prepared in 1842 by reacting molten boric acid with potassium cyanide. However, it was not produced commercially until 1950s. Many methods have been used for the production of BN such as reacting boron oxide or boric acid with ammonia or urea at 900-1500°C. Other methods are metallothermic or carbothermic reduction of boron oxide and nitridation [1, 2].

This study has two principle aims. The first aim is to produce BN powder by carbothermic and mechanochemical methods. The second aim is to

S. Guceri et al. (eds.), Nanoengineered Nanofibrous Materials, 115-120.

investigate the formation of BN nanotubes by utilizing both the produced BN powder and from the B_2O_3 by mechanochemical method. In literature it has been reported that boron nitride nanotubes could be produced by this method, which involves high energy milling of boron powder under NH_3, or BN powder under N_2 gasses and annealing in N_2 atmosphere [3].

2. FORMATION OF BORON NITRIDE BY CARBOTHERMIC METHOD

Formation of boron nitride by carbon reduction of boron oxide in the presence of nitrogen gas has been studied in our department [4]. In this method, process proceeds according to the following overall reaction:

$$B_2O_3(l) + 3C(s) + N_2\,(g) = 2BN(s) + 3CO(g) \qquad (1)$$

Experiments were conducted at 1100, 1200, 1300, 1400 and 1500°C for two hours and also at 1500°C for 15, 30, 60, 90, 180 and 240 minutes. Products were examined by X-ray, SEM and chemical analysis.

Results of the experiments are presented as a function of temperature and duration in Figures 1 and 2, respectively. It is seen from *Figure 1* that the quantity of B_2O_3 in the reaction products of the experiments conducted for 2 hours decreases and amount of BN increases with temperature. The amount of un-reacted carbon shows a significant decrease after 1300°C. It is also seen in *Figure 2* that the amount of B_4C increases after 1200°C up to 1400°C and then decreases, reaching zero at 1500°C.

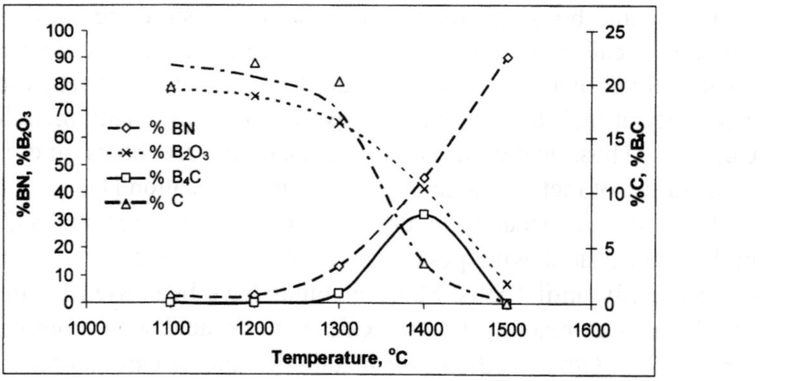

Figure 1. Variation of constituents of the products in the experiments conducted for 2 hours with temperature.

It is seen from *Figure 2* that the amount of B_2O_3 and carbon in the reaction product decreases with time and carbon is completely consumed in two hours at 1500°C. The quantity of BN forming increases with time up to two hours and remains constant afterwards. It is also seen in the experiments conducted at 1500°C that the amount of B_4C forming increases up to 30 minutes and then decreases reaching zero at 2 hours.

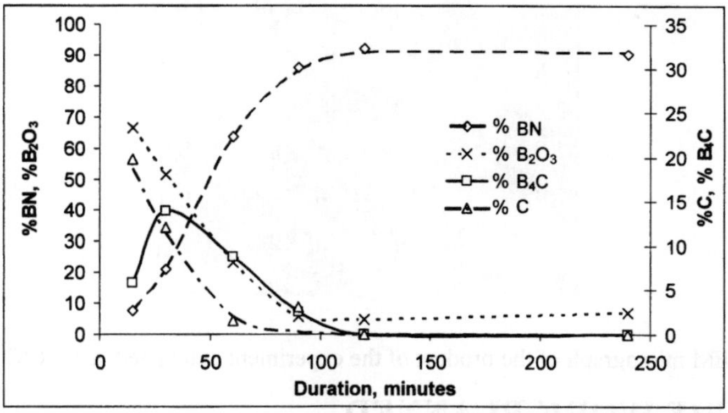

Figure 2. Variation of the constituents of the products in the experiments conducted at 1500°C with time.

Experiments have been carried out in order to investigate the role of B_4C in formation of BN. To check the possibility of formation of BN proceeding through the intermediate species B_4C, a pellet was prepared from a mixture of B_2O_3 and B_4C and subjected to N_2 at 1500°C for two hours. According to the results of this experiment, it may be stated that formation of BN from B_2O_3+B_4C mixture is slower than that from B_2O_3+C mixtures. But this may be attributed to the specific surface area of the C being much larger than that of B_4C powders. Therefore, the role of B_4C as an intermediate product has not been clearly understood and it has been proposed that the formation mechanism may proceed either by reduction of B_2O_3 by carbon or through the formation and reaction of B_4C as an intermediate product.

Experiments conducted by use of different geometrical arrangements indicate that B_2O_3 and carbon need not be in contact for the formation of BN under N_2 atmosphere and that BN formation reaction proceeds through boron containing gaseous intermediate species, which is probably gaseous B_2O_3. Rod-like particles and equiaxed particles are observed in *Figure 3*, from the SEM micrograph of the product of the experiment conducted at 1500°C for 1 hour. Energy dispersive analysis by X-ray (EDAX) energy profiles obtained from spot analysis of small and large particles on the rod, indicated as 1 and 3, were found to exhibit only B and N peaks; these particles were concluded to be BN. The analyses on the rod-like particles suggest that they are as yet un-reacted carbon on which BN particles, unreacted B_2O_3 and perhaps also B_4C exist. The results of the macroscopic and microscopic examinations indicated that C is necessary for formation of BN and that BN forms on carbon.

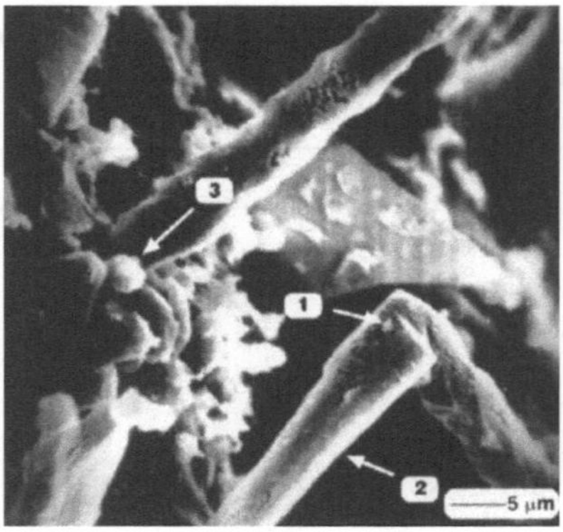

Figure 3. SEM micrograph of the product of the experiment conducted at 1500°C for 1 hour.

3. FUTURE WORK PLANNED

3.1 Mechanochemical Method for BN Synthesis

High energy ball milling is a mechanochemical process of intimate mixing, grinding, fracturing, high speed plastic deformation, fragmentation, cold welding and micro diffusion occurring within a layer of powder trapped between two colliding surfaces during impact [3, 5]. This mechanical treatment intensifies transport of the reactants and increases the rate of diffusion, which is the limiting factor for many solid-state reactions. Reactions with various mechanisms have been performed with this technique such as gas-solid reactions [6], mechanical alloying [5], phase transformations [7] and mechanical activation of solid-solid reactions [8].

The effect of mechanical activation on the carbothermic reduction of minerals such as ilmenite and rutile was studied by Chen et al., and it was reported that ball milling led to a significant lowering of the temperature range in which the carbothermic reductions occur [9]. Therefore, similarly, the formation process of boron nitride by carbothermic reduction and nitridation of boron oxide, which was conducted in the previous part of this study, will be performed with the aid of a high energy planetary ball mill. The effect of grain size reduction and mechanical activation on the reaction kinetics will be investigated. It is also known that a reaction between B_2O_3 and ammonia takes place at lower temperatures than carbothermic reduction and nitridation [1]. The reaction between ammonia and boron oxide will be conducted in the high energy ball mill and the possibility of the formation of boron nitride via mechanochemical process will be examined.

3.2 Nanotube Formation

Nanoscale materials such as nanoparticles and nanocomposites have been produced by the mechanochemical process [10, 11]. It has been reported that boron nitride nanotubes could also be produced by this method [3]. Other methods of BN nanotube formation are arc discharge, laser ablation and chemical reaction [12-14]. The geometry of the formed tubes depends on the starting materials and the method used. In mechanochemical formation of BN nanotubes, elemental boron or BN powder is milled in a high energy ball mill for 100-150 hours under ammonia or nitrogen atmosphere, respectively. High energy milling transfers a large amount of mechanical energy into powder particles, leading to morphological, structural and chemical changes. BN nanotubes are formed during the following annealing process, at 1200-1300°C for 6-16 hours under nitrogen atmosphere [3, 15]. High energy ball milling has two fundamental roles on BN nanotube formation. First, it produces a high density of structural defects and a high level of microstrain on the milled powder, as a result of which the nitriding reaction can take place at this relatively low temperature. Secondly, it introduces nanostructures to the powder such as Fe nanoparticles and BN crystallites, even when the starting material is elemental boron powder. These nanostructures act as seeds and catalyze the BN nanotube formation [3]. It has been proposed that the formation of the nanotubes takes place by solid-state nucleation and growth mechanism through a thermally activated process of surface self diffusion [16]. The advantage of this method is that a large quantity of materials containing BN nanotubes can be produced and the formation process takes place at temperatures far below that of the arc discharge or laser ablation [15]. However, Fe contamination may cause problems in some applications and control of the geometry of the formed nanotubes may be difficult with this method. BN nanotubes have the potential to be used in composite materials, as filters, or as thermal or acoustic insulators. But none of these applications is at a stage of commercialization, yet. Unlike carbon nanotubes or carbon containing nanotubes, the semiconducting properties of pure BN nanotubes are supposed to be independent of tube geometry, with a gap of 5.5 eV; therefore they can be candidate materials for practical applications [13].

In the present study, a mechanochemical method is planned to be utilized in the production of boron nitride. Nanotube formation upon extended high energy ball milling and annealing of the produced powder will also be investigated.

References

1. Popper, P., *Special Ceramics*. London: Heywood & Company, 1960.
2. Pikalov, S. N. Mechanism of formation of graphite-like boron nitride in the carbothermal process. Soviet Powder Metallurgy and Metal Ceramics 1988; 27:404-406
3. Chen, Y., Fitzgerald, J., Williams, J. S. and Bulcock, S. Synthesis of boron nitride nanotubes at low temperatures using reactive ball milling. Chemical Physics Letters 1999; 299:260-264
4. Aydoğdu, A. and Sevinç, N. Carbothermic formation of boron nitride. Journal of the European Ceramic Society 2003; 23:3153-3161
5. Calka, A. and Radlinski, A. P. Universal high performance ball-milling device and its application for mechanical alloying. *Materials Science and Engineering* 1991; A134:1350-1353
6. Chen, Y., Li, Z. L. and Williams, J. S. The evolution of hydriding and nitriding reactions during ball milling of titanium in ammonia. Journal of Materials Science Letters 1995; 14:542-544

7. Chen, Y., Williams, J. S. and Wang, G. M. High-temperature phase transformations of iron anhydrous ammonia system realized by ball milling at room temperature. Journal of Applied Physics 1996; 79:3956-3962
8. Welham, N. J. Mechanical activation of the solid-state reaction between Al and TiO_2. Materials Science and Engineering 1998; A255:81-89
9. Chen, Y., Hwang, T., Marsh, M. and Williams, J. S. Study on mechanism of mechanical activation. Materials Science and Engineering 1997; A266-228:95-98
10. Welham, J. N. Formation of Nanometric TiB_2 from TiO_2. Journal of the American Ceramic Society 2000; 83:1290-1292
11. Matteazzi, P. and Le Caer, G. Synthesis of nanocrystalline alumina-metal composites by room temperature ball milling of metal oxides and aluminum. Journal of the American Ceramic Society 1992; 75:2749-2755
12. Loiseau, A., Willaime, F., Demoncy, N., Hug, G. and Pascard, H. Boron nitride nanotubes with reduced numbers of layers synthesized by arc discharge. Physical Review Letters 1996; 76:4737-4740
13. Golberg, D., Bando, Y., Eremets, M., Takemura, K., Kurushima, K. and Yusa, H. Nanotubes in boron nitride laser heated at high pressure. Applied Physics Letters 1996; 69:2045-2047
14. Gleize, P., Schouler, M. C., Gadelle, P. and Caillet, M. Growth of tubular boron nitride filaments. Journal of Materials Science 1994; 29:1575-1580
15. Chen, Y., Conway, M. and Williams, J. S. Large-quantity production of high-yield boron nitride nanotubes. Journal of Materials Research 2002; 17:1896-1899
16. Chadderton, L. T. and Chen, Y. Nanotube growth by surface diffusion. Physics Letters 1999; A263:401-405

STRUCTURE AND PROPERTIES OF SILICON CARBIDE FIBERS AS FUNCTION OF THEIR SYNTHESIS CONDITIONS

K.L. VYSHNYAKOVA, L.N. PERESELENTSEVA, PhD
Frantsevich Institute for Problems of Materials Science of NAS of Ukraine, 3, Krzhyzhanovsky Str.,03142 Kiev, Ukraine

Abstract: Silicon carbide fibers were synthesized from hydrate cellulose precursors by means of their carbonization in the presence of silica; carbothermal reduction applying carbonized hydrate cellulose fibers; liquid-phase siliconizing of carbonized hydrate cellulose fibers. By methods of thermal analysis, IR-spectroscopy and x-ray diffraction mechanisms of synthesis processes of silicon carbide fibers were studied. Phase composition and structures of obtained SiC-fibers were investigated and their mechanical properties were examined. It was found the maximal strength characterizes fibers synthesized by the method of liquid-phase siliconizing of carbonized hydrate cellulose fibers. The synthesized fibers are destined for serving as reinforcing agents of polymer, metal, ceramic and carbon matrices.

1. INTRODUCTION

Creation of composite materials based on polymer, metal, ceramic and carbon matrices reinforced by high-strength and high-modulus fibers is still one of the most promising trends of the contemporary materials science. The most appropriate fibers for this purpose are silicon carbide fibers which excel by high strength and heat resistance, high elasticity modulus, chemical stability under aggressive media and when contacting with matrix materials having different nature in solid or liquid state. The most frequently used reinforcing agents for different kinds of matrices are SiC composite fibers "Nicalon" and thread-like crystals (whiskers) which are obtained from fibrous organosilicon precursors [1-7]. However, the whiskers processing is quite complicated. There are also certain limitations when applying the composite fibers at high temperatures. So, it is necessary to create SiC fibers of simpler, cheaper and of more ecologically safe manufacturing. According to [7], to produce high-temperature silicon carbide fibers applying porous activated carbon fibers

S. Guceri et al. (eds.), Nanoengineered Nanofibrous Materials, 121-129.

subjected to siliconizing in Si or Si-containing compounds vapors while adding components of boron or its compounds activating consolidation at high-temperature treatment. As precursors of carbon fibers they are mainly polymer synthetic artificial or natural fibers. General mechanisms of carbon fibers obtained from different precursors have already been sufficiently studied and the processes of destruction and formation of a carbon framework could describe physical-chemical reactions of transformation of polymer fibers into carbon fibers.

It is of interest to produce silicon carbide fibers from hydrate cellulose. The prerequisites for the problem statement are physical-chemical properties of hydrate cellulose, which is widely applied in producing of carbon fibrous materials. The capacity of hydrate cellulose to swell and to absorb considerable amounts of water and aqueous solutions is a property which allows to obtain ceramic fibers because it makes it possible to saturate hydrate cellulose fibers by any kind of soluble compounds, e.g. by aqueous salts solutions. Then, using further thermal treatment under controlled conditions, it becomes possible to obtain fibrous materials saturated by oxides, metals or carbides. Therefore, the purpose of the present paper is to study basic mechanisms of hydrate cellulose transformation into silicon carbide under its chemical and thermal treatment.

2. METHODS

For experiments hydrate cellulose fibers of purity not less than 99.0% (the remainder contains calcium oxide, silica, iron oxides etc.) were used. There were applied two kinds of viscose fibers of diameter of 10-15 μm - cord thread of spherical configuration and silk textile fiber of cut configuration. Furthermore, silk textile fiber, in contrast to wholly crystalline cord thread, contains a substantial amount of amorphous phase (~ 30%). Two different ways of silicon carbide fibers synthesis were used. According to one of them the fibers were subjected to impregnation by sodium silicate solutions with further precipitation of silicon acid, then heat treated at up to 2000 °C under nonoxidizing media. The amount of impregnated silica in cord fibers was about 10%, and in textile fibers about 35%. In other cases, as-carbonized fibers of hydrate cellulose were siliconized by liquid-phase method or else in Si-containing vapor.

Structure and composition studies at different steps of the transformation of hydrate cellulose into silicon carbide were conducted using methods of IR-spectroscopy, thermogravimetry and differential thermal analysis, x-ray analysis and electron microscopy. Derivatography investigations were carried out by means of the instrument "Derivatograph-K" of the Hungarian company "MOM". Heating of the samples under investigation was conducted in high-purity argon current with the heating rate of 3 grad/min. The phase composition of fibers was examined using x-ray diffractometer "DRON-1.5" (CuK_α). IR-spectroscopy studies were conducted using spectrometer "Specord-75IR" within the absorption frequency range of 400 to 4000 cm^{-1}. Electron microscopy studies were carried out using the instrument "CAMECA-50" under scanning and translucence conditions combined with local x-ray analysis and microdiffractometry. Specific surface, pore specific volume and pores sizes were evaluated by adsorption-structural method using the device ASAP-2000M.

3. RESULTS AND DISCUSSION

3.1. Obtaining Silicon Carbide Fibers by Means of Carbonization of Hydrate Cellulose Saturated with Silicon Acid Gel

Studying the pyrolysis process of hydrate cellulose saturated by silicon acid gel by methods of differential thermal and thermogravimetric analyses, IR-spectroscopy and x-ray diffractometry has allowed discovery of basic mechanisms of the transformation of hydrate cellulose fibers into silicon carbide fibers. All the investigation results represented below refer to the fibers samples made of textile hydrate cellulose. Nevertheless these basic mechanisms are also typical for cord fibers. Figure 1 shows the results of differential thermal analysis and thermogravimetric analysis of hydrate cellulose fibers samples saturated by silicon acid gel. The maximum mass loss rate is observed at 160-265 °C (TG and DTG curves) and the mass loss maximum occurs at 600 °C. An exothermal effect corresponds to the temperature range of 160 to 265 °C (DTA curve). IR-spectroscopy investigations of the samples in question suggest transformations occurring in the system of hydrate cellulose and silicon acid when it is subjected to annealing at 100-2000 °C (Figure 2). At 100-150 °C the system of two polymers, i.e. hydrate cellulose and silicon acid, is being formed. The IR-spectra of the samples contain all the absorption bands that are characteristic for this system [8]. As the annealing temperature increases up to 400 °C the representative absorption bands of hydrate cellulose disappear, and when the temperature reaches 600 °C absorption bands which represent hydration water of silicon acid gel disappear. Absorption bands (ν=940-960 cm^{-1}) that could represent newly formed Si-O-C bonds are starting to appear on the spectra at 200 °C and it should testify an interaction in the system and formation of intermediate compounds in the likeness of organosilicon compounds. Obviously, the beginning of this interaction conforms to the start of cellulose dehydration. The exothermal peak appearance on the DTA curve (Figure 1) also agrees with this. The absorption bands ascribed to Si-O-C

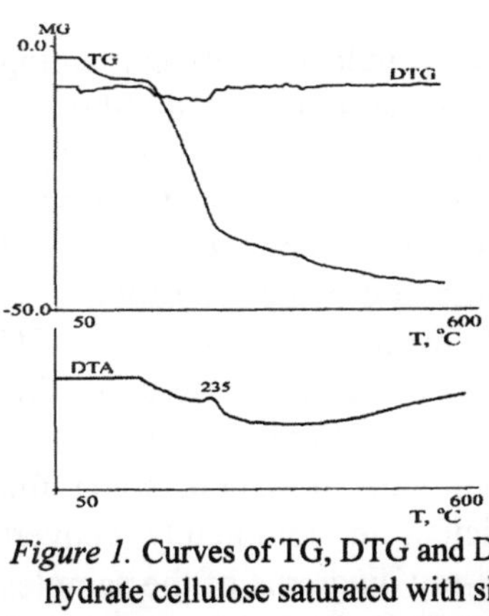

Figure 1. Curves of TG, DTG and DTA of hydrate cellulose saturated with silica

bonds of the intermediates register in the spectra up to the temperature of 600 °C [9-10]. Besides at the annealing temperatures above 600 °C changing of the absorption bands of Si-O bonds occur confirming formation of amorphous silica and its crystallization. It was found that at 800-1200 °C there appear absorption bands of Si-O-C and Si-C bonds (930-900 cm^{-1}) that could be ascribed to the inorganic phase SiOC. Starting from the temperature of 1200 °C absorption bands appear which are ascribed to Si-C bonds of β-SiC formation. At above 1600 °C the IR-spectra contain only the absorption bands which are characteristic for Si-C bonds. Figure 3 shows x-ray diagrams of hydrate cellulose fiber samples saturated with silicon acid gel after annealing in the temperature range of 1000-2000 °C. According to these data, crystalline silica (cristobalite) appears at the temperature of 1000 °C and exists up to 1400 °C at temperatures above 1200 °C appears silicon carbide of β-type; at 1500 °C and above there are peaks of β-SiC, amorphous carbon halo, and a peak which could be ascribed to graphite-like carbon. Therefore it could be established, that the conversion of hydrate cellulose saturated by silica into silicon carbide is a complicated process which includes different physicochemical reactions, i.e. pyrolysis and carbonization of hydrate cellulose; silicon acid gel dehydration, formation of organosilicon - like intermediates and their destruction; formation of amorphous silica and its crystallization in the form of cristobalite; and the formation of intermediate phase of SiOC, silicon carbide synthesis.

3.2 Carbothermal Method of Silicon Carbide Fibers Synthesis

To synthesize silicon carbide fibers there hydrate cellulose fibers containing not less than 95% of carbon which werpreliminarily carbonized at 800-900 °C were used. Silicon acid gel was precipitated on the fiber surface with SiO_2/C ratio no less than 1/3. The fibers were subjected to calcination to remove hydration water and then to isothermal annealingin the temperatures range of 1400-2000 °C under nitrogen or argon media. Silicon carbide formation took place in correspondence with the total scheme of chemical reaction $SiO_2 + 3C = SiC + 2CO$ starting at 1400 °C. The reaction completeness at this temperature depended on annealing duration that required at least 2 hours. As the temperature was increased the synthesis duration shortened, so at the temperature of 1800 °C, 20 minutes are sufficient to completely convert carbonized fibers into silicon carbide fibers. On x-ray diagrams of the samples annealed at 1400 °C there are mainly peaks of α-cristobalite, and also several peaks which could be ascribed to β-SiC. Whereas x-ray diagrams of the samples annealed at 1800 °C contain peaks that could only belong to β-SiC (figure 4). Formation of volatile substances (SiO, CO) as well as oxide film presence on the fibers surface facilitates initiation and growth of SiC thread-like crystals (whiskers). It is possible to control whiskers growth by means of creation of adjustable synthesis media.

3.3 Liquid-phase Siliconizing Method of SiC-fibers Synthesis

The process of hydrate cellulose fibers siliconizing was combined with their carbonization by means of placing silicon close to fibers so that when melted silicon could come upon the fiber surface. The heating temperature of the siliconizing process was within the range of 1700 to 2000 °C. The resultant fibers consisted mainly of β-SiC and of trace quantity of α-SiC.

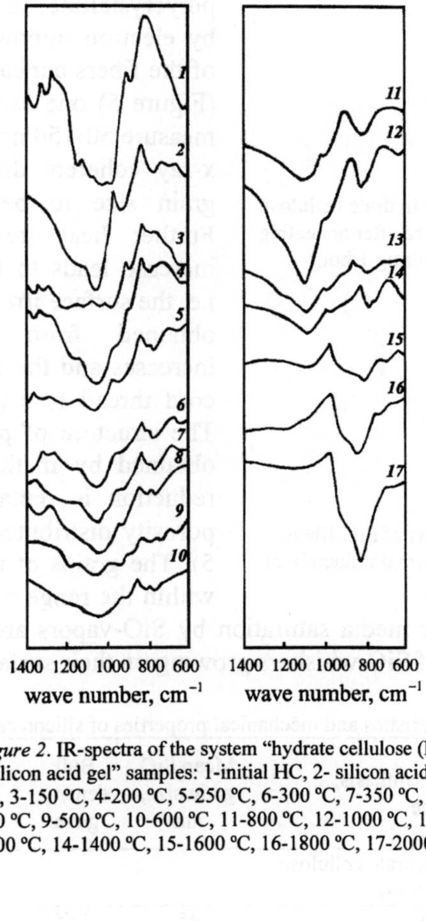

Figure 2. IR-spectra of the system "hydrate cellulose (HC) - silicon acid gel" samples: 1-initial HC, 2- silicon acid gel, 3-150 °C, 4-200 °C, 5-250 °C, 6-300 °C, 7-350 °C, 8-400 °C, 9-500 °C, 10-600 °C, 11-800 °C, 12-1000 °C, 13-1200 °C, 14-1400 °C, 15-1600 °C, 16-1800 °C, 17-2000 °C

3.4. Structure and Properties of Synthesized SiC Fibers

Figure 5 represents silicon carbide structures obtained in different ways. The structure of fibers made by means of carbonization of hydrate cellulose saturated by silica depends on heat treatment temperature. At the annealing temperature of 1200 °C the SiC fibers result having a smooth surface, whereas after annealing at 1400 °C and above the fibers structure is polycrystalline. In the picture obtained by electron microscope in scan mode of the fibers annealed at 1500-1800 °C (Figure 5) one can see the grains that measure 50-150 nm, and the method of x-ray coherent dissipation shows the grain size to be about 40-50 nm. Further heat treatment temperature increase leads to the fibers changing, i.e. the surface irregularity of the fibers obtained from textile precursors increases and the fibers obtained from cord thread lose their roundish form. The structure of polycrystalline fibers obtained by method of carbothermal reduction is remarkable for its high porosity distributed by volume (figure 5). The grains of these fibers measure within the range of 50 to 150 nm. The fibers obtained under media saturation by SiO-vapors are characterized by a substantial quantity of SiC-whiskers growing on their surface.

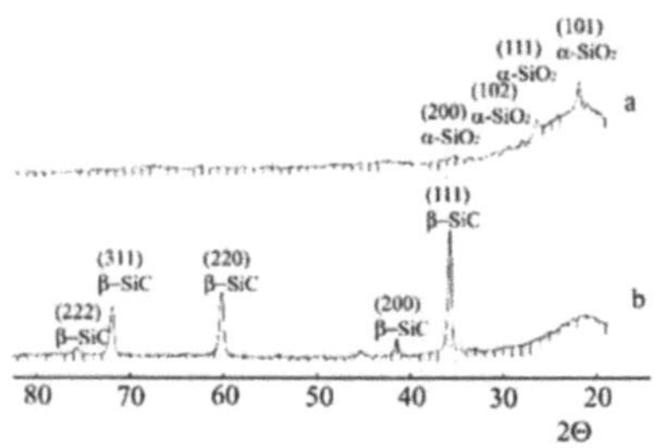

Figure 3. XRD patterns of hydrate cellulose samples saturated with silica after annealing at: a-1200 °C, b-1400 °C during 1 hour

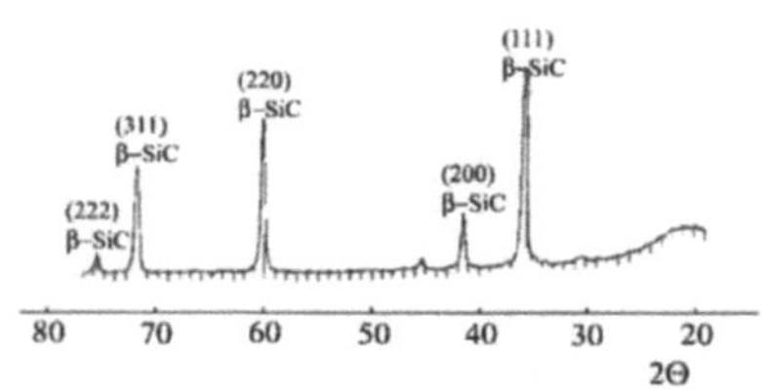

Figure 4. XRD patterns of SiC-fibers synthesized by carbothermal reduction at 1800 °C

Table 1. Structure characteristics and mechanical properties of silicon carbide fibers

Method of fibers synthesis, fibers type	Mean SiC grain size, nm	Bulk density, g/cm^3	Specific surface, m^2/g BET	Strength, MPa
Thermal treatment of hydrate cellulose saturated by SiO_2				
T=1600 °C	45	2.55	19.2	<100
Liquid-phase siliconizing, T=2000 °C	100	2.82	2	200
Carbothermal synthesis, 1800 °C	50-100	2.657	25	150

Since the growth centers of whiskers are the surface grains of the polycrystalline fibers, the whiskers cross sections are determined by these fiber grain sizes.

The fibers obtained by means of liquid-phase siliconizing, in contrast to all above considered fibers, keep dimensions and the shape of carbonized precursor (figure 5). They distinguish from the fibers obtained by other methods due to the densest and smoothest surface. It corresponds to the results of measurements of bulk density, specific surface and specific pore volume as well as the trials results of mechanical strength of fibers obtained in different ways (table 1). To estimate the mechanical properties of fibers studied, composite specimens based on polymer matrix were prepared. As it is shown in the table, fibers obtained by liquid-phase siliconizing have maximal strength about 2000 MPa. The rather high strength values have whiskerized fibers obtained by means of carbothermal reduction. Taking into consideration high values of specific surface and pores specific volume, it may be concluded that these high strength properties are conditioned by the presence of whiskers. The lowest strength characterizes fibers synthesized by carbonization of hydrate cellulose fibers saturated by silica. Summarizing the results, it is possible to make conclusion that the reached level of mechanical properties of synthesized silicon carbide fibers justifies the chosen way of high-strength fibers elaboration.

4. CONCLUSIONS

Using methods of thermal analysis, IR-spectroscopy and x-ray diffraction mechanisms of obtaining silicon carbide fibers from hydrate cellulose precursors by means of their carbonization in the presence of silica were studied; carbothermal reduction by applying carbonized hydrate cellulose fibers and by liquid-phase siliconizing of carbonized hydrate cellulose fibers. Phase composition and structures of obtained SiC-fibers were investigated and their mechanical properties were examined. It was found that the maximal strength characterizes fibers synthesized by the method of liquid-

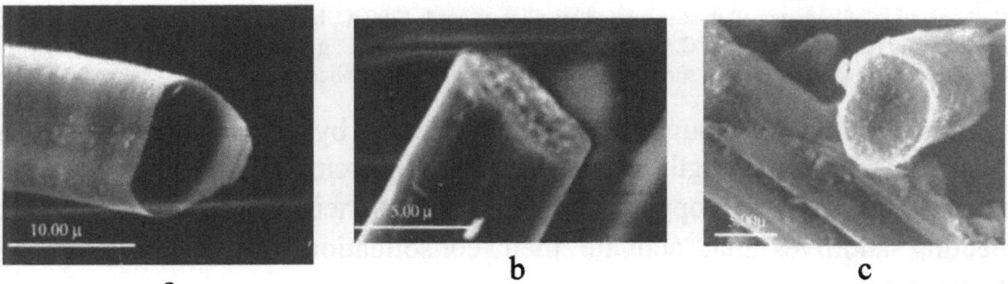

a b c

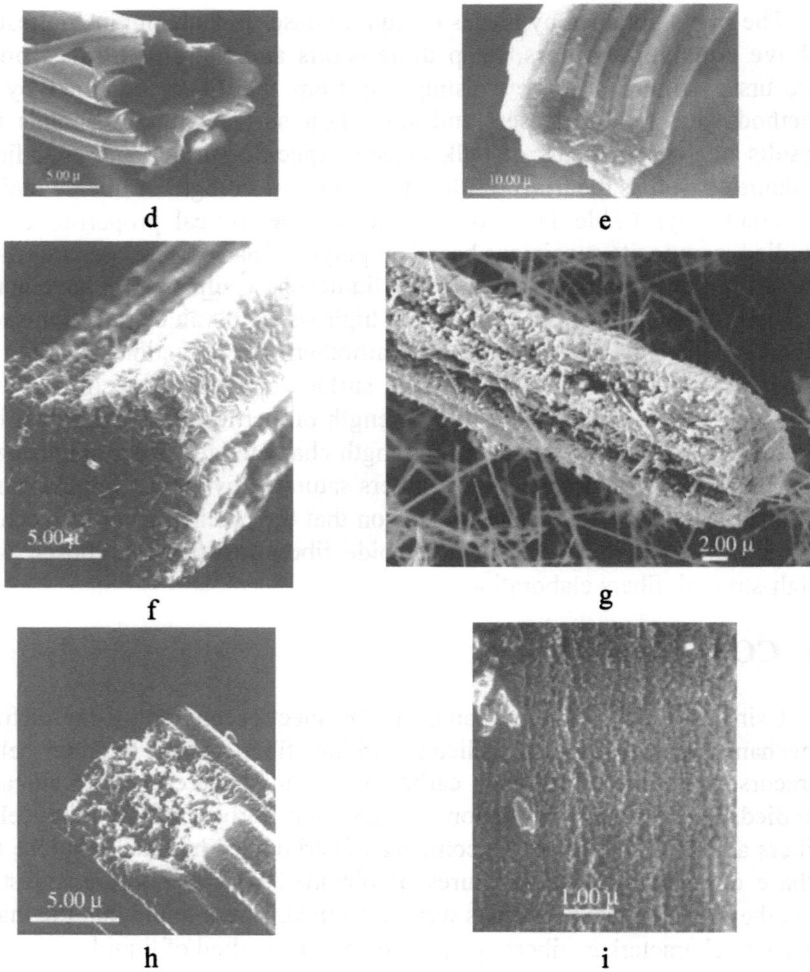

Figure 5. Microstructure of SiC obtained by different methods.
Carbonization of hydrate cellulose saturated by silica: a-cord, 1200 °C, b-cord, 1400 °C, c-cord 1500 °C, d-textile, 1200 °C, e-textile, 1400 °C. f-carbothermal reduction, 1850 °C, g-whiskerized fiber, 1750 °C, h-liquid-phase siliconizing, i- side surface fiber made by liquid-phase siliconizing

phase siliconizing. Though the fibers obtained by other methods possess insufficiently high strength to use them as reinforcing fillers for constructive materials, there is an opportunity to increase their strength by means of injecting additives that activate their consolidation and sintering at high temperatures.

Acknowledgement: We would like to thank Dr. G.S. Oleinik for his help with the electron microscopy experiments.

References

1. Hasegawa Y., Iimura M., Yajima S. Synthesis of Continuous SiC Fibers. J. Mater. Sci. 1980; 15: 720-28.
2. Yajima S., Okamura K., Omori M. Synthesis of Continuous SiC Fibers with High Tensile Strength. J. Am. Ceram. Soc. 1976; 59: 324-27.
3. Wang H., Fischman G.S. In Situ Synthesis of Silicon Carbide Whiskers from Silicon Nitride Powders. J. Am. Ceram. Soc. 1991; 74: 1519-22.
4. Raman V., Parashar V.K., Bahl O.P. Influence of Boric Acid on the Synthesis of Silicon Carbide Whiskers from Rice Husks and Polyacrilonitrile, J. Mat. Sci. Lett. 1997; 16:1252-54.
5. Chen S.H., Lin C.I. Effect of Processing Temperature on the Morfology of Silicon Carbide Whiskers. J. Am. Ceram. Soc. 1991; 74:447-49.
6. Raman V., Parashar V.K., Dhakate S. Bahl O.P. Synthesis of Silicon Carbide through the Sol-Gel Process from Rayon Fibers. J. Am. Ceram. Soc. 2000; 83: 952-54.
7. United States Patent 6,316,051, Okada, November 13, 2001, Process for producing silicon carbide fiber.
8. Bikles N., Segal M., *Cellulose and its derivatives*, Moscow: MIR, 1974.
9. Vyshnnyakova K.L., Tomila T.V. Studying of hydrate cellulose carbonization in the presence of silica by IR-spectroscopy, Ukrainian Chemical Journal, 2003, 69:85-90.
10. Kakimoto K., Wakai F., Bill J., Aldinger F. Synthesis of Si-C-O Bulk Ceramics with Various Chemical Compositions from Polycarbosilane. J. Am. Ceram. Soc. 1999; 82: 2337-41.

SELECTIVE OXIDATION OF HIPCO SINGLE-WALL CARBON NANOTUBES

S.N. BOKOVA[1], E.D. OBRAZTSOVA[1], A.V. OSADCHY[1], H. KUZMANY[2], U. DETTLAFF-WEGLIKOWSKA[3], S.ROTH[3]

[1] *Natural Sciences Center of A.M. Prokhorov General Physics Institute, RAS, 38 Vavilov str., 119991 Moscow, Russia;* [2]*Institute of Material Physics, University of Vienna, Strudlhofgasse, 4, A-1090, Vienna, Austria;* [3]*Max-Planck-Institute fur Festkorperforschung, Heisenbergstr. 1, D-70569 Stuttgart, Germany*

Abstract. The structure of HipCO single-wall carbon nanotubes was controlled by Raman scattering during the material heating in three ways: in a laser beam *("in situ")*, in an optical oven *("in situ")* and after a preliminary heating in the oven followed by Raman measurements. All experiments have demonstrated a diameter-selective oxidative etching of the smallest nanotubes (8 – 10 Å) at temperatures 400 – 450 °C.

1. INTRODUCTION

Carbon nanotubes, discovered 10 years ago, are still the focus of scientific interest due to their unique physical and chemical properties and quantum dimension.

The majority of methods provide less than 20% content of nanotubes in a raw material. Recently, a new method for the SWNT production – catalytic decomposition of CO under high pressure (HipCO) – has been proposed [1]. The method provides a material purity up to 99 weight % after a simple chemical purification. The structure of HipCO material is different from that of nanotubes produced by other methods. A wide diameter distribution of the nanotubes, including very small diameters ~7 Å, is a characteristic feature of HipCO material. Since the quantum properties of the nanotubes are best seen for nanotubes with small diameters, such nanotubes are most interesting for investigations.

S. Guceri et al. (eds.), Nanoengineered Nanofibrous Materials, 131-136.

Up to now very few works devoted to the thermal modification of single-wall carbon nanotubes have been published [2,3]. An advantage of our experiment is a possibility of *"in-situ"* observation of thermal modifications of nanotubes induced by different sources [4].

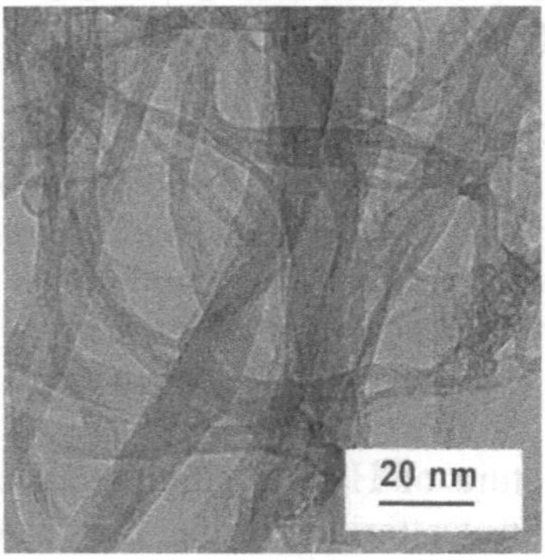

Fig.1. HRTEM view of the pristine HipCO material.

2. EXPERIMENTS AND DISCUSSION

In the course of this work a thermal dependence of the position and form of the Raman bands has been revealed. The sample was heated in a laser beam and in an oven on air. For excitation, the radiation of Ar^+ (514 nm) and Ar^+-Kr^+ (676 nm) was used. The material was synthesized by a method of catalytic decomposition of CO under high pressure (HipCo) (*Fig. 1*). The Raman spectra were registered at the consecutive values of the laser power density in the range 0,06-1,4 KW/cm^2.

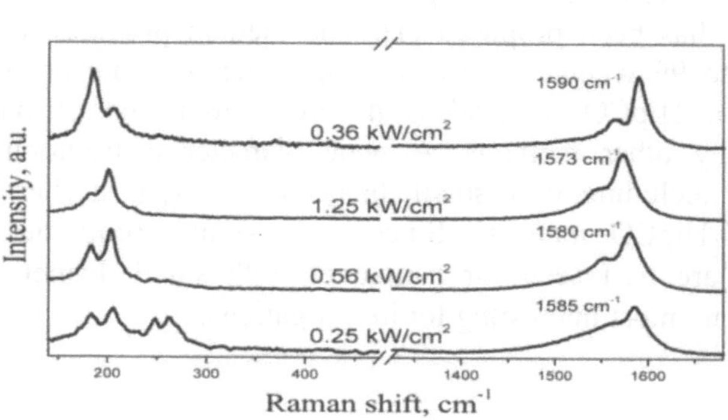

Fig.2. Raman spectra of SWNT at different laser power densities.

In Fig. 2 the Raman spectra of the single-wall carbon nanotubes (SWNT) are shown. The most essential changes of the form of the Raman spectrum were observed at laser power density 0,15-0,45 kW/cm^2 [5]. In a low-frequency spectral range, the Raman spectrum demonstrated four peaks. They correspond to the nanotubes with diameters 9.1, 9.8, 11.8, 13.3 Å. The intensities of the peaks of the smallest tubes (with diameters 9.1 and 9.8 Å) decreased while the laser power density increased. The most intensive peaks were observed with laser power density 0,2 kW/cm^2. The peaks for tubes of 9.1 and 9.8 Å have disappeared completely at 0,45 kW/cm^2. This change appears to be irreversible. While the laser power density decreased back to the initial value only peaks corresponding to the nanotubes with diameters 11.8 and 13.3 Å were observed in the Raman spectrum. The tangential mode shifted toward low frequencies at elevated temperature. Also, the tangential mode shape changed from a metallic-like type to a semiconductor-like type due to annealing of the smallest tubes.

To determine the value of the linear thermal shift of the tangential mode a special set of measurements of nanotubes in an optical oven has been performed. To exclude the material modification, a low power density of the laser beam was used for all measurements in the optical oven. The value of the thermal shift has been estimated as 0.03 cm^{-1}/grad (*Fig.3*).

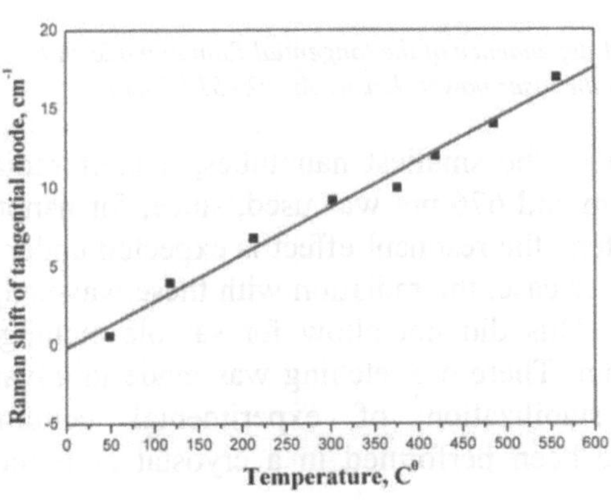

Fig.3. A dependence of the tangential Raman mode shift on

A dependence of the tangential mode shift on the laser power density is shown in *Fig.4*. In the graph point, which corresponds to a disappearance of the Raman signal from nanotubes with diameters 9.1 and 9.8 Å, the dramatic graph slope change is observed. At the lower laser power density the slope angle is bigger than at the higher power density. Hence, a presence of small nanotubes makes heating more effective.

After thermal oxidation of metallic nanotubes with diameters 9.1 and 9.8 Å, the material becomes porous. This breaks the integrity of the sample surface and results in a decrease of the area of effective absorption of the laser beam. Therefore, the further heating proceeds slower.

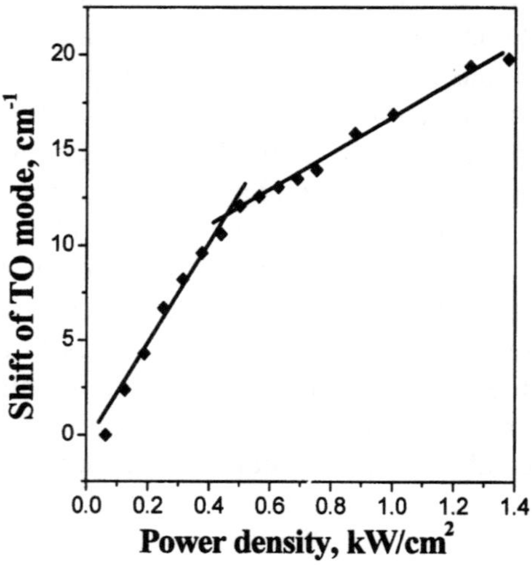

Fig.4. A dependence of the tangential Raman mode shift on the laser power density Ar^+ (λ=514.5 нм).

For observation of the smallest nanotubes, a laser radiation with wavelengths 647 nm and 676 nm was used, since, for nanotubes with the smallest diameters, the resonant effect is expected under excitation of ~1.8 - 2 eV. In our case, the radiation with these wavelengths had a low power output. This did not allow for sample etching *"in situ"* under the laser beam. Therefore, etching was made in advance in the oven. For the stabilization of experimental conditions, the measurements have been performed in a cryostat at liquid nitrogen temperature.

A series of experiments on heating the initial sample to different temperatures followed by Raman measurements have been carried out. Heating was done in the furnace in air at various temperatures ranging from 300 to 500°C with the step 50°C. After etching, the samples were fixed on a substrate and were placed in cryostat at 95-97K. Samples were investigated at three wavelengths of stimulating laser radiation (Ar^+-Kr^+ laser – 514 nm, 647 nm and 676 nm).

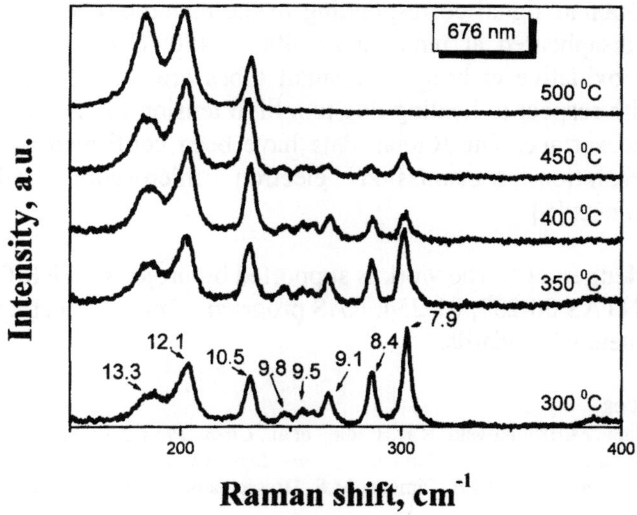

Fig.5 Raman spectra of the single-wall carbon HipCO nanotubes after etching at different temperatures.

In *Fig. 5* the Raman spectra of single-wall carbon HipCO nanotubes after etching at various temperatures are presented. The diameter of nanotubes contained in a sample was calculated. After minimal etching, the spectrum demonstrates peaks corresponding to the nanotubes with diameters 7.9, 8.4, 9.1, 9.5, 9.8, 10.5, 12.1 and 13.3 Å. A mutual comparison of Raman spectra, obtained at various temperatures, has revealed the irreversible changes. An intensity of the "breathing" mode corresponding to the nanotubes with the smallest diameters decreased while the temperature increased. At temperature higher than 450 °C, these modes have disappeared completely.

Thus, using laser radiation with energy 1.8 eV, we had an opportunity to observe thermo-induced change in Raman spectra of HipCO nanotubes with the smallest diameters. A diameter-selective oxidation effect has been observed in this case too.

3. CONCLUSION

In this work a selective oxidation of the smallest HipCO nanotubes has been observed by Raman spectroscopy. Three types of heating have been used: *"in-situ"* laser heating, *"in-situ"* heating in the optical oven with a temperature control and a preliminary heating in the oven followed by Raman measurements. The results of all experiments show that the Raman signal corresponding to the nanotubes with diameters 8-10 Å disappeared at temperature 400 – 450° C due to diameter-selective oxidative etching. The small tubes are assumed to be less stable with respect to heating due to a high tension of chemical bonds on curved surface. The Raman data have been confirmed by data of high resolution transmission electron microscopy (HRTEM) measurements [6].

Acknowledgement: The work is supported by projects RFBR 01-02-17358, INTAS 00-237, 01-254, RAS program "Optical Spectroscopy and Frequency Standards."

References

1. P.Nikolaev, M.Bronikovski, R.K.Bradley et al., *Chem. Phys. Lett.* ***313,*** *91 (1999).*
2. M. Yudasaka, H. Kataura, T.Ichihashi , *Nano Letters* ***9,*** *487 (2001).*
3. P. Corio, P.S. Santos, M.A. Pimena, M.S. Dresselhaus, *Chem. Phys. Lett.* ***360,*** *557 (2002).*
4. S.V. Terekhov, E.D. Obraztsova, A.S. Lobach, V. I. Konov, *Appl. Phys. A* ***74,*** *393 (2002).*
5. S.N. Bokova, V.I. Konov, E.D. Obraztsova et al., *Quantum Electronics, 33, 645 (2003).*
6. E.D. Obraztsova, S.N. Bokova, V.L. Kuznetsov, A.N. Usoltseva et al., *AIP Conferences Proceedings (XVII Int. Winterschool on Electronic Properties of Novel Materials, Kirchberg, Austria),* ***685,*** *215 (2003) .*

PHYSISORPTION OF OXYGEN MOLECULE ON CARBON NANOTUBES

Sefa Dag,[1,*] Oğuz Gülseren,[1] Salim Ciraci,[1] and, Taner Yildirim,[2]

*1) Department of Physics, Bilkent University, Ankara 06800, Turkey; 2) NIST Center for Neutron Research, Gaithersburg, MD 20899, USA,; *) Email: sefa@fen.bilkent.edu.tr*

Abstract: Our work is based on the first-principle pseudopotential plane wave calculations and involves the physisorption of oxygen molecule together with its equilibrium binding geometry and corresponding electronic energy structure. The triplet state of the physisorbed oxygen molecul is found to be a ground state. For different adsorption sites and pattern of coverage, selfconsistent field electronic energy structure calculations using the optimized physisorption structure corresponding to triplet ground state result in a small energy gap between unoccupied oxygen levels and the top of the valence band of the semiconducting carbon nanotube. In contrast to earlier studies, our results invalidate themetallization or hole doping of semiconducting carbon nanotube upon the physisorption of oxygen.

Keywords: Carbon nanotubes, adsorbate structure, magnetic materials, Ab-initio calculations of adsorbates

1. INTRODUCTION

Recent studies have shown that the physical properties of single wall carbon nanotubes (SWNT's) can be modified by the adsorption of foreign atoms or molecules [1, 2, 3, 4, 5]. Collins *et. al.*[1] found that exposure to air or oxygen dramatically influences electrical resistance and the thermoelectric power of a semiconducting SWNT. A semiconducting SWNT, which can be reversibly converted to a conductor by a small concentration of adsorbed oxygen has been proposed as a candidate for chemical sensor devices.[1] Several theoretical works aimed at the understanding of physical and chemical mechanisms underlying the enhanced conduction upon O_2 exposure.[6, 7, 8, 9, 10, 11, 12] Jhi *et. al.*[6] were first, who investigated the

S. Guceri et al. (eds.), Nanoengineered Nanofibrous Materials, 137-144.

effect of adsorbed O_2 on the semiconducting (8,0) SWNT by using pseudopotential plane wave method. Furthermore, they found the empty energy bands derived from oxygen states are overlap with the valence band of SWNT and give rise to a finite density of states at the Fermi level, E_F . In view of these results, they attributed the observed reduced resistance of the semiconducting SWNT upon O_2 exposure [1] to the hole-doping [6]. In contrast to this finding, the experiments by Derycke *et. al.*[13] have showed that the main effect of oxygen physisorption is not to dope the bulk of the tube, but to modify the barriers of the metal-semiconductor contact. The hole-doping picture for the O_2 physisorbed on the zigzag semiconducting tube has been refused also by recent first-principle studies [14, 15]. This paper investigates the atomic and electronic structureof the O_2 physisorbed zigzag SWNTs and further clarifies the effect of the adsorbed O_2 on the electronic structure.

2. METHOD OF THE CALCULATIONS

We performed first-principles total energy and electronic structure calculations using pseudopotential planewave method within spin-polarized, generalized gradient approximation (GGA) [16]. Calculations were carried out using periodically repeating tetragonal supercell with in-plane lattice constants $a_{SC} = b_{SC}$ and c_{SC} along the axis of the tube. To reduce the O_2-O_2 coupling, we used $c_{SC} = 2c$, i.e. double cell (c is the one dimensional lattice constant of the tube). Ultrasoft pseudopotentials [17] for C and O, the kinetic energy cut-off up to 400 eV and 6-12 Monkhorst-Pack [18] special **k**-points were used. For all systems we studied all atomic positions of adsorbate and SWNT, as well as c were fully optimized by using the conjugate gradient method. We studied the adsorption of O_2 on the (8,0) zigzag SWNT. The bare (8,0) tube is a semiconductor with a band gap of 0.64 eV.

3. PHYSISORPTION OF OXYGEN MOLECULE

We studied the bonding of O_2 by placing the molecule at different sites on the SWNT, and by calculating the binding energy corresponding to the optimized structure. Different physisorption positions described in Fig 1, *i.e.* on top of the axial C–C bond (A-site), above the center of the hexagonal carbon rings (H-site), on top of the zigzag C–C bond (Z-site), and perpendicular to the axis of the tube and above two adjacent zigzag bond (T-site) were considered. The binding energy involves short (E_s) and long (E_{vdW}) range interaction, *i.e.* $E_b = E_s + E_{vdW}$. By definition, $E_b > 0$ corresponds to a stable and exothermic bonding. In some cases, $E_s < 0$ is endothermic, but it corresponds to a local minimum where the desorption of O_2 from SWNT is prevented by a barrier. The contribution of the short range (chemical) interaction to the binding energy is calculated by using the expression

$$E_s = E_T[SWNT] + E_T[O_2] - E_T[O_2 + SWNT] \quad (1)$$

in terms of the total energies of the fully optimized bare SWNT, E_T [SWNT], the individual molecule, $E_T[O_2]$, and O_2 adsorbed on the SWNT, $E_T[O_2 + SWNT]$, which are calculated by using the same supercell, and the same parameters of calculation. The van der Waals energy is calculated by using the asymptotic form of the Lifshitz expression[19].

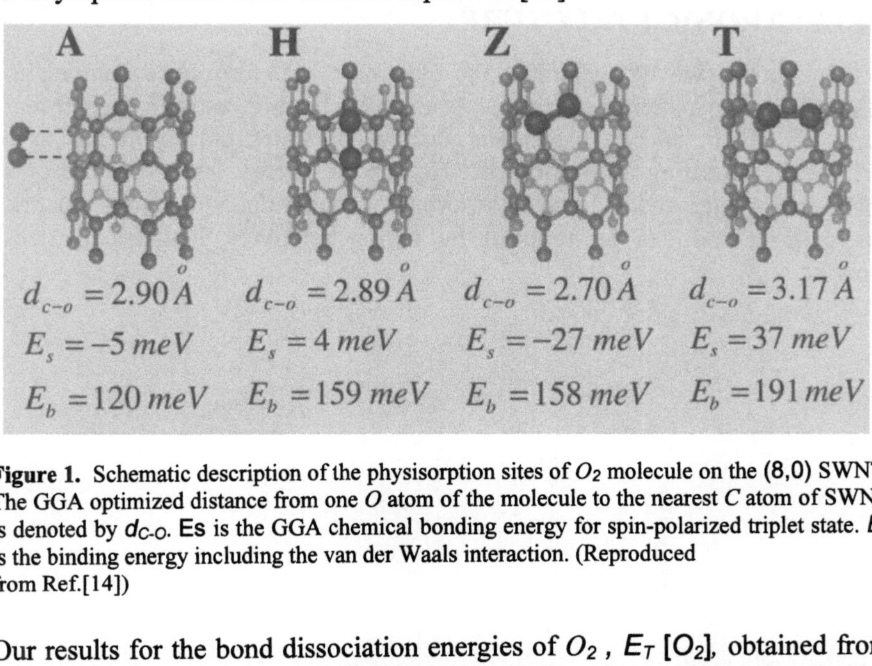

Figure 1. Schematic description of the physisorption sites of O_2 molecule on the (8,0) SWNT. The GGA optimized distance from one O atom of the molecule to the nearest C atom of SWNT is denoted by $d_{C\text{-}O}$. Es is the GGA chemical bonding energy for spin-polarized triplet state. E_b is the binding energy including the van der Waals interaction. (Reproduced from Ref.[14])

Our results for the bond dissociation energies of O_2, $E_T[O_2]$, obtained from the double supercell is 8.60 eV and 5.86 eV for spin-unpolarized and spin-polarized triplet state calculations, respectively. The triplet state is the ground state of O_2. The experimental bond energy of O_2 in the triplet state, 5.2 eV[20], is in fair agreement with the calculated value of 5.86 eV found in this study. The small binding energies in Fig. 1 are characteristics of physisorption. Spin-polarized calculations yield relatively lower (stronger) total energies, $E_T[O_2 + SWNT]$ and hence set the triplet state as the ground state with a net magnetic moment of ~2μ$_B$ (Bohr Magneton) per unitcell. In this case, GGA chemical bonding energies are generally weakened and at A- and Z- site; they become even negative with E_s =-5 and -27 meV, respectively. The spin-polarized calculations yield that T-site is energetically most favorable site with E_s=37 meV. Then the binding energy E_b is found to be 191 meV by adding E_{vdW}=154 meV to E_s. This binding energy is in agreement with the recent measurement by Ulbricht *et. al.*[21] We note that these binding energies are small, and become exothermic mainly owing to the long range vdW interaction. Relatively higher binding energy at T-site is attributed to relatively higher number of nearest neighbors of physisorbed O_2 and also to two strained zigzag *C-C* bonds deviating from sp^2-character.

4. ELECTRONIC STRUCTURE

In this section, we will examine the electronic structure corresponding to these adsorption patterns. Because of supercell method used in the present study we obtain energy bands and density of states corresponding to a periodically repeating adsorption pattern. For the energy level structure of a single adsorbate, the adsorbate-adsorbate interaction indigenous to the supercell method can be reduced by taking relatively longer cell sizes allowing

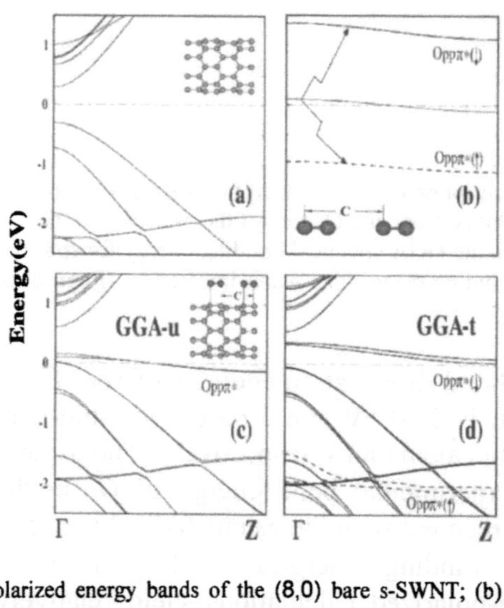

Figure 2. (a) Spin unpolarized energy bands of the (8,0) bare s-SWNT; (b) Insulating, spinpolarized $O_{pp\pi^*}$(↑) and $O_{pp\pi^*}$(↓) bands derived from metallic, spin-unpolarized bands of the individual O_2 chain having the lattice parameter *c*; (c) spin unpolarized energy bands of the O_2 physisorbed on the (8,0) tube with $O_{pp\pi^*}$ state pinning the Fermi level. (d) Spin-polarized bands corresponding to (c). Zero of energy is taken at the Fermi level shown by dash-dotted line. Up-spin and down-spin bands are indicated by corresponding arrows. Here c_{sc} = c. (Reproduced from Ref.[14])

longer nearest neighbor distances. Under these circumstances the bands are flatten and represent the energy level of the dopant. In Fig 2(a) we present the band structure of the bare (8,0) SWNT, which is a semiconductor with a band gap E_g ~ 0.64 eV at Γ-point between the bottom of the conduction band E_C and the top of the valence band E_V. The π^*-σ^* hybridized singlet state is in the conduction band. [23, 22] The free linear chain of O_2 molecules, in principle, has a weak bonding state with an equilibrium lattice parameters slightly larger than c. In Fig 2(b), the free linear chain of O_2 in registry with the (8,0) tube has half-filled, doubly degenerate $pp\pi^*$ bands. For the triplet state, (which is energetically more favorable than the spin unpolarized state as well as singlet state) these bands split into two doubly degenerate bands [$pp\pi^*(\uparrow)$ and $pp\pi^*(\downarrow)$]. The $pp\pi^*(\uparrow)$ band is filled and separated from the empty $pp\pi^*(\downarrow)$ bands by an energy gap of ~2 eV. In Fig 2(c) spin unpolarized GGA calculations with c_{sc} = c yield doubly degenerate, half-filled $pp\pi^*$ bands in the band gap for the physisorption of O_2 at A-site. The Fermi level touching the top of the valence band E_V makes the system metallic. The above situation is, however, changed in the spin polarized calculations, which yields the triplet state as the ground state. Under these circumstances two $O_{pp\pi^*}$ bands are split into four bands, two of them occur ~2eV below the valence band of s-SWNT. Remaining two empty bands rise 350 meV above E_V at the Γ-point and make a band gap of 90 meV [see Fig 2(d)]. Note that the bands here are only the artifact of the supercell method, and hence in the absence of band dispersion the energy level due to the single physisorbed O_2 and the top of the s-SWNT valence band, E_V shall be relatively larger than the calculated band gap. Our arguments have been justified by double cell calculations in which the O_2-O_2 coupling is reduced due to the large distance (~7 °A) between nearest molecules. In this case, sizable energy band gap of 0.2-0.4 eV occurs between E_V and the unoccupied spin-down bands [$O_{pp\pi^*}(\downarrow)$ states] of oxygen molecule. This situation eliminates the hole doping picture. The dispersionless $O_{pp\pi^*}(\downarrow)$ bands indicate that the coupling between O_2 molecules are negligible.

The effects of band formation are examined by studying the electronic energy structure corresponding to various patterns of physisorbed O_2. In Fig 3 we show the band structures and density of states calculated for the zigzag chain and row of O_2 physisorbed on (8,0) s-SWNT in the triplet ground state. In the zigzag chain, O_2 molecules are placed initially above the adjacent axial C-C bonds (*i.e.* A-sites), but upon relaxation they are tilted by 40° and inclined side ways to increase O-O separation of nearest molecules. We calculated the chemical bonding energy E_s=56 meV/molecule. Since the number of O_2 molecules per single cell is doubled, the $O_{pp\pi^*}(\uparrow)$ and $O_{pp\pi^*}(\downarrow)$ bands are doubled as compared to those illustrated in Fig 2(d). The system is a semiconductor with a direct band gap E_g ~0.4 eV. As clearly seen from the total and partial density of states the $O_{pp\pi^*}$ bands are split by ~ 2.5 eV upon spin polarization. The band structure of the O_2 row in Fig 3(b) displays slightly different situation. Here the gap between the lowest $O_{pp\pi^*}(\downarrow)$ band and E_V is reduced to 25 meV at the Z-point; it is so far the smallest band gap we obtained from spin polarized calculations. While this band complies with the hole doping picture since its is in the range of thermal excitation at room temperature, the metallization of s-SWNT is still too far.

Electronic energy structure of higher O_2 coverage was further analyzed by a system where initially O_2 molecules are physisorbed to all A-sites of the (8,0) tube. Upon relaxation of the system, the distance between O_2 molecule and SWNT surface has increased continuously. This indicates that under increased O_2–O_2 coupling the chemical bonding is weakened and eventually molecules start to escape from the surface of s-SWNT. The vdW interactions was needed to keep the molecules attached to s-SWNT. At each step of ionic relaxation, the bands derived from the bare s-SWNT and molecular oxygen

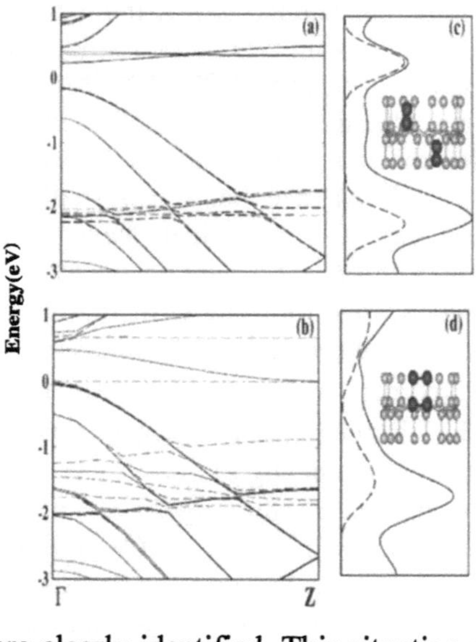

Figure 3. (a) Spin polarized electronic energy band structure of the zigzag chain of O_2 adsorbed above the adjacent axial *C-C* bonds along the axis of a (8,0) s-SWNT as shown by inset. Solid and dashed lines are the spin-down and spin-up bands. The zero of energy is taken at the Fermi level E_F indicated by dash-dotted line. (c) Corresponding total density of states of s-SWNT+O_2 and partial density of states on the oxygen atoms are shown by solid and dashed lines. (b) Band structure of the row of O_2 physisorbed at T-sites as shown by inset. (d) Same as (c). (Reproduced from Ref.[14])

are clearly identified. This situation suggests a negligible mixing between O_2 and s-SWNT states. In fact, the bands of bare s-SWNT remain practically unchanged, and the states of O_2 are broadened into bands owing to the intermolecular coupling. Several bands derived from $O_{pp\pi^*}(\uparrow)$ states forma sharp peak at ~2 eV below E_V . Empty bands derived from $O_{pp\pi^*}(\downarrow)$ states occur always above E_V, and appear as another peak. Even in full coverage we see that empty O_2 states and highest filled s-SWNT states are separated by a gap of ~100 meV, so that the electron transfer from SWNT to empty O_2 states and hence the metallization of the tube is prohibited. Under these circumstances, the s-SWNT+O_2 system can be viewed by two semiconductors forming a two-liquid system. The first one is due to O_2; the second one is a due to s- SWNT. The excitation of the electrons from s-SWNT bands to O_2 states may mediate intermixing of these two liquids.

5. RESULTS AND DISCUSSION

In this paper we studied the adsorption of oxygen molecule on semiconducting SWNT and resulting effects on the electronic structure. (*i*) The ground state of the adsorbed O_2 molecule is the triplet state with a net magnetic moment of $2\mu_B$. The binding energies calculated by including short and long range interactions are small and characteristics of physisorption. (*ii*) Stable chain formation and also uniform coverage of O_2 molecules in the triplet ground state on a s-SWNT are also possible. Since physisorbed O_2 has net magnetic moment in the ground state, new magnetic molecules (or nanomagnets) can be constructed as a result of O_2 decoration. (*iii*) GGA electronic structure calculations performed for optimized atomic geometry corresponding to the triplet ground state indicate that the unoccupied $O_{pp\pi^*}(\downarrow)$ bands (states) of O_2 occur 0.2–0.4 eV above the top of the valence band of the s- SWNT. Only the row of O_2 among different patterns studied in this paper yields a band gap as small as 25 meV.

References

1. P. G. Collins, K. Bradley, M. Ishigami, and A. Zettl, Science, **287** (2000) 1802.
2. J. Kong, N. R. Franklin, C. Zhou, M. G. Chapline, S. Peng. K. Cho, and H. Dai, Science **287** (2000) 622.
3. J. Kong, J. Cao, H. Dai, and E. Anderson, Appl. Phys. Lett. **80** (2002) 73.
4. O. Gulseren, T. Yildirim, and S. Ciraci, Phys. Rev. B **66** (2002) 121401.
5. O. Gulseren, T. Yildirim, and S. Ciraci, Phys. Rev. Lett. **87** (2001) 116802.
6. S. H. Jhi, S. G. Louie, and M. L. Cohen, Phys. Rev. Lett. **85** (2000) 1710.
7. S. M. Lee, Y. H. Lee, Y. G. Hwang, J. R. Hahn, and H. Kang, Phys. Rev. Lett. **82** (1999) 217.
8. D. C. Sorescu, K. D. Jordan, and P. Avouris, J. Phys. Chem. B **105** (2001) 11227.
9. X.Y. Zhu, S.M. Lee, Y.H. Lee, and T. Frauenheim, Phys. Rev. Lett. **85** (2000) 2757.
10. C-Y, Moon, Y-S, Kim, E-C, Lee, Y-G, Jin, and K. J. Chang, Phys. Rev. B **65** (2002), 155401.
11. D. J. Mann and M. D. Halls, J. Chem. Phys. **116** (2002) 9014.
12. A. Ricca and J. A. Drosco, Chem. Phys. Lett. **362** (2002) 217.
13. V. Derycke, R. Martel, J. Appenzeller, and Ph. Avouris, Appl. Phys. Lett. **80** (2002), 2773.
14. S. Dag, O. G¨ulseren, T. Yildirim and S. Ciraci, Phys. Rev. B **67** (2003) 165424.

15. P. Giannozzi, R. Car, and G. Scoles, J. Chem. Phys. **118** (2003) 1003.
16. J. P. Perdew and Y. Wang, Phys. Rev. B **46** (1992) 6671.
17. D. Vanderbilt, Phys. Rev. B **41** (1990) 7892.
18. H.J. Monkhorst and J.D. Pack, Phys. Rev. B **13** (1976) 5188.
19. E. M. Lifshitz, Zh. Eksp. Teor. Fiz. **29**, 94 (1956) (Sov. Phys. JETP **2**, 73 (1956)).
20. K.P. Huber and G. Herzberg, *Molecular Spectra and Molecular Structure. IV Constants of Diatomic Molecules* (Van Nostrand Reinhold, New York, 1979).
21. H. Ulbricht, G. Moos, and T. Hertel, Phys. Rev. B **66**, 075404 (2002).
22. X. Blase, L. X. Benedict, E. L. Shirley, and S. G. Louie, Phys. Rev. Lett. **72**, 1878 (1994).
23. O. Gulseren, T. Yildirim, S. Ciraci, and C. Kilic, Phys. Rev. B **65**, 155410 (2002); O.Gulseren, T. Yildirim, and S. Ciraci, Phys. Rev. B **65**, 153405 (2002).

ELECTRONIC STRUCTURE OF THE FLUORINATED HIPCO NANOTUBES

L.G.BULUSHEVA[1], A.V.OKOTRUB[1], T.A.DUDA[1], E.D.OBRAZTSOVA[2], A.L.CHUVILIN[3], E.M.PAZHETNOV[3], A.I.BORONIN[3], U.DETTLAFF-WEGLIKOWSKA[4]

[1]Nikolaev Institute of Inorganic Chemistry SB RAS, pr. Ak. Lavrentieva 3, Novosibirsk 630090, Russia
[2]Natural Sciences Center of A.M. Prokhorov General Physics Institute RAS, st. Vavilova 38, Moscow 119991, Russia
[3]Boreskov Institute of Catalysis SB RAS, pr. Ak. Lavrentieva 5, Novosibirsk 630090, Russia
[4]Max-Planck-Institute for Solid State Research, Heisenbergstr.1, Stuttgart 70569, Germany

Abstract: Single-wall carbon nanotubes (SWNTs) produced by high pressure CO decomposition (HiPco) process has been fluorinated using a volatile mixture of BrF_3 and Br_2. The composition of the fluorinated material was estimated by X-ray photoelectron spectroscopy to be C_4F. Raman spectroscopy showed the narrower tubes are more readily fluorinated. Comparison between CKα-spectrum of the fluorinated SWNTs and theoretical spectra obtained for the (8,8) tube models with different fluorine pattern revealed the fluorine atoms preferably form the chains around a tube circumference.

Keywords: Single-wall carbon nanotubes; fluorination; X-ray spectroscopy; semi-empirical calculations.

1. INTRODUCTION

Among the possible methods of sidewall functionalization of carbon nanotubes, fluorination attracts a special attention due to a possibility to covalently attach a large amount of fluorine [1]. The fluorinated single-wall carbon nanotubes (SWNTs) are solvated in alcohols [2] that dramatically enhances their reactivity and manipulation [3]. The first extensive fluorination of laser-ablated graphite-grown SWNTs has been carried out using elemental fluorine at the elevated temperatures [4]. It was found that if the reaction is performed at 325°C, the fluorination is nondestructive for tubular structure yielding overall stoichiometry of about C_2F. Determination of the distribution of functional groups over tube surface represents a serious problem. Scanning tunneling microscopy

S. Guceri et al. (eds.), Nanoengineered Nanofibrous Materials, 145-151.

(STM) image of the fluorinated SWNTs revealed the fluorinated regions form bright bands around the tube circumference [5].

Quantum-chemical calculations of thermodynamic stability of fluorinated carbon nanotubes show the fluorine atoms tend to bind close to each other on the tube surface [6]. With a C_2F stoichiometry the armchair (n,n) and zigzag (n,0) tubes were found to have the lowest energy when the conjugated π-bonds are parallel to the tube axis or helically wrapped around the tube [7]. The fluorination of narrower SWNTs was predicted to be more exothermic that is connected with an energetic gain due to the reduced pyramidalization of the π-system [8]. Band electronic structure calculations on the fluorinated carbon nanotubes showed the electronic properties of tubes can range from insulating over semiconducting to a metallic-like behavior depending on the fluorine decoration [7, 9]. The metallic conductivity of the fluorinated SWNT is realized with π-bond chains parallel to the tube axis although these predictions are still not supported experimentally. Electric transport measurements on the bulk materials containing SWNTs detected the fluorination increases the sample resistance by 4–7 orders of magnitude [4, 10].

Recently developed large scale production method (HiPco) yields SWNTs [11] with average tube diameter of 11 Å being smaller than diameter of SWNTs produced by other processes [12]. In the present work the HiPco material was fluorinated by exposing to the BrF_3 vapor at room temperature. Previously, this procedure was applied for the fluorination of arc-produced SWNTs [10] and multiwall carbon nanotubes (MWNTs) [13, 14]. The influence of fluorination on the electronic structure of HiPco nanotubes is examined by means of X-ray spectroscopy and quantum chemical calculations.

2. EXPERIMENTAL DETAILS

HiPco SWNTs were produced at CNI, Houston, USA. Fluorination was carried out at room temperature. The samples placed in a teflon flask was held in the vapor over a solution of BrF_3 in Br_2 for one day. Thereafter the flask content was dried by a flow of N_2 until the termination of Br_2 evolution (~48 hours). The samples were characterized by applying transmission electron microscopy (JEM-2010 microscope), Raman spectroscopy (Ar^+ laser with λ = 0.5145 μm), and X-ray photoelectron spectroscopy (VG ESCALAB HP spectrometer with Al Kα radiation).

X-ray fluorescent spectra of HiPco material before and after fluorination were recorded with a "Stearat" spectrometer using ammonium biphthalate (NH_4AP) single crystal as an analyzing crystal. This crystal has nonlinear reflection efficiency, which is corrected by the procedure described elsewhere [15]. The sample was deposited on the copper support and cooled down to liquid nitrogen temperature in the vacuum chamber of the X-ray tube. The X-ray tube with copper anode operated at

0.5 A, 6 kV. The accuracy of determination of X-ray band energy was ±0.15 eV with the spectral resolution of 0.5 eV.

3. CALCULATION

Three models of (8,8) tube with different fluorine pattern were calculated using the semi-empirical MNDO method within the GAMESS package [16]. The (8,8) tube has a 10.9 Å diameter coincident with the average diameter of HiPco tubes. The fluorinated carbon rows and polyacetylene-like chains are alternated around tube circumference in the model **I** and along tube axis in the model **II**. The model **III** is derived from model **II** by placing every other fluorine atom inside the tube. The endohedral covalent binding of fluorine was shown to be thermodynamically possible for armchair tubes [8]. Fluorine atoms were used to saturate dangling bonds at the cluster boundary. The composition of models is $C_{1.8}F$ that is close to a limiting stoichiometry of C_2F achieved for SWNTs. The geometry of fluorinated tube models was fully optimized by standard BFGS procedure to the gradient value of 10^{-4} Ha/Bohr.

The theoretical CKα-spectra were plotted for the central atoms to eliminate the boundary states effect. X-ray transition intensity was calculated by summing the squared coefficients with which carbon 2p-atomic orbitals (AOs) involved in the concrete occupied molecular orbital (MO). The energy location of intensity corresponded to the MO eigenvalue. Calculated intensities were normalized by maximal value and broadened by Lorenzian functions with half width at half maximum of 0.7 eV.

4. RESULTS AND DISCUSSION

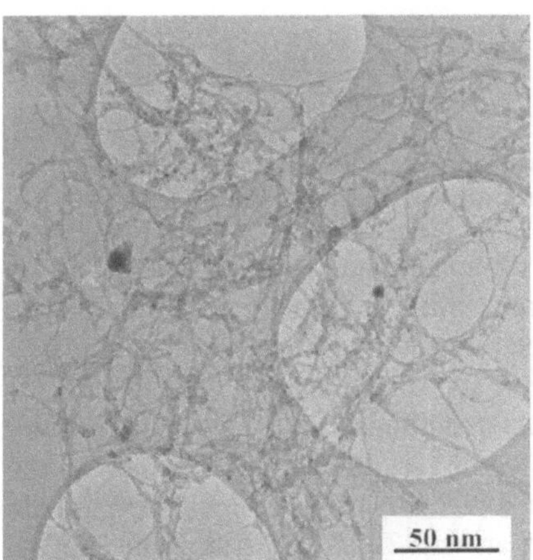

Figure 1. TEM image of the fluorinated HiPco material.

A transmission electron microscopy (TEM) image of the fluorinated HiPco material is presented in Fig. 1. As can be seen from the image, the applied fluorination procedure is not destructive to nanotubes. The ropes of SWNTs are sometimes covered by amorphous carbon. In addition, the fluorinated material contains encapsulated iron particles.

The Raman spectroscopy was used to probe a structural change of the starting material with fluorination. Fig. 2(a) shows well-defined radial breathing modes (RBM) in the region of 170–280 cm^{-1}, a narrow tangential G band at 1591 cm^{-1} and a disorder mode D with low intensity around 1336 cm^{-1}. The pattern is typical for the HiPco SWNTs [17]. Raman spectroscopy on the fluorinated product yields a spectrum being very different from the starting material (Fig. 2(b)). First, the relative intensity of the D mode is greatly increased after fluorination. Second, the RBM region is drastically changed showing two broad bands around 205 cm^{-1} and 273 cm^{-1}; the reason for the changes in the overall symmetry and bonding structure of the SWNTs with fluorination [18]. The low intensity of the band located at 273 cm^{-1} probably reflects the preferable fluorination of the smallest diameter SWNTs. Moreover, fluorination shifts the spectra in both regions towards the higher frequencies. This may be related to an increase in the inter-tube spacing due to the fluorine on the tube sidewalls. The all mentioned changes in the Raman spectra have been observed with nanotubes fluorinated by other methods [2, 18, 19].

The C 1s X-ray photoelectron spectrum (XPS) of the fluorinated HiPco material is shown in Fig. 3. The fluorinated material has a low conductivity and one need to take into account sample charging during the

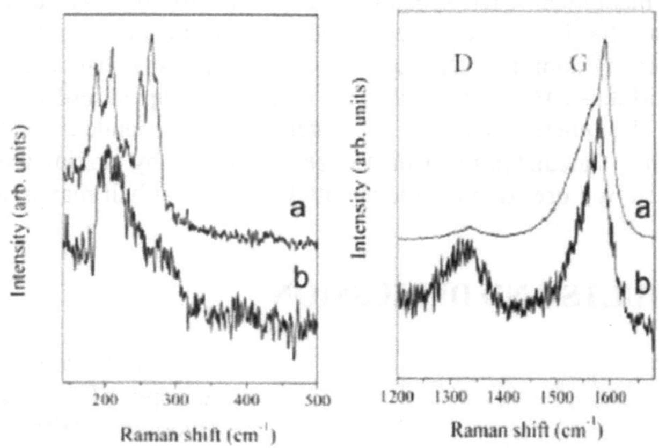

Figure 2. Raman spectra of raw HiPco material (a) and that after fluorination (b). The right-side spectra were normalized using the main tangential mode at 1591 cm^{-1}.

measurement. the binding energies were calibrated with respect to the F 1s line. The spectrum was fitted to three Gaussian-shaped curves that gave a C_4F composition of the product. The reduced fluorinated content could be due to amorphous carbon in the starting material, which remains unchanged with the applied synthetic conditions [10]. The binding energy centered at 289.5 eV corresponds to the carbon atoms bonded to the fluorine ones and the signal obtained at 286.1 eV is assigned to the bare carbon. The low-intensity shoulder around 291.8 eV is attributed to the CF_2 groups.

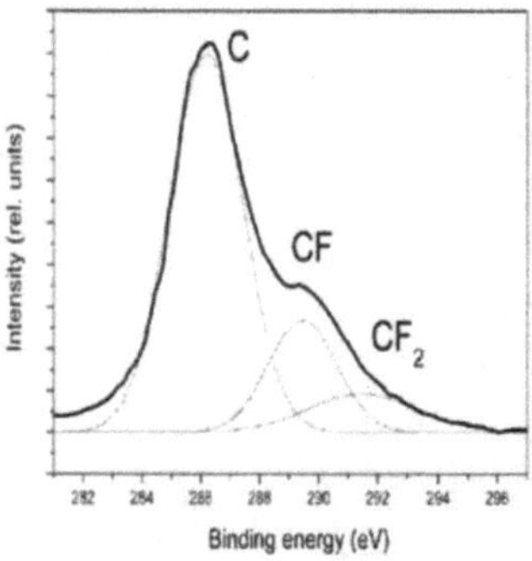

Figure 3. C 1s spectrum of the fluorinated HiPco material.

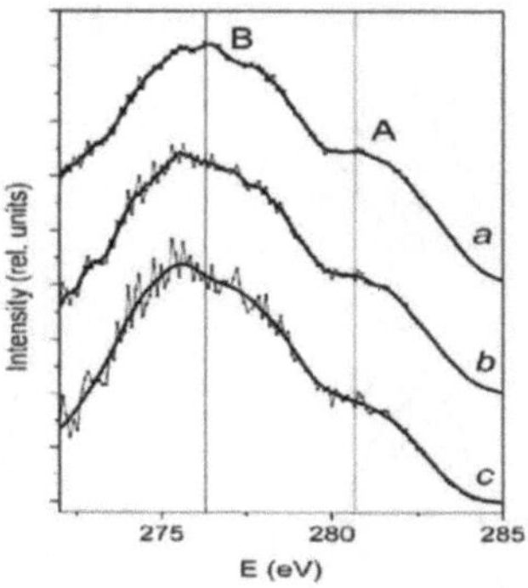

Figure 4. CKα-spectra measured for the raw HiPco material (a) and fluorinated HiPco material (b). The spectral profile (c) was obtained by subtraction of half intensity (a) from the spectrum (b)

Valence electronic structure of HiPco nanotubes before and after fluorination was compared using X-ray fluorescence spectroscopy. CKα-spectrum of the raw HiPco material shows two features: the main maximum B around 276.5 eV and less intense maximum A at 281 eV (Fig. 4(a)). Fluorination slightly decreases intensity of maximum A and shifts position of maximum B by about 1 eV towards the lower energy region (Fig. 4(b)). With C_4F composition of the fluorinated HiPco material we suggested the product contains equal portions of non-fluorinated carbon and C_2F components. To recognize the electronic states of carbon constituted the walls of the fluorinated HiPco nanotubes, half intensity of the starting material CKα-spectrum was subtracted from the intensity of CKα-spectrum measured for the product. The resultant spectral profile is presented in Fig. 4(c). This spectrum is different from the CKα-spectrum of fluorinated sample by noticeable reduction of the maximum A intensity.

The more preferable arrangement of the fluorinated HiPco tubes was determined from the comparison of the resultant spectral profile with theoretical CKα-spectra plotted for the models **I–III** (Fig. 5). The thin dashed and dotted lines in Fig. 5 (a–c) correspond to the density of 2p-occupied states for carbon atoms constituted respectively the polyacetylene-like and fluorinated carbon chains. One can see the change in the fluorine pattern has noticeable effect on both calculated components. The density for the first component was held at the calculated energies while the density for carbon atoms bonded to fluorine was moved 3.4 eV towards the high energies correspondingly with the XPS data. A sum of these two components represents the theoretical CKα-spectrum of a model (the thin solid line in Fig. 5). The theoretical spectra were aligned to the experimental spectral profile by the position of main maximum B. The spectrum calculated for the model **I** showed the best agreement with the experiment (Fig. 5(a)). Alternation of the

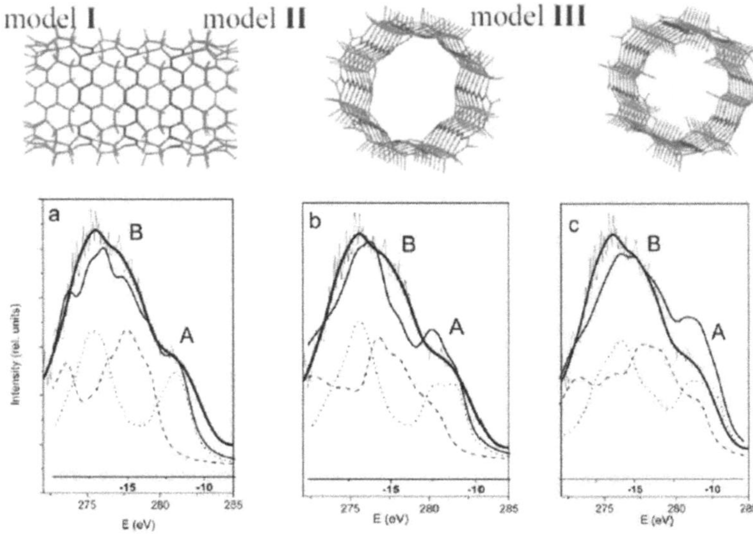

Figure 5. Comparison of the resultant spectral profile for fluorinated HiPco nanotubes (thick solid line) with the theoretical CKα-spectra (thin solid line) plotted for model **I** (a), model **II** (b), and model **III** (c). The dashed lines correspond to the density of states of the fluorinated carbon and the dotted lines correspond to the density of states of the bare carbon.

fluorinated carbon rows and acetylene-like chains along tube axis results in the overestimation of the intensity of high-energy maximum A (Fig. 5(b, c)). This intensity strongly increases when half of the fluorine atoms are located inside the tube (model **III**). The reason is reduction of electron repulsion in the model **III** that is reflected also in the lowest calculated total energy.

In summary, the low temperature fluorination of the HiPco material yielded a product of about C_4F composition. Comparison of experimental CKα-spectrum with theoretical ones plotted by the results of quantum-chemical calculations on the (8,8) tube models with different fluorine decoration revealed that the fluorine atoms are likely to form chains around the tube circumference.

Acknowledgments: The work was financially supported by the INTAS (project 01-254) and the RFBR (grant 03-03-32286a).

References

1. Bahr, J.L. and Tour, J.M. (2002) Covalent chemistry of single-wall carbon nanotubes, *J. Mater. Chem.* **12**, 1952-1958.
2. Mickelson, E.T., Chiang, I.W., Zimmerman, J.L., Boul, P.J., Lozano, J., Liu, J., Smalley, R.E., Hauge, R.H., and Margrave, J.L. (1999) Solvation of fluorinated single-wall carbon nanotubes in alcohol solvents, *J. Phys. Chem. B* **103**, 4318-4322.
3. Khabashesku, V.N., Edward Billups, W., and Margrave, J.L. (2002) Fluorination of single-wall carbon nanotubes and subsequent derivatization reaction, *Acc. Chem. Res.* **35**, 1087-1095.

4. Mickelson, E.T., Huffman, C.B., Rinzler, A.G., Smalley, R.E., Hauge, R.H., and Margrave, J.L. (1998) Fluorination of single-wall carbon nanotubes, *Chem. Phys. Lett.* **296**, 188-194.
5. Kelly, K.F., Chiang, I.W., Mickelson, E.T., Hauge, R.H., Margrave, J.L., Wang, X., Scuseria, G.E., Radloff, C., and Halas, N.J. (1999) Insight into the mechanism of sidewall functionalization of single-walled nanotubes: an STM study, *Chem. Phys. Lett.* **313**, 445-450.
6. Bauschlicher Jr., C.W. (2000) Hydrogen and fluorine binding to the sidewalls of a (10,0) carbon nanotube, *Chem. Phys. Lett.* **322**, 237-241.
7. Kudin, K.N., Bettinger, H.F., and Scuseria, G.E. (2001) Fluorinated single-wall carbon nanotubes, *Phys. Rew. B* **63**, 045413.
8. Bettinger, H.F., Kudin, K.N., and Scuseria, G.E. (2001) Thermochemistry of fluorinated single wall carbon nanotubes, *J. Am. Chem. Soc.* **123**, 12849-12856.
9. Seifert, G. Kohler, T., and Frauenheim, T. (2000) Molecular wires, solenoids, and capacitors by sidewall functionalization of carbon nanotubes, *Appl. Phys. Lett.* **77**, 1313-1315.
10. Okotrub, A.V., Romanenko, A.I., Yudanov, N.F., Bulusheva, L.G., Asanov, I.P., Anikeeva, O.B., Chuvilin, A.L., Boronin, A.I., and Bonard J.-M. (2002) Fluorine effect on the structure and electrical transport of arc-produced carbon nanotubes, in H. Kuzmany, J. Fink, M. Mehring, and S. Roth (eds.), *Structural and Electronic Properties of Molecular Nanostructures,* AIP Conference Proceedings **633**, pp.263-266.
11. Nikolaev, P., Bronikowski, M.J., Bradley, R.K., Rohmund, F., Colbert, D.T., Smith, K.A., and Smalley, R.E. (1999) Gas-phase catalytic growth of single-walled carbon nanotubes from carbon monoxide, *Chem. Phys. Lett.* **313**, 91-97.
12. Chiang, I.W., Brinson, B.E., Huang, A.Y., Willis, P.A., Bronikowski, M.J., Margrave, J.L., Smalley, R.E., and Hauge, R.H. (2001) Purification and characterization of single-wall carbon nanotubes (SWNTs) obtained from the gas-phase decomposition of CO (HiPco process), *J. Phys. Chem. B* **105**, 8297-8301.
13. Okotrub, A.V., Yudanov, N.F., Chuvilin, A.L., Asanov, I.P., Shubin, Yu.V., Bulusheva, L.G., Gusel'nikov, A.V., and Fyodorov I.S. (2000) Fluorinated cage multiwall carbon nanoparticles, *Chem. Phys. Lett.* **322** 231-236.
14. Yudanov, N.F., Okotrub, A.V., Shubin, Yu.V., Yudanova, L.I., Bulusheva, L.G., Chuvilin, A.L., and Bonard J-M. (2002) Fluorination of arc-produced carbon material containing multiwall nanotubes, *Chem. Mater.* **14** 1472-1476.
15. Okotrub, A.V. and Bulusheva, L.G. (1998) CKα- spectra and investigation of electronic structure of fullerene compounds, *Fullerene Science and Technology* **6**, 405-432.
16. Schmidt, M.W., Baldridge, K.K., Boatz, J.A., Elbert, S.T., Gordon, M.S., Jensen, J.H., Kosoki, S., Matsunaga, N., Nguyen, K.A., Su, S.J., Windus, T.L., Dupluis, M., and Montgomery, J.A. (1993) *J. Comput. Chem.* **14**, 1347-1363.
17. Cinke, M., Li, J., Chen, B., Cassell, A., Selzeit, L., Han, J., and Meyyappan, M. (2002) Pore structure of raw and purified HiPco single-walled carbon nanotubes, *Chem. Phys. Lett.* **365**, 69-74.
18. Gu, Z., Peng, H., Hauge, R.H., Smalley, R.E., and Margrave, J.L. (2002) Cutting single-wall carbon nanotubes through fluorination, *Nanoletters* **2**, 1009-1013.
19. Pehrsson, P.E., Zhao, W., Baldwin, J.W., Song, C., Liu, J., Kooi, S., and Zheng, B. (2003) Thermal fluorination and annealing of single-wall carbon nanotubes, *J. Phys. Chem. B* **107**, 5690-5695.

TITANIUM COVERAGE ON A SINGLE–WALL CARBON NANOTUBE: MOLECULAR DYNAMICS SIMULATIONS

H. Oymak and S. Erkoc
Department of Physics, Middle East Technical University, Ankara 06531 Turkey

Abstract: The minimum energy structures of titanium covered finite-length C(8,0) singlewall carbon nanotubes (SWNT) have been investigated. We first parameterized an empirical potential energy function (PEF) for the CTi system. The PEF used in the calculations includes two- and three-body atomic interactions. Then, performing molecular dynamics simulations, we obtained the minimum-energy configurations for titanium covered SWNTs. The reported configurations include low and high coverage of Ti on SWNTs. We saw that one layer of Ti did not distort the nanotube significantly, whereas two-layer coverage showed an interesting feature: the second layer of Ti pushed the first layer inside the wall, but the general shape of the nanotube was not affected so much.

Key words: nanotubes; molecular-dynamics; emprical potentials.

1. INTRODUCTION

Carbon nanotubes are ideal for investigating molecular scale wires. Especially stable metal wires with diameters of the order of nanometers are indispensable in the development of nanoelectronics and nanodevice technologies [1]. In order to explore the intrinsic electrical properties of nanotubes and to obtain functional electronic devices with desired characteristics, low resistance metal-tube ohmic contacts are vital [2]. Several studies [3,4] revealed that low resistance metal-tube ohmic contacts could be achieved with Ti, Nb, and Ni metals. It was reported that the resistance of a Ti contacted metallic single-wall carbon nanotube (SWNT) was the lowest (as low as 12 kΩ) measured with individual SWNTs among all reported results [2,4]. A continuous Ti coverage of varying thicknesses, and quasi continuous coverage of Ni and Pd were obtained using electron-beam evaporation techniques [2,5]. It was pointed out by several researchers that the coverage of any metal on a SWNT could be mediated by first depositing titanium as a buffer layer [2,5,6]. It follows that a deep understanding of metal-tube

S. Guceri et al. (eds.), Nanoengineered Nanofibrous Materials, 153-157.

interactions and the metal coverage mechanism is indispensably necessary. In a recent *ab initio* calculation [6] aluminum-covered C(8,0) SWNTs were studied. Inspired from that work we have investigated titanium adatom and titanium coverage on the same C(8,0) nanotube.

2. METHODS OF CALCULATIONS

In order to obtain the minimum-energy structures of Ti adsorbed (Ti-SWNT) systems, we conducted molecular dynamics (MD) simulations based on an empirical PEF parameterized for CTi binary system. It is parameterized to give experimental data for the system under study. The PEF used in this study contains only two- and three-body interactions:

$$\Phi = \sum_{i<j}^{N} U_{ij}(\mathbf{r}_i, \mathbf{r}_j) + \sum_{i<j<k}^{N} W_{ijk}(\mathbf{r}_i, \mathbf{r}_j, \mathbf{r}_k) \quad (1)$$

where $\mathbf{r}_i$ is the position vector for the atom i and N is the total number of atoms in the system. The Lennard-Jones potential has been taken for the two-body part, and the Axilrod-Teller triple-dipole potential for the three-body part:

$$U_{ij} = \varepsilon_0 \left[\left(\frac{r_0}{r_{ij}} \right)^{12} - 2 \left(\frac{r_0}{r_{ij}} \right)^{6} \right], \quad W_{ijk} = \frac{Z(1 + 3\cos\theta_i \cos\theta_j \cos\theta_k)}{(r_{ij}\, r_{ik}\, r_{jk})^3} \quad (2)$$

where $r_{ij} = |\mathbf{r}_i - \mathbf{r}_j|$ is the interatomic distance between the atoms i and j; r_0 is the equilibrium distance of the pair potential, where the pair energy assumes its minimum value, ε_0; θ_i, θ_j, and θ_k are the angles of the triangle formed by the three atoms i, j, k; and Z is the three-body parameter.

In this work the equations of motion for the CTi system have been solved by employing a MD technique based on the Nordsieck-Gear algorithm [7], through the seventh-order predictor-corrector method. The simulations were performed in such a way that each system under study was considered as an isolated system. By taking the MD temperature as 1 K and 300 K, we considered both a low-temperature and a room-temperature study. For the calculation of each system 50000 time steps (each time-step was 2.6×10^{-16} s) were enough to reach thermal equilibrium. This criterion was checked by looking at the variation of the total potential energy of the system with respect to the number of time steps. The structures of the systems discussed in this work are snapshots of the final MD step.

In order to determine some of the PEF parameters for which there were no experimental data, we have first investigated the microclusters of type C_kTi_l

$(k+l=2,3)$ theoretically by performing density functional theory (DFT) calculations. The exchange and correlation potentials contributions have been considered at B3LYP level [8]. The compact effective potential (CEP) basis functions with ECP triple-split basis, namely, CEP-121G, have been used in the calculations. DFT calculations were performed by using the GAUSSIAN 98 package [9].

3. RESULTS AND DISCUSSIONS

Making use of the available experimental data and the quantum mechanical DFT calculation results for dimer and trimers of C and Ti, we have determined the parameters, (r_0, ε_0, Z) of the PEF, given in above equations, for the CTi binary system. The binding energy (i.e. pair energy) and the interatomic equilibrium distance for two-atom systems, as well as the binding energy (i.e. formation energy) and the geometry for three-atom systems, have been taken into account during the parameterization process. The main objective was to re-obtain the geometry and the total interaction energies of the dimers and trimers of the elements under consideration as close as those resulting from experiments and, at the same time, those resulting from previous and current *ab initio* results. The thus-obtained parameter set for the CTi binary system is:

ε_0 (C–C) = 6.21 eV, r_0(C–C) = 1.48 Å, Z(C–C–C) = 292.1 eV Å^9;

ε_0 (Ti–Ti) = 2.0 eV, r_0(Ti–Ti) = 2.3 Å, Z(Ti–Ti–Ti) = 1764.0 eV Å^9;

ε_0 (C–Ti) = 5.17 eV, r_0(C–Ti) = 1.638 Å, Z(C–C–Ti) = 60.0 eV Å^9,
Z(C–Ti–Ti) = 160.0 eV Å^9.

We now have come to the process of the Ti coverage on a SWNT. We considered a C(8,0) SWNT containing 176 C atoms. After several trials of different initial configurations we found that the optimum configuration for a single Ti atom on the SWNT was the hollow site (H-site: above the center of hexagons). We next prepared a ”Ti-ring”, containing 8 Ti atoms, uniformly distributed 5 Å above the H sites around the circumference near the middle of the tube. We saw after the optimization of this system that the Ti-ring placed itself exactly at the H sites with neither clustering nor a significant distortion in the SWNT both at 1 K and 300 K [Figures 1(a) and 1(b)]. Now the trend is obvious; we tried to fully cover the SWNT by forming ”one-layer” of Ti containing 10 Ti rings (containing totally 80 atoms) by placing each Ti atom 5 Å above the SWNT again. Optimization at 1 K and 300 K led to the Figure 1(c) and 1(d), respectively. We obtained perfectly Ti-covered nanotubes and at both temperatures with again no any significant distortion in the SWNT and no clusterization of Ti atoms. This is one of the most important results of this study. It is highly tempting to say that the thus-obtained SWNT with ”one-layer" Ti can be used as a template (or substrate) for a second type of metal atom, e.g. Au, to obtain a stable metal nanowire. Finally, we tried to cover the SWNT with a "two-layer" Ti (containing 160 Ti atoms). To do so, we followed two different ways. First we placed a second layer of Ti (80 Ti atoms) 4 Å above the previously relaxed (i.e., optimized) first layer in such a way that each Ti atom of second layer was exactly between two neighbor Ti atoms of the first layer. It is found that after the MD relaxation the second layer Ti atoms pushed the first layer Ti atoms through the nanotube wall toward the inside so that the SWNT has become covered by Ti atoms both from inside and outside. This is another important outcome of this study. In Figures 2(a) and 2(b) we give thus-optimized structures of SWNTs at 1 K and 300 K, respectively. It is seen that the structure of SWNT is no longer intact; it undergone a nonnegligible, although not drastic, distortion in both cases. We again saw no any clusterization of Ti atoms. In the second way of covering the nanotube by two Ti layers, we placed both first and second layers above the nanotube at the same time. First layer placed 5 Å above the nanotube and the second one 3 Å above the first one in such a way that each Ti atom of second layer was again between two neighbor Ti atoms of the first layer. Relaxed structures are shown in Figures 2(c) and 2(d). The result was the similar the previous one: The SWNTs have been covered both from inside and outside. The distortions in the SWNTs at these times are now more severe but the general tube shape is still remained. Furthermore, the temperature made this time a significant difference in the final shape of the system. In passing, although we have chosen a tube of finite length (with ten hexagons along the tube axes), the results obtained are valid for the tubes of any size.

Acknowledgements: The authors would like to thank TUBITAK for partial support through the project TBAG-U/41 (102T007), and METU (Middle East Technical University) for partial support through the project METU-BAP-2003-07-02-00- 46. The authors would like to thank also Prof. Dr. Salim Ciraci for drawing their attention to the subject of this work.

References

1. S. Ciraci, A. Buldum, I. Batra, J. Phys.: Condens. Matter 13 (2001) 537.
2. Y. Zhang, N.W. Franklin, R.J. Chen, H. Dai, Chem. Phys. Lett. 331 (2000) 35.
3. H.T. Soh, C.F. Quate, A.F. Morpurgo, C.M. Marcus, J. Kong, H. Dai, Appl. Phys. Lett. 75 (1999) 627.
4. C. Zhou, J. Kong, H. Dai, Phys. Rev. Lett. 84 (2000) 5604.
5. Y. Zhang, H. Dai, Appl. Phys. Lett. 77 (2000) 3015.
6. V.M.K. Bagci, O. G¨ulseren, T. Yildirim, Z. Gedik, S. Ciraci, Phys. Rev. B 66 (2002) 045409.
7. C.W. Gear, Numerical Initial Value Problems in Ordinary Differential Equations, NJ: Prentice-Hall, Englewood Cliffs, 1971.
8. R.G. Parr, W. Yang, Density Functional Theory of Atoms and Molecules, Oxford University Press, Oxford, 1989.
9. Gaussian 98, Revision A.9, Gaussian, Inc., Pittsburgh PA, 1998.

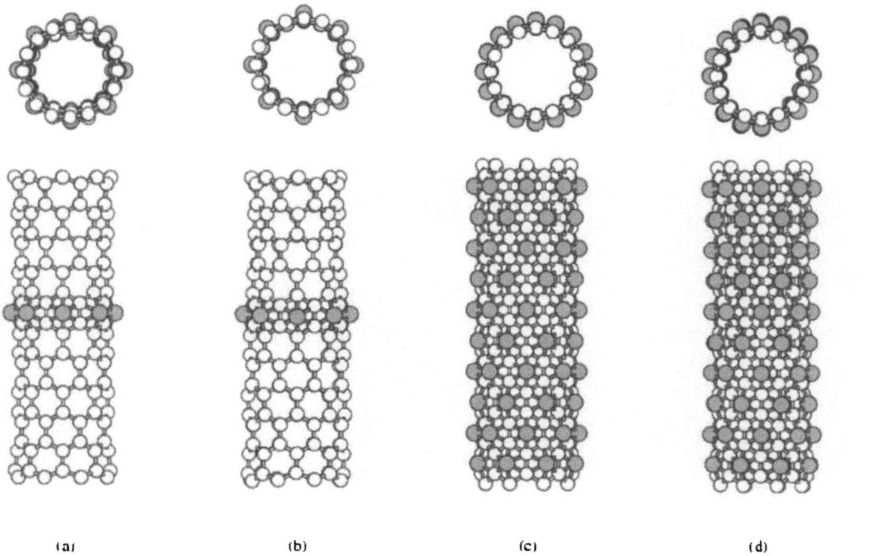

Figure 1. Optimized structure of SWNT (a) at 1 K with one ring of Ti atoms; (b) at 300 K with one ring of Ti atoms; (c) at 1 K with one layer of atoms; (d) at 300 K with one layer of atoms.

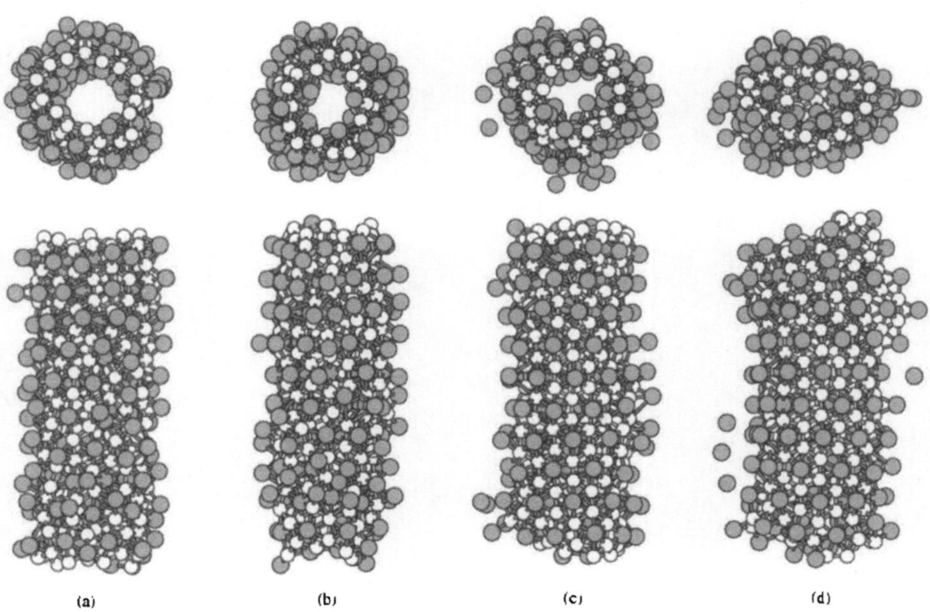

Figure 2. Optimized structure of SWNT (a) at 1 K with two layers of Ti atoms (the second layer is placed on the previously optimized one-layer Ti-SWNT system); (b) 300 K with two layers of Ti atoms (the second layer is placed on the previously optimized one-layer Ti-SWNT system); (c) at 1 K with two layers of Ti atoms (both first and second layers are placed on the SWNT at the same time); at 300 K with two layers of Ti atoms (both first and second layers are placed on the SWNT at the same time).

USING SUPERCRITICAL WATER MANIPULATES THE STRUCTURES OF POROUS MATERIALS & NANO-SCALE PARTICLES

J.C. Li [a,b], R. Aswani [a], X.L. Gao [a], S. King [c] and A.I. Kolesnikov [d]

[a] *Department of Physics, UMIST, P.O. Box 88, Manchester, M60 1QD, UK;* [b] *Institute of Physics, Chinese Academy of Science, Po Box 603, Beijing 100080, China;* [c] *ISIS facility, Rutherford Appleton Laboratory, Chilton, Didcot, Oxon, QX11, 0XQ, UK;* [d] *IPNS, Argonne National Laboratory, Argonne, Illinois, 60439, USA*

Abstract: Supercritical water has attracted a growing interest due to its applications in disposing toxic waste and its uses as a medium for chemical synthesis [1]. Properties of supercritical water are, however, rather peculiar compared to normal water. Its dielectric constant is rather small and its ionic product is three orders of magnitude higher than ambient water. Thus it is a very corrosive medium that attacks a range of materials and chemical substances. In this paper, we report series of small angle neutron scattering measurements on supercritical water, in confined porous Vycor glass, which showed that exposure of the latter to water at elevated temperatures and pressures results in loss of its original pore structure, producing nano-scale SiO_2 particles.

1. INTRODUCTION

The use of supercritical water (SCW) in oxidation processes is an emerging technology that is being developed for the treatment of toxic organic wastes, such as pesticides, herbicides, PCBs, PAHs, obsolete munitions, chemical warfare agents and for the synthesis of new materials under hydrothermal conditions [1]. The process is carried out at temperatures and pressures above the critical point of water (374°C and 221bar) at which organic chemicals become completely miscible. When oxygen is introduced oxidisable materials are converted readily to carbon dioxide, water, and other innocuous compounds. The effluent from such waste processing can be evaluated with respect to compliance with applicable discharge regulations, thus ensuring protection of the environment.

The behaviour of SCW is still imperfectly understood. One suggestion is that its reactivity may be due to changes of the intermolecular structure at

S. Guceri et al. (eds.), Nanoengineered Nanofibrous Materials, 159-164.

elevated temperatures, and especially to changes in the hydrogen bond network. An alternative suggestion is that at high temperature, a large mount of OH and H species (split from the natural water molecules) give rise to the high-lever oxidation activity [2].

In the recent years, there considerable amount of work has been done for the structure [3-5] and dynamics [6-12] of SCW and the SCW in confine volume [13], aiming to have better understanding the H-bonding at below and above the supercritical state. Because of the large incoherent cross-section of hydrogen, neutron scattering is the most appropriate technique for studies of the dynamics of hydrogen containing materials, such as water and the water of biological systems.

Our recent neutron scattering work shows that the properties of SCW change drastically when it is in confined space [13]. In order to better understand the behaviour of the SCW in confined space, we have used Vycor glass (produced by spinodal decomposition process which provides a well-inter-connected spaghetti-like porous structure) as the confinable media [14]. In this paper, we report series of small angle neutron scattering measurements, aim to exam the changes of the porous structure of Vycor glass in water under high pressures and temperatures.

2. EXPERIMENTAL DETAILS

Two small angle neutron scattering (SANS) instruments, LOQ on ISIS at Rutherford Appleton Laboratory (UK) and SAND on IPNS at Argonne National Laboratory (USA), were used in this study. Because of the use of time of flight technique on pulsed neutron sources, both instruments have very similar design with two dimensional detector arrays, which can utilise wavelengths from 2 to 10 Å which provide a wide Q coverage from 0.005 to 0.2 $Å^{-1}$. This corresponds a coverage of real-space measurement of 10-1000 Å. Therefore, any structural variation below 1000 Å can be detected by this type of facility. The high pressure (up to 2000 bar) and temperature (up to 450°C) cell is designed specifically for this work (see Figure 1), it is made of ZrTi alloy which has zero combined coherent neutron scattering cross-section, hence the background in the measured spectra is from the incoherent scattering only which gives a smooth background in measured scattering intensity, S(Q). The sample volume in the cell is 7 mm in diameter and 30 mm in height. The heaters and temperature sensors were all inserted in the cell and a 1 mm steel pipe leads to an external water-pressuriser (see Figure 1) which provided the required pressure while the experiment was running.

The sample used in this experiment was porous Vycor glass (corning 7930). It was produced by allowing a borosilicate glass to undergo spinodal decomposition and by then etching out the B_2O_3 component. This process yields a porosity of 28% and an average pore diameter of 30 Å [14]. The samples were cut into 6 mm in diameter and 25 mm in height rods, preinserted in the cell, which then was connected to the water pump. The water was used as pressure transfer media and we used double distilled H_2O.

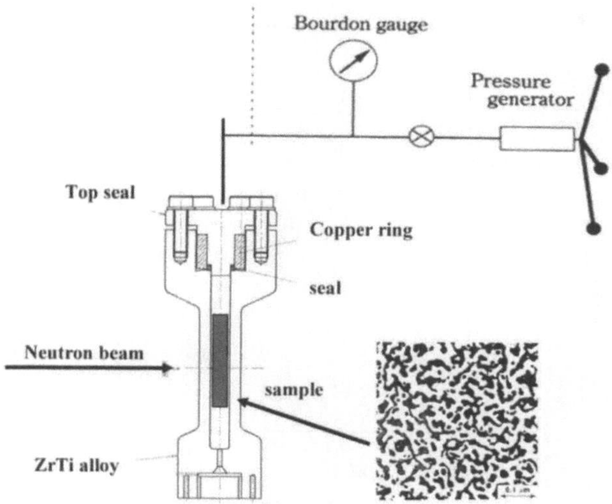

Figure 1. Schematic illustration of the supercritical water pressure system and the experimental set-up.

3. RESULTS AND DISCUSSION

Figure 2 shows the first experiments carried out on LOQ at ISIS. The upper curve (data series 1) in the figure was for a dry Vycor glass at room temperature and ambient pressure, a structural spectrum which is similar to what we have studied in the past [15,16]. The lower curve (data series 2) is for the same sample, which has reacted with SCW for a few hours using the SCW high pressure cell shown in Figure 1. The curve shows a straight line in the log-log plot across almost the entire Q range measured from 0.005 to 0.2 $Å^{-1}$ (i.e. one and half orders of Q), it shows a behaviour very much similar to the aggregated particles with a gradient of –3.5 in the log-log plot (i.e. the scattering intensity $S(Q) \sim Q^{-3.5}$). It indicates that the structure of the Vycor glass was completely destroyed by the SCW, resulting a near close-packed particles.

Further investigation of the dynamic process of the reaction between SCW and Vycor, in-situ measurement was then set up on SAND at IPNS using the same pressure system illustrated in Figure 1. The experiment was first set at 350°C and 1000 bar after an initial measurement for the Vycor glass at room temperature, the measurements were taken for every 15 mins per step at the fixed temperature and pressure. However the experiment shows that the reaction was very rapid and the 15 mins per data set was not possible to catch up the process (less than 15 mins per data set would have led to poor statistics and hence not very useful data).

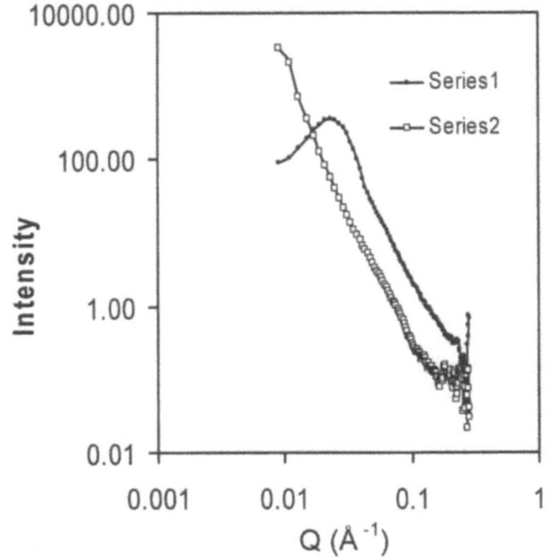

Figure 2. Plot of two series of SANS measurements on LOQ at ISIS for porous Vycor glass: Series 1: Vycor glass before reacts with supercritical water; Series 2: After reaction a few hours with supercritical water using the system shown in Figure 1.

At the same temperature, 350°C, we reduced the pressure to 500 bar in order to reduce the reaction speed. As result, we were able to follow the process in a much slow speed over 10 hours. Figure 3 shows the second series measurements as function of time. Again we collected the data every 15 mins (i.e. in Figure 3, the data set 1 is at 15 mins, the data set 2 is at 30 mins, the data set 3 is at 45 mins, *etc*). As one can see, during the ten hours, we were able to see the small changes of the porous structures step by step, which were never seen before.

The main changes of the scattering intensity in Figure 3 can be seen in two areas in the spectra: at the spinodal peak position Q = 0.025 Å^{-1}, the scattering intensity decreases with time while at the position Q = 0.01 Å^{-1} the intensity increases. After a few hours the main peak at 0.025 Å^{-1}, a typical feature for the Vycor glass, completely disappeared, being replaced by a straight line in the log-log plot with gradient of –3.0 for the entire measured Q range from Q = 0.003 to 0.05 Å^{-1}. This feature has often been seen in the fractal-like aggregated particles (silica aerogel for instance [16]). The gradient is the fractal dimension, a measurement of packing density, i.e. lower the dimension, and lower the packing density of the aggregates. The Q range for this experiment is limited due to the limitation of the instrument. Further investigation to extend lower Q region is much desirable in order to define the size distribution of the particles.

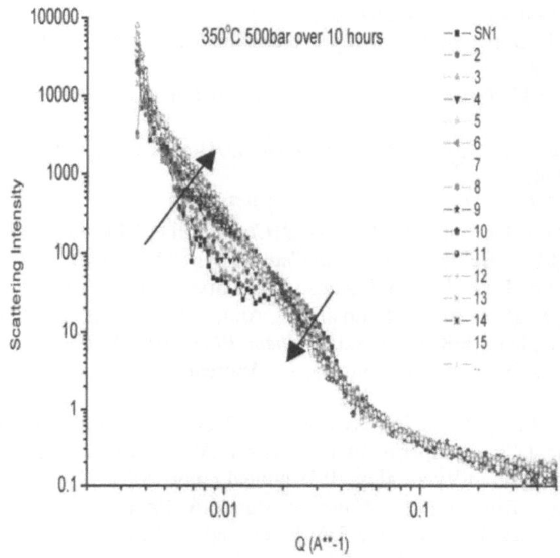

Figure 3. Plot of series of SANS spectra measured at temperature T = 350°C and pressure P = 500 bar on SAND at IPNS for period of 10 hours using the high pressure cell shown in Figure 1. The data set 1 is at 15 mins, the data set 2 is at 30 mins, the data set 3 is at 45 mins, etc. The main changes are in two areas: at the spinodal peak position, $Q = 0.025$ Å^{-1}, the intensity decreases with time while the intensity at $Q = 0.01$ Å^{-1} increases. After a few hours the main peak at 0.025 Å^{-1} completely disappeared.

4. SUMMARY

The SANS measurements revealed that water has slowly "melt" the Vycor glass near to the SC point (T = 350°C and P = 500 bar) and alters the porous structure of Vycor glass drastically. The SANS measurements give us better understanding of the reaction processes between SCW and the Vycor glass as function of time and pressure. The process has provided a valuable technology for manipulating nanostructures of a range of SiO_2 based materials. Controlling the reaction process can lead to manufacture glasses with defined pores sizes and volumes. It can also have further implications towards the synthesis of new nano-materials from solutions of silicates in supercritical water and use the supercritical water as a coating technique.

Acknowledgments: We would like to thank the Engineering and Physical Sciences Research Council (UK) for financial support (GR/M74689), ISIS at Rutherford Appleton Laboratory and IPNS at Argonne National Laboratory for the use of neutron beam time. The first author wishes to acknowledge the

Chinese National Science Foundation (Grant No: 10228511) for their financial support of the programme.

References

1. P. Savage, *Chem. Rev.*, 99 (1999) 603.
2. C.H.Uffindell, "Supercritical water and oxidation technology", UMIST, PhD thesis, 2000, Manchester, UK
3. P. Postorino, R.H. Tromp, M-A. Ricci, A.K. Soper and G.W. Neilson, *Nature,* 366 (1993) 668.
4. M-C. Bellissent-Funel, T. Tassaing, H. Zhao, D. Beysens and Y. Guissani, *J. Chem. Phys.*, 107 (1997) 2942.
5. A.K. Soper, *J. Phys.: Condens. Matter*, 9 (1997) 2717.
6. W.J. Lamb, G.A. Hoffman and J. Jonas, *J. Chem. Phys.*, 74 (1981) 6875.
7. T. Tassaing, M. C. Bellissent-Funel, *J. Chem. Phys.*, 113 (2000) 3332.
8. E. Franck and K. Roth, *Farday Discuss. Chem. Soc.,* 43 (1967) 108.
9. Y. Ikushima, K. Hatakeda, N. Saito and M. Arai, *J. Chem. Phys.*, 108 (1998) 5855.
10. D. M. Carey and G. M. Korenowski, *J. Chem. Phys.*, 108 (1998) 2669.
11. M.A. Ricci, M. Nardone, A. Fontana, C. Andreani and W. Hahn, *J. Chem. Phys.*, 108 (1998) 450.
12. I.A. Beta, J.C. Li and M.-C. Bellissent-Funel, *Chem. Phys. (2003) in press.*
13. I.A. Beta and J.C. Li, experimental report on eVS measurements of water in Sub- and Supercritical region in Vycor glass, ISIS annual report 2003.
14. J.C. Li and D.K. Ross, *J. Phys.: Condens. Matter,* 6 (1994) 351.
15. J.C. Li, D.K. Ross, L.D. Howe, J.P.A. Fairclough, K.L. Stefanopoulos, R.K. Heenan and K. Ibel, *Phys. Rev. B, 49* (1994) 5911.
16. D.W. Schaefer and D. Keefer, *Phys. Rev. Lett.,* 56 (1986) 2199.

FUNCTIONALIZATION OF CARBON NANOTUBES: SINGLE ATOM ADSORPTION

Oğuz Gülseren,[1,*] Sefa Dag,[1] Engin Durgun,[1] Taner Yildirim,[2] and, Salim Ciraci,[1]
*1) Department of Physics, Bilkent University, Ankara 06800, Turkey; 2) NIST Center for Neutron Research, Gaithersburg, MD 20899, USA, *) Email: gulseren@fen.bilkent.edu.tr*

Abstract: Unusual mechanical and electronic properties of carbon nanotubes promise a new generation of materials with potential applications in nanotechnology. The physical and chemical properties can also be efficiently engineered by the adsorption of atoms or molecules so that carbon nanotubes can be functionalized. Using state-of-the-art first principles total energy calculations, we show the single-wall carbon nanotubes (SWNTs) can be functionalized and the mechanical deformation can control the electronic and chemical properties of SWNTs. We studied the adsorption of single atom on a semiconducting and metallic single-wall carbon nanotubes for a large number of foreign atoms. The stable adsorption sites, binding energy and the resulting electronic properties are analyzed. The character of the bonding and associated physical properties exhibit dramatic variations depending on the type of the adsorbed atom. While the atoms of good conducting metals, such as Cu and Au, form very weak bonding, atoms such as Ti, Sc, Nb and Ta are adsorbed with relatively high binding energy. Most of the adsorbed transition metal atoms excluding Ni, Pd and Pt have a magnetic ground state with a significant magnetic moment. In summary, our results suggest that carbon nanotubes can be functionalized in different ways by their coverage with different atoms, showing interesting applications such as one-dimensional nanomagnets or nanoconductors and conducting connects etc.

Keywords: Carbon nanotubes, adsorbate structure, magnetic materials, Ab-initio calculations of adsorbates

1. INTRODUCTION

Single-wall carbon nanotubes [1-3] (SWNT) can be functionalized by adsorption of atoms or molecules, which can induce dramatic changes in the physical and chemical properties of the bare tube. Functionalization of

S. Guceri et al. (eds.), Nanoengineered Nanofibrous Materials, 165-173.

SWNTs has been attracting our interest for two possible, insofar technologically important applications; namely fabrication of metallic nanowires and nanomagnets. Single wall carbon nanotubes can serve as templates to produce reproducible, very thin metallic wires with controllable sizes. [4,5] These metallic nanowires can be used as conducting connects and hence are important in nanodevices based on molecular electronics. Recently, Zhang et al. [6,7] have shown that continuous Ti coating of varying thickness, and quasi continuous coating of Ni and Pd can be obtained by using electron beam evaporation techniques. Metal atoms such as Au, Al, Fe, Pb were able to form only isolated discrete particles or clusters instead of a continuous coating of SWNT. Low resistance contacts to metallic and semiconducting SWNTs were achieved by Ti and Ni ohmic contacts.[8] Most recently, *ab-initio* density functional calculations [9] have indicated that stable rings and tubes of Al atoms can form around a semiconducting SWNT. It is argued that either persistent currents through these conducting nanorings, or conversely very high magnetic fields can be induced at their center[9]. It is expected that novel molecular nanomagnets and electromagnetic devices can be generated from these metallic nanostructures formed by adatom adsorption on SWNTs. As an example, one can contemplate to generate a nanodevice by the modulating adsorption of adatoms on a bare (8,0) SWNT, which is a semiconductor [10] with an energy gap of ~0.64 eV. This band gap can increase to 2 eV by the adsorption of hydrogen atom [11]. Then, a quantum well (or dot) can form between two barriers at the hydrogen covered sections of the (8,0) tube. This structure is connected to the metallic reservoirs through metal coated ends of SWNT. This way a resonant tunneling device with metal reservoirs and connects at both ends can be fabricated on a single SWNT.

Clearly, the study of adsorption of atoms on nanotube surfaces is essential to achieve low resistance ohmic contacts to nanotubes, to produce nanowires with controllable size, and to fabricate functional nanodevices. This paper presents an extensive study of the adsorption of individual atoms on the surface of a semiconducting (8,0) and also a metallic (6,6) SWNT. The binding geometry and binding energy and resulting electronic structure of various (ranging from alkali and simple metals to group IV elements, and including most of the transition metal as well as hydrogen and oxygen) atoms have been investigated. We believe that the results of this work is important for further studies related with the functionalization and coating of carbon nanotubes.

2. METHOD OF THE CALCULATIONS

Binding geometry and binding energy, and resulting electronic structure of 26 different atoms (Na, Al, Cu, Au, Ni, Fe, Ti, W, Nb, Mo, Pd, Pb, C, Si, Cr, Co, Sc, V, Zn, Ag, Pt, Ta, Mn, S, H and O) adsorbed on a (8,0) zigzag SWNT and four different atoms (Au, Mn, Mo, Ti) adsorbed on a (6,6) armchair SWNT have been calculated by using the pseudopotential plane wave method

[12] within the generalized gradient approximation (GGA) [13]. Spin-unpolarized and spin-polarized (relaxed) calculations have been carried out for single atom, bare SWNT, and single atom absorbed SWNT in a periodically repeating tetragonal supercell with lattice constants, $a_{sc} = b_{sc} \sim 15$ Å and c_{sc}. To minimize the adsorbate-adsorbate interaction, the lattice constant along the axis of the tube, c_{sc}, is taken to be twice the 1D lattice parameter of the bare tube, $c_{sc} = 2c$. Ultra soft pseudopotentials [14] and plane waves up to an energy cutoff of 300 eV are used. The Brillouin zone of the supercell is sampled by (1,1,11) **k**-points within the Monkhorst-Pack special **k**-point scheme [15]. For the adsorption of individual atoms we considered four possible sites (*i.e.* H-site, above the hexagon; Z-, and A-sites above the zigzag and axial C-C bonds; and T-site above the carbon atom) as described in Fi;

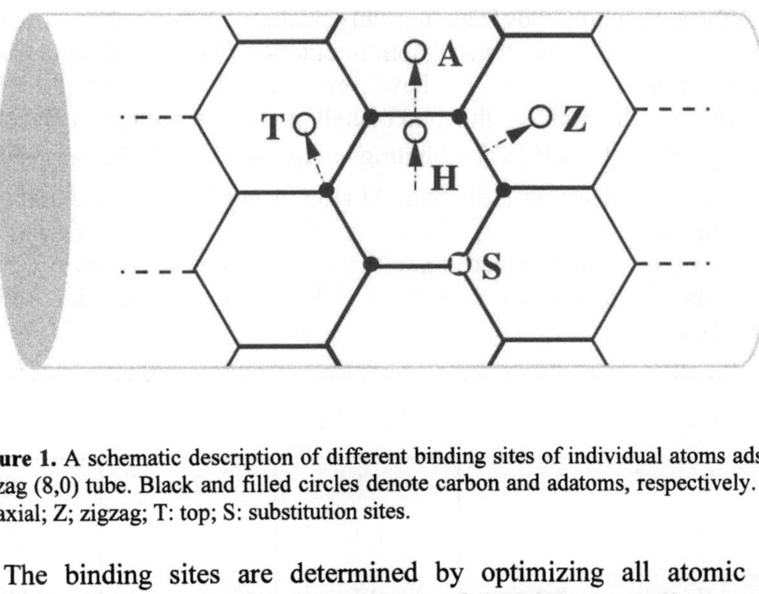

Figure 1. A schematic description of different binding sites of individual atoms adsorbed on a zigzag (8,0) tube. Black and filled circles denote carbon and adatoms, respectively. H: hollow; A: axial; Z; zigzag; T: top; S: substitution sites.

The binding sites are determined by optimizing all atomic positions (adsorbate atom and 64 carbon atoms of SWNT), as well as c using the conjugate gradient (CG) method. Binding energies are obtained from the expression,

$$E_b = E_T[SWNT] + E_T[A] - E_T[A+SWNT] \quad (1.1)$$

in terms of the total energies of the fully optimized bare nanotube ($E_T[SWNT]$), free atom A ($E_T[A]$), and the atom A adsorbed on a SWNT ($E_T[A+SWNT]$). The binding energies E_b are obtained from the lowest ground state total energies (either magnetic or non-magnetic) of both $E_T[A+SWNT]$ and $E_T[A]$; a bare nanotube has a non-magnetic ground state with zero net spin. $E_b > 0$ corresponds to a CG optimized stable structure and indicates the bonding (a local or global minimum on the Born-Oppenheimer surface).

3. BINDING GEOMETRY AND BINDING ENERGY

In Table I and II, we present our results. The atoms that were observed to form continuous and quasi continuous coating on the SWNT (Ti, Ni, and Pd) have relatively higher binding energies as compared to those atoms (Au, Fe, Pb) that form only discrete particles on the surface of the tube [4,5]. We also note that in forming a good coverage not only adatom-SWNT interaction, but also other factors, possibly adatom-adatom interaction play crucial role. Good conductors such as Au, Ag, and Cu have very weak binding. On the other hand, Na with *3s* electron on the outer shell is bound with a significant binding energy ($E_b = 1.3\ eV$). The binding energy of Zn with the $(4s)^2$ outer shell is almost zero. While an individual Al atom is not bound to the graphite surface, its binding on the (8,0) SWNT is relatively strong. This can be explained by the curvature effect, since the binding was found to be even stronger at the high curvature site of SWNT under uniaxial radial deformation[16].

Atom	Site	$\bar{d}_{C-A}$ (Å)	E_b^{sp} (eV)	E_b^{su} (eV)	μ (μ_B)
Ti	H	2.2	1.8	2.6	1.68
Mn	H	2.5	0.1	3.1	5.60
Mo	H	2.3	0.1	4.3	3.61
Au	T	2.3	0.3	0.4	0.79

Table 1. Calculated binding energies (E_b^{sp} spin-polarized and E_b^{su} spin-unpolarized) of individual atoms adsorbed on the armchair (6,6) SWNT, the most favorable binding site, the average carbon-adatom bond distance ($\bar{d}_{C-A}$), and the net magnetic moment (μ) of the adatom+SWNT system.

We note that specific adsorbate-SWNT (A+SWNT) systems are found to be in a magnetic ground state, hence $E_T^{sp}[A+SWNT] < E_T^{su}[A+SWNT]$. No matter what the value of the binding energy is, a stable binding of a particular A+SWNT geometry is meaningful if it belongs to a ground state. Hence, a positive difference between the spin-unpolarized and spin-polarized total energies,

$$\Delta E_T = E_T^{su}[A+SWNT] - E_T^{sp}[A+SWNT] > 0$$

indicates that magnetic ground state with a net spin is favored. We can extract following useful information from the results of calculations listed in these Tables. In general, the binding energies calculated for non-magnetic state are higher than those corresponding to the magnetic ground state. This is partly due to the reference of energies in Eq. 1.

Atom	Site	$\bar{d}_{C-A}$ (Å)	E_b^{sp} (eV)	E_b^{su} (eV)	μ (μ_B)
Na	H	2.3	-	1.3	-
Sc	H	2.2	1.9	2.1	0.64
Ti	H	2.2	2.2	2.9	2.21
V	H	2.2	1.4	3.2	3.67
Cr	H	2.3	0.4	3.66	5.17
Mn	H	2.4	0.4	3.4	5.49
Fe	H	2.3	0.8	3.1	2.27
Co	H	2.0	1.7	2.8	1.05
Ni	A	1.9	1.7	2.4	0.04
Cu	A	2.1	0.7	0.8	0.53
Zn	H	3.7	0.04	0.05	0
Nb	H	2.2	1.8	3.9	2.86
Mo	H	2.2	0.4	4.6	4
Pd	A	2.1	1.7	1.7	0
Ag	A	2.5	0.2	0.3	0.6
Ta	H	2.2	2.4	2.9	3.01
W	H-A	2.1	0.9	3.4	2.01
Pt	A	2.1	2.4	2.7	0
Au	A-T	2.2	0.5	0.6	1.02
Al	H	2.3	-	1.6	-

C	Z	1.5	-	4.2	-
Si	H	2.1	-	2.5	-
Pb	H	2.6	0.8	1.3	0
H	T	1.1	-	2.5	-
O	Z	1.5	-	5.1	-
S	A	1.9	-	2.8	-

Table 2. Calculated binding energies (E_b^{sp} spin-polarized and E_b^{su} spin-unpolarized) of individual atoms adsorbed on the zigzag (8,0) SWNT, the most favorable binding site, the average carbon-adatom bond distance ($\overline{d}_{C-A}$), and the net magnetic moment (μ) of the adatom+SWNT system.

Most of the transition metal atoms adsorbed on the (8,0) and (6,6) SWNT have magnetic ground state with $\Delta E_T > 0$, and hence they give rise to the net magnetic moment ranging from 5.49 μ_B (for Mn) to zero magnetic moment (for Pd and Pt). While Ni adsorbed SWNT has very low magnetic moment 0.04 μ_B, the adsorbates such as Au, Ag or Cu have magnetic moment in the range of 0.4-0.6 μ_B. Our spin-polarized and spin-unpolarized calculations show that these transition metal atoms in Table 2 have also magnetic ground state when they are free. Since a bare SWNT having a non-magnetic ground state, the net spin of the A+SWNT system originates from the magnetic moment of the adsorbed atom. On the other hand, atoms, such as Na, Al, C, Si, Pb, O, S, H, favor non-magnetic ground state when adsorbed on the (8,0) SWNT. The magnetic moment generated upon the adsorption of individual transition atoms has important implications, and points to an issue, whether molecular magnets or nanomagnets) can be produced from carbon nanotubes.

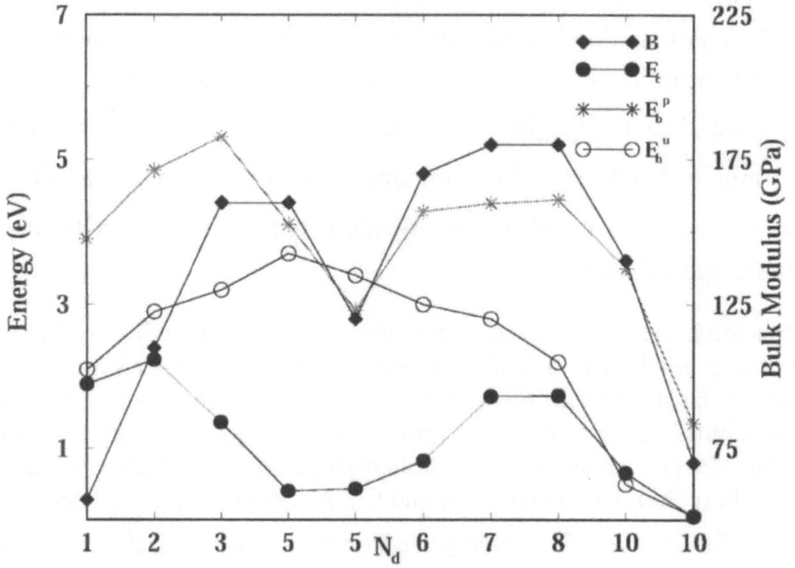

Figure 2. Variation of the calculated spin-unpolarized E_b^{su} and spin-polarized E_b^{sp} binding energy of transition metal atoms with respect to the number of d-electrons N_d. The bulk cohesive energy E_c and the bulk modulus B from Ref. [17] is included for the comparison of the trends.

Figure 2 presents the variation of the ground state properties [18,19] (such as the cohesive energy E_C and bulk modulus B), binding energies E_b^{su} and E_b^{sp} of the first row transition metal elements with respect to the number of *d*-electrons N_d. $E_c(N_d)$ and $B(N_d)$ curves show a minimum (at $N_d = 5$ for the $3d^5 4s^2$ configuration of Mn atom) between two maxima of equal strength; first maximum occurs at $N_d = 3$ or 4, the second one at $N_d = 7$. This behavior of bulk properties were explained by the Friedel model [20,21]. Although the overall shape of the variation of the binding energies of first row transition metal atoms with N_d, $E_b^{sp}(N_d)$, is reminiscent of the $B(N_d)$ and $E_c(N_d)$, there are some differences. The binding energy of Ti ($N_d = 2$) is highest, and hence the first maximum is higher than the second one $E_b^{sp}(N_d = 7)$ and $E_b^{sp}(N_d = 8)$ corresponding to the binding energies of Co and Ni. While the binding energy of Sc($N_d = 1$) is close to that of Ti at the first maximum, the binding energy of Cu($N_d = 10$) is small, and it eventually decreases to almost zero for Zn, which has a filled valence shells (*i.e* $3d^{10}4s^2$). Interestingly, Cr and Mn atoms which have the same $3d^5 4s^2$ configuration, have similar binding energies forming the minimum between double maximum, but different E_c and B. The variation of E_b^{su} and E_b^{sp} with N_d, however, shows dramatic differences. While $E_b^{sp}(N_d)$ mimics $E_c(N_d)$ and has a minimum for $N_d = 5$, as E_c and B do, E_b^{su} passes through a maximum. This situation confirms that magnetic states correspond to ground state.

The transition metal atoms with a few *d*-electrons, such as Sc, Co, Ti, Nb, Ta form strong bonds with a binding energy ranging from 2.4 eV to 1.8 eV, and hence can be suitable for metal coating of SWNT. These metals can also be used as a buffer layer to form uniform coating of good conductors such as Au, Ag, Cu. Most of the adatoms we studied yield strongest binding at the H-site. Ni, Pd, Pt (column VIII elements) and Cu, Ag, Au (column I-B elements) seem to prefer A-site. The average adsorbate-C distance, $\bar{d}_{C-A}$, ranges between 1.9 Å (minimum) and 3.7 Å (maximum); most of them occur at ~2.1 Å. On the other hand, the binding energies E_b^{su}, as well as E_b^{sp} and magnetic moments of the adatom adsorbed on the (8,0) SWNT came out to be consistently lower for the adatom adsorbed on the (6,6) tube. Perhaps, this trend can also be explained by the curvature effect[16,22].

4. DISCUSSIONS AND CONCLUSIONS

Our study shows that interesting physical properties can be generated by the adsorption of single atom on a SWNT. Changes in the physical properties depend on the valency of the adsorbate. Na having *3s*-valence state with small ionization potential donates this valence electron to the empty conduction band of semiconducting SWNT. Therefore, adsorption of Na results in the metallization of the (8,0) tube. Similar effects occur upon the adsorption of Al. However, adsorption of individual transition metal atoms gives rise to dramatically different results. Because of their occupied *d*-states, the transition metal adsorbed SWNT has a magnetic ground state. In most of the cases, adsorbed individual transition atom gives rise to a band gap between spin-up and spin-down bands within the supercell geometry. Ti, that leads to a metal, appears to be an exception. Another important property that is specific for Ti is its continuous coating of SWNT. This should be related to both Ti-Ti and Ti-SWNT interaction. Whether the Ti covered SWNT leads to regular atomic structure, and the nanowire produced from Ti covered SWNT allows ballistic quantum transport are important issues requiring further investigations. The adsorption of transition metal atoms on the (6,6) tube exhibits similar trends as the (8,0) tube, except that the binding energies in the former are consistently smaller. We attribute this behavior to the relatively smaller curvature of the (6,6) tube.

Higher coverage and decoration of adsorbed foreign atoms can produce nanostructures (such as nanomagnets, nanometer size magnetic domains, 1D conductors and thin metallic connects, and electronic devices) which may find interesting technological application, such as spintronics and high density data storage, and interconnects between devices. The *d*-orbitals of the transition metal atoms are responsible for relatively higher binding energies, which display an interesting variation with the number of filled *d*-states.

Acknowledgments: This work was partially supported by the National Science Foundation under Grant No. INT01-15021 and TUBITAK under Grant No. TBAG-U/13(101T010).

References

1. S. Iijima, Nature (London) **354**, 56 (1991).
2. S. Iijima, T. Ichihashi, and Y Ando, Nature (London) **356**, 776 (1992).
3. R. Saito, G. Dresselhaus, and M. S. Dresselhaus, *Physical Properties of Carbon Nanotubes*, Imprerial College Press, London (1998).
4. H. Dai, E.W. Wong, Y.Z. Lu, S. Fan, C.M. Lieber, Nature(London) **375**, 769 (1995).
5. W.Q. Han, S.S. Fan, Q.Q. Li, Y.D. Hu, Science **277**, 1287 (1997).
6. Y. Zhang and H. Dai, Appl. Phys. Lett. **77**, 3015 (2000).
7. Y. Zhang, N.W. Franklin, R.J. Chen, and H. Dai, Chem. Phys. Lett. **331**, 35 (2000).
8. C. Zhou, J. Kong, and H. Dai, Phys. Rev. Lett. **84**, 5604 (2000).
9. V.M.K. Bagci, O. Gülseren, T. Yildirim, Z. Gedik, and S. Ciraci, Phys. Rev. B **66**, 045409 (2002).
10. O. Gülseren, T. Yildirim, and S. Ciraci, Phys. Rev. B **65**, 155410 (2002); ibid **65**, 153405 (2002).
11. T. Yildirim, O. Gülseren, and S. Ciraci, Phys. Rev. B **64**, 075404 (2001); O. Gülseren, T. Yildirim, and S. Ciraci, Phys. Rev. B **66**, 121401 (2002).
12. M.C. Payne, M.P. Teter, D.C. Allen, T.A. Arias, and J. D. Joannopoulos, Rev. Mod. Phys. **64**, 1045 (1992).
13. J.P. Perdew, J.A. Chevary, S.H. Vosko, K.A. Jackson, M.R. Pederson, D.J. Singh and C. Fiolhois, Phys. Rev. B **46**, 6671 (1992).
14. D. Vanderbilt, Phys. Rev. B **41**, 7892 (1990).
15. H.J. Monkhorst and J.D. Pack, Phys. Rev. B **13**, 5188, (1976).
16. O. Gülseren, T. Yildirim and S.Ciraci, Phys. Rev. Lett. **87**, 116802, (2001).
17. C. Kittel, Introduction to Solid State Physics. (John Wiley and Sons, New York, 1996).
18. S. Dag, O. Gülseren, T. Yildirim, and S. Ciraci, Phys. Rev. B **67**, 165424 (2003)
19. E. Durgun, S. Dag, V.M.K. Bagci, O. Gülseren, T. Yildirim, S. Ciraci, Phys. Rev. B **67**, R201401 (2003).
20. P. Soderlind, O. Eriksson, J.M. Wills, and A.M. Boring, Phys. Rev. B **48**, 5844 (1993).
21. J. Friedel, The Physics of Metals, edited by J. M. Ziman (Cambridge University Press, New York, 1969); D.G. Pettifor, in Solid State Physics, edited by H. Ehrenreich and D.Turnbull (Academic, New York, 1987). Vol. **40**, p43.
22. X. Blase, L.X. Benedict, E.L. Shirley and S.G. Louie, Phys. Rev. Lett. **72**, 1878 (1994)

TOWARDS FIBER-BASED MICRO- AND NANOFLUIDICS

Karin KEIS, Konstantin G.KORNEV, Yash K.KAMATH, and Alexander V.NEIMARK
Center for Modeling and Characterization of Nanoporous Materials, TRI/Princeton, 601 Prospect Avenue, P.O.Box 625, Princeton, NJ, 08542.
aneimark@triprinceton.org
http://www.triprinceton.org/aneimark

Abstract: We describe a new method of controlled manipulation of micro-quantities of fluids by using a conduit formed by two parallel fibers. We show that a droplet of a wetting fluid, which is deposited on the fiber rails, can be either pinned at the point of deposition or driven to spread along the inter-fiber channel. The principle of controlled manipulation is that the droplet spreads out or comes back spontaneously due to the action of capillary forces, which critically depend upon the interfiber spacing. Compared to conventional microchannel design, the suggested fiber-based conduits significantly reduce the viscous drag. We present a quantitative analysis of the phenomenon of droplet spreading and the critical conditions for droplet pinning. Depending on the fiber diameter, the fiber-based conduits provide an opportunity of fluid manipulation in micro- and nanofluidic devices.

Keywords: Microfluidics, nanofluidics, capillarity, fibers, Lucas-Washburn law, Bosanquet law.

1. INTRODUCTION

In conventional microfluidic devices the fluid is pumped by applying pressure drop, temperature drop or voltage [1-6]. In this paper, we present a new principle of controlled fluid transport at micro- and nanoscale. As an alternative to the existing microfluidic platforms based on photo-lithographically fabricated channels, we suggest to exploit the fiber-formed conduits. Recent advances in manufacturing of nanofibers [7-9] make possible the fabrication of building blocks for fiber-based micro- and nanofluidics. The fiber-based conduits possess unique features that significantly ease the manipulation of fluids in a controllable manner. In particular, capillarity succeeds in governing the droplet self-propulsion without the need for an additional external means. The viscous drag, which is the major obstacle in

S. Guceri et al. (eds.), Nanoengineered Nanofibrous Materials, 175-182.

conventional microfluidics, can be significantly reduced due to a relatively small area of the fluid-solid contact surface.

This paper is aimed at the explanation of the physical principle of manipulation of fluids by fiber rails, a conduit composed of two parallel fibers. For these purposes, the details of the fluidic device are not important. For the sake of simplicity, we take aside the design of fluidic devices exploiting this principle and focus on the specifics of droplet spreading over two parallel fibers to demonstrate the advantages of this new technique.

2. SPREADING/PINNING OF DROPLETS ON FIBER RAILS

Droplets on fibers take on the equilibrium configurations, which cannot be realized on planar substrates. Remarkably, if a droplet of a wetting fluid could spread completely over a planar substrate, it would never spread on a fiber to form a sheath-like film. Typically, a spherical droplet deposited on a fiber will

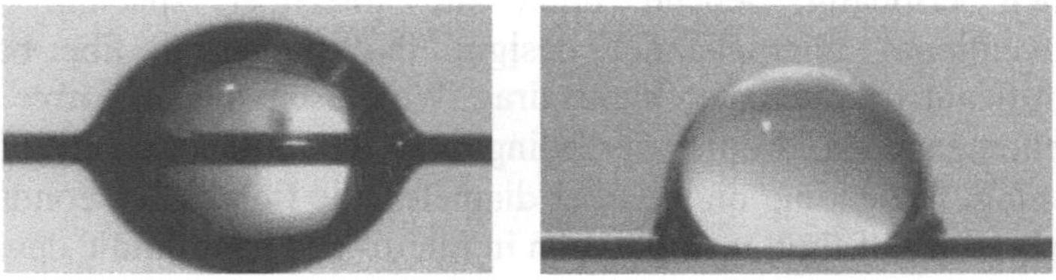

Figure 1. The left is the barrel (hexadecane) and the right is the clam-shell (water) configurations of droplets on a single wire.

equilibrate to a barrel (Fig.1) [10]. If the droplet size is smaller than the fiber diameter or the fluid is partially wetting, the droplet takes on a clam-shell configuration shown in Fig. 1. Droplets deposited on fiber rails attain similar configurations. However, these configurations can become unstable if the fibers are sufficiently close one to the other. For the given liquid-fiber system, there is a critical inter-fiber spacing, d_c, which controls the droplet stability [11-13]. If the inter-fiber spacing is narrower than d_c, the capillary forces facilitate wicking into the channel between the fibers. If the inter-fiber spacing is greater than d_c, the spreading is hindered. That is the droplet takes on a barrel-like configuration and does not come into the inter-fiber channel. Thus, the spreading/pinning transition is characterized by the critical inter-fiber spacing,

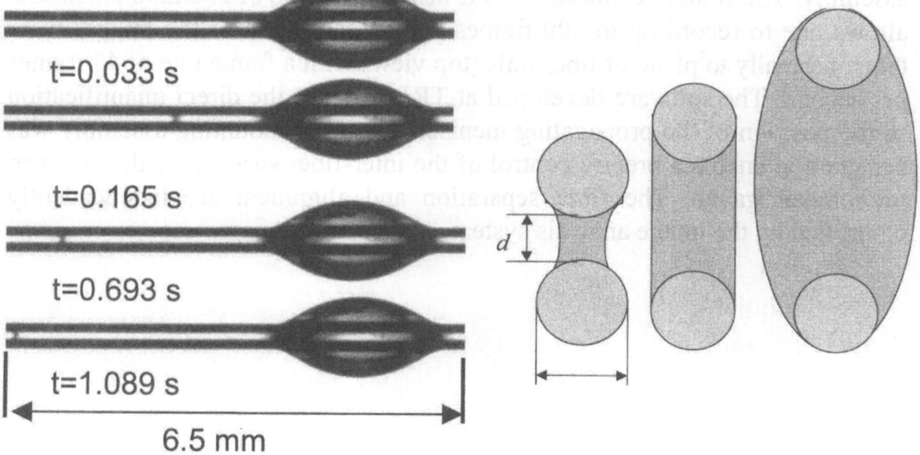

Figure 2. Droplet spreading over the fiber rails. Left: absorption of a hexadecane droplet into the inter-fiber channel formed by stainless steel fibers. The propagating meniscus is clearly seen. The spacing is d = 0.0955mm. Right: the cross-sectional shape of liquid columns confined between the fibers: a) $d < d_c$ wicking occurs at the negative curvature of the liquid-gas interface; b) $d = d_c$; the critical spreading condition corresponds to zero curvature of the liquid-gas interface; c) d_c, droplet forms a stable barrel-like configuration without wicking into the inter-fiber channel. For hexadecane on the rails made of stainless steel fiber, the critical fiber diameter-to-spacing ratio is d/D = 0.83, i.e., for fibers of diameter 0.1524 mm, the critical spacing is d_c = 0.1275mm

which determines the equilibrium droplet state. As shown in Fig.2, the shape and the cross-sectional area of the liquid column in the fiber channel depend on the inter-fiber spacing. This distance determines both the capillary pressure and the hydraulic resistance.

Thus, changing the inter-fiber spacing, we can manipulate with the droplet. Fluid can be forced to come into the fiber channel or it can be collected into a droplet. The principal point is that the droplet spreads or shrinks back spontaneously. Moreover, compared to the conventional microfluidic channels, the fiber channels reduce the viscous drag significantly. Indeed, at the surfaces of menisci the fluid slips freely, thus accelerating the fluid flow. The quantitative analysis of rate of droplet spreading and critical condition characterizing the criterion of droplet spreading/pinning are done below.

3. EXPERIMENTAL

3.1. Imaging System

The imaging system consists of a high-speed digital CCD camera (Dalsa CA-D1-0256A) with a video zoom lens (Navitar Zoom 6000), a power supply, an image acquisition board, a light source and a fiber mounting assembly. The camera connected to PC has a resolution of 256x256 pixels and allows one to record up to 200 frames per second. Most of the images were taken normally to plane of fiber rails (top view) with a frame rate of 45 frames per second. The software developed at TRI provides the direct quantification of the position of the propagating menisci. The fiber mounting assembly was designed to enable a precise control of the inter-fiber spacing by the attached micrometer gauge. The fiber separation and alignment are independently controlled by the image analysis system.

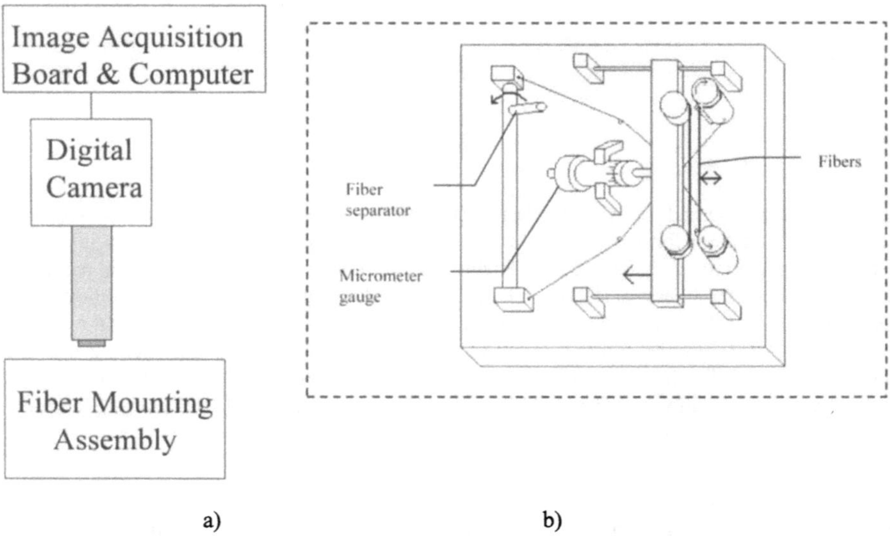

Figure 3. a) Experimental setup; b) Schematic diagram of the fiber mounting assembly.

3.2. Experimental Protocol

The experimental procedure was as follows. The stainless steel wires of D = 0.1524 mm diameter were taken to form the rails. Spreading experiments were performed with hexadecane (Aldrich). The fibers were cleaned with acetone and mounted providing a desired inter-fiber separation. In order to prevent the droplet from spreading immediately upon deposition, the fiber mounting assembly was equipped with a fiber separator (see Figure 3). The separator temporarily keeps the fibers apart beyond the critical spreading separation. The fiber separator is activated manually. In the normal position, the fiber separation is determined by a micrometer gauge. A droplet of approximately constant volume (1.4 μl) is placed on the fibers kept apart by the fiber separator. At this stage, a stable barrel-shaped droplet is formed and the necessary adjustments of the camera position and setting are done. The recording process is started simultaneously with the release of the handle thus bringing the fibers instantly to the desired distance. At this moment, the droplet begins to spread. At least 10 runs were conducted for each experimental condition and the average values were reported. All experiments were conducted at the room temperature.

4. RESULTS AND DISCUSSION

4.1. Spreading/Pinning Criterion

In Fig. 4 we show the complete wicking time as a function of the fiber separation for hexadecane droplets. The complete wicking time is the time interval counted from the start of droplet spreading (i.e. the moment of release of the handle) to the droplet disappearance. For closely spaced fibers the spontaneous wicking was found to be very slow. In particular, at the inter-fiber spacing of 0.0145 mm, the wicking time was about 4 minutes. However, the wicking time gradually decreased with the increase of the inter-fiber spacing. At the inter-fiber spacing of 0.105 mm, the wicking time was about 20 s. The fastest wicking occurred within ~10 s at the spacing of 0.121 mm. Beyond this distance, the wicking time increased with the increase of the inter-fiber spacing. At the inter-fiber spacing of 0.1238 mm (that is about d = 0.83 D), the liquid wicked out very slowly, but still completely depleting the droplet reservoir. Intermediate situation with a slight movement of liquid column (partial wicking) was observed instantly before reaching the critical d_{cr}. The critical inter-fiber spacing, at which hexadecane droplet remained stable without wicking, was found to be 0.1275 ± 3x10^{-3} mm. For $d>d_{cr}$, the menisci did not move and the droplet remained in its stable barrel-shaped configuration. Schematic pictures of the cross-sectional shapes of liquid columns between parallel fibers at different d are shown in Figure 2.

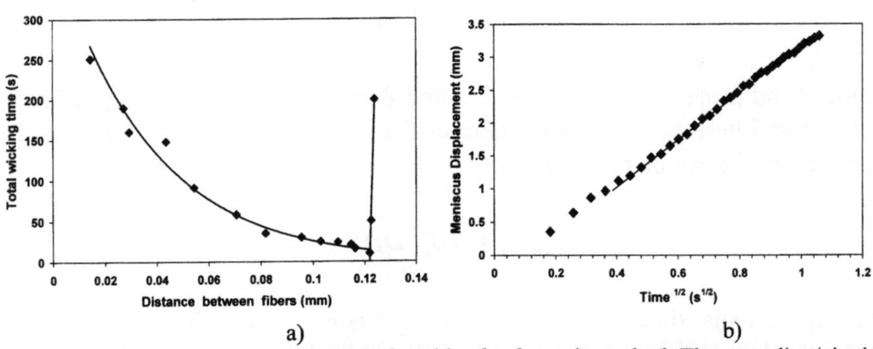

Figure 4. a) Critical spreading condition found by the dynamic method. The spreading/pinning criterion for hexadecane is $d_c \approx 0.1275$ mm. As d approaches the fiber diameter, D = 0.1524 mm, the wicking time increases steeply. b) Meniscus displacement versus square root of time. The inter-fiber spacing is 0.0955 mm.

4.2. Wicking Kinetics

The selected images in Figure 2 show the process of spreading of hexadecane droplet along the fiber rails. The velocity of the meniscus propagation is determined by the ratio of the driving force (capillary pressure) to the viscous drag. This velocity is generally presented by the Lucas-Washburn-type (LW) equation [14,15]:

$$\frac{dL}{dt} = \frac{k}{\eta L}(P_d - P_c) \qquad (1)$$

Here the L is the length of liquid column, k is the permeability of the inter-fiber channel, η is the viscosity of the liquid, P_d is the pressure in the droplet, P_c is the capillary pressure at the meniscus, and t is time. The permeability k is almost independent of the length of the liquid column. It is proportional to the cross sectional area of the liquid column. The capillary pressure at the meniscus is much larger than the pressure in the droplet. Thus, the right hand side of Eq.1 is proportional to $\sim 1/L$. Therefore, in the intermediate stage of wicking, a plot of the meniscus displacement (L) versus square root of time ($t^{1/2}$) should be linear. It is linear indeed, see Fig. 4b). Although the LW equation describes the intermediate and late stage of wicking fairy well, the initial velocity is finite in accord with the Bosanquet law [16-19]. According to Bosanquet, in the very beginning of spreading the wetting force, which is proportional to the surface tension times the fiber diameter, $F_w = 2\pi d\sigma\cos\theta$ are balanced by the inertial forces, proportional to the velocity squared times the column cross-section, $F_{in} \infty U^2 dD\rho$, where ρ is the fluid density. Thus, the meniscus velocity scales as

$$U_{Bosanquet} \propto (\sigma / d\rho)^{1/2}, \qquad (2)$$

where σ and ρ are the surface tension and density of the liquid, respectively. In the most interesting regime of near-critical inter-fiber spacing, since $d_c \sim D$, the estimate (2) can be rewritten as

$$U_{Bosanquet} \propto (\sigma / D\rho)^{1/2}. \qquad (3)$$

The experiments show that the droplet spreading along fibers follows the Bosanquet law at the very beginning of its run and the Lucas-Washburn laws at late times. To double check the applicability of the Bosanquet and Lucas-Washburn models, we tested the flow patterns by adding TiO_2 particles into hexadecane. As expected, the streamlines are monotonous lines starting from the droplet surface and ending at the menisci. At the transition zone near the droplet edges, the streamlines condense to form a stream tube in the inter-fiber channel. No vortices were observed.

5. CONCLUSIONS

As shown, the fiber rails can be used for controlled delivery or collection of micro-quantities of wetting fluids in the form of droplets. Determined by the inter-fiber spacing, the rate of fluid transport through the inter-fiber channels can be controlled by external fields or forces. Compared to flow through the micropipes and microchannels, the viscous drug is reduced due to a partial replacement of the supporting walls by the free surfaces of menisci. The viscous drag can be further decreased by addition of specific surfactants that form the lubricating layers at the liquid/solid interfaces. Although we demonstrated the principle of droplet manipulation by fiber rails on the submillimeter scale, the capillary mechanism should be effective at the micrometer and submicrometer scales as well. The fiber rails can be built of any fibers and scaled down by using, in particular, composite carbon nanotube-polymer fibers [7,20], carbon nanopipes [21], or electrospun polymer nanofibers [8,9]. The proposed principle of exploiting fiber-based conduits opens new perspectives in micro- and nanofluidic technologies.

Acknowledgement: This work was supported in parts by a group of TRI corporate members and the NIH grant R21 EB002889-01 "Nanofibrous Supports for Biomedical Sensors."

References

1. Beebe, D. J., Mensing, G. A. & Walker, G. M. Physics and applications of microfluidics in biology. *Annual Review of Biomedical Engineering* **4**, 261-286 (2002).
2. Lindner, D. Microsystems for chemical and biological applications. *Mrs Bulletin* **26**, 333-336 (2001).
3. Geng, X., Yuan, H., Oguz, H. N. & Prosperetti, A. Bubble-based micropump for electrically conducting liquids. *Journal of Micromechanics and Microengineering* **11**, 270-276 (2001).
4. Voldman, J., Gray, M. L. & Schmidt, M. A. Microfabrication in biology and medicine. *Annual Review of Biomedical Engineering* **1**, 401-425 (1999).
5. Jones, T. B. Liquid dielectrophoresis on the microscale. *Journal of Electrostatics* **51**, 290-299 (2001).
6. Darhuber, A. A., Valentino, J. P., Davis, J. M., Troian, S. M. & Wagner, S. Microfluidic actuation by modulation of surface stresses. *Applied Physics Letters* **82**, 657-659 (2003).
7. Vigolo, B. et al. Macroscopic fibers and ribbons of oriented carbon nanotubes. *Science* **290**, 1331-1334 (2000).
8. Fong, H. & Reneker, D. H. in *Structure formation in polymeric fibers* (ed. Salem, D.) 329-358 (Hanser, Munich, 2000).
9. MacDiarmid, A. G. et al. Electrostatically-generated nanofibers of electronic polymers. *Synthetic Metals* **119**, 27-30 (2001).

10. Carroll, B. J. Accurate Measurement of Contact-Angle, Phase Contact Areas, Drop Volume, and Laplace Excess Pressure in Drop-on-Fiber Systems. *Journal of Colloid and Interface Science* **57**, 488-495 (1976).

11. Dyba, R. V. & Miller, B. Evaluation of Wettability from Capillary Rise between Filaments. *Textile Research Journal* **39**, 962-& (1969).

12. Princen, H. M. Capillary Phenomena in Assemblies of Parallel Cylinders .3. Liquid Columns between Horizontal Parallel Cylinders. *Journal of Colloid and Interface Science* **34**, 171-& (1970).

13. de Lazzer, A. & Langbein, D. Liquid surface in regular N-pods. *Journal of Fluid Mechanics* **358**, 203-221 (1998).

14. Lucas, R. Ueber das Zeitgesetz des kapillaren Aufstiegs von Flussigkeiten. *Kolloid Zeitschrift* **23**, 15-22 (1918).

15. Washburn, E. W. The dynamics of capillary flow. *Physical Review* **17**, 273-283 (1921).

16. Bosanquet, C. H. On the flow of liquid into capillary tubes. *Philosophical Magazine* **45, Series 6**, 525-531 (1923).

17. Szekely, J., Neumann, A. W. & Chuang, Y. K. The rate of capillary penetration and the applicability of the Washburn equation. *Journal of Colloid & Interface Science* **35**, 273-278 (1971).

18. Quere, D. Inertial capillarity. *Europhysics Letters* **39**, 533-538 (1997).

19. Kornev, K. G. & Neimark, A. V. Spontaneous penetration of liquids into capillaries and porous membranes revisited. *Journal of Colloid & Interface Science* **235**, 101-113 (2001).

20. Neimark, A. V. et al. Hierarchical Pore Structure and Wetting Properties of Single Wall Carbon Nanotube Fibers. *Nano Letters* **3**, 419-423 (2003).

21. Gogotsi, Y., Libera, J. A. & Yoshimura, M. Hydrothermal synthesis of multiwall carbon nanotubes. *Journal of Materials Research* **15**, 2591-2594 (2000).

THEORETICAL MODELS FOR NANODEVICES & NANOMAGNETS BASED ON CARBON NANOTUBES

S. CIRACI, O. GÜLSEREN, S. DAG and E. DURGUN
Department of Physics, Bilkent University, Ankara 06800, Turkey
T. YILDIRIM
NIST Center for Neutron Research, National Institute of Standards and Technology, Gaithersburg, Maryland 20899

Abstract: Carbon nanotubes, in which two-dimensional hexagonal lattice of graphene is transformed to a quasi one-dimensional lattice conserving the local bond arrangement, provide several structural parameters to engineer novel structures with desired properties allowing ultimate miniaturization. Recent interest in nanoscience and nanotechnology has driven a tremendous research activity in carbon nanotubes. Most of the effort has gone to modify various physical properties of nanotubes and to functionalize them by external agents, such as deformation, chemisorption, doping, coating etc. In this paper, we exploit various modified properties of functionalized tubes to develop theoretical models for nanodevice and nanomagnets. Our study is based on the first-principles pseudopotential plane wave calculations within the density functional theory.

1. INTRODUCTION

Nanoelectronics based on carbon nanotubes have been considered as a new frontier aiming at the ultimate miniaturization of electronic circuits with ultra high density components and new functionalities. Researchers, who can touch and relocate atoms, have been challenged to discover the novel properties of these strange materials in order to transform them into new devices or other technological applications. Several devices fabricated so far with different functionalities have kept the interest in the field alive.

Carbon nanotubes [1] are unique materials which offer a variety of structural parameters to engineer their physical and chemical properties. [2,3] They can be synthesized as single wall (SWNT) or multiple wall (MWNT)

S. Guceri et al. (eds.), Nanoengineered Nanofibrous Materials, 183-197.

nanotubes; they can form ropes or even crystals. Even an ultimate one-dimensional (1D) carbon chain at the center of a MWNT (and stabilized by the innermost SWNT) has been discovered in the cathode deposits. [4] SWNTs are basically rolled graphite sheets, which are characterized by two integers *(n,m)*. The rolling (or chiral vector) $\mathbf{C} = n\mathbf{a}_1 + m\mathbf{a}_2$, is defined in terms of the two-dimensional (2D) hexagonal Bravais lattice vectors of graphene, $\mathbf{a}_1$ and $\mathbf{a}_2$. Then the radius of the tube is given in terms of *(n,m)* by the relation, $R = a_0(n^2 + m^2 + nm)^{1/2}/2\pi$, where $a_0 = |\mathbf{a}_1| = |\mathbf{a}_2|$. SWNTs exhibit different electronic structures depending on *n* and *m* (*i.e.* on their chirality and radius). In the last decade, extensive research has been carried out on SWNTs aiming at the modification of electronic structure for desired device operations.

The mechanical properties of carbon nanotubes are striking. [5] They are flexible and can sustain large elastic deformations radially, at the same time are very strong with high yield strength.[6,7] Their strength far exceeds that of any other fibre. Even more striking is the response of electronic structure to the radial deformation leading dramatic changes.[8] Moreover, it has been shown experimentally, as well as theoretically that the physical properties of SWNTs can be dramatically changed by the adsorption foreign atoms, molecules or by generating defects. [9,10]

Here we have taken a theoretical approach and developed various models for nanodevices and nanomagnets using first-principles calculations within the density functional theory. The predictive power of the method in previous studies holds the promise that the present theoretical models can lead to novel nanodevices despite technical difficulties.

2. METHOD

The first principles total energy and electronic structure calculations have been performed using the pseudopotential plane wave method and the generalized gradient approximation within the density functional theory. We carried out both spin-unpolarized and spin-polarized calculations using a periodically repeating tetragonal supercell with lattice constants, a_{sc}, b_{sc} and c_{sc}. The lattice constants, a_{sc} and b_{sc}, are chosen such that the interaction between nearest neighbor tubes is negligible. We used ultra soft pseudopotentials and plane waves up to an energy cutoff of 400 eV. For all systems we studied all atomic positions of adsorbate and SWNT, as well as *c* are fully optimized by using the conjugate gradient method. Further details for the method of calculations can be obtained from Ref [11].

3. ATOMIC AND ELECTRONIC STRUCTURE

The electronic band structure of single wall carbon nanotubes are deduced from graphene by mapping its band structure corresponding to the 2D hexagonal lattice on a cylinder. [3,12,13,14,15,16] In this respect, SWNT presents an interesting example, in which dimensionality is reduced from two to one. The analysis based on the band folding indicates that the *(n,n)* armchair nanotubes are always metal with π^*-conduction, and π-valence bands crossing at the Fermi level, and exhibit 1D quantum conduction. [16,17] The *(n,0)* zigzag SWNTs are generally semiconductor and only are metal if *n* is an integer multiple of three. Although the overall electronic structure of SWNTs has been described by this simple picture, recent studies [11,18] have shown much more complicated structural dependence. For example as predicted by first-principle calculations [11] and confirmed by experimental studies [18], (9,0) tube is, in fact, a small band gap semiconductor.

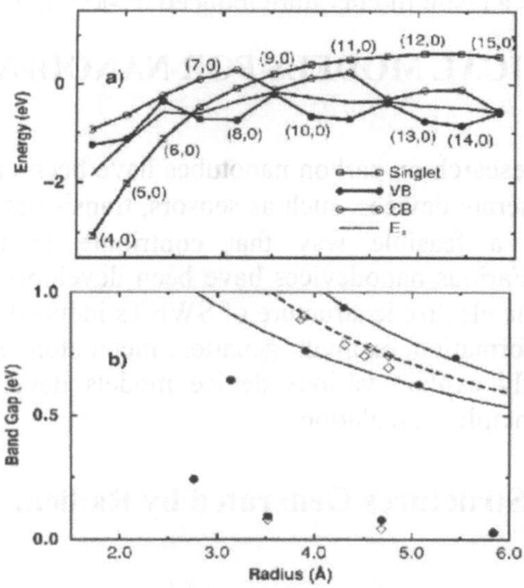

Figure 1. (a) Energies of the double degenerate π-states (VB), the double degenerate π^*-states (CB) and the singlet π^*-state as a function of nanotube radius. Each data point corresponds to *n* ranging from 4 to 15 consecutively. (b) The calculated band gaps as a function of the tube radius shown by filled symbols. Dashed lines is the plots of the fit $E_g = \gamma_0 d_0/R$. The experimental data are shown by open diamonds.[18,19,20] (Repreduced from Ref[11])

First-principles calculations [11] result in small but non-zero energy band gaps of 93, 78 and 28 meV for (9,0), (12,0) and (15,0) SWNTs, respectively. Recently, these gaps are measured by Scanning Tunneling Spectroscopy (STS) experiments [18] as 80, 42 and 29 meV, in the same order.

In Fig. 1, the double degenerate π-states (which are the valence band edge at the Γ-point), the double degenerate π*-states (which become the conduction band edge at Γ for large *R*), and the singlet π-state (which is in the conduction band for large *R*) are shown. As seen, the shift of the singlet π*-state is curvature dependent and below a certain radius it determines the band gap. For tubes with radius greater than 3.3 A° (*i.e.* n > 8), the energy of the singlet π*-state at the Γ-point of the BZ is above the doubly degenerate π*-states (*i.e.* bottom of the conduction band), while it falls between the valence and conduction band edges for n=7,8, and eventually dips even below the double degenerate, π-states of the valence band for the zigzag SWNT with radius less than 2.7 A° (*i.e.* n < 7). Therefore, all the zigzag tubes with radius less than 2.7 A° are metallic. For n=7,8, the edge of the conduction band is made by the singlet π*-state, but not by the double degenerate π*-state. The band gap derived from the zone folding scheme is reduced by the shift of this singlet π*-state as a result of curvature induced σ*-π* mixing [21].

4. THEORETICAL MODELS FOR NANODEVICES

Much of the research on carbon nanotubes have been carried out with the motivation to generate devices, such as sensors, transistors, nanomagnets etc, and hence find a feasible way that contribute to the objectives of miniaturization. Various nanodevices have been developed by exploiting the modification of the electronic structure of SWNTs induced by various effects, such as radial deformation, exohydrogenation, metal atom adsorption. In what follows we briefly explain various device models developed theoretically based on first-principles calculations.

4.1 Quantum Structures Generated by Radical Deformation

It has been shown theoretically that the band gap of semiconducting (zigzag) SWNT's decreases, and eventually is closed under increasing radial deformation. This property has been proposed to generate quantum structures. [8,22,23] The idea is simple and originates form the numerous work done on semiconductor heterostructures A_nB_m or quantum dots. It is known that owing to the band off-sets of the semiconductor heterostructures, the energies of the band states of one semiconductor *B* may fall into the band gap of the adjacent semiconductor *A*. According to the Effective Mass Approximation, the height (depth) of the conduction (valence) band edge of *A* from that of *B*, ΔE_c (ΔE_V) behaves as a potential barrier for electrons (holes). For example, *m* layers of *B* between *n* layers of two *A*'s form a quantum well yielding confined electronic states. The depth of the well and the width of the barrier and well (in terms of number of layers *n* and *m*, respectively) are crucial parameters to monitor the resulting electronic properties. By repeated formation of (A_nB_m) one generates (multiple quantum well structure) MQWSs or (resonant tunneling double barrier) RTDB by the combination $(A_nB_mA_n)$. Similar what one achieves in semiconductor heterostructures, periodic MWQSs or finite RTDB structure can be realized on an individual s-SWNT, and their electronic properties can be variably and reversibly monitored. Performing empirical tight binding and also first-principle calculations, Kiliç *et al.* [8] have shown that MQWS is generated by applying different radial deformations at adjacent parts of a (7,0) semiconducting SWNT, and repeating them periodically. They considered a supercell consisting of 16 unitcell of the (7,0) SWNT, the *n* unitcell of it were kept undeformed, and the remaining $m=16-n$ were radially deformed. Tight-binding charge densities of the band edge states for superlattices corresponding to n = 4,8,12, have clearly demonstrated the confinement of states and quantum structure formation. Tombler *et al.* [22] have shown that a quantum dot can be produced by bending (hence inducing radial deformation) SWNT at a given point. Metallization of a semiconducting (7,0) SWNT upon radial deformation and formation of MQWS is shown in Fig 2.

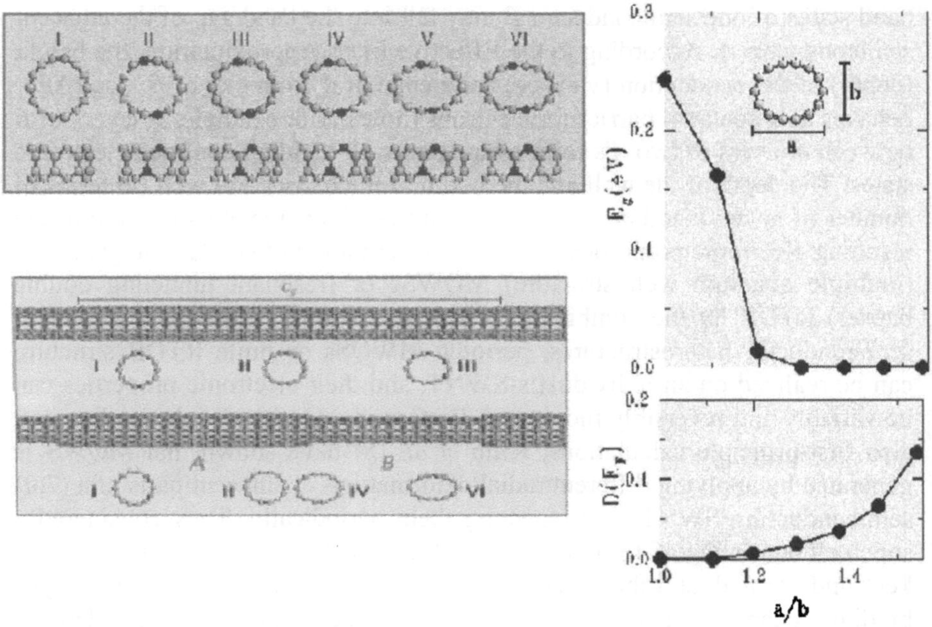

Figure 2. Top panel: Cross section of SWNTs under various degree of radial strain. Bottom panel: Schematic description of MQWs generated on a (7,0) SWNT. Side panels: Variation of band gap and density of states at Fermi level, as a function of radial deformation expressed as a/b.

4.2 Formation of MQWs by Modulating Hydrogenation

Earlier it has been shown that the electronic structure of SWNTs is modified upon exohydrogenation, which is resulted with the opening of a wide band gap. [24] Using this effect, a model of electronic device based on a similar band gap modulation and hence MQWS formation as discussed above has been proposed recently. [25] In this case the band gap modulation is provided by the modulating hydrogen adsorption as described schematically in Fig 3.

It has been shown that the zone of the SWNT which is kept clean and the zone which is hydrogenated, are expected to have different band structure.[24] This is the most crucial aspect that we have to explore. Normally, in the band continua, states in $(C_{32})_l$ (*i.e.* specified by *l*) zone are extended to the adjacent $(C_{32}H_{32})_q$ (*i.e.* specified by *q*) zone. These are propagating states and have comparable integrated probability (in both adjacent zones. Defining $\rho_l = \int\limits_{l} |\Psi_{n,k}(\mathbf{r})|^2 d\mathbf{r}$ and $\rho_q = \int\limits_{q} |\Psi_{n,k}(\mathbf{r})|^2 d\mathbf{r}$, $\rho_l \sim \rho_q$ for propagating states. On the other hand, if the energy of a state in one (say in the hydrogen-free l-) zone of the superlattice coincides with the band gap of the adjacent hydrogenated q- zone, this particular state cannot propagate, and has to be confined to

(a)

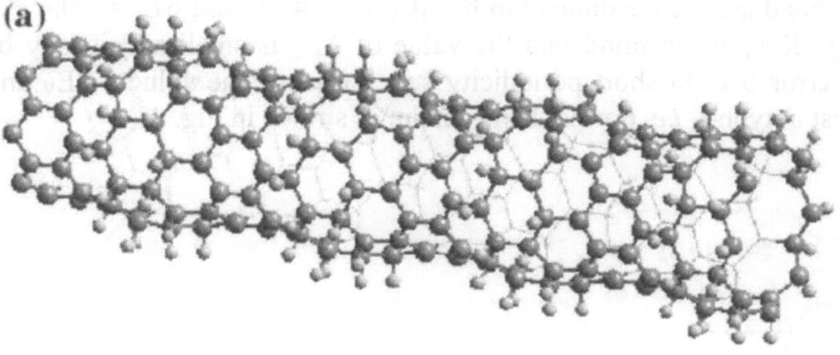

(b)

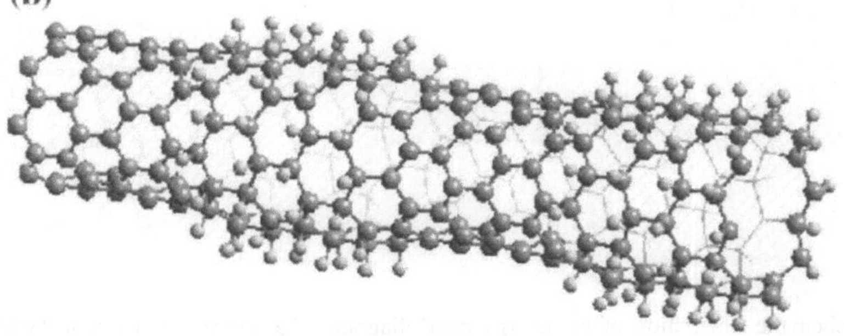

Figure 3. Superlattice structures, $[(C_{32})_l(C_{32}H_{32})_q]$ formed on a single (8; 0) SWNT by modulating adsorption of hydrogen atoms. (a) $l = q = 1$; (b) $l = q = 2$. The positions of all H and C atoms, as well as superlattice parameter c_s are calculated from full structure optimization by using conjugate gradient method. (Reproduced Ref. [25])

the zone it belongs to. Accordingly, $\rho_l \gg \rho_q$, are very much different. As a result, electrons or holes can be confined in a hydrogen free region of SWNT sandwiched between two regions which are uniformly hydrogenated.

We can now estimate band off-sets by taking a direct and practical approach and deduce the band offsets and resulting energy band diagram from the calculated electronic structure of the superlattice l = q = 2. [25] We note that the band off-sets are incorporated in the calculated state densities. The energy band diagram in direct space is obtained by analyzing the characters and the energy positions of the band structure in *l* and *q* zones and by examining the gaps of the local density of states calculated from the first principles. The band gaps at the hydrogen-free and exohydrogenated regions of the superlattice are slightly different from the values corresponding to $l = \infty$ and $q = \infty$, and are deduced from the local density of states to be $E_{g,l=2} \sim 0.5$ eV and $E_{g,q=2} \sim 2.3$ eV, respectively. The band-offsets at the valence and conduction band are then estimated to be $\Delta E_V =\sim 1.4$ eV and $\Delta E_C =\sim 0.4$ eV, respectively. Keeping in mind that the value of ΔE_C is small and it may be within the error bars in short periodicity superlattices, The values, ΔE_V and ΔE_C, suggest a type-I, *i.e.* normal band-lineup as shown in Fig. 4.

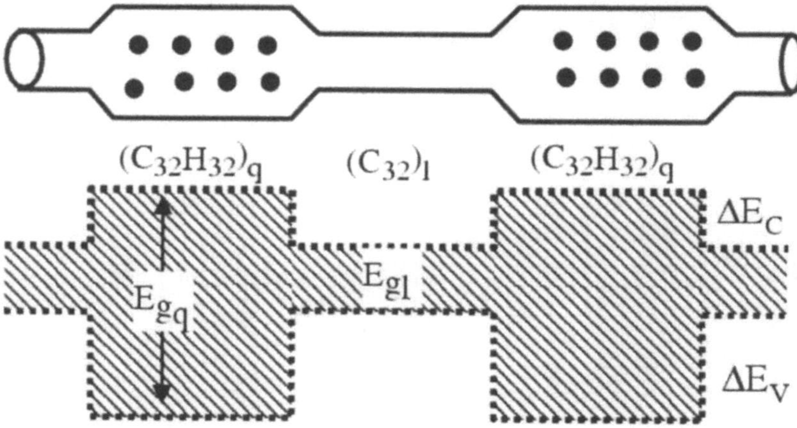

Figure 4. Schematic illustration of the energy band diagram of a resonant tunnelling double barrier structure generated by a modulating adsorption of H atoms on a s-SWNT superlattice indicating a type-I, normal band-lineup. (Reproduced from Ref. [25])

The present discussion for the periodically repeating quantum wells can be extended to finite systems, for example to the resonant tunnelling double barriers, made by a zone $(C_{32})_l$ placed between two zones of $(C_{32}H_{32})_q$, *i.e.* $(C_{32}H_{32})_q(C_{32})_l(C_{32}H_{32})_q$. Metallic reservoirs adjacent to both barriers, *i.e.* $(C_{32}H_{32})_q$, and metallic connects of the device can be achieved by metal coating [26,27,28]. Also a metal-semiconductor heterostructure can be formed by modulating hydrogenation of a *(n,n)* armchair SWNT. This way metallic connects are provided by the hydrogen free ends of the tube.

4.3 SWNT-Metal Electrode Contact

The switching of the current in a semiconducting single wall carbon nanotube s-SWNT) at room temperature by an external electric field has been utilized to fabricate a new field effect transistor (FET). [29, 30, 31] The interaction between s-SWNT and metal electrode, and the Schottky barrier, Φ_B formed thereof have been proposed as the origin of FET operation. Model calculations have been carried out to provide further understanding of experimental I-V characteristics. [32, 33] The SWNT-electrode interaction and resulting electronic structure are crucial for the electron transport and hence for all device properties, but they are not thoroughly investigated theoretically. The origin of the potential barrier formed between a semiconducting SWNT and a specific metal electrode has been investigated recently by using first-principles calculations. [10, 34]

The interaction and binding energy between individual Au atom and (8,0) tube is essential for the contact. The contact on the gold electrode is formed by placing the s-SWNT on the Au(100) surface, so that a surface Au atom faces the center of the

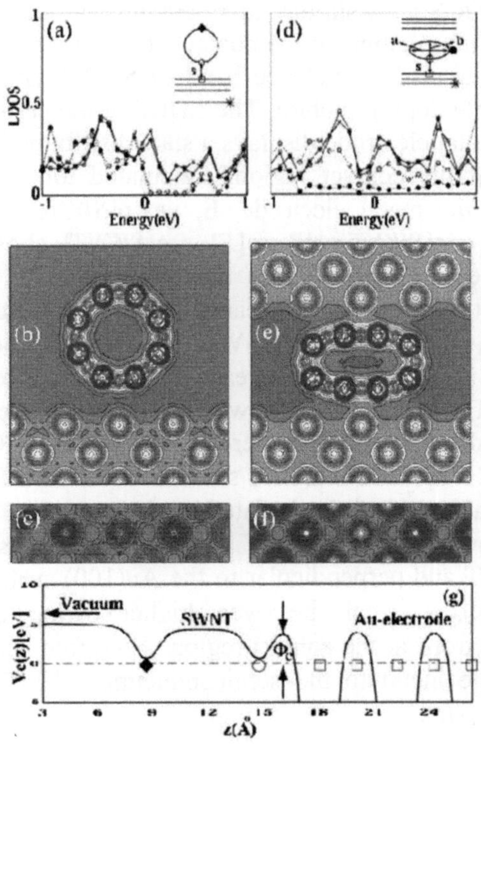

(a) Local densities of states (LDOS) at different Au and C atoms of the (8,0) zigzag SWNT side bonded to the Au(100) surface. The location of atoms is described by inset. The zero of energy is taken at the Fermi energy E_F. State densities shown by empty circles and filled diamonds correspond to 6 neighboring carbon atoms. Other densities are for single atoms. (b) Contour plots of the SCF electronic potential, $V_e(\mathbf{r})$, on a vertical plane. (c) Same as in (b) on a horizontal plane bisecting **s**. In the dark gray regions $V_e(\mathbf{r}) > E_F$. (d), (e) and (f) correspond to the radially deformed (8,0) SWNT pressed between two 3-layer Au(100) slabs with b/a=0.47 as shown by inset. State densities shown by empty circles in (d) correspond to 6 neighboring carbon atoms. Other densities are for single atoms. (g) Variation of $V_e(\mathbf{z})$ on a perpendicular line passing through the center of SWNT. (Repreduced from Ref [34])

hexagon. The Au-SWNT equilibrium distance, **s** occurred at 3.18 A° as a result of full structure optimization. This implies a weak Au-SWNT interaction which is characterized by physisorption. The local densities of states (LDOS) calculated for two Au atoms at different sites (one is at the contact just below the SWNT hexagon; the second being farthest to SWNT mimics the clean Au surface) in Fig. 5a have finite state density at the Fermi level. These two LDOS's are similar; there are only minute changes. Whereas the LDOS of the carbon atoms at the contact still has a band gap. The Fermi level lies near the top of the valence band of SWNT and complies with a small Φ_B in the hole doping picture. The LDOS at the carbon atom which is farthest from the metal electrode displays a state distribution similar to that of the carbon atom at the contact region. Calculated total charge density of SWNT bonded to the metal electrode, E, i.e. $\rho[SWNT+E]$, and difference charge density, $\Delta\rho = \rho[SWNT + E] - \rho[E] - \rho[SWNT]$, shows minute charge rearrangement. These results indicate that the weak Au electrode-SWNT interaction does not induce any significant changes in the electronic structure. The SCF electronic potential between SWNT and Au electrode, $V_e(\mathrm{r}) = V_{ion}(\mathrm{r}) + V_H(\mathrm{r}) + V_X(\mathrm{r})$, is presented on a vertical plane and also on a horizontal plane bisecting **s**. The shaded area shows that the electronic potential energy at the contact yields a potential barrier $\Phi_C = V_e - E_F > 0$. At the contact midway between SWNT and Au(100) surface Φ_C is calculated to be ~ 3.9 eV, that is comparable with the calculated work function (Φ ~ 5 eV) of Au-slab. Fig. 5(g) shows the variation of $V_e(z)$ on a line passing through the center of the SWNT and perpendicular to the Au(100) surface. The effective potential barrier $\Phi_{c,eff}$, can be even higher owing to the increased confinement of electrons at the contact region. Therefore, electrons which are transferred from the metal to the semiconducting SWNT has to tunnel a potential barrier $\Phi_{c,eff}(\mathrm{r})$.

Under radial deformation, the spacing **s** has decreased to 2.6 A°, and eventually the barrier has collapsed (i.e. $V_e < E_F$) at specific sites at the contact (see Fig. 5(e) and (f)). Under these circumstances, the electron from the SWNT can be ballistically transferred to the gold electrodes.

4.4 Metal Nano-Rings

Metal nanorings are of particular interest, because they may lead to the realization of nanosolenoids on SWNTs. The small radius of a metallic nanoring around the carbon nanotube may lead to interesting electromagnetic properties. The magnetic field B at the center of the ring can be expressed in terms of the quantized angular momentum L_z of the electrons in the direction parallel to the tube axis

$$B = \frac{\mu_0 e L_z}{4\pi m r^3}$$

where r is the radius of the nanoring. Taking the lowest possible value for L_z and r = 5.9 A° we estimate B to be at the order of 100 Gauss.[26] The current in the metal ring that can induce such a high magnetic field is comparable to the current attained in the suspended, monoatomic gold chains. Relatively higher magnetic fields at the order of Tesla can be induced by higher current passing through a thick Ti based metal coating around the SWNT, or by increasing the number of turns and hence by forming a nanocoil. Miyamoto *et al.*[35] have examined the chiral conductivity in bare BC_2N nanotubes. They estimated that magnetic field of a few tenths of Tesla can be induced at the center of the tube by assuming relaxation time of carriers ~50 times larger than that in Cu and homogeneous chiral current density confined to the tubule wall.

Metal nanorings winding the SWNT are also important, because they may allow the generation of persistent current at the nanoscale. Persistent currents in the nanoring can start by sudden application of an external magnetic field. In this way it is possible to use the nanotube, with a ring at its end, as a local magnetic probe at nanoscale. The persistent currents induced around a SWNT have been proposed as possible qubit in quantum computation. A superconducting ring may also be used for Schrödinger's cat experiments where one deals with superposition of macroscopic quantum states. [36] The two supercurrent quantum states (clockwise and counterclockwise flow) sit in two separate quantum wells. It has been observed that a weak microwave, which does not break Cooper pairs, can cause quantum tunnelling between these two macroscopic states. In this kind of experiments the main problem is to isolate the superconducting quantum interference device (SQUID) from the outside (nonquantum) environment and that is why isolated carbon nanotubes can be very useful.

Motivated with these exciting future applications, Bagci *et al.* [26] studied the formation of Al-nanoring around a (8,0) SWNT. Since aluminium atoms have strong Al-Al interaction, but not so strong Al-SWNT interaction, they tend to form clusters or nanoparticles (*i.e.* 3D islands) on the surface of SWNT. However, Bagci *et al.* showed that at special configuration stable nanoring can be formed. Al-nanoring coverage is obtained by placing Al atoms on top of carbon atoms (T-site), forming a zigzag ring. This structure includes 64 C and 16 Al atoms in the double unit cell. In this initial configuration the Al-Al distance is 2.33 A° and the angle of Al-Al-Al bond is ~137^0. After structure optimization, the Al-Al bond length is increased to 2.56 A°, and the Al-Al-Al bond angle is decreased to 124°, yielding the radius of the nanoring to be 5.9 A°. A side view of the optimized structure of the Al-nanoring wrapping the (8,0) SWNT is illustrated in Fig. 6. The binding energy of the Al nanoring is calculated to be 0.85 eV. The stability of the Al nanoring around the nanotube can be understood from the stable structures of planar Al monoatomic chains. Recent studies on the low dimensional structures of metals have revealed several stable atomic structures in one dimension (1D). [37, 38, 40] The firstprinciples calculations predicted linear chain, planar zigzag, triangular, ladder and non-planar dumbbell and pentagonal structures as stable structures for Al wires. [38, 39, 40]. More interestingly, it was found that by going from bulk to a chain structure the character of bonding in Al wires changes and acquires directionality.[40] Among a number of these 1D stable structures of Al predicted by first-principles calculations [40] was the planar zigzag monoatomic chain of Al with a bond angle 139° and bond distance 2.53 A°. This zigzag structure is only a local minimum on the Born-Oppenheimer surface, and hence its binding energy (1.92 eV) is intermediate between the binding energies of bulk and linear structures. An energy barrier of ~0.1 eV prevents the transition from the planar zigzag structure to other relatively more stable 1D structures.[40] The final optimized structure of the zigzag Al nanoring around the (8,0) nanotube has structural parameters similar to this planar zigzag structure, except that it is rolled on a cylinder. The SWNT initially serves as a template in the formation of the ring structure and also increases the stability of the ring by preventing the transitions to other relatively more stable structures. Therefore, the Al nanoring around the SWNT is expected to be stable at room temperature. Interestingly, the nanoring is also stable by itself, since the position of Al atoms do not change significantly upon discarding the underlying carbon nanotube.

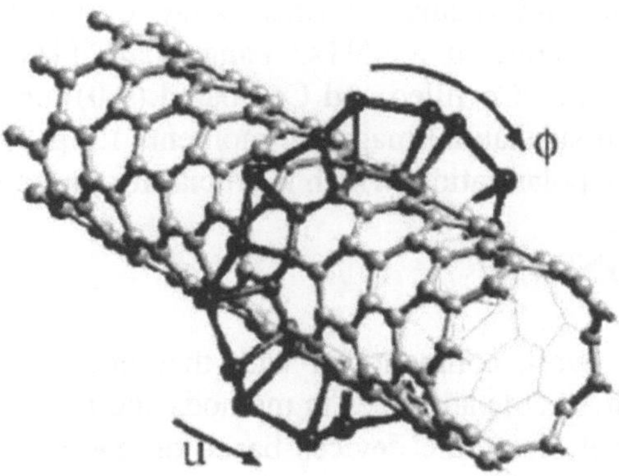

Figure 6. A view of the optimized structure of the Al zigzag nanoring formed on a (8,0) SWNT.

The electronic properties of the flat zigzag Al nanoring system are also quite interesting. The electronic energy bands of the Al metal ring (without SWNT) are derived from the dispersive bands of the flat zigzag Al chain. [40] When the flat zigzag Al chain is rolled into a ring, its bands are zone folded at the Γ-point and they appear as a number of discrete energy levels. For the case of the Al nanoring wrapping the nanotube, these states are mixed with the states of the nanotube and give rise to the bands and density of states, which attributes a metallic behavior to the combined Al-nanoring and (8,0) nanotube system. According to Mulliken analysis 0.15 electrons are transferred from each Al atom to the SWNT.

4.5. Nano-magnets

Earlier it was shown that most of the transition metals adsorbed on SWNT have magnetic ground state. [27, 28] It is expected that a nanomagnet can be formed once SWNT is fully covered or coated by this specific atoms. As a matter of fact, supercell calculations predicting the magnetic ground state of an individual atom adsorbed SWNT actually correspond to the chain of atoms adsorbed on the SWNT. Therefore, first-principles calculations by Durgun *et al.* [27, 28] present evidence that one can generate nanomagnets from SWNTs covered by transition metal atoms (such as Ti, Ta, etc). We found that single Ti atom adsorbed on the (8,0) SWNT give rise to 2.27 μ_B (Bohr magneton) net magnetic moment. This value rises to ~13 μ_B/unit cell when the SWNT is fully covered by Ti atoms.

From the spintronics and molecular magnetism point of view, the subject is extremely important, and requires further study of the magnetization of transition metal atom covered SWNTs. Yang *et al.* [41] performed first-principles calculations of Co filled and Co coated (9,0) tube, and found that these systems exhibit substantial magnetic moment (1.4 μ_B per Co atom) and have a very high spin polarization, which is crucial for the spin-valve effect.

5. CONCLUSIONS

Intensive laboratory work compounded with theoretical studies during last decade has made remarkable advances in methods and techniques which were able to produce novel electronic devices based on carbon nanotubes. These devices have kept the great expectations from these materials vivid and have continued to motivate further research. It appears that one still needs new paradigms in the integration of several nanodevices. In addition to their contribution for the better understanding of various fundamental effects, first-principles calculations are crucial for discovering new properties and for designing novel nanodevices. Many features of nanostructures and their I-V characteristics can be revealed accurately and easily in the computer medium by using first-principles calculations. Specific features of nanostructures, which cannot be easily probed in laboratory environment, can be provided also by first-principles and empirical calculations. In this respect, the role of theoretical work and numerical simulations in the development of the field is indispensable.

References

1. S. Iijima , Nature (London) **354**, 56 (1991); S. Iijima, T. Ichihashi, and Y. Ando, Nature (London) **356**, 776 (1992).
2. M.S. Dresselhaus, G. Dresselhaus and P.C. Eklund, *Science of Fullerenes and Carbon Nanotubes*; Academic Press, San Diego (1996).
3. R. Saito, G. Dresselhaus and M. S. Dresselhaus, *Physical Properties of Carbon Nanotubes*; Imperial College Press, London (1998).
4. X. Zhao, Y. Ando, Y. Liu, M. Jinno, and T. Suzuki, Phys. Rev. Lett. **90**, 187401 (2003).
5. T. Cagin, (paper in this proceeding).
6. M.F.Yu, *et al.*, Science, **287**, 637 (1998).
7. J.P. Salvetat *et al.* Appl. Phys. A **69**, 255 (1999).
8. Ç. Kılıç,, S. Ciraci, O. Gülseren and T. Yildirim, Phys. Rev. B **62**, 16345 (2000).
9. T. Yildirim *et al* .(paper in this proceeding)
10. O. Gülseren *et al.* (paper in this proceeding)
11. O. Gülseren, T. Yildirim and S. Ciraci, Phys. Rev. **65**, 153405 (2002)
12. M.S. Dresselhaus, G. Dresselhaus and R. Saito, Phys. Rev. B **45**, 6234 (1992).

13. N. Hamada, S. Sawada and A. Oshiyama, Phys. Rev. Lett. **68**, 1579 (1992).
14. J.W. Mintmire, B.I. Dunlap and C.T. White, Phys. Rev. Lett. **68**, 631 (1992).
15. C.T. White, D.H. Robertson and J.W. Mintmire, Phys. Rev. B **47**, 5485 (1993).
16. S. Ciraci, A. Buldum and I. Batra, J. Phys: Condens. Matter **13**, 537 (2001).
17. S. Frank, P. Poncharal, Z. L. Wang and W. A. Heer, Science **280**, 1744 (1998).
18. M. Ouyang, J. Huang, C.L. Cheung and C.M. Lieber, Science **292**, 702 (2001).
19. T.W. Odom, J. Huang, P. Kim and C.M. Lieber, Nature **391**, 62 (1998).
20. T.W. Odom, J. Huang, P. Kim and C.M. Lieber, J. Phys. Chem. B **104**, 2794 (2000).
21. X. Blase, L. X. Benedict, E. L. Shirley, and S. G. Louie, Phys. Rev. Lett. **72**, 1878 (1994).
22. T.W. Tombler, C. Zhou, L. Alexseyev, J. Kong, H. Dai, L. Liu, C.S. Jayanthi, M. Tang, and S. Wu, Nature **405**, 769 (2000).
23. C. J. Park, Y. H. Kim, K. J. Chang, Phys. Rev. B **60**, 10656 (1999).
24. T. Yildirim, O. Gülseren and S. Ciraci, Phys. Rev. B **64**, 075404 (2001).
25. O. Gülseren, T. Yildirim, S. Ciraci, Phys. Rev. B **68**, 115419 (2003).
26. V. M. K. Bagci, O. Gülseren, T. Yildirim, Z. Gedik, and S. Ciraci, Phys. Rev. B **66**, 045409 (2002).
27. E. Durgun, S. Dag, V. M. K. Bagci, O. Gülseren, T. Yildirim and S. Ciraci, Phys. Rev. B **67**, 201401(R) (2003).
28. E. Durgun, S. Dag, S. Ciraci, and O. Gülseren, J. Phys. Chem. B **108**, 575-582 (2004).
29. S.J. Tans, A. R. M. Verscheuren and C. Dekker, Nature **393**, 49 (1998); A Bachtold, P. Hadley, T. Nakanishi, and C. Dekker, Science **294**, 1317 (2001);T. Nakanishi, A. Bachtold and C. Dekker, Phys. Rev B**66**, 073307 (2002).
30. H. T. Soh, C. Quate, A. F. Morpurgo, C.M. Marcus, J. Kong and H. Dai, Appl. Phys. Lett. **75**, 627 (1999);C. Zhou, J. Kong, and H. Dai, Appl. Phys. Lett. **76**, 1597 (2000).
31. R. Martel, V. Derycke, C. Lavoie, J. Appenzeller,K. K. Chan, J.. Tersoff, and Ph. Avouris, Phys. Rev. Lett. **87**, 256805; Ph. Avouris, R. Martel, V. Derycke and J. Appenzeller, Physica, B**323**, 6 (2002).
32. A. Rochefort, M. Di Vientra, P. Avouris, Appl. Phys. Lett. **78**, 2521 (2001).
33. S. Heinze, J. Tersoff, R. Martel, V. Derycke, J. Appenzeller and Ph. Avouris, Phys. Rev. Lett **89**, 106801 (2002).
34. S. Dag, O. Gülseren, S. Ciraci, T. Yildirim, Appl. Phys. Lett. **83**, 3180 (2003).
35. Y. Miyamoto, S. G. Louie, and M. Cohen, Phys. Rev. Lett. **76**, 2121 (1996); Y. Miyamoto, Phys. Rev. B **54**, 11149, (1996); Y. Miyamoto, A. Rubio, S. G. Louie, and M. L. Cohen, Phys. Rev. B **60**, 13885 (1999).
36. J. R. Friedman, V. Patel, W. Chen, S. K. Tolpygo, and J. E. Lukens, Nature **406**, 43 (2000).
37. H. Mehrez and S. Ciraci, Phys. Rev. B **56**, 12632 (1997).
38. O. Gülseren, F. Ercolessi, and E. Tosatti, Phys. Rev. Lett. **80**, 3775 (1998).
39. D. S. Portal, E. Artacho, J. Junquera, P. Ordejon, A. Garcia, and J. M. Soler, Phys. Rev. Lett. **83**, 3884 (1999).
40. P. Sen, S. Ciraci, A. Buldum and I. P. Batra, Phys. Rev. B **64**, 195420 (2001).
41. C-K. Yang, J. Zhao and J. P. Lu, Phys. Rev. Lett. **90**, 257203 (2003).

INTIMATE RELATIONSHIP BETWEEN STRUCTURAL DEFORMATION AND PROPERTIES OF SINGLE-WALLED CARBON NANOTUBES AND ITS HYDROGENATED DERIVATIVES

TANER YILDIRIM *, OGUZ GULSEREN AND SALIM CIRACI**

** NIST Center for Neutron Research, Gaithersburg, MD 20899, USA; **Physics Department, Bilkent University, 06533, Ankara, Turkey*

Abstract: Carbon nanotubes continue to surprise scientists with their novel properties. Recently we have discovered many intimate relationships between structural deformation and the properties of single-walled nanotubes (SWNT) that could be important in technological applications. From the first-principles we show that by using pressure, carbon nanotubes can be covalently joined to form one and two-dimensional networks of interlinked nanotubes. We also found that the band gap of an insulating nanotube can be engineered by elliptical distortion, which is found to be in the elastic range. This could allow the fine-tuning of the properties of SWNTs via reversible deformation and ultimately lead to variable quantum devices. Finally, we have shown that the chemical reactivity of nanotubes can be tuned by elliptical deformation, which may provide a way to attach various atoms such as H and metals to a specific location on a nanotube. In particular, we have studied hydrogenated carbon nanotubes for a large number of configurations and hydrogen coverage. We show that the electronic and atomic structure of carbon nanotubes undergo dramatic changes with hydrogen chemisorption. The first principle calculations indicate that selective bonding of hydrogen to nanotubes can give rise to a number of potentially useful applications in the emerging field of molecular electronics.

Keywords: Carbon nanotubes, SWNT, first-principles calculations, absorption, nanodevices, hydrogenated SWNT

S. Guceri et al. (eds.), Nanoengineered Nanofibrous Materials, 199-211.

1. INTRODUCTION

Carbon nanotubes, originally discovered as by products of fullerene synthesis [1], are now considered to be the building blocks of future nanoscale electronic and mechanical devices [2]. It is, therefore, desirable to have a good understanding of their electronic and mechanical properties and the interrelations between them. In particular, single wall carbon nanotubes (SWNTs) provide a system where the electronic properties can be controlled by the structure of the nanotubes and by various deformations of their geometries [3-12]. SWNTs are basically rolled graphite sheets, which are characterized by two integers *(n,m)* defining the rolling vector of graphite. Therefore, the electronic properties of SWNTs, to the first order, can be deduced from that of the graphite by mapping the band structure from a 2D hexagonal lattice onto a cylinder. Such an analysis indicates that *(n,n)* armchair nanotubes are always metallic and *(n,0)* zigzag nanotubes are metallic only if n is an integer multiple of three. Using the state-of-the-art first-principles total energy calculations, we have shown that these electronic properties can be significantly modified by structural deformation of single-walled nanotubes (SWNT) [9-13]. Here we present a brief review of our findings.

2. PRESSURE-INDUCED INTERLINKING OF CARBON NANOTUBES

In this section we discuss the effect of the external pressure on intertube interactions between nanotubes packed in hexagonal lattices, as so called "nanoropes"[10]. The intertube interactions in nano ropes can be probed by applying external pressure to vary the intertube distance. For fullerenes, such high-pressure studies have yielded many interesting results including new compounds such as the pressure-induced polymeric phases of C_{60}. It is, therefore, of interest to inquire if similar covalent-bonding can occur between the nanotubes in a rope. This could have important consequences for nanoscale device applications and composite materials that require strong mechanical properties since nano ropes consisting of inter-linked SWNT will be significantly stronger than nano ropes composed of van der Waals (vdW) packed nanotubes.

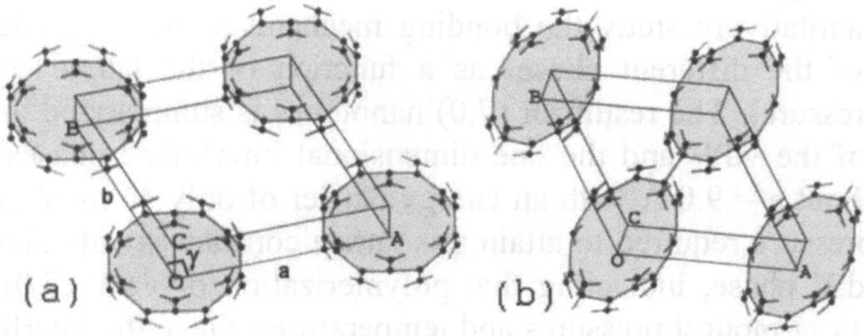

Figure 1. Optimized structures of the vdW (7,0) (a), and one dimensional interlinked (7,0) (b) nanotube lattices. The interlinked structure shown in (b) has lower energy than vdW packed (7,0) nanotubes shown in (a).

We investigated possible new pressure-induced ground state structures for (n,0) nanotube ropes from first-principles total energy calculations using the pseudopotential method within the generalized gradient approximation (GGA)[13]. For simplicity, we model the nano ropes as a hexagonal lattice of nanotubes with one nanotube per unit cell. The pressure dependence of the lattices of nanotubes was determined by calculating the total energy as a function of nanotube separation (i.e. a and b) while the other parameters, including atom positions, c, and γ are optimized. We observe that (7,0) nanotubes become elliptically distorted with applied pressure (i.e. decreasing nanotube-nanotube distance). At a critical pressure, we observe a structural phase transformation from the van der Waals nanotube lattice (as shown in Fig.1 a) to a new lattice in which the nanotubes are interlinked along the [110] direction, where the strain of the nanotube is largest (Fig.1(b)). The covalent bonding between nanotubes is therefore the result of curvature-induced re-hybridization of the carbon orbitals. The same structural transformation was observed for the other (n, 0) nano ropes.

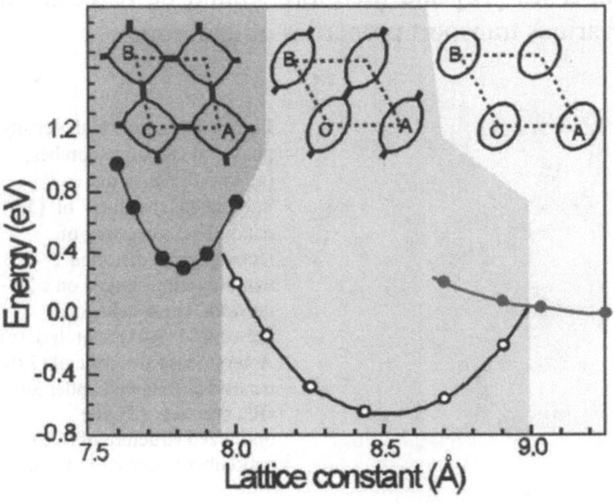

Figure 2. Planar lattice constant variation of the total energy of (7,0) nanotube ropes in three different phases. Inset shows the view of the structures along c-axis. The zero of energy was taken to be the energy of vdW packing of the nanotubes.

To quantitatively study the bonding mechanism, we calculated the total energies of the different phases as a function of the lattice constant (i.e. applied pressure). The result for (7,0) nanotubes is summarized in Fig.2. The energies of the vdW and the one dimensional interlinked phases cross each other at about a = 9.0 Å with an energy barrier of only 46 meV/unitcell (552 K). The pressure required to attain this lattice constant is only about 0.3 GPa for the vdW phase, indicating that polymerization of vdW (7,0) nanoropes could occur at modest pressures and temperatures. Once the interlinked phase is reached, the energy barrier required to break the bonds and obtain free nanotubes is about 0.7 eV (25 meV/atom), which is comparable to that of 1D polymerized C_{60} molecules (20 meV/atom).

Fig.2 also shows that another interlinked phase of (7,0) nanotubes becomes the ground state for lattice parameter smaller than 8.0 Å. In this new phase the nanotubes are interlinked along both a- and b-axes (see Fig.3(a)). This two dimensional interlinked structure is about four times stiffer than the 1D interlinked phase and sixteen times stiffer than the vdW nanoropes.

We observe that applying even higher pressures yields more complicated and denser phases for many of the nano ropes studied here (see Fig.3). For (9,0) nanoropes, we find that the nanotubes are interlinked along three directions, forming a hexagonal network. The length of the intertube bond, d_{C-C} = 1.644 Å, is significantly elongated for a sp^3 C-C bond. The two dimensional interlinked phase of (7,0) nanotubes is further transformed to a denser structure at 30 Gpa with a band gap of 2 eV (Fig.3(c)). By comparison, (6,6) nanotubes do not form an interlinked structure up to a pressure of 60 GPa. Rather the nanotubes are hexagonally distorted such that the local structure of the nanotube faces is similar to that in graphite sheets (Fig.3(d)). Furthermore, releasing the pressure yields the original structure, indicating that the distortion is purely elastic. The structural changes clearly have strong effects on the electronic properties [10] and therefore should be detected in the pressure dependence of various transport properties of nanoropes.

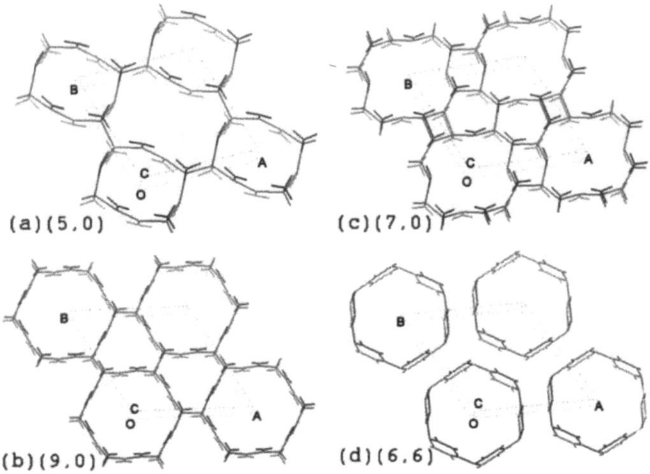

Figure 3. Various high density phases of carbon nanotubes; (a) Two dimensional interlinked structure of (5,0) nanotubes, consisting of rectangularly distorted nanotubes interlinked on a 2D network. (b) A hexagonal network of (9,0) nanotubes, (c) A very dense structure of (7,0) nanotubes obtained under 30 GPa pressure. (d) The optimized structure of (6,6) nanotubes under P=53 GPa.

The new pressure-induced, high density phases[10] reported here may provide a way of synthesizing novel carbon base materials with interesting physical properties. For example, interlinking of the nanotubes may improve the mechanical performance of composites based on these materials. The change in the band gap of a SWNT with applied pressure can be exploited to realize various quantum devices on a single nanotube with variable and reversible electronic properties [10]. Finally, we note that several months after our predictions [10] were published, similar structures of interlinked nanotubes were experimentally observed [14-15], indicating the predicting power of the first-principles computational technique.

3. BAND GAP ENGINEERING

In the previous section we showed that at modest pressures the circular cross-section of the nanotubes in a rope is distorted to elliptical one. This raises an interesting question; namely what is the effect of such an elliptical deformation on the electronic and chemical properties of an individual carbon nanotube? Recently, we have addressed this issue using first-principles calculations [9]. Our results are summarized in Figure 4.

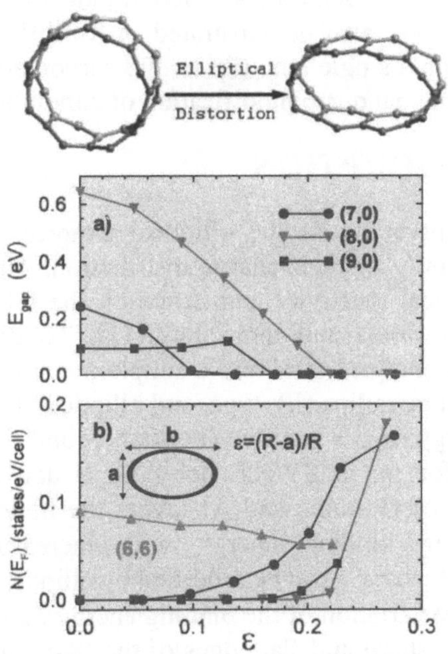

Figure 4. Variation of the energy band gap E_{gap}(a) and density of the states at the Fermi energy (b) as a function of elliptical deformation .

We find that the electronic properties of SWNTs can be significantly modified by radial strain, which distorts the circular cross section to an elliptical one. Figure 4 shows that the energy gap of an insulating SWNT decreases and eventually vanishes at an insulator-metal transition with increasing applied elliptical strain. The density of states at the Fermi level $N(E_F)$, of a metallized SWNT increases with increasing deformation. More interestingly, the elliptical deformation necessary to induce metallicity is found to be in the elastic range. Therefore, all strain induced changes in electronic and also in mechanical properties are reversible.

The band gap engineering in carbon nanotubes by reversable deformation discussed above can be used to construct variable and reversible quantum structures In Ref.[9], we explored this idea, namely that various quantum structures can easily be realized on an individual carbon nanotubes and the properties of these structures can be controlled by applied transverse compressive stress. If such a deformation is not uniform but has different magnitude at different positions along the nanotube then the different regions will have different band gap and quantum wells of the desired electronic character can be formed. In this respect, the present scheme is quite different from previous constructions of nanotube heterostructures or quantum dots, where one had to fabricate each time a different junction to satisfy the desired electronic character. We also demonstrated in Ref.[9] various nanodevices and band-alignment on a single tube (up to 300 carbon atoms) using this idea of deformation induced band gap modification of nanotubes.

4. TUNABLE ABSORPTION

Another significant effect of the elliptical distortion of SWNTs is the change in the uniformity of their charge distribution. This, in turn, imposes changes in the chemical reactivity and hence on the interaction of the tube surface with foreign atoms and molecules [11]. To demonstrate this, we calculate the binding energies of H and a simple metal atom, Al, on SWNTs as functions of nanotube radius (R), type, and elliptical deformation. We find that the binding energy, E_B, is positive (i.e. stable) and follows a remarkably simple $1/R$ scaling for *(n, 0)* SWNTs for both H and Al, even though H prefers to sit on top of C atoms and Al favors the hollow sites (i.e. top of hexagons). Increasing binding energy with increasing curvature (i.e. decreasing R) suggests that E_B can be modified by elliptical deformation.

Figure 5 shows the variation of the binding energy E_B of a single hydrogen atom adsorbed on the sharp and flat edges of the (8,0) surface with elliptical distortion. With increasing distortion, the binding energy of the sharp site increases, while it decreases for the flat site, creating an energy difference of

≈ 1.1 eV at the distortion value $\varepsilon = 0.3$. This value is 44 % of the binding energy of H on the undeformed SWNT.

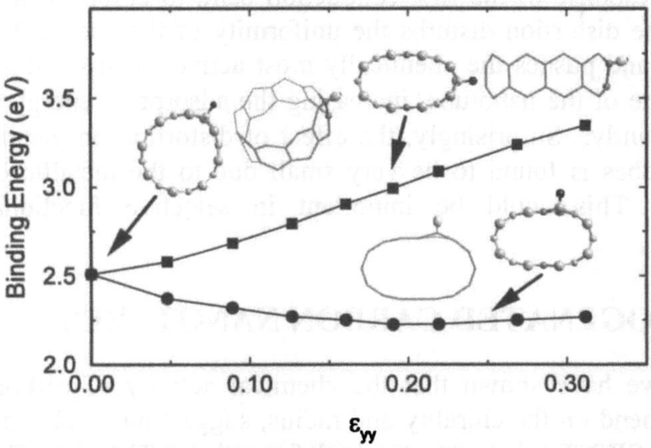

Figure 5. Binding energies of the sharp (top) and the flat sites (bottom) versus elliptical deformation for a hydrogen adsorbed on a (8,0) SWNT.

The binding energy of Al exhibits a behavior similar to that of H, despite H and Al favoring different sites on the (8,0) tube. Elliptical deformation increases the binding energy of the both flat and sharp sites for Al, given an energy difference of 0.5 eV at the deformation $\varepsilon = 0.3$. It is remarkable that Al, which is not bound to the graphite, can be adsorbed at the sharp site of the distorted SWNT with a binding energy of 1.8 eV. Hence, we conclude that not only band gap engineering but also chemical reactions taking place on the surface of a SWNT can be engineered through radial deformation. Our findings from first-principles calculations [11] are in good agreement with the idea of "Kinky Chemistry", which was first proposed by Srivastava *et al.* using a many-body empirical potential [16].

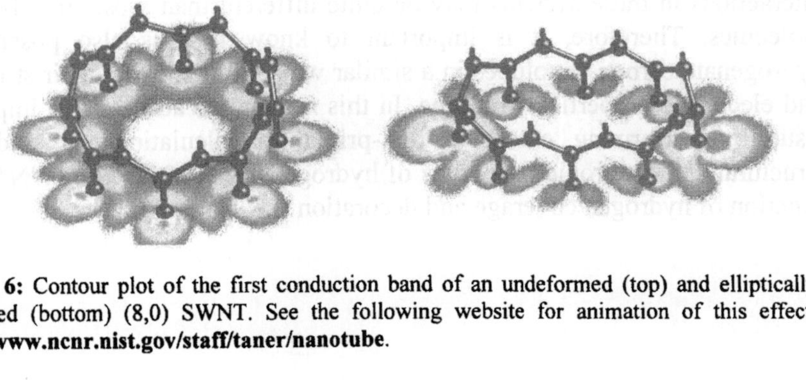

Figure 6: Contour plot of the first conduction band of an undeformed (top) and elliptically deformed (bottom) (8,0) SWNT. See the following website for animation of this effect: **http://www.ncnr.nist.gov/staff/taner/nanotube**.

An explanation of this remarkable change of the binding energy with elliptical deformation is sought in the electronic structure of SWNTs. Figure 6 shows the response of the first conduction band to elliptical distortion. It is clear that the distortion disturbs the uniformity of the charge distribution of the SWNT and pushes the chemically most active electrons from the flat to the sharp site of the nanotube, increasing the adsorption energy of the sharp site significantly. Surprisingly, the effect of distortion on metallic armchair *(n,n)* nanotubes is found to be very small due to the metallic bands of the tubes [11]. This could be important in selective functionalization of nanotubes.

5. HYDROGENATED CARBON NANOTUBES

Above we have shown that the chemical activity of carbon nanotubes strongly depend on the chirality and radius, suggesting tunable absorption of atoms on SWNTs by structural deformation. The interplay between adsorption and electromechanical properties can give rise to novel physiochemical properties. The experimentally observed sensitivity of the electronic properties of SWNTs to the presence of oxygen and hydrogen is clear evidence for the importance of this interplay [18]. Furthermore the understanding of the interaction between hydrogen and carbon nanotubes is particularly important due to potential application of nanotubes for hydrogen storage where the large effective surface area promises a large absorption capacity.

Hydrogen-carbon interactions have been studied extensively both theoretically and experimentally for many interesting polyhedral molecules, such as cubane C_8H_8 [19,20], and dodecahedrane ($C_{20}H_{20}$) [21] (see Fig.7). These novel polyhedral molecules, which represent the zero dimensional case, exhibit many interesting properties. However, due to the one dimensional nature and the curvature of carbon nanotubes, the hydrogen-carbon interactions in these systems may be quite different than those in polyhedral molecules. Therefore, it is important to know if it is also possible to hydrogenate carbon nanotubes in a similar way and if so what their structural and electronic properties would be. In this section we address this important issue by performing extensive first-principles calculations to study the structural and electronic properties of hydrogenated SWNTs (H-SWNT) as a function of hydrogen coverage and decoration (i.e. isomers).

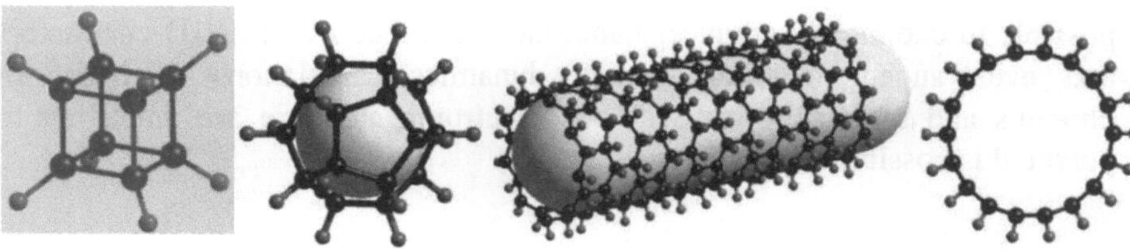

Figure 7. Three different polyhedra of carbon and hydrogen. The molecules shown at the left (i.e. cubane) and middle (i.e. dodecahedrane) have been synthesized successfully. The right panel shows the fully exo hydrogenated carbon nanotube (a side and top view) studied in this work.

We have considered zigzag ((7,0), (8,0), (9,0), (10,0), (12,0)) and armchair ((6,6), (10,10)) SWNTs, which are hydrogenated at two different coverages. For full coverage, we consider two isomers; namely (*i*) **exohydrogenation** where each carbon atom is bonded to a hydrogen atom from outside of the nanotube and (*ii*) **endo-exohydrogenation** where each carbon atom is bonded to a hydrogen from inside and outside of the tube alternatively. For half coverage, we consider the three most interesting isomers; namely (*i*) **uniform** pattern where every other carbon atom is bonded to a hydrogen from outside, (*ii*) **chain** pattern where every other carbon zigzag chain is saturated by hydrogen, and (*iii*) **dimer** pattern where every other carbon dimmer rows perpendicular to the zigzag carbon chains are saturated by hydrogen. We find that geometric and electronic structures and binding energies of these different H-SWNT's strongly depend on the pattern of hydrogenation. The optimized structures and their binding energies are summarized in Fig. 8. From this figure we see that for both zigzag and armchair nanotubes, hydrogenation of each carbon atom from inside and outside alternatively yield the most stable isomer with a very weak curvature dependence and a large band gap.

Our calculations [17] indicate that strategically placing hydrogen on the exterior of carbon nanotubes leads to dramatic changes in their electronic and atomic structures. One of the most important results [17] is that upon hydrogenation at uniform half coverage, the zigzag (n,0) SWNTs are metallized with high density of states at the Fermi level as shown in Fig. 9. Surprisingly, the cross section of the nanotube —initially being cylindrical — becomes rectangular, with a diamond-like carbon atom at each corner (see Fig. 9). These chemically passive carbon atoms isolate the four conducting faces of the H-SWNT, yielding a *"four-wire nanocable"*. Because of the high density of conduction electrons in this particular configuration, it may be possible to chemically engineer nanotube wires that are ideal 1D conductors and even superconducting. Lattice dynamics calculations of hydrogen phonons and their coupling with the electronic structure are underway to pursue this possibility further.

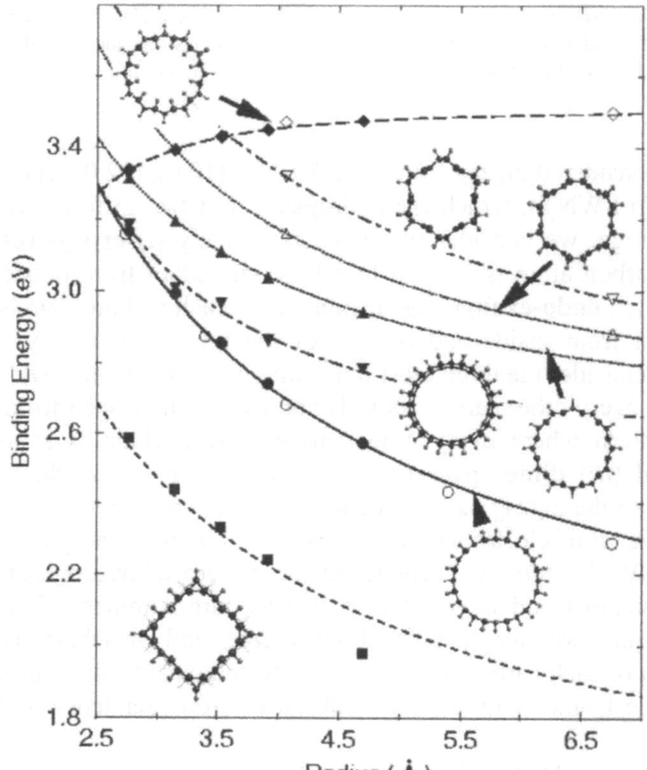

Figure 8. Average binding energies, E_b, of hydrogen atoms adsorbed on various zigzag and armchair SWNTs versus bare tube radius R. Filled and open symbols are for zigzag and armchair nanotubes, respectively. Circles and diamonds are for exo- and endo-exohydrogenation at full coverage, respectively. The filled squares show the zigzag nanotubes uniformly exohydrogenated at half coverage. The chain and dimmer patterns of adsorbed hydrogen atoms at half coverage are shown by down- and up-triangles, respectively. Insets show top view of several H-SWNT isomers.

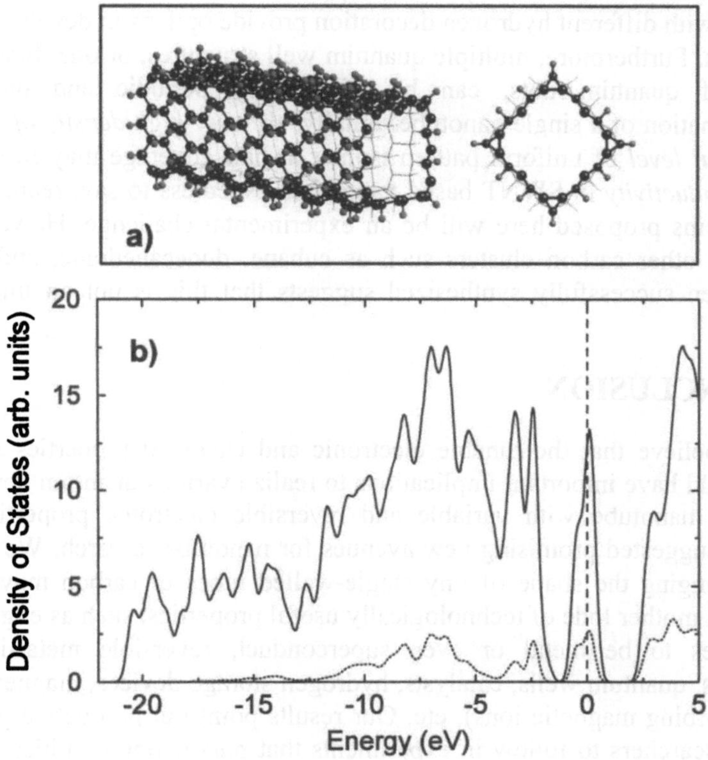

Figure 9. TOP: A side and top view of a hydrogenated (8,0) SWNT at half coverage. The black and gray spheres represent carbon and hydrogen atoms, respectively. Note that the fully optimized structure has a rectangular cross section. BOTTOM: The electronic density of states (DOS), indicating a very large number of states at the Fermi level. Hence hydrogenation of an insulating (8,0) SWNT at half coverage induce metalization. The dotted line shows the contribution to the DOS from hydrogen atoms.

For other isomers at half coverage as well as exo- and endo-exohydrogenations at full coverage, the cross sections remain quasi-circular and electronic structures vary greatly among the resultant materials. For example, the chain pattern H-SWNTs exhibit small band gap of ~2eV while the dimer pattern H-SWNTs exhibit a band gap of ~4 eV. Our results indicate that selective bonding of hydrogen to nanotubes can give rise to a number of potentially useful applications in the emerging field of molecular electronics.

Our study of hydrogenated carbon nanotubes reveals many important and novel effects of hydrogen adsorption on SWNTs, and brings a number of new problems and issues to be explored [22]. For example, one can argue that the band gap of a SWNT can be engineered by the controlled hydrogenation of a single nanotube as in the alloy of Si_xGe_{1-x}. A number of isomers which can be tailored with different hydrogen decoration provide options in developing new materials. Furthermore, multiple quantum well structures, or one dimensional chain of quantum dots, can be tailored by periodic and modulated hydrogenation of a single nanotube. Finally, *the very high density of states at the Fermi level* of uniform pattern isomer at half coverage may result in to *superconductivity* in SWNT based nanowires. Needless to say, realization of the systems proposed here will be an experimental challenge. However, the fact that other carbon clusters such as cubane, dodecahedrane, and $C_{60}H_{32}$ have been successfully synthesized suggests that this is not an impossible task.

6. CONCLUSION

We believe that the tunable electronic and chemical properties reported here could have important implications to realize various quantum devices on a single nanotube with variable and reversible electronic properties. Our studies suggested promising new avenues for nanotube research. We showed how changing the shape of tiny single-walled tubes of carbon may open a potential mother lode of technologically useful properties, such as engineering nanotubes to be metal or even superconduct, reversible metal-insulator junctions, quantum wells, catalysts, hydrogen storage devices, magnetic tubes (by absorbing magnetic ions), etc. Our results point out productive paths for other researchers to follow in experiments that pursue opportunities to make new materials and technologies with nanotubes.

Acknowledgments: This work was partially supported by the National Science Foundation under Grant No.'s INT97-31014, INT01-15021 and TUBITAK under Grant No. TBAG-U/13(101T010).

References

1. S. Iijima, Nature 354, 56 (1991).
2. R. Saito, G. Dresselhaus, and M. S. Dresselhaus, *Physical Properties of Carbon Nanotubes*, Imprerial College Press, London (1998).
3. X. Blase *et al.* Phys. Rev. Lett. **72**, 1878 (1994).
4. Bezryadin *et al.* Phys. Rev. Lett. **80**, 4036 (1998).
5. Rochefort *et al.* Chem. Phys. Lett. **297**, 13824 (1998).
6. C-J. Park *et al.* Phys. Rev. B **60**, 10656 (1999).
7. L. Liu *et al.* Phys. Rev. Lett. **84**, 4950 (2000).
8. L. Yang and J. Han, Phys. Rev. Lett. **85**, 154 (2000).
9. Kilic, S. Ciraci, O. Gulseren, and T. Yildirim, Phys. Rev. B **62**, R16345 (2000).
10. T. Yildirim, O. Gulseren, C. Kilic, and S. Ciraci, Phys. Rev. B **62**, 12648 (2000).
11. O. Gulseren, T. Yildirim, and S. Ciraci, Phys. Rev. Lett. **87**, 116802 (2001).
12. T. Yildirim , O. Gulseren, and S. Ciraci, Phys. Rev. B **64**, 075404 (2001).

13. First-principles calculations were performed by the pseudo potential plane wave code CASTEP; M. C.Payne *et al.* Rev. Mod. Phys. **64**, 1045 (1992).
14. M. J. Lopez *et al.* Phys. Rev. Lett. **86**, 3056 (2001).
15. S. Rols *et al.* Phys. Rev. B **64**, 153401 (2002).
16. Srivastava *et al.* J. Phys. Chem. B **103**, 4330 (1999).
17. O. Gulseren, T. Yildirim, and S. Ciraci Phys. Rev. B **66**, 121401 (2002).
18. P. G. Collins et. al. Science **287**, 1801 (2000); J. Kong et. al. Science **287**, 622 (2000).
19. P. E. Eaton and T. W. Cole, Jr., J. Am. Chem. Soc **86**, 962 (1964).
20. T. Yildirim, P. M. Gehring, D. A. Neumann, P. E. Eaton, and T. Emrick, Phys. Rev. Lett. **78**, 4938 (1997).
21. B.S. Hudson, D. A. Braden, S. F. Parker, and H. Prinzbach, Angew Chem. Int. Edit. **39**, 514 (2000).
22. O. Gulseren, T. Yildirim, and S. Ciraci, Phys. Rev. B (in press, 2003).

GEOMETRY EFFECT ON ONE-ELECTRON DENSITY OF STATES OF BORON NITRIDE NANOTUBES

A.V. OSADCHY, E. D. OBRAZTSOVA
Natural Sciences Center of A.M. Prokhorov General Physics Institute, Russian Academy of Sciences, 38 Vavilov street, 119991, Moscow, Russia

Abstract: In this work one-electron density of states (DOS) for single-wall boron nitride nanotubes of different configurations has been calculated numerically with the computer program specially developed. The result has been obtained by numerical integration of dispersion curves of hexagonal boron nitride. The allowed values of wave vector have been derived using "zone folding" model. The data calculated have been plotted on "Kataura"-type graph demonstrating a dependence of the gap value on BN nanotube diameter. No metallic BN nanotubes have been revealed.

1. INTRODUCTION

Recently synthesized [1] boron nitride single-wall carbon nanotubes possesses a variety of interesting physical properties. A nanotube is a hexagonal boron nitride monolayer rolled up so as to form a cylinder with diameter of 10-20 Å. The appearance of circular periodicity, which is a multiple of the cylinder circumference, gives rise to additional boundary conditions for the quantum dimension of the electronic states.

Boron nitride nanotubes are unique due to their one-dimensional structure. A very small diameter is a reason of a specific view of electron density of states (DOS) demonstrating a set of singularities [2-4]. At the moment, there is a very few number of experimental works on this material. One of the most interesting fields is optics of BN nanotubes. Experimentally, the information about electronic structure may be obtained by scanning tunneling microscopy (STM), optical absorption and resonance Raman spectroscopy. For understanding the optical properties knowledge about electronic structure of this material is necessary.

S. Guceri et al. (eds.), Nanoengineered Nanofibrous Materials, 213-218.

This work is devoted to development of the computer program for numerical calculations of one-electron DOS for single-wall boron nitride nanotubes [5,6], applicable to nanotubes of any geometry. The data calculated have been plotted on "Kataura"-type graph demonstrating a dependence of the gap value on the BN nanotube diameter. No metallic BN nanotubes have been revealed.

2. ALGORITHM OF DOS CALCULATION

DOS of single-wall boron nitride nanotubes has been calculated by Monte-Carlo numerical integration method. 2D-dispersion for boron nitride was integrated over all allowed values of wave vectors. Allowed values of wave vector depend on geometry of nanotube. The obtained density of states also depends on nanotube geometry.

The 2D-dispersion surface of boron nitride has been deduced on the basis of published one-dimensional dispersion relations for few directions in Brillouin zone of hexagonal BN [7]. An inequality of B and N atoms has been taken into account. A view of 2D-dispersion for boron nitride is show in Fig.1. The corresponding equation is:

$$E(K_x,K_y)=\frac{\tilde{\varepsilon}\pm\sqrt{\tilde{\varepsilon}^2+4\cdot\tilde{\gamma}_0\cdot\omega(K_x,K_y)}}{2}\text{, where} \qquad (1)$$

$$\omega(K_x,K_y)=\left(1+4\cdot\cos\left(\frac{\sqrt{3}\cdot K_x\cdot\tilde{a}_0}{2}\right)\cdot\cos\left(\frac{K_y\cdot\tilde{a}_0}{2}\right)+4\cdot\cos^2\left(\frac{K_y\cdot\tilde{a}_0}{2}\right)\right)^2$$

and $\tilde{a}_0$, $\tilde{\gamma}_0$, $\tilde{\varepsilon}\equiv\varepsilon_{2p1}-\varepsilon_{2p2}$ - constants

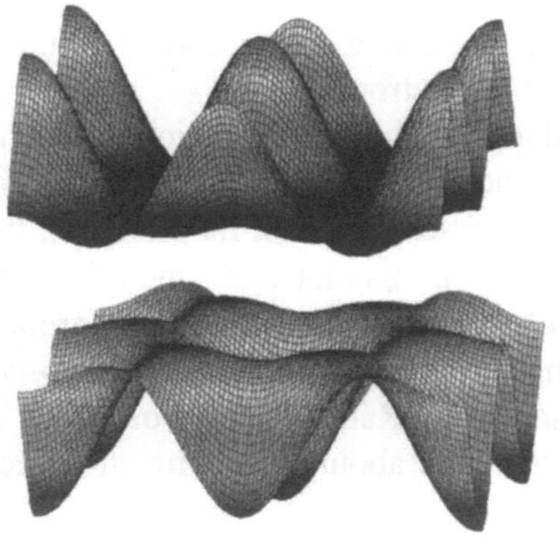

FIGURE 1. A view of two-dimensional (2D) dispersion relations for hexagonal boron nitride.

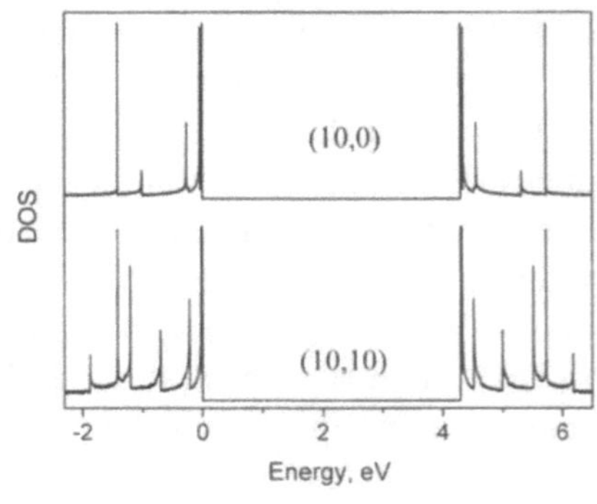

FIGURE 2. The calculated one-electron density of states for BN single-wall nanotubes of different geometry.

The constant values have been optimized to reach a minimal deviation from the dispersion along known directions [6]. The constant values obtained are:

$$\widetilde{a}_0 = 0.249 \text{ nm}, \ \widetilde{\gamma}_0 = 2.85 \text{ eV}, \ \widetilde{\varepsilon} = 4.3 \text{ eV} \tag{2}$$

The numerical differentiation should be performed for all allowed values of the wave vector within the first Brillouin zone. In case of single-wall nanotubes, the additional boundary conditions come into play. As a result, not all values of the wave vector K are allowed. Limitations appear because of rolling up the boron nitride plane into a cylinder with a diameter of about 10 Å; i.e., they are a consequence of the tube nanodimension.

The allowed values of the wave vector are determined as follows [8]: we introduce a Cartesian coordinate system XY, which has the origin at the point K of the first Brillouin zone and is rotated clockwise with respect to the initial coordinate system (Fig. 1) through an angle three times as great as the nanotube chirality angle. The latter is expressed as:

$$\alpha = \arctan\left(\frac{\pi}{6} - \frac{1}{\sqrt{3}} \cdot \frac{n-m}{n+m}\right), \tag{3}$$

where n and m are the nanotube parameters determining the width of the graphite plane strip and the way it is rolled up to form the nanotube (according to the adopted classification [9]).

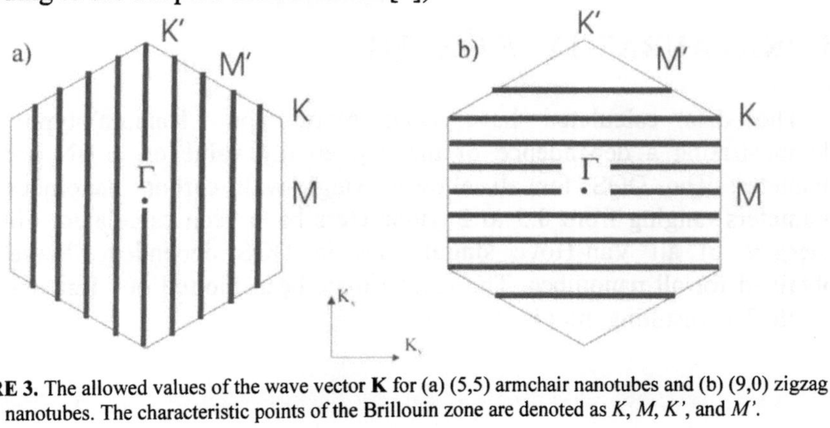

FIGURE 3. The allowed values of the wave vector **K** for (a) (5,5) armchair nanotubes and (b) (9,0) zigzag nanotubes. The characteristic points of the Brillouin zone are denoted as *K*, *M*, *K'*, and *M'*.

In the new coordinate system XY, the projection of the wave vector on the Y axis takes any values, while the projection on the X axis takes only the quantized values.

$$K_{X'} = \frac{2}{3 \cdot D} \cdot (3 \cdot Q - n + m), \tag{4}$$

where Q is an integer and D is the nanotube diameter in nanometers. The diameter of the nanotube as a function of its parameters n and m is expressed as

$$D = \frac{a_0}{\pi} \cdot \sqrt{n^2 + m^2 + n \cdot m} \qquad (5)$$

By performing the inverse transformation of coordinates from the rotated coordinate system to the system used in Eq. (1), we obtain all allowed values of the wave vector K.

Using the formulas presented above, it is possible to perform the numerical differentiation of the dispersion relation with an allowance for the boundary conditions.

For the practical realization of the problem, we developed a computer program [6] that provides the calculation of the density of states for any type of single-wall boron nitride nanotubes (the nanotube type should be preset before the calculation). The numerical differentiation was performed by the Monte Carlo method. Depending on the required accuracy of calculation, the execution time varied from 60 s to 1 h. The results of the calculation were entered into the Microcal Origin package.

The examples of DOS calculated for BN nanotubes are shown in Fig. 2. For all nanotube geometries DOS demonstrates a big (more than 4 eV) first mirror gap (E11). It means that there are no BN single-wall nanotubes with metallic properties.

3. "KATAURA"-TYPE GRAPH

The data calculated have been plotted on "Kataura"-type graph demonstrating a dependence of the gap energy value on a BN nanotube diameter. The DOS for all allowed single-wall carbon nanotubes with diameters ranging from 0.5 to 2 nanometers have been calculated. Then the energies of all Van-Hove singularities in DOS dependence have been obtained for all nanotubes. The results have been plotted on "Kataura"-type graph. The resulting plot is shown in Fig. 4.

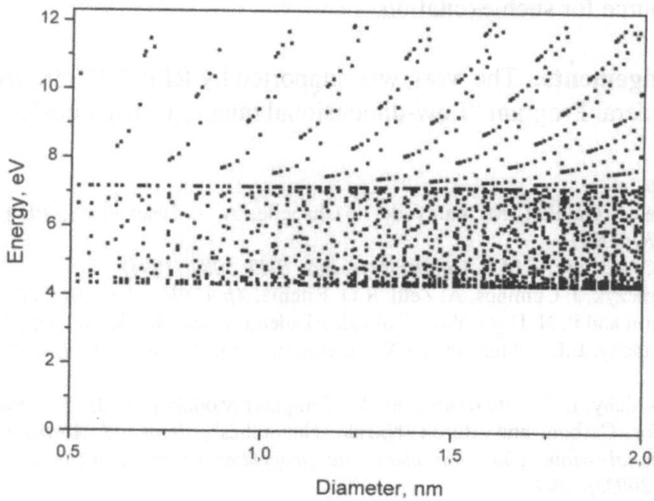

FIGURE 4. *"Kataura"-type* plot for single-wall boron nitride nanotubes.

Unfortunately *"Kataura"-type* plot for single-wall boron nitride nanotubes shows no dependence informative for investigation of electronic structure of nanotubes. This plot shows no strong correlation between the diameter and gap energy (contrary to the case of carbon nanotubes). It is obvious from this plot that there are no metallic BN nanotubes. All nanotubes have big energy gap. Also it is possible to note that energy less than 7.2 eV can't be use for investigation of resonance optical effects in boron nitride nanotubes. It means that resonance behavior in Raman and absorption spectra may be observed under excitation energies much exceed the visible photon energies, being efficient for carbon nanotubes.

4. CONCLUSIONS

Modeling of the electron density of states has been performed for single-wall boron nitride nanotubes of all allowed geometries. The calculation was based on the use of the original computer program that performs numerical integration by Monte-Carlo method for two-dimensional dispersion relations for hexagonal boron nitride. The integration is performed for all allowed values of the wave vector using π-electron approximation.

The data calculated have been plotted on *"Kataura"-type* graph demonstrating a dependence of the gap value on the BN nanotube diameter. This plot shows a very big gap value for nanotubes of all geometries. No metallic nanotubes have been revealed. It was shown that a possibility to observe the resonance effects in Raman and absorption spectra of BN nanotubes exists only under excitation energies higher than 7 eV. It is difficult to find a source for such excitation.

Acknowledgements: The work was supported by RFBR-17358, INTAS-01-254 and Federal Program "Low-dimensional quantum structures".

References

1. R.S. Lee, J. Gavillet, M. Lamy de la Chapelle, A. Loiseau et al., *Phys. Rev. B* ***64**, 121405(R)(2001).*
2. A.Rubio, J.L. Corkill, M.L. Cohen, *Phys. Rev. B 49, 5081 (1994).*
3. B.G. Demczyk, J. Cumings, A. Zettl, R.O. Ritchie, *Appl. Phys. Lett.* ***78**, 2772 (2001).*
4. D. V. Kirin and P. N. D'yachkov, Dokladi Akademii Nauk, Ser. Khim. **373**, 344 (2000).
5. A.V. Osadchy, E.D. Obraztsova, S.V. Terekhov, V.Yu. Yurov, *JETP Lett. 77 (2003) 405-410*
6. A. V. Osadchy, E. D. Obraztsova, et al. "Computer Modeling of One-Electron Dencity of States for Carbon and Boron Nitride Nanotubes", *Book of Abstract of "XVIIth International winterschool on electronic properties of novel materials", Kirchberg, Austria, 2003, p. 394.*
7. Yong-Nian Xu, W. Y. Ching, Phys Rev. B **44**, 7787 (1991).
8. *S. Reich and C. Thomsen, Phys. Rev. B* ***62**, 4273 (2000).*
9. *M.S. Dresselhaus, G. Dresselhaus, P.C. Eklund,* in. Science of Fullerenes and Carbon Nanotubes, New York, NY, San Diego, CA: Academic Press, 1996.

STRUCTURAL STABILITY OF CARBON NANOCAPSULES: MOLECULAR-DYNAMICS SIMULATIONS

O. B. Malcioglu, V. Tanriverdi, A. Yilmaz, S. Erkoc
Department of Physics, Middle East Technical University, Ankara 06531 Turkey

Abstract: Structural stability of empty and endohedrally doped carbon nanocapsules have been investigated by performing molecular–dynamics computer simulations. Calculations have been realized by using an empirical many–body potential energy function for carbon. It has been found that empty carbon nanocapsules are relatively more stable with respect to endohedrally doped ones against heat treatment.

Key words: carbon nanocapsules; molecular-dynamics; empirical potential.

1. INTRODUCTION

Nanotechnology is a highly active field today, not only due to the possibility of incredible improvement in the technology at hand, but also to the array of new technologies that may emerge. Carbon based nano structures constitute an important part in the area, because of their very unusual structural and electronic properties.

Nano capsule structure is quite popular among the people working in the area, and a number of applications using its unusual properties are possible [1]. In fact, most of the reported nanotube structures in the literature have at least one closed cap. Two possible examples of applications may be given as a novel memory device [2], and controlled medicine delivery [3]. In general, one end of the capsule can be opened by exposure to carbon dioxide, and they can be doped by a variety of materials [4].

2. THE PEF AND THE MD SIMULATION

The empirical many–body potential energy function (PEF) developed for carbon [5] is used in the calculations. This PEF describes the structural properties and energetics of carbon relatively accurately. The total interaction energy of a system of particles is taken to be the sum of total two–body and total three–body contributions [6]

$$\Phi = \phi_2 + \phi_3 \tag{1}$$

Total two–body and three–body energies are expressed, respectively, as

S. Guceri et al. (eds.), Nanoengineered Nanofibrous Materials, 219-223.

$$\phi_2 = A\sum_{i<j}^{N} U_{ij}^{(1)} \quad , \quad \phi_3 = -B\sum_{i<j}^{N} U_{ij}^{(2)}\left[1+\beta^n\left(\sum_{k\neq i,j}^{N} W_{ijk}\right)^n\right]^{-1/2n} \tag{2}$$

here U_{ij} and W_{ijk} represent the two–body and three–body interactions, respectively.

$$U_{ij}^{(1)} = f_c(r_{ij})\exp(-\lambda_1 r_{ij}) \tag{3}$$

$$U_{ij}^{(2)} = f_c(r_{ij})\exp(-\lambda_2 r_{ij}) \tag{4}$$

$$W_{ijk} = f_c(r_{ik})g(\theta_{ijk}) \tag{5}$$

where

$$g(\theta_{ijk}) = 1+\frac{c^2}{d^2}-\frac{c^2}{d^2+(h-\cos\theta_{ijk})^2} \tag{6}$$

$$f_c(r) = \begin{cases} 1 & \text{for} \quad r < R-D \\ \frac{1}{2}-\frac{1}{2}\sin[\frac{\pi}{2}(r-R)/D] & \text{for} \quad R-D < r < R+D \\ 0 & \text{for} \quad r > R+D \end{cases} \tag{7}$$

The parameters of the PEF for carbon are as follows [5]: $A = 1393.6\, eV$, $B = 346.74\, eV$, $\lambda_1 = 3.4879\, Å^{-1}$, $\lambda_2 = 2.2119\, Å^{-1}$, $\beta = 1.5724\times10^{-7}$, $n = 0.72751$, $c = 38049$, $d = 4.3484$, $h = -0.57058$, $R = 1.95\, Å$, and $D = 0.15\, Å$.

The equations of motion of the particles are solved by considering the Verlet algorithm [7]. The canonical NVT ensemble molecular–dynamics [7] is used. The temperature scaling is taken into account at every MD step and the temperature of the system is kept constant at a given temperature. One time step is taken as $10^{-16}\, s$. The initial velocities of the particles are determined from the Maxwell distribution at the given temperature.

3. RESULTS AND DISCUSSION

There are six models considered in this work, the first capsule (NC1) is the empty Carbon nano capsule with armchair Carbon nano tube, C(5,5), capped at both ends by C_{60} hemispherical structure , the second capsule NC2 is same as NC1 but doped with C_{20} buckyball at the middle, the third capsule (NC3) is again the initial NC1 capsule but doped with C_{20} at one of the ends. The next three are prepared in same fashion as previous three capsules, that is NC4 is the empty one, NC5 is the one doped with C_{20} at the middle, and NC6 is the one doped close to one of the ends, but this time the tubular section of the capsule is formed from C(9,0) zigzag Carbon nano tube. The tubular sections of the capsules were generated using programs given in [8].

The simulations are carried out starting at low temperature (1 K) and the temperature of the system is increased by a predetermined step (300 K) up to the system considered collapses. Each system collapsed at a different temperature. At every temperature rise the system is relaxed for about 50000 time steps. This number of steps were enough to reach equilibrium at every temperature rise. The empty and doped nano capsule structures at various temperatures are displayed in Figure 1.

The simulation results for NC1 are shown in Figure 1(a). NC1 is considered to be collapsed at 5700 K, which is a significantly higher temperature when compared with that of an armchair carbon nanotube with no caps in a similar simulation (4400 K) [9]. This may be due to absence of loosely connected atoms at the ends due to caps, and the added stability due to increased average coordination number. One point of interest may be the structure of the tubular section becomes more elliptical at central region, with increasing temperature. This effect was also observed in open ended tubes[9].

The simulation results for NC2 are shown in Figure 1(b). There is an enlargement of the tubular section at the location of dopant. This enlargement rapidly decays away from the location. Although there were no explicit constraints, the dopant C_{20} did not move inside the capsule through the course of the simulation. Flattening of the tubular section is again observable, in the area between the dopant and the caps. Capsule was considered to be collapsed at 2700 K, less than half the temperature initial structure persisted. Nano capsule collapsed at the position of the dopant, along with the dopant.

The simulation results for NC3 are shown in Figure 1(c). In the case of NC3 the nano capsule persisted up to 3900 K. This capsule is considerably more stable than NC2, however it is less stable than NC1. Again, there is an enlargement in the tubular section at the position of dopant, but this time flattening effect is not clearly observable. The capsule was collapsed at the region of dopant, while the other end was virtually unaffected. Again dopant did not move through the capsule.

The simulation results for NC4 are shown in Figure 1(d). The empty NC4 reaches up to 5100 K, a temperature which is considerably higher than that reported for its open-ended counterpart (4000 K) [9]. The flattening of the tubular section is again observable, but with the introduction of caps, the resulting geometry is not elliptical but somewhat triangular, that is there are three flattened regions instead of two.

The simulation results for NC5 are shown in Figure 1(e). When the dopant is placed at the middle, there is an enlargement through all of the tubular section, so the enlargement at the position of the dopant is not as large as seen in NC2. However, the capsule collapses at only 1500 K, not even a third of the temperature NC4 structure collapsed. Collapse is rather violent, the capsule is bent at the middle, there are very loosely bonded atoms, and the dopant structure is shattered. The dopant again did not move through the capsule, and the capsule was collapsed at the region of dopant. The flattening effect is also observable with the increasing temperature.

The simulation results for NC6 are shown in Figure 1(f). When the dopant is introduced at the side, however, there is a dramatic increase in the temperature the capsule resists without collapsing, approximately 4200 K. Although the collapse temperature is not as high as the NC4 structure, it is nearly tripled when compared to NC5. The overall enlargement when the dopant was introduced at the middle is not present, but there is a significant but rapidly decaying enlargement at position of the dopant. Likewise, in NC3, the flattening effect is not clearly observable unlike the initial structure, N4. Capsule collapses at the tubular section near the dopant.

As an overall conclusion we can state the following: The capsule structures are more stable than their open-ended counterparts under heat treatment. There is a decrease in diameter through the tubular section in the relaxed capsule structures, when compared with the ideal structures of open–ended carbon nanotubes, but there is flattening along the tubular section, away from the caps, in the empty capsules, as predicted by previous theoretical work [9]. Introduction of an endohedral dopant significantly lowers the temperature the structure may resist, and the capsules collapse at the region of dopant indicating a possible strain. The caps are more resistant to this possible strain, and when the dopant is placed close the caps structure becomes more stable thermally. The middle of the capsule is more prone to the thermal instability introduced by dopant. The C_{20} dopant does not move through the tubular section of the capsule without external influence, possibly due to close diameters. The diameter of ideal C(5,5) tube is about 6.8 Å, the diameter of ideal C(9,0) tube is about 7.1 Å, whereas the diameter of ideal C_{20} ball is about 4.0 Å. When a dopant is introduced at one of the ends, the flattening effect in the tubular section is not clearly observable.

Acknowledgments: The authors would like to thank METU (Middle East Technical University) for partial support through the project METU-BAP-2003-07-02-00-47, and TUBITAK for partial support through the project TBAG–U/41 (102T007).

References

1. http://www.ucl.itri.org.tw/eng/research/scope/polymer/polymer/polymer_6_e13.html
2. Y-K Kwon, D. Tománek, and S. Iijima, Phys. Rev. Lett. **82**, 1470(1999).
3. http://pubs.acs.org/subscribe/journals/mdd/v04/i04/html/MDD04FeatureVogelson.html
4. P.J.F. Harris, *Carbon Nanotubes and Related Structures*, (Cambridge Univ. Press, 2002).
5. J. Tersoff, Phys. Rev. Lett. **61**, 2879(1988).
6. J. Tersoff, Phys. Rev. B **38**, 9902(1988).
7. D.W. Heermann, *Computer Simulation Methods in Theoretical Physics*, Springer–Verlag, 1990.
8. R. Saito, G. Dresselhaus, and M.S. Dresselhaus, *Physical Properties of Carbon Nanotubes*, Imperial College Press, Singapore, 1998.
9. S. Erkoc, and O.B. Malcioglu, Int. J. Mod. Phys. C **12**, 865(2001).

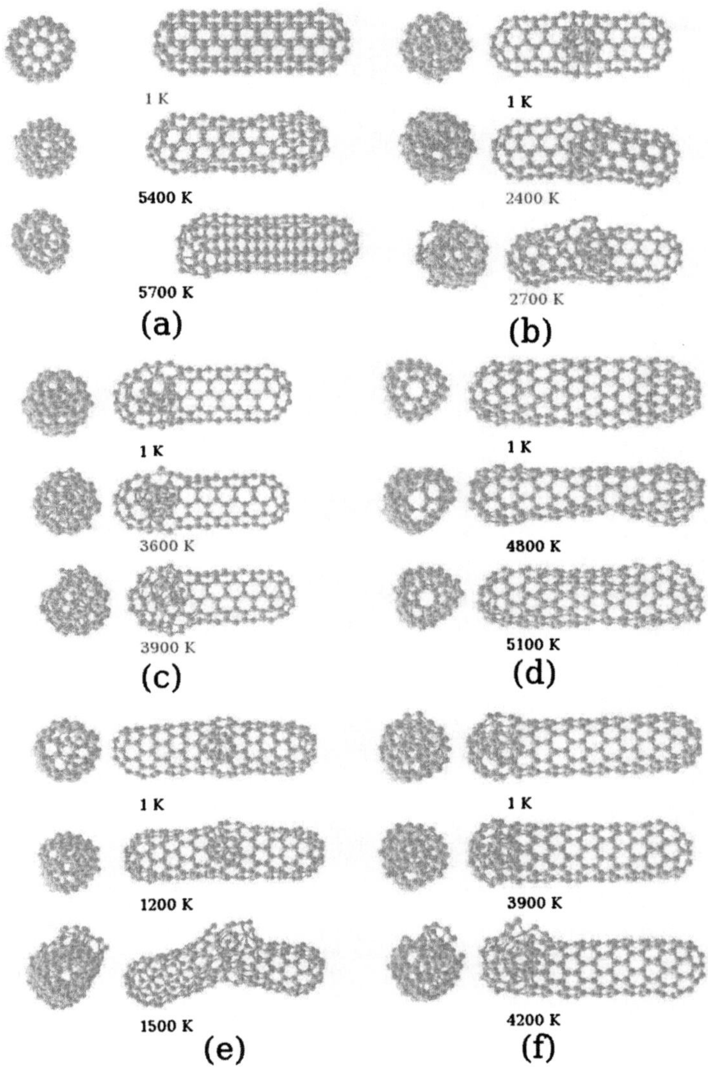

Figure 1: (a) Capsule NC1 formed from C(5,5) carbon nanotube capped with C_{60} buckyball, shown at various temperatures. NC1 collapses at about 5700 K. (b) Capsule NC2 formed from NC1 encapsulating C_{20}. C_{20} is located at the middle of the capsule. NC2 collapses at about 2700 K. (c) Capsule NC3 formed from NC1 encapsulating C_{20}. C_{20} is located at one end of the capsule. NC3 collapses at about 3900 K. (d) Capsule NC4 formed from C(9,0) carbon nanotube capped with C_{60} buckyball, shown at various temperatures. NC4 collapses at about 5100 K. (e) Capsule NC5 formed from NC4 encapsulating C_{20}. C_{20} is located at the middle of the capsule. NC5 collapses at about 1500 K. (c) Capsule NC6 formed from NC4 encapsulating C_{20}. C_{20} is located at one end of the capsule. NC6 collapses at about 4200 K.

CARBON NANOTUBE MULTI-TERMINAL JUNCTIONS: STRUCTURES, PROPERTIES, SYNTHESIS AND APPLICATIONS

L.A.CHERNOZATONSKII, I.V. PONOMAREVA
Institute of Biochemical Physics, Russian Academy of Sciences, Moscow 119991, Russian Federation

Abstract: We summarize recent studies on the structures, properties, methods of preparation and possible applications of carbon nanotube T-, Y- and X-junctions in which all carbon atoms remain sp^2-coordinated throughout. We propose energetically efficient pathways for the process of junction formation in which all atoms maintain their sp^2 arrangements throughout.

Key Words: carbon nanotubes, Y-junction, electronic properties, CVD method, topological defects.

1. INTRODUCTION

The synthesis of carbon nanotubes (CNT) with branched structures is an important step in the development of carbon based nano-electronic devices, because these materials are potentially able to bring in new mechanical or electrical properties [1]. Recently CNT junctions have attracted much attention because of their unique transport properties [2]. It has stimulated many theoretical works [3-5]. Formation of three-terminal nanotube junctions by connecting different CNTs was first proposed in 1992 [6,7]. Such multi-terminal junctions have been first observed in 1995 [8] in arc discharge synthesized material. These junctions have been proposed as promising building blocks for nano-scale device applications [9]. But these earlier experimental observations did not attract much attention due mainly to the difficulties associated with their synthesis and the complexities of their structures. Nevertheless, experimentalists have recently succeeded in achieving controlled growth of Y-junctions [10-17]. In addition to previously reported Y-junctions, H-junctions and 3D CNT, webs have also been produced recently using a special CVD method [18] and X-shaped molecular connections by welding electron-beam irradiation of crossing single-wall nanotubes (SWNT) [19]. Recently, energetically efficient pathways for formation of such SWNT multi-terminal junctions have been proposed [20].

S. Guceri et al. (eds.), Nanoengineered Nanofibrous Materials, 225-235.

Contrary to the CNT case, the study of transport properties of multi-terminal junctions received the most attention when compared with mechanical properties investigations. However, the mechanical properties of carbon nanotube multi-terminal junctions are expected to be even more intriguing than those of carbon nanotubes because of the variety in their forms and geometries. Using an empirical many body-potential, we have shown [21] that the deformation of Y-junction occurs under constant external load applied to branch ends of the junction and leads to a new metastable state of junction with parallel branches.

Due to the complex structure, multi-terminal nanotube junctions exhibit a wider spectrum of transport properties in comparison with those of straight nanotubes.

2. STRUCTURE

Carbon SWNT's constitute molecularly perfect materials with many interesting properties. The electronic structure of these nanotubes can be either metallic or semiconducting, depending on both the diameter and chirality, which can be uniquely determined by the chiral vector *(n,m)*, where *n* and *m* are integers [1]. Thus a SWNT junction can be constructed from different nanotube sets. The resulting structure of multi-terminal junctions still contains three-fold coordination for all carbon atoms, and so it requires the presence of topological defects in the form of pentagons, heptagons, and octagons. This is essential for maintaining sp^2 configuration for all carbon atoms in order to maximize stability. Note that the introduction of a pentagon transforms a hexagonal sheet into a *convex* shape. On the other hand, a heptagon (or octagon) makes a *saddle* shape [1]. Crespi [22] transformed Euler's well-known formula by considering only nonhexagonal polygons on the surface of a closed polyhedron in terms of the faces, vertices, and edges. He proposed a consideration for the formation of complex multiple junctions in terms of local bond surplus at the junction - number of polygonal sides due to nonhexagonal polygons. For example, the presence of a pentagon in a hexagonal sheet contributes to a bond surplus of -1, while heptagons and octagons contribute to bond surpluses of +1 and +2 each, respectively. Thus a junction made of *N* nanotubes has a bond deficit of 6(*N*-2). For all 3-terminal junctions (such as Y- or T- junctions) this gives a bond surplus of +6, and for X-junctions this gives a bond surplus of +12, and so one. Crespi's formula can be simply transformed into a modified equation for junction with *N* open nanotube ends:

$$N_7 + 2N_8 - N_5 = 6(N-2), \qquad (1)$$

where N_5, N_7, N_8 are numbers of pentagons, heptagons and octagons. We have classified planar 3-terminal junctions (Fig.1), which contain tube axes of both branches and stem in the same plane [23], i.e. the Y-, and T- SWNT structures generally observed in experiments – (Fig.2). The latter can be constructed only from "a" - armchair (n,n) or "z" - zigzag (n,0) nanotubes. Thus these junctions are considered only with "a,aa", "a,az", "a,zz", "z,zz", """z,za", "z,aa" sets of nanotubes. The classification is based on the geometry of different junction forms [4, 24]. Figure 1 shows 3 possible groups of Y-junctions with an acute angle between branches.

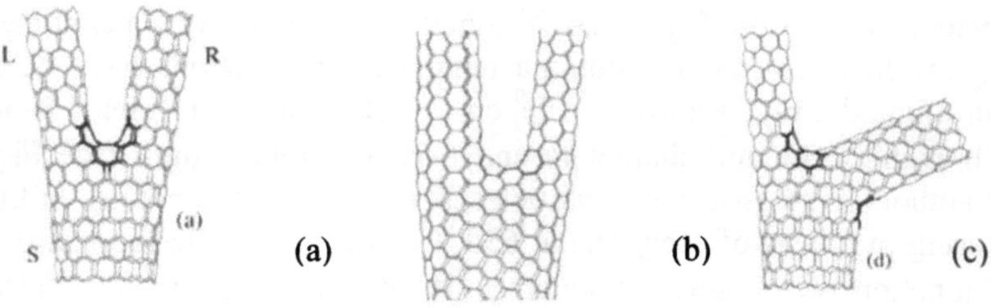

Fig.1 Y-junctions with an acute angle: (a) symmetrical "z,zz" fork with (14,0)-stem and two (7,0)-branches (indexes S, L and R denote stem, left, and right branches); (b) nonsymmetrical "a,aa" fork with (10,10)-stem, and (5,5), (4,4) branches; (c) "z,zz" bough configuration with (17,0) –stem, (7,0) and (10,0)-branches. The heptagons and octagons are shown in dark.

In Figs.1 (a) and (b) six heptagons are situated in the bifurcation area forming acute angle junctions with the same ("a,aa" or "z,zz") structure of nanotubes. The "Bough" must contain stem and main branch (stem continuation) with the same tube structure, but bough-branch may have another structure when one heptagon is situated in area of passage of stem into bough. The "slingshot" group is shown at the Fig.2.

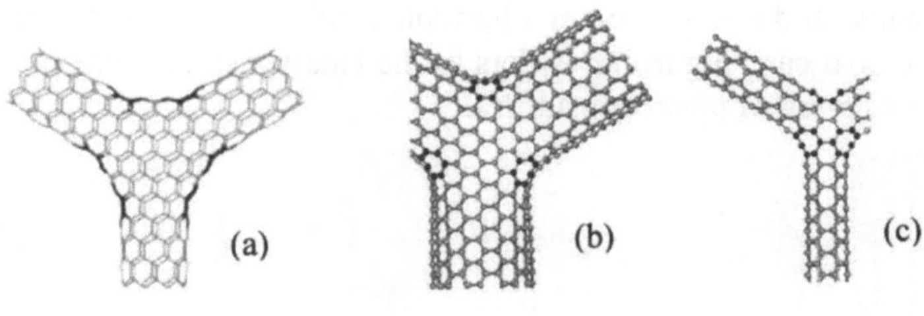

Fig.2 Y-junctions with an obtuse angle: D_{3h} (a) and C_{3h} (b) symmetry "slingshots" with all (8,0) and (9,9) branches respectively of the Y junction; D_{2h}(c) symmetrical slingshot with all (4,4)-branches. The heptagons and octagons are shown in dark.

The T- junction consists of a stem perpendicularly connected with another nanotube (the same or another tube structure [6,8]). Other multi-terminal junctions (MTJ), for example, X- and perpendicular-crossed junctions containing different nanotube types of branches, can also be classified.

3. METHODS OF PREPARATION

In order for the multi-terminal junctions (MTJ) to be useful from device perspective, controlled and macroscopic scale production of these junctions is required. Today there are several methods of CNT Y-junction synthesis that allow for the reproducible and high-yield fabrication [2,11-19]. We consider four main methods from them. The first method is the template-based chemical vapor deposition [2] when Y-junctions of multiwall nanotubes (MWNT) grow in nano-channel alumina templates. Y-junctions have been produced in aligned arrays (density ~10^{10} cm^{-2}) with adjustable "stem" (~50 nm) and "branch" (~35 nm) diameters and 6-10 μm total length (see Fig. 3(a)). The authors have used cobalt particles as catalysts in the bottom of the channels during pyrolysis of acetylene at 650 °C. This allows the fabrications of CNT Y-junction with controlled length (~10 μm) and tube branches (15-100 nm) with an acute angle between them resembling "tuning forks" – Fig.3(a). However, this method requires the use of a special template.

The second method is the pyrolysis of methane with N_2 over cobalt supported on special calcined magnesium oxide substrate. The growth normally lasted for 1 h after reducing Co catalysts at 1000 °C in flowing gases of H_2 (40 sccm) with N_2 (100 sccm) at a pressure of 200 torr, after replacing the N_2 with CH_4 (10 sccm). The Y junctions have very straight arms with uniform (~200nm) diameters, and the angles between the three arms are close to 120° - Fig. 3(b). Some CNTs branched several times to form multiple Y junctions, which still keep their arms straight.

The third method is synthesis in addition to the previously reported *Y*-junction, also *H*-junction and 3D web with CNT's ~50nm in diameter (see Fig. 3(c)). The authors used Si substrate, which was first scratched, cleaned ultrasonically, and then placed in a horizontal tube of a thermal CVD furnace. A ceramic boat carrying iron powders as the catalyst source was positioned at a distance of 5 cm of upstream substrate.

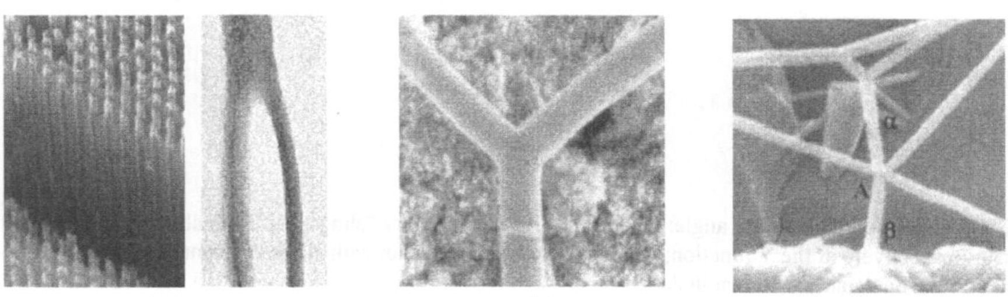

(a) (b) (c)

Fig.3 Carbon nanotube Y-junctions grown by CVD methods. (a) Y-junction CNT array as viewed from the large diameter stem side (scanning electron micrograph - left); acute angle "fork" junction with stem 100 nm and branched ~50 nm in diameter removed from the template (transmission electron micrograph - right) [2]. (b) Y- junction with uniform 200 nm diameter arms, and the angles between them close to 120° [16]; (c) 3D five-terminal junction of CNT's ~50nm in diameter [18].

Very recently Terrones et al. [19] have obtained stable SWNT junctions of various geometries (see, for example, electronic micrographs in insets of Fig. 4, 6) by an electron beam welding after a few minutes of irradiating at specimen temperatures of 800°C. The nanotube behavior was monitored under usual imaging conditions in a transmission electron microscope: 1.25 MeV electron energy beam with intensity of ~ 10 A/cm^2. The junctions were created via vacancies and interstitial defects, induced by the focused electron beam, that promote the formation of intertube links. The authors have supposed that electron beam exposure at high temperatures induces structural defects, which promote the joining of SWNTs via cross-linking of dangling bonds. They have modeled this process by molecular dynamics simulations, but did not describe energetic process of transformation of separate tubes into their junction.

4. ENERGETICALLY EFFICIENT PATHWAYS OF MTJ FORMATION FROM SWNT'S

Different growth mechanisms are responsible for multi-terminal junction formation in synthesis methods described above. For example in work [24] multi-terminal junction growth during CVD synthesis was explained in terms of interaction of nanotube surface with catalyst particle.

Here we consider the energetically efficient process of MTJ under electron beam welding of SWNT's [19]. Using tight-binding as well as classical molecular dynamics [25] we simulate the formation of single-wall carbon nanotube T-, Y- and X-junctions via the fusing of two nanotubes. We propose energetically efficient pathways for this process in which all atoms maintain their sp^2 arrangements throughout. Recent experimental advances have greatly increased the plausibility of synthesizing multi-terminal junctions as proposed in the simulations. At first, we consider different steps leading to the formation of the (5, 5)–(10, 0)–(5, 5) T-junction. The 8-step process in which *sp*^{2} coordination is maintained throughout for all the atoms in the junction region is shown in Fig. 4. An armchair (5,5) nanotube wall contains defects (shown in dark) which could arise due to localized e-beam or ion-beam irradiation-induced localized heating/welding at the location of the junction point [19]. In the first step, the capped end of the (10,0) nanotube is near the defected region of the (5,5) nanotube. In the second step, the inter-nanotube connectivity is through 4 bonds forming the sides of 4 octagons. This structure also contains defect rings in the form of 4 pentagons and 2 heptagons. In step three, two of the octagons are annihilated, and, additionally, the tructure contains 4 pentagons and 6 heptagons in the junction region. In step four, the rearrangement of atoms causes the neck to widen. The defects are identical to step three, but their positioning is different. The subsequent arrangements of defects as we proceed toward the final structure are: 6 pentagons and 12 heptagons (step five); 4 pentagons, 6 heptagons and 2 octagons (step six); and 2 pentagons and 8 heptagons (step seven). The final structure (step eight) contains 7 heptagons and no pentagons.

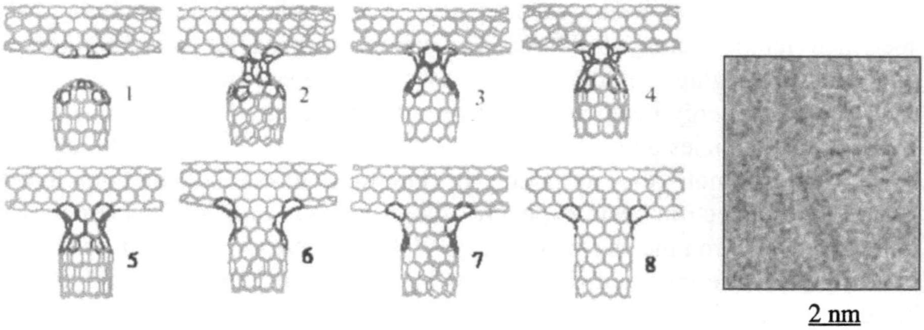

Fig.4. Steps of an energetically efficient pathway for coalescence of the (10,0) tube with the (5, 5) tube: (5,5) –(10, 0)–(5, 5) T-junction formation. The defect rings are shown in dark and include 5-, 7- and 8-fold rings. Inset – HREM image of T-junction obtained be electron beam welding [19]. The shape of junction obtained experimentally is similar to the junction shape in the proposed pathway.

All structures are fully relaxed at each step and the total energy is calculated using the GTBMD scheme [25]. As seen in the Figure 5, an increase in energy (indicating a barrier to the process in steps 1–3) is followed by a drop in energy (in steps 3–8) without any significant barrier. Once the initial barrier is overcome, the formation of the junction in steps 3–8 can be driven by the lowering of energy.

The formation pathway for X-junctions by bringing one (6, 6) armchair tube near another (6, 6) tube has been studied. Since the number of atoms involved is rather large, we perform relaxations using Brenner's three-body reactive potential [26], and the GTBMD scheme is used only to check the stability of the final structure. Here, the first step does not require any a priori existence of defects on the sidewalls of the nanotubes. The five-step process leading to the formation of a four-terminal X-junction is shown in Fig. 6. The final structure obtained in step five contains topological defects in the form of 12 heptagons only according to equation (1). This structure was also found to be stable under GTBMD relaxation. During the considered process of MTJ formation no structural defects are generated and the transformation is achieved through creation/annihilation of topological defects.

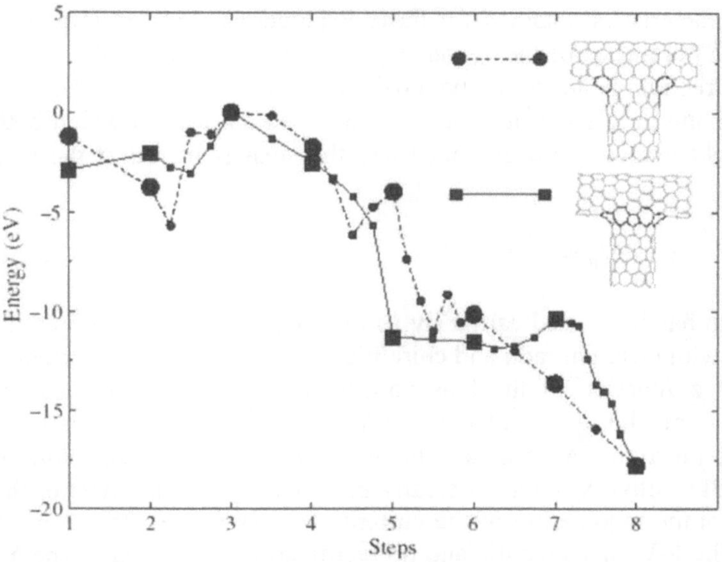

Fig. 5. Relative energies at each step for (5,5)–(10,0)–(5,5) and (9,0)–(10,0)–(9,0) T-junctions. Intermediate energy points are represented by small circles and squares.

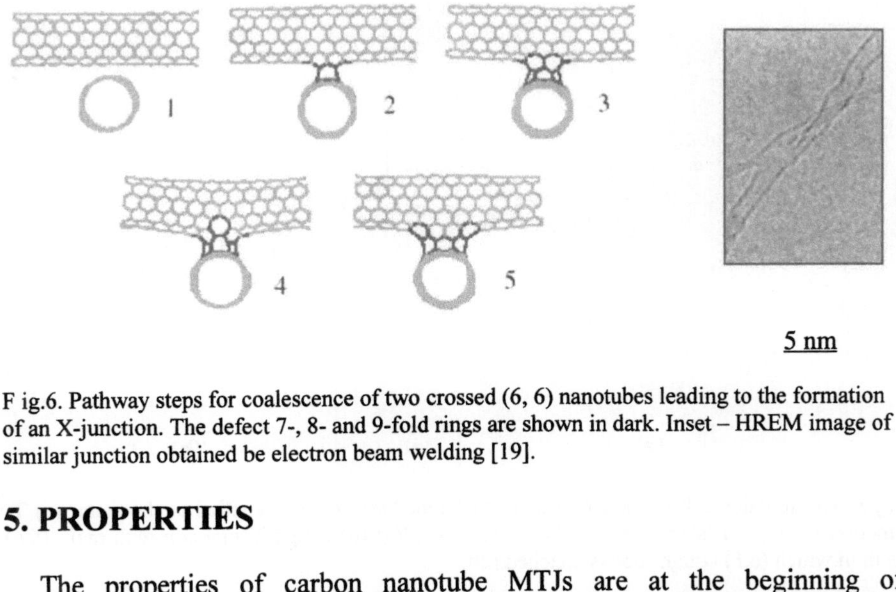

F ig.6. Pathway steps for coalescence of two crossed (6, 6) nanotubes leading to the formation of an X-junction. The defect 7-, 8- and 9-fold rings are shown in dark. Inset – HREM image of similar junction obtained be electron beam welding [19].

5. PROPERTIES

The properties of carbon nanotube MTJs are at the beginning of investigations. So far the conductance measurements on template prepared Y-"fork" junctions have shown intrinsic nonlinear and asymmetric *I–V* behavior at room temperature [2]. Theoretical calculations have supported these experimental findings [3, 4]. Recently, the quantum conductances of T- and "bough" type Y-junction have been computed using a Green's function embedding scheme [27,20]. The Green function was constructed using the same Hamiltonian as used for obtaining structural GTBMD relaxations. Fig.7

shows I–V characteristics obtained for these Y-junctions. The left (L), right (R) and stem (S) branches of the T and Y-junctions are shown in the insets. The current direction is taken to be positive when flowing towards the junction region and negative otherwise. First we used a set-up in which both the left and right branches are grounded and the stem is biased at the bias voltage V_b, i.e.

$$V_L = V_R = 0;\ V_S = V_b. \tag{2}$$

A similar set-up has been used earlier by us to study the transport properties of Y-junctions with different radii and chiralities [4]. In Fig. 7(a) the current I_s in the stem as a function of the bias voltage V_s was calculated also for different values of the gate voltage V_L=V_g (V_R = 0) for the case of symmetrical T-junction. Asymmetry in the *I-V* behavior with current saturation for all positive V values is clearly indicated. The main effect of the variation in V is the modulation of the current. As seen in Fig.7(b), there is asymmetry in the I–V characteristic and no rectification for "bough" type Y-junction . This is consistent with our finding reported earlier- that no rectification at all is possible for asymmetric Y-junctions [4].

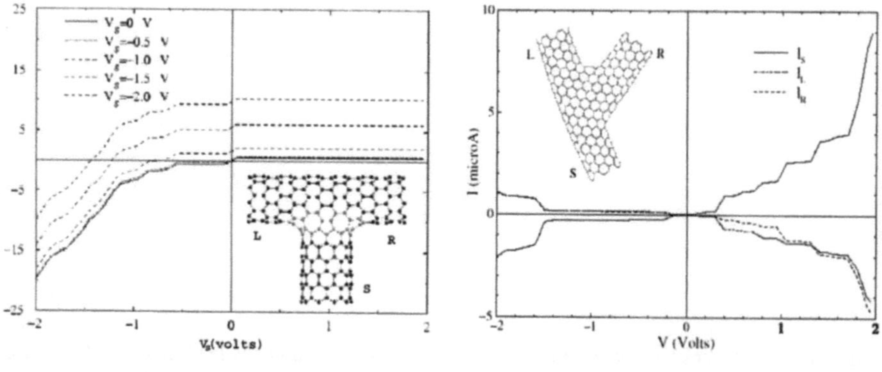

(a) **(b)**

Fig.7. The junction *I-V* characteristics for different bias voltages applied at the stem: (a) T-junction with (10,0) stem tube and (9,0) crossbar [26]; (b) bough Y-junction with main (6,6) stem in which (6,6) bough-tube is attached [20].

Unusual MTJ mechanical properties should be connected with their geometric forms. Here we demonstrate one of them [21]. Let's consider Y-'fork" junction with (20,20) tube stem and the angle between (13,13) tube branches of 23° under external load – Fig. 8(a). We have shown by molecular dynamics (with Brenner potential [26]) simulation that the deformation of Y-junction form occurs under constant external load and at the end of the deformation process the branches of junction are parallel with the distance between them corresponded to the interlayer distance in graphite (Fig.8(b)). External load was simulated by placing the Y-junction between two graphite planes moving toward each other during simulation. The new state is stable up to temperature 2000K. The transition between two states has been investigated.

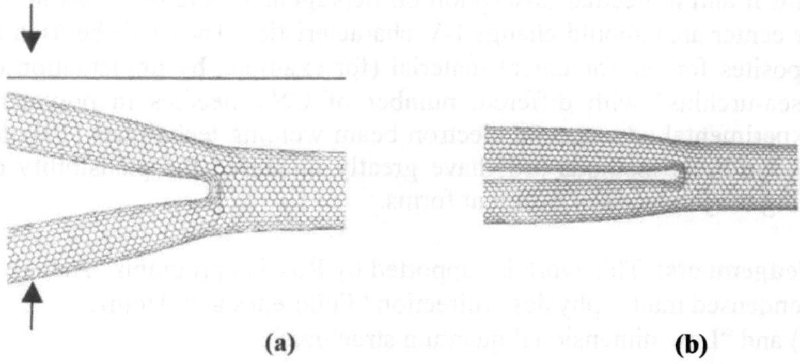

Fig.8. (a) Y-junction of (20,20) tube-stem and (13,13) tubes-branches (arrays show the force action on the "fork"; the heptagons are shown in dark); (b) "fork" with stick together branches is in metastable state after unloading.

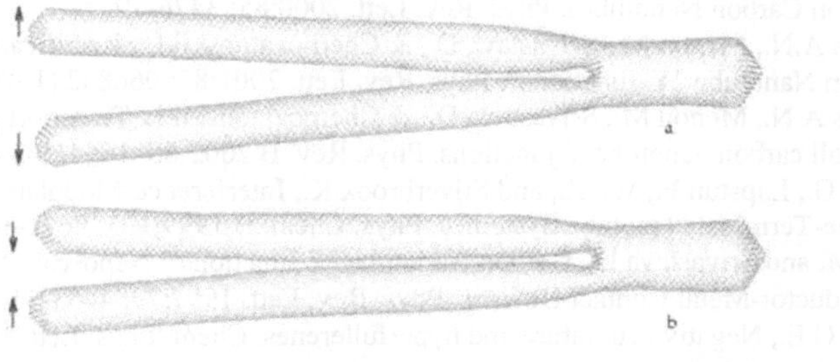

Fig. 9. Behavior of Y-junction of tune fork type with long (~21 nm) branches after unloading in intermediate action time: (a) distance between branches of 2.5 nm, it can be seen that branches overlap area is only about 25% of branches length; (b) distance between branches of 1.4 nm, the van der Waals overlap area is more than a half of branch length. Arrows show the direction of branch movement during MD simulation.

We have observed that if the angle between branches exceeds some critical value van der Waals interaction between them acts like a zipper-mechanism forcing branches to stick together – Fig.9(b). So within the empirical many-body potential scheme it was shown that Y-junction with acute angle between branches has at least two local energy minimums.

6. CONCLUSION

Modelling of new multi-terminal junctions, their forming mechanisms, interpretation of the experimental data on Y-junction transport properties and prediction new MTJ properties open new areas of carbon nanotube structure investigation. MT carbon nanotube junctions can be used as new elements of nanomechanical systems and NEM: springs, vibrators (0.01-1 THz frequencies), mixer elements, memory elements, nanoactuators; nanoelectronic devices: rectifiers, switches, antennas & so on; as nanosensors (gas chemical and molecular adsorption on heptagons or/and octagons in the connector center area should change I-V characteristic). They will be used in nanocomposites for reinforcement material (for example, by implantation of carbon "sea-urchins" with different number of CNT needles in polymer). Recent experimental advances in electron beam welding techniques [19] and new MTJ synthesis methods [18] have greatly increased the plausibility of synthesizing the junctions of different forms.

Acknowledgements: This work is supported by Russian programs "Actual lines in condensed matter physics" (direction "Fullerenes and Atomic Clusters") and "Low dimensional quantum structures".

References

1. Saito R., Dresselhaus G., and Dresselhaus M. S., *Physical Properties of Carbon Nanotubes* Imperial College Press, London, 1998.
2. Papadopoulos C., Rakitin A., Li J., Vedeneev A.S., and Xu. J.M. Electronic Transport in Y-Junction Carbon Nanotubes. Phys. Rev. Lett. 2000;85: 3476-79.
3. Andriotis A.N., Menon M., Srivastava D., & Chernozatonskii L., Rectification Properties of Carbon Nanotube "Y-Junctions". Phys. Rev. Lett. 2001;87: 066802 (1-4).
4. Andriotis A.N., Menon M., Srivastava D., & Chernozatonskii L. Transport properties of single-wall carbon nanotube Y junctions. Phys. Rev. B 2002;65: 165416-24.
5. Treboux G., Lapstun P., Wu Z., and Silverbrook K., Interference-Modulated Conductance in a Three-Terminal Nanotube System. J. Phys. Chem. B 1999;103: 8671-86.
6. Menon M. and Srivastava D., Carbon Nanotube "T Junctions": Nanoscale Metal-Semiconductor-Metal Contact Devices. Phys. Rev. Lett. 1997;79: 4453-56.
7. Scuseria G.E., Negative curvature and hyperfullerenes. Chem. Phys. Lett. 1992;195: 534-36.
8. Chernozatonskii L.A., Phys. Lett. A. Carbon nanotube connectors and planar jungle gyms 1992;172: 173-76.
9. Zhou D. and Seraphin S., Complex branching phenomena in the growth of carbon nanotubes. Chem. Phys. Lett. 1995;238: 286-290 .
10. Li J., Papadopoulos C., and Xu J., Nanoelectronics: Growing Y-junction carbon nanotubes. Nature 1999;402: 253-57.
11. Satishkumar B.C., Thomas P.J., Govindraj A., and Rao C.N.R., *Y*-junction carbon nanotubes .Appl. Phys. Lett. 2000;77: 2530-32.
12. .Nagy J.B, Ehlich R., Biro L.B., & Gjulai J.. Appl. Phys. A 2000;70: 481-83.
13. Gan B., Ahn J., Zhang A., Yoon S. F., Rusli, Huang Q. F., Yang H., Yu M. B., and Li W. Z., Branching carbon nanotubes deposited in HFCVD system. Diamond Relat. Mater. 2000;9: 897-900.
14. Gan B., Ahn J., Zhang Q., Huang Q.-F., Kerlit C., Yoon S. F., Rusli, Ligachev V. A., Zhang X.-B., and Li W.-Z., Topological structure of Y-junction carbon nanotubes. Mater. Lett. 2000;45: 315-19.
15. Deepak F.L.. Govindraj A., Rao C.N.R., Synthesis strategies for Y-junction carbon nanotubes. Chem. Phys. Lett. 2001;345: 5-10.
16. Li W. Z.,. Wen J. G, and Ren Z. F., Straigt carbon nanotube Y-junctions. Appl. Phys. Lett. 2001;79: 1879-81.

17. Osvath Z., Koos A.A., Horvath Z.E., Gyulai J., Benito A.M., Martinez M.T., Maser W.K., Biro L.P., Arc-grown Y-branched carbon nanotubes observed by scanning tunneling microscopy (STM). Chem. Phys. Lett. 2002;365: 338-42.
18. Ting J.-M. and Chang C.-C., Multijunction carbon nanotube network. Appl. Phys. Lett. 2002;80: 324-25.
19. Terrones M., Banhart F., Grobert N.,. Charlier J.-C, Terrones H., and Ajayan P. M., Phys. Rev. Lett. Molecular Junctions by Joining Single-Walled Carbon Nanotubes. 2002;89:075505 (1-4).
20. Ponomareva I.V., Chernozatonskii L.A., Andriotis A. N., Menon M.. New Journal of Physics 2003;5: 1.1-1.12 (http://www.njp.org/).
21. Chernozatonskii L.A., Ponomareva I.V . Sticking of Carbon Nanotube Y-junction branches. JETP Lett. 2003;78: 327-31.
22. Crespi V.H., Relations between global and local topology in multiple nanotube junctions. Phys. Rev. B 1998;58: 12 671.
23. Chernozatonskii L.A. and S.V. Lisenkov, Classification of three-terminal nanotube junctions. Fullerenes, Nanotubes and Carbon Nanostructures 2003; (in press).
24. Chernozatonskii L.A, J. Nanopartical Res. 2003; (in press).
25. Menon M, Richter E and Subbaswamy K R. Structural and vibrational properties of fullerenes and nanotubes in a nonorthogonal tight-binding scheme. J. Chem. Phys. 1996;104 : 5875-87.
26. Brenner D. W., Empirical potential for hydrocarbons for use in simulating the chemical vapor deposition of diamond films. Phys. Rev. B. 1990;42: 9458-9471.
27. Menon M., Andriotis A. N., Ponomareva I.V., Chernozatonskii L.A.. Carbon nanotube ``T-junctions": Formation pathways and conductivity. Phys. Rev. Lett. 2003;91: 115501(1-4).

SIMULATION OF CARBON NANOTUBE JUNCTION FORMATIONS

E. Tasci, O.B. Malcioglu and S. Erkoc
Department of Physics, Middle East Technical University, Ankara 06531 Turkey

Abstract: In this work we have examined the possible formation of a junction between two identical C(10,0) carbon nanotubes. One of the tubes was rotated 90 degrees with respect to the other. Simulation have been performed by means of a molecular-dynamics technique at 1K. For this purpose, we have introduced two stiff layers of graphite positioned above and below the nanotubes. By moving these layers we have created an effective force pushing the tubes closer to each other. In this simulation we have used a semi-empirical many-body potential for carbon.

Key words: carbon nanotubes; molecular-dynamics, empirical potential; nanojunction.

1. INTRODUCTION

Since Iijima's 1991 dated discovery of carbon nanotubes [1], many potential applications have been developed. Many of these applications, such as fluidic and electronic devices, require a junction between nanotubes, and there is still much work to be done in order to find an effective way to create such junctions. Our aim in this work is to investigate the possibility of formation of a junction under pressure. For this, the system was considered to be at 1 K through the simulation.

The system consisted of two identical C(10,0) nanotubes, rotated 90 degrees with respect to each other (Figure 1b) so that their axis reside on x and y directions. The separation in the z axis between the tubes at the beginning of the simulation was larger than the cutoff distance of the potential used , thus all the atoms in each nanotube was outside the effective influence range of the other.

Then, two stiff graphene layers were introduced above and below the two-tube system (Figure 1a). These layers were large enough so that the nanotubes cannot escape from the sides, and the entire tube is under the influence of the particular graphene layer. The purpose of these layers was to create an effective pressure without directly interfering with the tube atoms. Periodical boundary conditions were applied to the system considered.

In the course of the simulation, the stiff layers of graphite was moved towards each other in the z axis, followed by a relaxation run. This process continued until the nanotubes mended.

S. Guceri et al. (eds.), Nanoengineered Nanofibrous Materials, 237-240.

2. THE PEF AND THE MD SIMULATION

The empirical many-body potential energy function (PEF) developed for carbon is used in the calculations [2]. This PEF describes the structural properties and energetics of carbon that is relatively accurate. The total interaction energy of a system of particles is taken to be the sum of total two-body and total three-body contributions:

$$\Phi = \phi_2 + \phi_3 \tag{1}$$

Total two–body and three–body energies are expressed, respectively, as

$$\phi_2 = A\sum_{i<j}^{N} U_{ij}^{(1)} \quad , \quad \phi_3 = -B\sum_{i<j}^{N} U_{ij}^{(2)}\left[1+\beta^n\left(\sum_{k\neq i,j}^{N} W_{ijk}\right)^n\right]^{-1/2n} \tag{2}$$

here U_{ij} and W_{ijk} represent the two–body and three–body interactions, respectively.

$$U_{ij}^{(1)} = f_c(r_{ij})\exp(-\lambda_1 r_{ij}) \tag{3}$$

$$U_{ij}^{(2)} = f_c(r_{ij})\exp(-\lambda_2 r_{ij}) \tag{4}$$

$$W_{ijk} = f_c(r_{ik})g(\theta_{ijk}) \tag{5}$$

where

$$g(\theta_{ijk}) = 1 + \frac{c^2}{d^2} - \frac{c^2}{d^2 + (h - \cos\theta_{ijk})^2} \tag{6}$$

$$f_c(r) = \begin{cases} 1 & \text{for} \quad r < R - D \\ \frac{1}{2} - \frac{1}{2}\sin[\frac{\pi}{2}(r-R)/D] & \text{for} \quad R - D < r < R + D \\ 0 & \text{for} \quad r > R + D \end{cases} \tag{7}$$

The parameters of the PEF for carbon are as follows: $A = 1393.6\,eV$, $B = 346.74\,eV$, $\lambda_1 = 3.4879\,\AA^{-1}$, $\lambda_2 = 2.2119\,\AA^{-1}$, $\beta = 1.5724\times10^{-7}$, $n = 0.72751$, $c = 38049$, $d = 4.3484$, $h = -0.57058$, $R = 1.95\,\AA$, and $D = 0.15\,\AA$.

The equations of motion of the particles are solved by considering the Verlet algorithm. The canonical NVT ensemble molecular–dynamics is used. The temperature scaling is taken into account at every MD step and the temperature of the system is kept constant at a given temperature. One time step is taken as $10^{-16}\,s$. The initial velocities of the particles are determined from the Maxwell distribution at the given temperature.

3. RESULTS AND DISCUSSION

Nanotubes and corresponding graphene layers were considered to be of infinite length, so that the deformation of the nanotubes at the ends, and various structural instabilities due to discontinuity are removed. In order to simulate infinite length, periodic boundary conditions in the x and y directions were introduced. One side effect of this was that the system became a grid of carbon nanotubes instead of isolated two. In order to compansate this, the length of the simulated region was chosen to be large enough so that the area where possible formation of a junction is expected is not effected by "ghost" atoms due to periodic boundary conditions.

Systematical relocation of the graphite layers each by 0.25 A towards each other and the relaxation of the system for 30000 time steps at 1 K for every relocation showed that, the tubes indeed moved closer but also, rotated in their longitudinal axes.

Since the graphite atoms were stiff (i.e., they are involved in potential calculations but do not move), the graphene layer was not deformed. This may also be the case in the presence of a strong macroscopic graphitic crystal.

At the end of the simulation, it has been observed that the two tubes successfully merged with each other (Figure 2). Although the shape of the carbon nanotubes are greatly distorted, it is known that carbon nanotubes can withstand an enormous amount of structural deformation that can be reversed. Simulations are still been carried out if the resulting structure resumes tubular form when the layers are eventually removed.

The effect of temperature may also be interesting. Increased temperature may result in mending before the nanotubes become this much distorted, increasing the chance of resuming tubular form after the formation.

Using two different tubes with different electrical properties may yield a transistor-like component whereas a weaving of such cross-nanotubes may form a net-like system. Simulations of both of these aspects are in progress. Since this process is easier to conduct with respect to other processes like arc melding [3,4], it may be more convenient in engineering applications.

Acknowledgements: The authors would like to thank METU (Middle East Technical University) for partial support through the project METU-BAP-2003-07-02-00-19, and TUBITAK for partial support through the project TBAG–U/41 (102T007).

References:

1. Iijima, S. (1991) Helical Microtubules of Graphitic Carbon, Nature, 354, 56
2. Tersoff, J. (1988) Empirical Interatomic Potential for Carbon, with Applications to Amorphous Carbon, Phys. Rev. Lett., 2879, 61
3. Ebbesen, T.W., (1994) Carbon Nanotubes, Ann. Rev. Mater. Sci., 24, 235
4. Ebbesen, T.W., (1997) in Carbon Nanotubes: preparation and properties, ed. T.W. CRC Press, Boca Raton, 139

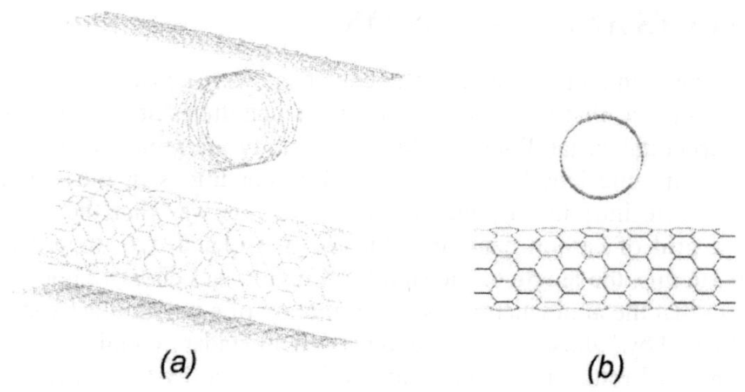

Figure 1: (a) The system prepared after addition of graphene layers and relaxation. (b) The initial position of nanotubes.

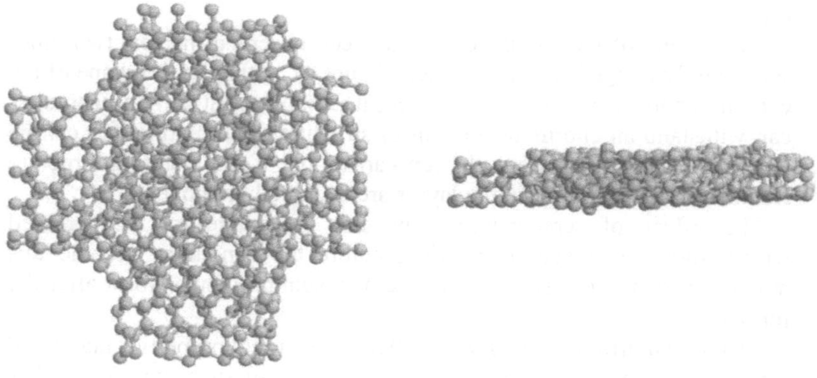

Figure 2: Mended carbon nanocapsules.

STABILITY OF CARBON NANOTORI

E. Yazgan, E. Tasci, O.B. Malcioglu and S. Erkoc
Department of Physics, Middle East Technical University, Ankara 06531 Turkey

Abstract: The structural stability of carbon nanotori have been investigated by performing molecular--dynamics simulations. The systems considered are C_{170}, C_{360}, C_{520}, and C_{750} tori, which have been constructed using an algorithm developed in our laboratory based on Fonseca's idea. Calculations have been realized by using an empirical many-body potential energy function for carbon.

Key words: carbon nanotorus; molecular-dynamics; empirical potential.

1. INTRODUCTION

The discovery of fullerenes [1] and carbon nanotubes [2] opened a new era in nanoscience. Much attention has been given to nanotubes and fullerenes. However, only a relatively small number of studies on the carbon nanotoroidal cage structures exist. Various methods have been suggested to model toroidal structures [3-7]. A molecule in the shape of a torus is expected to have special chemical and physical properties due to its multi-connectivity and its negative curvature regions [8,9]. In this work, we examine the structural stabilities of some nanotori formed using the method of Fonseca et al. [5].

Fonseca--type nanotori consist of nanotubes, joined by 'knee regions'. Curvature of the knee regions are given by pentagon-heptagon pairs. The heptagons give negative curvature and the pentagons positive curvature. Thus at the knee regions stress is reduced. The tori constructed have 5-fold symmetry and are composed of armchair and zigzag type nanotubes. An algorithm to generate Fonseca-type tori is explained in [10].

2. THE PEF AND THE MD SIMULATION

The empirical many-body potential energy function (PEF) developed for carbon [12] is used in the calculations. This PEF describes the structural properties and energetics of carbon relatively accurate. The total interaction energy of a system of particles is taken to be the sum of total two-body and total three-body contributions:

$$\Phi = \phi_2 + \phi_3 \qquad (1)$$

S. Guceri et al. (eds.), Nanoengineered Nanofibrous Materials, 241-244.

Total two-body and three-body energies are expressed, respectively, as

$$\phi_2 = A\sum_{i<j}^{N} U_{ij}^{(1)} \quad , \quad \phi_3 = -B\sum_{i<j}^{N} U_{ij}^{(2)}\left[1+\beta^n\left(\sum_{k\neq i,j}^{N} W_{ijk}\right)^n\right]^{-1/2n} \tag{2}$$

here U_{ij} and W_{ijk} represent the two-body and three-body interactions, respectively.

$$U_{ij}^{(1)} = f_c(r_{ij})\exp(-\lambda_1 r_{ij}) \tag{3}$$

$$U_{ij}^{(2)} = f_c(r_{ij})\exp(-\lambda_2 r_{ij}) \tag{4}$$

$$W_{ijk} = f_c(r_{ik})g(\theta_{ijk}) \tag{5}$$

where

$$g(\theta_{ijk}) = 1+\frac{c^2}{d^2}-\frac{c^2}{d^2+(h-\cos\theta_{ijk})^2} \tag{6}$$

$$f_c(r) = \begin{cases} 1 & \text{for} \quad r < R-D \\ \frac{1}{2}-\frac{1}{2}\sin[\frac{\pi}{2}(r-R)/D] & \text{for} \quad R-D < r < R+D \\ 0 & \text{for} \quad r > R+D \end{cases} \tag{7}$$

The parameters of the PEF for carbon are as follows: A=1393.6 eV, B=346.74 eV, λ_1=3.4879 Å^{-1}, λ_2=2.2119 Å^{-1}, β=1.5724×10^{-7}, n=0.72751, c=38049, d=4.3484, h=-0.57058, R=1.95 Å, and D=0.15 Å.

The equations of motion of the particles are solved by considering the Verlet algorithm. The canonical NVT ensemble molecular-dynamics is used. The temperature scaling is taken into account at every MD step and the temperature of the system is kept constant at a given temperature. One time step is taken as 10^{-16} s. The initial velocities of the particles are determined from the Maxwell distribution at the given temperature. The simulations are carried out starting at 1 K and the temperature is increased with steps of 100 K up to the temperature the system deforms. At every temperature rise the system is relaxed for about 50000 time steps.

3. RESULTS AND DISCUSSION

Four different toroidal structures have been considered in this study; C_{170}, C_{360}, C_{520} and C_{750}. The deformation temperatures of C_{170}, C_{360}, C_{520}, and C_{750} are found to be 3700 K, 4500 K, 4200 K and 4100 K respectively. Deformation of C_{170} is shown in Figure 1. The stability of the toroidal structures against heat treatment show a complex dependence on geometrical parameters. Each toroid deformed at a different temperature. Deformation on C_{170}, C_{520} and C_{750} toroidal structures started at a knee-region, however deformation on C_{360} started between two knee-regions.

In a similar work, thermal stability of nanotori with different geometry (different locations of pentagons and heptagons) but similar size have been investigated [12]. In that work, C_{120} nanotorus has deformed at about 4600 K with a little bit of distortion in the structure before deformation temperature. Two different isomers of C_{240}, had deformation temperatures of 4100 K and 5400 K. It may be argued that smooth geometry of these nanotori have increased the thermal stability considerably, since strain due to curvature is distributed along the whole structure, and not concentrated at the knee-regions like in the present case.

Thermal stability investigations on other carbon nanostructures such as nanotubes and fullerenes, showed that the deformation temperature varies depending on the size and the geometry of the nanostructures. Deformation temperature of nanotori considered are comparable with carbon nanotubes and nanorods of similar cross section radius. Deformation temperature of nanotubes, nanorods, and fullerenes change in the range 1500-7600 K [12-17]. Molecular-dynamics simulations show that nanotubes expand as they are heated [14]. This observation finds confirmation from experiment [18]. As a conclusion, in an isolated environment these structures seem to be quite stable.

Acknowledgements: The authors would like to thank METU (Middle East Technical University) for partial support through the project METU-BAP-2003-07-02-00-19, and TUBITAK for partial support through the project TBAG--U/41 (102T007).

References

1. Kroto H.W. C_{60}: Buckminsterfullerene. Nature **318**, 162 (1985).
2. Iijima S. Helical Microtubules of Graphitic Carbon. Nature **354**, 56 (1991).
3. Dunlap B.I. Connecting Carbon Tubules. Phys. Rev. B **46**, 1933 (1992).
4. Chernozatonskii L.A. Carbon Nanotube Elbow Connections and Tori. Phys. Lett. A **170**, 37 (1992).
5. Fonseca, A., Hernadi, K., Nagy, J. B. Nagy, PH. Lambin and Lucas, A.A. Model Structure of Perfectly Graphitizable Coiled Carbon Nanotubes. Carbon **33**, 1759 (1995).
6. Berger, J., and Avron, J.E. Classification Scheme for Toroidal Molecules. J. Chem. Soc. Faraday Trans. **91**, 4037 (1995).
7. Diudea, M.V., Silaghi-Dumitrescu, I and Parv, B., Toroidal Fullerenes from Square Tiled Tori. Internet Electronic Journal of Molecular Design. Perprint No 1, http://www.biochempress.com
8. Shapere A., and Wilczek F. Geometric Phases in Physics. Signapore: World Scientific (1989).
9. Avron J.E., Klein M., Pnueli A., and Sadun L. Hall Conductance and Adiabatic Charge Transport of Leaky Tori. Phys. Rev. Lett. **69**, 128 (1992).
10. Yazgan, E., Tasci , E. and Erkoc, S. An Algorithm to Generate Toroidal and Helical Structures using Pentagons, Hexagons and Heptagons, International Journal of Modern Physics: C (2003), in print.
11. Tersoff, J. Empirical Interatomic Potential for Carbon, with Applications to Amorphous Carbon. Phys. Rev. Lett. **61**, 2879 (1988).

12. Erkoc S. and Vural D.C., Molecular-Dynamics Simulations of Carbon Nanocage Structures: Nanoballs and Nanotoroids. Int. J. Mod. Phys. C **12,** 685 (2001).
13. Erkoc S. From Carbon Nanotubes to Carbon Nanorods. Int. J. Mod. Phys. C **11,** 1247 (2000).
14. Erkoc S. and Malcıoglu O.B., Effect of Chirality on the Stability of Carbon Nanotubes: Molecular-Dynamics Simulations. Int. J. Mod. Phys. C **12,** 865 (2001).
15. Erkoc S. and Turker L. Structural and Electronic Properties of Carbon Nanoballs: C_{20}, C_{60}, and C_{20}@ C_{60}. Int. J. Mod. Phys. C **12,** 1391 (2001).
16. Erkoc S. Stability of Carbon Nanoonion C_{20}@ C_{60} @C_{240}: Molecular-Dynamics Simulations. Nano Letters **2,** 215 (2002).
17. Malcıoglu O.B. and Erkoc S. Structural Properties of Diamond Nanorods: Molecular-Dynamics Simulations. Int. J. Mod. Phys. C **14,** 441 (2003).
18. Yudasaka M., Kataura H., Ichihashi T., Qin L.-C., Kar S., and Iijima S. Diameter Enlargement of HiPco Single-Wall Carbon Nanotubes by Heat Treatment. Nano Letters **1(9),** 487 (2001).

(a) (b) (c)

Figure 1. Carbon 170: Relaxed structure (a) at 1 K, (b) at 3600 K and (c) at 3700 K in which the structure considered deforms.

USE OF HIGH SURFACE NANOFIBROUS MATERIALS IN MEDICINE

S.V. Mikhalovsky[1], L.I. Mikhalovska[1], V.G. Nikolaev[3], V.I Chernyshev[2], V.V. Sarnatskaya[3], K.Bardakhivskaya[3], A.V. Nikolaev[4], L.A. Sakhno[3],

[1]*School of Pharmacy and Biomolecular Sciences, University of Brighton, Lewes Road, Brighton, BN2 4GJ, UK;* [2]*AV Palladin Institute of Biochemistry, National Academy of Sciences, Kiev, Ukraine;* [3]*RE Kavetsky Institute of Experimental Pathology, Oncology and Radiobiology, National Academy of Sciences, Kiev, Ukraine;*[4]*Moscow Institute of Physics and Technology, Russia.*

1. INTRODUCTION

Interest in potential biomedical applications of nanomaterials and nanostructured materials arises from the perception that they are able to interact with individual biomolecules, cells and their individual parts and other biological structures of similar size in the micrometre and nanometre range. They may also have properties different from micro- and macro-objects. Recently nanofibres (NF) were mentioned as a connection between the nanoscale world and the macroscale world, because at least in theory, having diameter of a fraction of a micron, nanofibres may be infinitely long [1]. Production of nanofibres by electrospinning has been extensively studied recently and a variety of porous NF made by different modifications of this method has been described. They include polylactide (PL), poly-glycolide (PG), polyethylene terephthalate, polyacrylonitrile, polyvinyl-carbazole, polycarbonate, polystyrene/ polymethylmethacrylate, PL/PEO, PL/polyvinylpyrrolidone and PL/(polyethylene oxide), [2]. Co-electro-spinning of colloidal particles such as hydroxyapatite with polymer solutions results in the production of NF with incorporated nanoparticles. Carbon nanotubes were co-electrospun with PEO and nanocomposite fibrils with very high mechanical strength were obtained [3]. Similarly, if polymer NF are produced from a drug containing solution, controlled drug release systems can be obtained [4]. Nanofibres of biodegradable polymers such as PL and PG are of particular interest for tissue engineering scaffolds, as they will completely dissolve after implantation thus leaving no foreign material in the body [5]. Using electrospinning technique, collagen fibres of 100 nm in diameter have been produced, which is similar to the size of natural collagen fibres [6]. The electrospun collagen NF were shown to promote cell growth and penetration into the engineered tissue scaffold.

Currently the main problem with a large scale production of nanofibres is an extremely slow rate of the electrospinning process. Non Woven

S. Guceri et al. (eds.), Nanoengineered Nanofibrous Materials, 245-256.

Technologies, Inc. Georgia, USA, has developed a thin-plate die technology that the company claims to have created a route to the cost effective production of NF using an alternative and more conventional meltblown technology [7]. A few other companies reported recently about achieving a large scale NF production using a modified electrospinning process. STAR Inc., NY, claim that they can produce electrospun nanofibres in bulk quantities and the mesh made from these NF could be used as a biodegradable anti-adhesion biomaterial preventing scar tissue formation during post-operative wound healing [8]. So far, the material has proven effective at reducing tissue adhesions in animal trials and the Phase III clinical trials (on humans) are being planned. The company also plans to use their nanomesh for drug delivery. eSpinTechnologies, founded in Chattanooga, TN are also commercialising nanofibres made of organic and biological polymers for use in biomedical devices and filtration systems [9]. Donaldson Co., Inc., Minneapolis, MN has patented its EON Nanofibers made with an electrospun process [10]. The most obvious and logical biomedical application of NF seems to be for filtration of biological media in order to remove pathogenic substances and micro-organisms. By producing filters from nanofibres it is possible to achieve high surface area with controlled pore size, and in this case NF have a strong advantage over microfibres. Filters made using microfibres have much smaller surface area – a few square metres per gram vs a few hundred square metres per gram in nanofibre-made filters. Nanofibrous filters may become an efficient means for water purification and thus could be used against bio-terrorism. Recently reported filters made from alumina NF by Argonide, Sanford, FL are capable of removing heavy metals, viruses and bacteria from water and protein separation [11]. These alumina whiskers produced by have an average diameter of 2 nm and surface area of 500-650 m^2/g. Alumina nanofibre-based filters are a result of successful collaboration between the American business, Los Alamos National Laboratories and Russian scientists from Special Design and Technology Centre, Tomsk, and Vector Virological Institute, Novosibirsk.

An interesting example of man-made self-assemblying peptide NF was reported in [12]. The authors synthesised and studied peptide-amphiphile molecules, which reversibly self-assembled into nanofibrous networks resulting in the formation of aqueous gels. These structures could be further reversibly polymerised to enhance their stability.

Electrospun polymer NF were used as a template for producing nanotubes by coating nanofibres with another material and then removing the polymer core [13]. Using this approach, aluminium and titanium dioxide hollow fibres were produced [14].

Nanomaterials and nanofibres in particular are not a recent invention of a human genius, as they are widespread in living nature. Such different materials as wood, silk, hair, connective tissue, bones, skin and blood vessels – to name just a few – are all formed by some regular and repetitive arrangement of NF, usually made up by proteins or polysaccharides. The molecule of DNA itself is an example of a nanofibre, about 2 nm in diameter and hundreds of nanometres long. A striking example of a nanostructured material used by nature is blue feathers, such as those of the bluebird. There is no blue pigment in the feather and its colour is due to its sophisticated nanostructure. Colour production in bluebird is caused by coherent light scattering owing to a nanostructural arrangement of keratin and air bubbles in the feather barbs [15,16]. One of the most studied systems containing natural nanofibres is that of fibrin(ogen) polymerisation followed by blood clot formation. It is an example of a self-assembling macro-molecular structure and it is the result of the final stages of the blood coagulation cascade. The clot assembly consists of several distinctive steps, which influence the rate of fibrin fibres assembly and the spatial arrangement of fibres in the final clot. The first step is the proteolytic cleavage of plasma protein fibrinogen (factor I in the blood coagulation system) by the enzyme thrombin. Thrombin cleaves low molecular peptide fragments known as fibrinopeptide A and fibrinopeptide B from the fibrinogen molecule. This leads to the formation of monomeric fibrin, which has exposed binding sites and immediately forms dimers and small oligomers. Oligomers grow to yield protofibrils [17]. After reaching the critical mass the protofibrils form fibrils and finally form a clot network [18]. *In vivo* fibrin monomers spontaneously form a solid polymer (fibrin clot), which is covalently cross-linked by the blood plasma enzyme trans-glutaminase into insoluble polymer. However, *in vitro* fibrin polymer could be polymerised reversibly and re-polymerised with a predictable structure. It is possible to form a fibrin network with a wide spectrum of fibre thickness, branching and finally a polymer with different porosity and pore size distribution. Several factors appear to play an important role in determining the final fibre diameter and network structure, such as ionic strength, pH, fibrinogen/thrombin ratio or the presence of other plasma proteins [19,20]. Fibrin self-assemblying occurs mainly via non-covalent intermolecular A-α and B-β binding between the complementary polymerisation sites localised in the central E-domains and peripheral D-domains of fibrin. The physicochemical nature of the noncovalent binding depends on the electrostatic interactions and hydrogen bonding between amino acid residues in the course of fibrin self-assembly. The significance of the electrostatic interactions became clear from observations of the of ionic strength effect on the polymerisation rate [21]. E.Mihalyi [22] suggested hydrogen bonding as the main mechanism of non-covalent binding for fibrin polymerisation. Later on it was shown that beside electro-static and hydrogen bonding, hydrophobic interactions also participate in fibrin polymerisation [23]. The study of fibrin network structure is of great academic and clinical interest. On the one hand, it is particularly relevant to such common diseases and conditions as bleeding, thrombosis, arterio-sclerosis and wound healing and on the other hand fibrin networks have a potential for use as a drug carrier, tissue scaffold or wound sealant. In this work an investigation of fibrin clot structure has been carried out using scanning and transmission electron microscopy.

Activated carbon fibres (ACF) is another interesting example of nanostructured fibrous materials that have high potential for medical applications. Although the fibre diameter is in the micrometre range, ACF have large internal surface area due to the presence of micro- (pore width <2 nm), meso- (2-50 nm) and macropores (>50 nm). This classification is based on the difference in mechanism of gas adsorption according to the IUPAC recommendations [24]. All three types of pores contribute to the nanostructure of ACF and should be regarded as nanopores as they have at least one dimension at nanometre scale. Granulated activated carbon (GAC) adsorbents have been used in extracorporeal blood purification (haemoperfusion) since 1970s [25]. Preliminary results of the comparative study of ACF vs. GAC in several potential medical applications are presented and discussed in this paper.

2. MATERIALS AND METHODS

All reagents and solvents were used as purchased without further purification.

2.1. Preparation of Fibrin Monomer and Fibrin Films

Fibrin monomer (F-1) solution was prepared from aseptically collected fresh blood containing trisodium citrate. Fibrin monomer concentration determined by turbidimetry was 22.5 mg/cm^3. The pH of F-1 solution was 4.4. 2% agarose gel was prepared using a mixture of 0.2M acetate/0/75M bicarbonate buffer (7:1 w/w) with pH 7.3. The agarose gel was used as hydrophilic surface. Polyethylene (Goodfellow, UK) was used as a hydrophobic surface. Fibrin films were prepared by spraying F-1 solution onto a surface from the distance of 5 cm. The samples were then incubated at ambient temperature for 30 min. in a humid chamber prior to fixation for electron microscopy studies. Afterwards the samples were fixed in 2% (v/v) glutaraldehyde solution in 0.1M cacodilate buffer, pH 7.4, followed by treatment with 1% (w/v) osmium tetroxide solution in 0.05M cacodilate buffer, pH 7.4. After fixation the samples were treated with aqueous 2% (w/v) tannic acid for 1 h and then the samples were dehydrated by sequential treatment with a series of water-acetone mixtures with increasing concentration of acetone (30, 50, 70, 90 and 100%, volume/volume). Finally, the samples were cut into two pieces. One piece was freeze-dried and sputter-coated with gold [26] and examined with a Scanning Electron Microscope, SEM (JEOL JCM-840, Japan). The other piece was embedded in epoxy resin (Araldite-Epon) at 60 C for 48 hours and 50-100 nm sections were cut with Ultratom LKB-V, Sweden. The sections were further coated with carbon by glow discharge procedure and negatively contrasted with 1% (w/v) uranyl acetate for 1 h followed by a 2-min treatment with a lead-containing solution (Reynolds) at ambient temperature [27]. These samples were examined with a Transmission Electron Microscope, TEM (Hitachi H-600, Japan).

2.2. Activated Carbons

Activated carbon fibres were produced from hydrated cellulose fibres using carbonisation and steam activation. Properties of ACF are given in Table 1. Pore size distribution in ACF was obtained from nitrogen adsorption isotherm at 77 K (Table 2). Activated carbon SCN commonly used in haemoperfusion was produced by pyrolysis of poly(4-vinyl-pyridine) cross-linked with di-vinyl-benzene followed by steam activation. As it is widely used in haemoperfusion, SCN has been chosen as a reference material to compare with ACF. Powdered activated carbon Norit A Supra was used as received (Table 1).

Table 1. Characteristics of activated carbons.

Activated carbon	Particle shape	Particle size, μm	Bulk density γ, g/cm^3	Total pore volume Vs, cm^3/g*
ACF	Fibres	(7.0–9.0) (D) x 50 (L)	0.07 – 0.09	0.80 - 1.50
SCN	Spherical beads	300-1000	0.19-0.41	0.85 - 2.15
Norit A Supra	Irregular particles	10 x 25	0.30	0.81

* Vs was estimated from benzene vapour adsorption.

2.3. Adsorption Experiments

Unconjugated bilirubin adsorption was studied from 0.18 mg/mL solution of bilirubin in 3.0% human serum albumin (HSA) in single-pass column experiments at flow rate of 2.5 ml/min for 4 hours at room temperature (RT). Column volume was 20 cm^3. Unconjugated bilirubin content was determined by Jendrassik-Grof method using diagnostic kit 605-C (Sigma, USA) [28].

Creatinine and vitamin B_{12} adsorption were studied in batch experiments from 0.3 g/L and 1.0 g/L solution, respectively, with adsorbent/adsorbate ratio (w/v) 1:200 g/mL, for 1 hour at RT. Creatinine and vitamin B_{12} content were determined with UV-Vis spectrophotometry.

Adsorption of bacterial endotoxin (BET) was carried out in batch from 10 mL of PBS (pH 7.2) containing 0.05 g/L of *E. coli* 055:B5 lipopoly-saccharide (L4005, Sigma, USA) for 60 min. BET concentration was measured by spectrophotometric determination of 3-deoxysugars [29].

Table 2. Pore surface area S (m^2/g) and total pore volume Vs (cm^3/g) of carbon samples and their bilirubin adsorption capacity (mg/g of carbon)

Sample	S_{micro}	S_{meso}	S_{macro}	S_{total}	Vs	Bilirubin adsorption
ACF1	610	620	14	1250	0.80	9.4
ACF2	820	670	14	1505	1.10	20.5
ACF3	1250	670	15	1930	1.50	50.1
SCN1	n.d.*	n.d.	n.d.	900	0.85	3.2
SCN2	n.d.	n.d.	n.d.	1200	1.20	4.8
SCN3	n.d.	n.d.	n.d.	1320	2.15	18.0

* not determined.

2.4. Experiments with Biospecific Adsorbents

Native double-strained (ds) DNA from calf thymus (Sigma, USA) was immobilised by physical adsorption on activated carbons Norit A Supra and ACF2 (**DNA-Norit** and **DNA-ACF** adsorbents, respectively) as in [30]. Single-strained (ss) DNA was obtained by thermal denaturaton of ds-DNA. Efficiency of DNA-immobilised adsorbents was studied in single-pass column experiments by perfusion of blood plasma of patients with systemic lupus erythematosus (SLE) through 2.5 cm^3 microcolumn at flow rate of 0.2 mL/min for 1h at RT. Anti-DNA antibodies concentrations were measured using ELISA.

Inactivated *Staphylococcus Aureus Cowan* (SAC) cells were immobilised by physical adsorption on ACF3 (**SAC-ACF** adsorbent). SAC suspension was inactivated by incubation in 0.5 % formaldehyde for 3 h at RT, washed 5 times in PBS (pH 7.4) and finally heat treated at 80° C for 10 min.

Plasma immunoadsorption (IA) was performed in 13 mongrel dogs with histologically confirmed malignant mammary tumours of third and fourth stages. Plasma was perfused through a 100-cm^3 column packed with 15 g of a SAC-ACF or uncoated ACF adsorbent at flow rate of 20 mL/min for 30-80 min. Anti-tumour effect of IA was estimated by tumour regression and ulcer healing [31]. IgG content was determined by rocket immunoelectrophoresis. Circulating immune complexes (CIC) were determined in plasma by precipitation with polyethylene glycol [32].

BET-ACF adsorbents were prepared by shaking carbon fibres and saline solution of *E. coli* 055:B5 for 3 h at RT. The mixture was centrifuged at 50g and the supernatant was analysed for BET concentration. Carbon fibres were washed three times with saline solution and packed in 2-cm^3 columns (0.27 g fibres per column).

Guerin carcinoma was induced in rats by hypodermic inoculation with 25% suspension of viable tumour cells (2 mL/kg). The animals were used to estimate the anti-tumour effect of direct haemoperfusion (DHP) through the column packed with BET-ACF adsorbent (n=8 animals) and with ACF (n=7) as well as of intravenous injection of 0.75 mg BET per kg of body weight (n=8). The control group of rats (n=9) were intravenously injected with saline solution. The rats were treated 24 h after operation on the primary tumour on the 6th day after inoculation of tumour cells.

2.5. Bactericidal Properties of ACF

ACF impregnated with metallic copper (**ACF-Cu**), 2.5-3.5% w/w, was obtained by magnetron sputtering in low temperature argon plasma [33]. To evaluate bactericidal properties of ACFs, 0.5 mL of one of the following microbial cultures – *P. mirabilis*, *E. coli*, *S. aureus* or *P. aeruginosa* was distributed on a Petri dish with beef-extract agar as a nutrient medium. Dishes were completely covered with gauze, ACF or ACF-Cu and incubated for 45 min at 37°C. After removing the covers, dishes were incubated for 24 h at 37°C and the numbers of bacteria were counted.

3. RESULTS AND DISCUSSION

3.1. Nanofibrous Fibrin Structures

Fibrin monomer was sprayed as droplets from the spray nozzle to the target site, and this flight took between 100 – 120 ms (data not presented). Confocal microscopy showed that the sprayed F-1 formed a heterogeneous film with high density boundary on the surface. Electron micrographs may be interpreted noting that in TEM mode the samples are sectioned and dark regions represent proteins with higher density than the background, hence they are seen as dark regions on micrographs. SEM, on the other hand, shows only the surface structures and on SEM micrographs dark regions correspond to the background or fluid filled space, whereas light regions correspond to proteins. The micrographs shown in Figures 1 and 2 are representative of a large number of images taken.

Investigation of fibrin clot structure using scanning and transmission electron microscopy revealed sequential steps of fibrin network formation (Figure 1, a-d). Formation of fibrin dimer (1a), protofibril formation (1b), branching of fibrils into fibrin network (1c) and finally the morphology and fibrin clot structure (1d) are clearly observed.

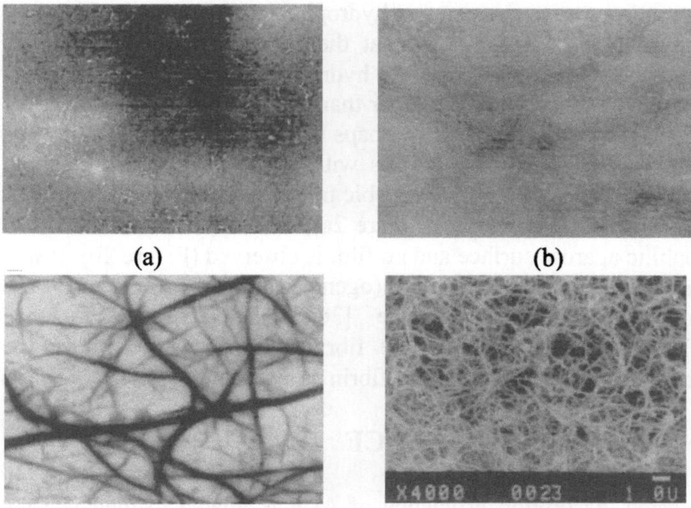

(a) (b)

(c) (d)

Figure 1. Different steps of fibrin network formation. a – fibrin dimer formation; b – protofibril formation; c – fibrin network; d – fibrin clot. a, b and c are TEM micrographs, x 60,000 (a) and x 35,000 (b and c); d is SEM micrograph, x 7,000.

Fibrin film produced on a hydrophobic polyethylene surface is easily detachable from the surface without being damaged. It retains contours of the surface on which it was polymerised and the micrograph shows that fibrin is concentrated at the interface (Figure 2a). Fibrin sprayed over the hydrophilic agarose surface behaves very differently. It does not form any film structure as shown in Figure 2b.

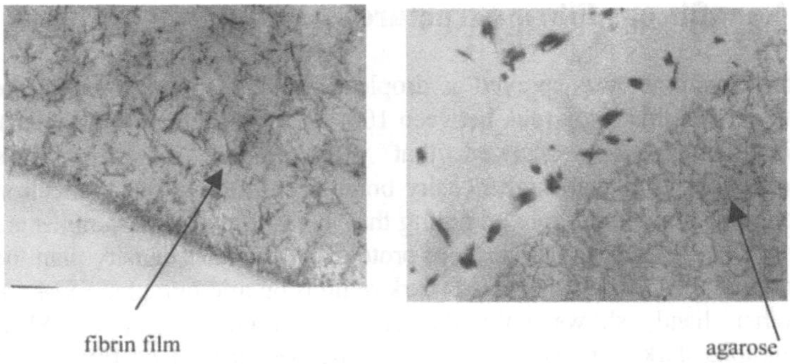

Figure 2. Polymerisation of fibrin polymer at (a) hydrophobic polyethylene surface and (b) hydrophilic agarose surface. TEM micrographs, x 17,500 (a) and x 35,000 (b). 22.5 g/L fibrin monomer sprayed on the surface and incubated for 30 min at RT. Black blobs – fibrin.

To explain these observations, the events at the air/water interface should be taken into account. The droplets of fibrin monomer during their short flight are exposed to air, which should cause concentration of the protein at the air/solution interface, as it was established earlier [34]. This observation allows us to suggest that hydrophobic media (air in this case) promote formation of a fibrin film at their interface which is further developed once the droplets reach a hydrophobic surface. Mobility of proteins at air/liquid interface is higher than at solid/liquid interface [35], however the time of flight is perhaps too short (in the order of milliseconds) to establish strong links within the fibrin network of the single droplet. Hence, on the hydrophobic polyethylene the fibrin network expands over the whole surface (Figure 2a), whereas it breaks down on the hydrophilic agarose surface and no film is observed (Figure 2b). It was shown that platelet adhesion to fibrin(ogen) coating is influenced by the nature of the underlayer surface [36]. The effect of surface hydrophilicity/hydrophobicity on the fibrin film formation should be considered in medical applications of fibrin as a sealant.

3.2. Medical Applications of ACF

As expected, adsorption efficiency of ACF is related to their surface characteristics and porosity. ACF samples with higher degree of burn-off have larger V_s, S_{micro} and S_{total}. As a result, adsorption capacity of ACF3 towards unconjugated bilirubin from protein-containing solutions is five times higher than that of ACF1 (Table 2). Analysis of the distribution of specific surface area for different ACF samples shows that unconjugated bilirubin is mostly adsorbed in micropores.

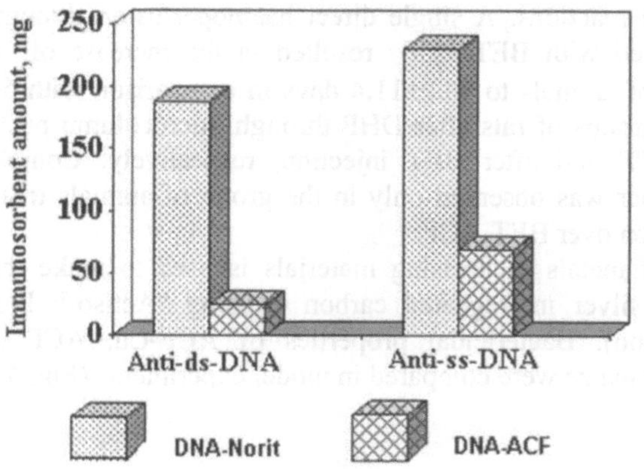

It should be noted that ACF showed excellent operating stability with no signs of saturation after 4 h perfusion. Projection of the experimental data on the treatment of patients with hepatic failure suggests that a 180-200 cm^3 column packed with highly activated ACF would be able to remove up to 600 mg of bilirubin from the blood plasma during 4 h perfusion at a flow rate of 25 mL/min.

In batch experiments DNA-ACF showed significantly higher adsorption efficiency than DNA-Norit towards anti-DNA antibodies removal from the blood plasma of SLE patients (Fig.). This result demonstrates high potential of DNA-containing ACF immunoadsorbents in the treatment of autoimmune diseases associated with elevated levels of anti-DNA antibodies.

Plasma IA performed in dogs with malignant tumours using SAC-ACF has resulted in a statistically significant reduction of the tumours. *Staphylococcus Aureus Cowan* suspension that has also been used in such cases is highly toxic and its use is associated with frequent and grave side effects. SAC immobilised on ACF showed little or no side effects. Plasma perfusion over SAC-ACF or ACF exhibits immunomodulatory effect withdrawing IgG and CIC from circulation, although IA seems to be more efficient (Table 3).

Table 3. IgG and CIC levels in blood plasma of dogs with mammary adenocarcinoma during plasma perfusion (B, pre-column level, P, post-column level) and one day after (A)

Solute	Adsorbent	Points and time of sample withdrawal (min) B-30	P-30	B-60	P-60	A
IgG (g/L)	ACF	4,76±0,4	4,15±0,3	4,92±0,36	4,61±0,34	5,22±0,51
	SAC-ACF	4,45±0,3	2,94±0,2	4,50±0,24	3,45±0,36	4,21±0,25
CIC (a.u.)	ACF	44,7±3,1	20,6±3,0	50,6±3,2	37,4±3,5	63,3±3,1
	SAC-ACF	77,5±2,5	34,0±2,4	62,1±3,4	45,6±3,1	61,3±3,7

Similarly, ACF-immobilised bacterial endotoxin is a safer option for cancer treatment than a free BET suspension. Despite well-known anti-tumour activity of BET, their therapeutic applications are discouraged by

BET-induced toxic reactions. A single direct haemoperfusion through a microcolumn packed with BET-ACF resulted in the increase of the average life span of animals to 76.2±11.4 days in comparison with 52.7 and 54.2 days in groups of rats after DHP through microcolumn packed with uncoated ACF and after BET injection, respectively. Complete regression of tumour was observed only in the group of animals treated with haemoperfusion over BET-ACF.

Incorporation of metals in dressing materials is used to make them bactericidal (e.g., silver impregnated carbon dressing "Actisorb Plus", Johnson & Johnson). Bactericidal properties of ACF-Cu, ACF and hygroscopic cotton gauze were compared in model experiments (Fig. 4).

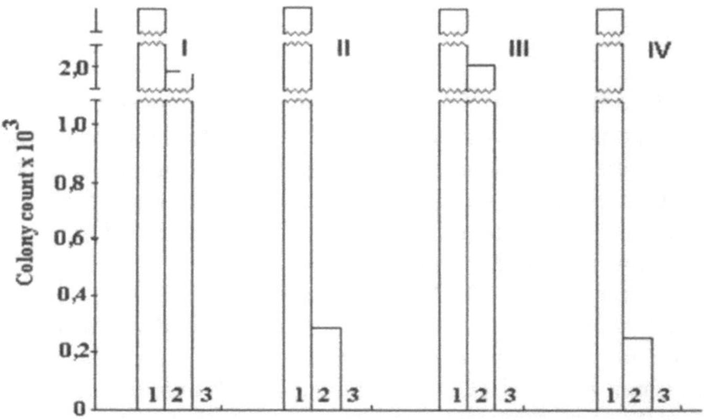

Figure 4. Bactericidal effect of ACF and ACF-Cu. 1 – Gauze, 2 – ACF, 3 – ACF-Cu. **I** – *P. mirabilis*; **II** - *E. coli*; **III** - *S. aureus*, **IV** - *P. Aeruginosa.*

After contact with the gauze the bacterial growth did not differ from that in the control. ACF exhibited a bacteriostatic effect with slight inhibition of *P. mirabilis* and *P. aeruginosa* growth and 3-fold reduction of *S. aureus* and *E. coli* colonies. In presence of ACF-Cu growth of all colonies was completely oppressed. It can be concluded that ACF-Cu has a strong bactericidal effect on different microorganisms combined with high adsorption potential.

The results presented in this paper confirm a large number of opportunities for medical applications of nanofibres such as fibrin or nanostructured fibres such as ACF. In case of natural fibrin fibres their properties can be significantly modified for a particular application by forming their network at different interfaces. Adsorptive properties of ACF can be significantly altered by their surface chemical or physical modification. It is important to realize that despite successful demonstration of these approaches in designing fibres, there is a lot of room for further improvement. A simple calculation allows estimation of nonporous nanofibrous carbon surface area as being between 200 and 2000 m^2/g for the fibre diameter 100 nm and 10 nm, respectively. (In this calculation the density of carbon fibre was taken as 2.0 g cm^{-3}). Such a large surface area is comparable with that of ACF but it does not require the activation stage, which significantly reduces mechanical strength of carbon materials. It is also an easily accessible external surface which has no intraporous diffusion limitation that may reduce the adsorption rate by porous carbons. Although carbon nanofibres are not produced on a large scale yet, recent advances in making polymeric nanofibres make possible production of carbon nanofibres by polymer pyrolysis in the near future, and that will open exciting opportunities for medical applications of these materials.

Acknowledgments: The authors are grateful to NATO for their support of this work via Collaborative Linkage Grant LST.CLG.978860 and to the Royal Society (VN and KB) for having sponsored their visit to Brighton. SM acknowledges NATO support for attending the ASI 'Nanoengineered Nanofibrous Materials'. Part of the work on fibrinogen was funded by Oxford Bioresearch Laboratory.

References

1. Frenot A., Chronakis I.S. Polymer nanofibers assembled by electrospinning. Current Opinion in Colloid and Interface Science 2003; 8:64–75.
2. Bognitzki M., Czado W., Frese T., *et al.* Nanostructured fibres via electrospinning. Adv Mater 2001; 13:70.
3. Ko F., Gogotsi Y., Ali A., *et al.* Electrospinning of continuons carbon nanotube-filled nanofiber yarns. Adv Mater 2003; 15:1161-5.
4. Kenawy E.R., Bowlin G.L., Mansfield K., *et al.* Release of tetracycline hydrochloride from electrospun poly(ethylene-co-vinyl acetate), poly(lactic acid) and a blend. J Contr Release 2002; 81:57–64.
5. Boland E.D., Wnek G.E., Simpson D.G., Pawlowski K.J., Bowlin G.L. Tailoring tissue engineering scaffolds y employing electrostatic processing techniques: a study of poly glycolic acid). J Macromol Sci 2001; 38:1231–43.
6. Matthews J.A., Wnek G.E., Simpson D.G., Bowlin G.L. Electrospinning of collagen nanofibers. Biomacromolecules 2002; 3:232–8.
7. USP 6,114,017 (2000).
8. Zong X., Kim K., Fang D., *et al.* Structure and process relationship of electrospun bioabsorbable nanofiber membranes. Polymer 2002; 43:4403–12.
9. Doshi J., Reneker D.H. Electrospinning process and applications of electrospun fibers. J Electrostatics 1995; 35:151-60.
10. Grafe T., Graham K. Polymeric nanofibers and nanofiber webs: A new class of nonwovens. Proceedings of the International Nonwovens Technical Conference (Joint INDA – TAPPI Conference); 2002 September 24-26; Atlanta, GA.
11. Tepper F., Lerner M., Ginley D. Nanosized Alumina Fibers. American Ceramic Society Bulletin 2001; 80:57-60.
12. Hartgerink,J.D., Beniash E., Stupp S.I. Peptide-amphiphile nanofibers: A versatile scaffold for the preparation of self-assembling materials, PNAS 2002; 99:5133–8.
13. Bognitzki M., Hou H., Ishaque M., *et al.* Polymer, metal, and hybrid nano- and mesotubes by coating degradable polymer template fiber (TUFT process). Adv Mater 2000; 12:637–40.
14. Caruso R.A., Schattka J.H., Greiner A. Titanium dioxide tubes from sol–gel coating of electrospun polymer fibers. Adv Mater 2001; 13:1577–9.
15. Prum R.O., Torres R.H., Williamson S., Dyck J. Constructive interference of light by blue feather barbs. Nature 1998; 396:28–9.
16. Shawkey M.D., Estes A.M., Siefferman L.M., Hill G.E. Nanostructure predicts intra-specific variation in structural plumage colour. Proc Roy Soc, Ser B 2003; 270:1455-60.
17. Doolittle R.F. Fibrinogen and fibrin. Ann Rev Biochem 1984; 53:195-229.

18. Henschen A. On the structure of functional sites in fibrinogen. Thromb Res 1983; Suppl 5:27-39.
19. Doolittle R.F., Blomback M., Henschen A., *et al.* N-terminal disulphide knot of human fibrinogen. Nature 1968; 218:130-4.
20. Kudryk B., Reuterby J., Blomback B. Adsorption of plasmic fragment D to thrombin modified fibrinogen-Sepharose. Thromb Res 1973; 2:297-303.
21. Ugarova T.P., Budzynski A.Z. Interaction between complementary polymerisation sites in the structural D and E domains of human fibrin. J Biol Chem 1992; 267:13687-93.
22. Mihalyi E. Transformation of fibrinogen into fibrin: II Changes in pH during clotting of fibrinogen. J Biol Chem 1954; 209:733-41.
23. Collen D., Vandereycken G., de Maeyer L. Influence of hydrostatic pressure on reversible polymerisation of fibrin monomer. Nature 1970; 220:669-71.
24. Sing K.S.W., Everett D.H., Haul R.A.W., Moscou L., Pierotti R.A., Rouquerol J., Siemieniewska T. Reporting physisorption data for gas solid systems with special reference to the determination of surface-area and porosity (recommendations 1984). Pure & Appl Chem 1985; 57:603-19.
25. Mikhalovsky S.V. Emerging technologies in extracorporeal treatment: focus on adsorption. Perfusion 2003: 18, Suppl 1:47-54.
26. Langer B.G., Weisel J.W., Dinauer P.A., *et al.* Deglycosylation of fibrinogen accelerates polymerisation and increases lateral aggregation of fibrin fibers. J Biol Chem 1988; 263:15056-63.
27. Weisel J.W., Stauffacher C.V., Bullit E., Cohen C. A model for fibrinogen: domains and sequence. Science 1985; 230:1388-91.
28. Varley Harold, Gowenlock Alan H., Bell Maurice. *Practical Clinical Biochemistry*, v. 1. London: Heinemann, 5th Ed., 1980.
29. Cynkin M., Ashwell G. Estimation of 3-deoxysugar by means of the malonaldehyde thiobarbituric acid reaction. Nature 1960; 186:155-6.
30. Beloglasov V.A., Snezhkova E.A., Nikolaev V.G. Hemoperfusion through DNA-coated and uncoated synthetic activated charcoals as an additive to the bronchial asthma traditional treatment. Artif Cell Blood Sub 1998; 26:191-7.
31. Terman D.S., Yamamoto T., Mattiolli M., *et al.* Extensive necrosis of spontaneous canine mammary adenocarcinoma after extracorporeal perfusion over *Staphylococcus aureus* Cowan I. J Immun 1980; 124:805-15.
32. Creighton W.D., Lambert P.H., Miescher P.A. Detection of antibodies and soluble antigen-antibody complexes by precipitation with polyethylene glycol. J Immun 1973; 3:1219-27.
33. Eretskaya E., Sakhno L., Pimonenko N., *et al.* Method for obtaining the metallised carbon dressing material. SU Patent 1,817,483.
34. Bagnall R.D. Adsorption of plasma proteins on hydrophobic surfaces. II - Fibrinogen and fibrin-containing protein mixtures. J Biomed Mater Res 1978; 12:203-17.
35. Baszkin A., Boissonnade M.M. Competitive adsorption of albumin and fibrinogen at solution-air and solution-polyethylene interfaces. *In situ* measurements. In: *Proteins at Interface II*, TA Horbett and JL Brash, eds. Washington, DC: Amer Chem Soc, 1995, Ch. 15, pp. 229-27.
36. Kiaei D., Hoffman A.S., Horbett T.A. Platelet adhesion to fibrinogen adsorbed on glow discharge-deposited polymers. In: *Proteins at Interface II*, TA Horbett and JL Brash, eds. Washington, DC: Amer Chem Soc, 1995, Ch. 32, pp. 450-62.

NANO-SCALE ENGINEERING OF SURFACES

Functionalized Nanoparticles as Versatile Tools for the Introduction of Biomimetics on Surfaces

V. PRASAD SHASTRI*, A.M. LIPSKI, J.C. SY, W. ZNIDARSIC, H. CHOI AND I-W. CHEN

Department of Materials Science and Engineering & Laboratory for Research in Structure of Matter, University of Pennsylvania, Philadelphia PA, USA 19104 and Joseph Stokes Research Institute, The Children's Hospital of Philadelphia, PA 19104

Abstract: Historically, advances in materials synthesis and processing have been pivotal in the emergence of advanced medical technologies. Tissue Engineering (TE) and other organ regenerative techniques depend on the creation of a mechanically well-defined environment (scaffold) that is rich in biomolecular signals to achieve its objectives, namely the growth of functional neo-tissue. The evolution of neo-tissue is governed by material-cellular interactions, which in turn are dictated by surface characteristics such as texture and chemical functionality. To enable predictive outcomes in TE, polymeric fibers and cellular solid scaffolds should be engineered to include the presentation of biomolecular signals. Since information that is introduced on the material surface is processed as biomechanical and biochemical signals through receptors, which are nano-meter sized entities on cell surfaces, it is therefore important that this information be presented on the same length scale as it occurs in nature. A robust, reproducible and simple way to assembly physical and biochemical information therefore needs to be developed. We hypothesized that an assembly of functionalized particles could serve as a versatile tool for imparting texture and chemical functionality on a variety of surfaces. We further hypothesized that nano-scale resolution of physical and chemical information can be achieved using functionalized nanoparticles (FNP) as building blocks. Pre-functionalization allows for the precise control over density and spatial distribution of information on the surface. Using functionalized inorganic oxides; surfaces of hard (stainless steel and titanium) and soft (polyurethane) substrates have been modified so as to impart reproducible texture and chemically derivatizable functionality. These surface modification coatings can be tuned to possess tethered or covalently adsorbed biomolecules such as peptides, growth factors and proteins and can serve as a platform for engineering biomimetic interfaces to modulate cellular behavior toward implants and in scaffolds for TE.

S. Guceri et al. (eds.), Nanoengineered Nanofibrous Materials, 257-264.

1. INTRODUCTION

Depending on the surface chemistry and physical characteristics of a biomaterial (metallic or polymeric) the cellular response to the material interface can be varied. Although local modulation of cellular events is influenced to a significant extent by the layer of adsorbed serum proteins at the interface, it is not the sole determinant in the long-term compatibility of a material. Several other factors play a key role and they include chemistry of the surface (lipophilic versus hydrophilic), surface texture, local pH and degradation products.

When a material is chemically inert or slow degrading the typical response entails the encapsulation of the implant within an avascular fibrous capsule. The fibrous capsule serves to sequester the implant from the body's immune system while localizing the undesirable effects of the implant. This isolation process is effective from the body's perspective, nevertheless it is a barrier to transport and can serve as an impediment to the functioning of an implanted biosensors or drug delivery device. Such a fibrotic response is typically observed around polymers like Teflon, Silastic and polyurethane [1].

However, when a degradable material is implanted, the response can vary from integration of the implant with the surrounding tissue, i.e., if the implant is porous or fibrous; to a strong inflammatory response, typically triggered by localized accumulation of degradation products [2]. The inflammatory response may be acute and get resolved favorably or be sustained resulting in the rejection of the implant. Degradable polymers elicit such responses depending on their chemical make-up and configuration, i.e., fiber, film, particles or solid. The mass of the material has a direct bearing on the local changes in pH and accumulation of degradation products, while the configuration and chemistry of the material plays an important role in dictating cell attachment and cell morphology [3]. The latter is of extreme importance in the development of tissue-augmentation devices (e.g., bone pins, bone screws) and in tissue engineering where cell recruitment, proliferation and differentiation in a localized and controlled manner is vital for directing tissue-morphogenesis [4].

Surface modification offers a means to alter interfacial properties without affecting the bulk characteristics of a material. Some of the commonly used processes for the modification of chemical composition or texture of a material include; (a) chemical etching, (b) oxidative plasma treatment (typically used for metals), (c) chemical and plasma enhanced vapor deposition and (d) physisorption and chemisorption of block co-polymers and biopolymers [5-8]. Common drawbacks with these approaches are that they are usually material specific, i.e., limited to modification of either a hard or soft material, and they do not provide a means of precisely and reproducibly controlling chemical functionality and texture. In addition, seldom can both texture and chemical functionality be controlled in a single step or process.

It occurred to us that one could affect both chemical composition and texture of surface via the assembly (or coating) or functionalized nanoparticles. The advantages of this paradigm over conventional methodologies are the following:

1. Particles allow for control over surface characteristics from the sub-micron to nanometer length scale,
2. Pre-functionalized particles enable a high degree of control over density and spatial distribution of functional moieties,
3. Spherical morphology of the particles lend themselves to self-assembly and hence a high degree of order and reproducibility in surface chemistry and texture,
4. Since the assembly process can be done from aqueous phase, it is compatible with biological systems,
5. Complex surface chemistries can be attained by varying the size and chemical composition of the particles, and
6. The approach can be applied to both hard and soft material and in any configuration (wires, fibers, mesh, porous body, solid).

We have demonstrated the suitability of this approach in the modification of both metal and polymer surfaces and shown that precise control of chemical composition and texture may be achieved by varying the chemical, physical and structural characteristics of the nanoparticles.

2. EXPERIMENTAL

2.1 Synthesis and Characterization of Amine-Functionalized Silica Nanoparticles (FSNP)

Silica nano-particles were prepared using the Stober process, via the controlled hydrolysis of tetraethylorthosilicate (Aldrich), using ammonia and alcohol as catalyst [10]. Functionalization of the silica nanoparticle surface with primary amines (Si-O-R-NH_2) was achieved by condensation with aminopropyltriethoxysilane (APS). The mean particle diameter and size distribution was ascertained using Dynamic Laser Light Scattering and Transmission Electron Microscopy (TEM). The presence of amine functionality on the surface of the particles was visualized using salisaldehyde. The surface charge was determined by Zeta Potential measurement.

2.2 Coating of Metal and Polymer Substrates

316-L stainless steel and titanium foil substrates (Goodfellow Corporation) were cut into rectangle pieces (0.7mm x 10mm), ultrasonically cleaned in organic solvents of increasing polarity, followed by rinsing in deionized water and then dried in a 60° C oven for 24 hours prior to use. FSNP were then assembled from an ethanolic solution onto metal substrates using spin coating (50 μL/deposition, 1 wt%, 2000 rpm, total of 10 sequential depositions). An identical procedure was employed to coat polyurethane (PU) thin films that were prepared by spin casting from a tetrahydrofuran solution onto borosilicate glass slides.

2.3. Characterization of FSNP Modified 316-L Stainless Steel and Titanium Surfaces

The modified metal surfaces were characterized using Scanning Electron Microscopy (SEM), Energy Disperssive X-ray Analysis (EDAX), Rutherford Back Scattering (RBS), X-ray Photoelectron Spectroscopy (XPS), and Atomic Force Microscopy (AFM, tapping mode image). The concentration of amine groups on the surface was determined by a quantitative assay as described in the literature [11]. The assay involves first reacting the amine group with sulfo-SDTB followed by hydrolysis of the adduct with perchloric acid to liberate a colored cation that is assayed spectrophotometrically at 498 nm.

3. RESULTS

Silica nanoparticles ranging from 50- 800 nm in mean diameter with polydispersity (PD) of less than 0.1 were prepared successfully using the modified-Stober process. Functionalization of the nanoparticles surface via the silanol groups with a silane coupling agent aminopropyltriethoxysilane yielded amine modified silica nanoparticles with no appreciable change to both particle size and PD (Table 1, Figure 1). The surface modification was

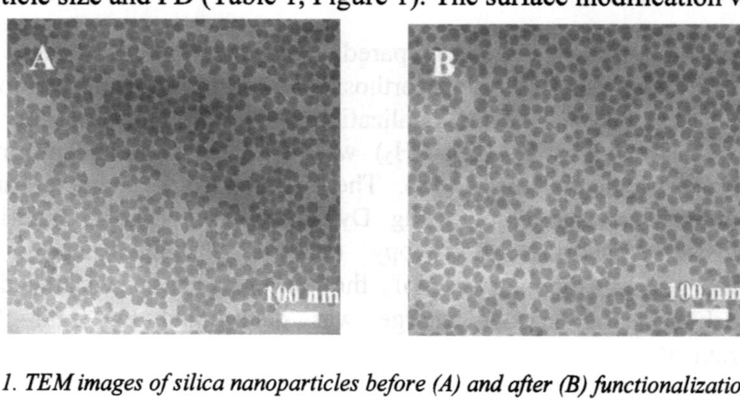

Figure 1. TEM images of silica nanoparticles before (A) and after (B) functionalization

Table 1. Characteristics of silica nanoparticles

	SiO_2	SiO_2-NH_2
Mean Diameter (nm)	56.0	60.2
PD	0.07	0.09
Zeta Potential (mV)	-30	+26

accompanied by reversal of surface charge, which was confirmed by zeta potential measurements (Table 1). 316-L stainless steel and titanium foil surfaces were successfully modified using non-functionalized and amine-functionalized silica nanoparticles (FSNP). SEM analysis of the surface confirmed the presence of a silica nanoparticle assembly on the surface (Figure 2). Quantification of chemically accessible amine groups on the modified metal surfaces indicates that amine group density as high as 0.6 NH_2 groups/$Å^2$ may be achieved. Analyses of modified metal surfaces using RBS indicate that nanoparticle assemblies are reproducible in thickness for given substrate and coating conditions (number of depositions and nanoparticle concentration). For 10 depositions, typical thickness range from 350 to 450 nm. AFM analyses reveal that although the titanium surface exhibits much

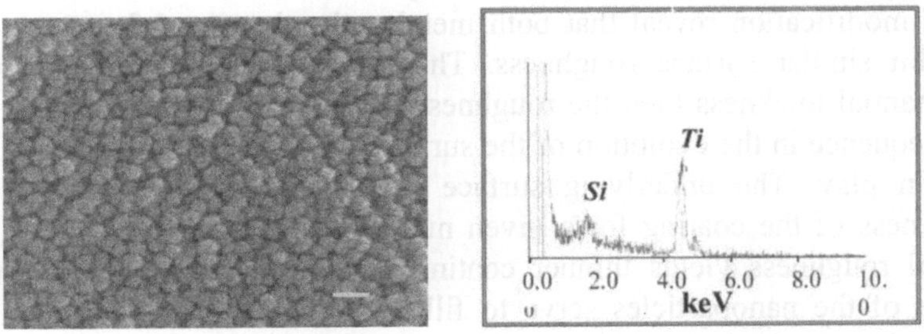

Figure 2. SEM image of amine-FNSP modified titanium surface (left, scale bar: 100 nm) and EDAX spectrum of the surface (right)

greater surface roughness than the stainless steel surface, after coating with amine-FSNP the surface roughness of both surfaces were comparable (Table 2). In fact, even more dramatic contrast can be seen in the case of PU surface, which has a initial surface roughness of ~ 3nm and a roughness post-coating of ~32 nm. Finally, the coatings appear to be quite robust and capable of withstanding handling by forceps and chemical environments that are typically used in the coupling of proteins, peptides and growth factors to surfaces.

Table 2. Roughness of metal and PU substrates before and after modification with functionalized silica nanoparticles

Substrate	**Average Surface Roughness (nm) (determined from AFM)**
Titanium	51.4
Titanium-Amine-FSNP	21.1
Stainless Steel	31.0
Stainless Steel-Amine-FSNP	27.3
PU	3.1
PU-Amine-FSNP	31.8

4. DISCUSSION

We have demonstrated that metal and polymer surfaces can be modified with functionalized nanoparticles to bear well-defined chemical functionality and topography. These coatings appear to be quite robust and resistant to mechanical abrasion due to typical handling procedures. Both titanium and stainless steel surfaces bear an oxide layer due to the spontaneous oxidation of titanium and, iron/chromium, respectively. It is plausible that this oxide layer facilitates the covalent binding of the silica nanoparticles to the metal surface via metal-silicate bonds. The AFM analyses of the metal surfaces before and after modification reveal that both metal and polymer surfaces post-coating exhibit similar surface roughness. This suggests that if the coating is of substantial thickness then the roughness of the underlying surface is of little consequence in the evolution of the surface topology, i.e., no epitaxial effects are in play. The underlying surface roughness has only bearing on the thickness of the coating for a given number of depositions. That is, greater initial roughness yields thinner coating for identical coating conditions as some of the nanoparticles serve to fill the crevices on the surface prior to achieving topological uniformity. The amine quantification studies reveal that the modified metal surfaces are rich in amine functionality and that the amine groups are chemically accessible and can be modified in a reproducible manner under reaction conditions similar to those used for protein coupling chemistry.

Thus these coatings are highly suitable for surface biomimetic engineering. We are currently exploring layer-by-layer assembly of biologically relevant information such as collagen, polysaccharides and apatite onto these functionalized nanoparticles and studying their assembly into 2-dimension and 3-dimensional hierarchies. It is hypothesized that these biomimetic particles can be assembled into 3-dimensional assemblies by either coating or dispersion in polymeric fibers (Figure 3) and porous media to present biologically relevant information in a spatially well defined manner for tissue regenerative and drug delivery applications.

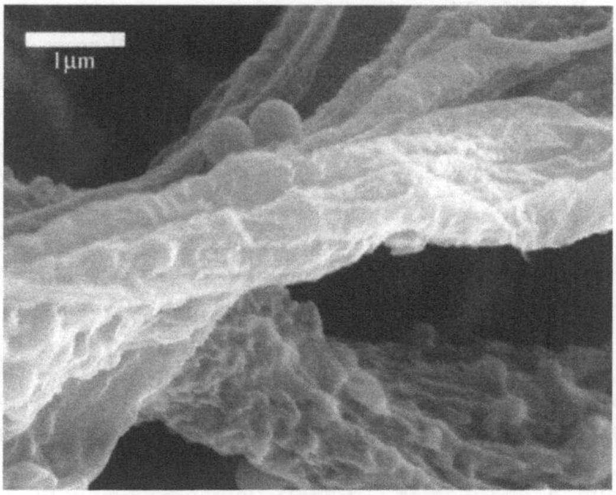

Figure 3. Poly(L-lactic acid) fibers containing poly(acrylic acid) functionalized silica nanoparticles on the fiber surface

5. CONCLUSIONS

Functionalized nanoparticles are excellent vehicles for the modification of chemical and topological characteristics of surfaces. Via these nanoparticle assemblies, roughness and texture of the surface, as well as spatial resolution of the chemical functionality can be simply controlled by varying the chemistry and size of the nanoparticles (colloids). It is envisaged that these functionalized nanoparticle assemblies will serve as a platform for the further assembly of biologically relevant information such as peptide-aggregates, enzymes, apatites and genetic material. This paradigm has been extended to include synthetic polymeric surfaces and polymeric nanoparticles. In addition, we are currently exploring the assembly of such nano-scale-functionalized hierarchies on collagen fibrils and electro-spun polymeric felts.

Acknowledgements: The authors wish to thank the National Science Foundation, National Institutes of Health, the State of Pennsylvania and Ben Franklin Technology Partners through the Nanotechnology Institute for funding of this work.

References

1. Jennings, D., Morykwas, M., DeFranzo, A. and Argenta, L.(1991), Analysis of silicon in human breast and capsular tissue surrounding prostheses and expanders, *Ann. Plastic Surg.*;**27**, 553–558
2. Martin, C., Winet, H. and Bao, J. (1996), Acidity near eroding polylactide-polyglycolide in vitro and in vivo in rabbit tibial bone chambers, *Biomaterials*, **17**, 2373-2380

3. Bergsma, E., Rozema, F., Bos, R. and de Bruijn, W. (1993), Foreign body reaction to resorbable poly(L-lactide) bone plates and screws used for the fixation of unstable zygomatic fractures, *J Oral Maxillofac. Surg.*, **51**, 666-670

4. Langer, R and Vcanti, J. (1993), Tissue Engineering, *Science*, **260**, 920-926

5. Nitschke, M., Schmack, G., Janke, A., Simon, F., Pleul, D. and Werner, C. (2002), Low pressure plasma treatment of poly(3-hydroxybutyrate): toward tailored polymer surfaces for tissue engineering scaffolds, *J Biomed Mater Res.* **59(4)**, 632-638

6. Gao, J., Niklason, L. and Langer, R., (1998), Surface hydrolysis of poly(glycolic acid) meshes increases the seeding density of vascular smooth muscle cells. *J Biomed Mater Res.* **42(3)**, 417.

7. Clarotti, G., Schue, F., Sledz, J., Ait Ben Aoumar, A., Geckeler, K.E., Orsetti, A. and Paleirac, G. (1992), Modification of the biocompatible and haemocompatible properties of polymer substrates by plasma-deposited fluorocarbon coatings, *Biomaterials*, **13(12)**, 832-840

8. Park, S., Bearinger, J.,P., Lautenschlager, E.P., Castner, D.G. and Healy KE. (2000), Surface modification of poly(ethylene terephthalate) angioplasty balloons with a hydrophilic poly(acrylamide-co-ethylene glycol) interpenetrating polymer network coating, *J. Biomed. Mater. Res.* **53(5)**, 568-576

9. Nitschke, M., Menning, A. and Werner, C. (2000), Immobilization of PEO-PPO-PEO triblock copolymers on PTFE-like fluorocarbon surfaces, *J Biomed Mater Res.* **50(3)**, 340-343

10. Choi, H. and Chen, I-W. (2002), Surface-modified silica colloid for diagnostic imaging, *J. Colloids. Interf. Sci.* **258**, 435-437

11. Gaur, R.K. and Gupta, K.C. (1989), A spectrophotometric method for the estimation of amino groups on polymer supports, *Anal. Biochem.* **180(2)**, 253-258

CATALYTIC FILAMENTOUS CARBONS (CFC) AND CFC–COATED CERAMICS FOR IMMOBILIZATION OF BIOLOGICALLY ACTIVE SUBSTANCES

G.A. KOVALENKO, D.G. KUVSHINOV,
O.V.KOMOVA, A.V. SIMAKOV, N.A. RUDINA
Boreskov Institute of Catalysis, Novosibirsk 630090, Russia

Abstract: Adsorption properties of granulated catalytic filamentous carbons (CFC) and CFC-coated macrostructured ceramics (honeycomb monoliths, foams) have been studied for adsorptive immobilization of biologically active substances including enzyme glucoamylase, invertase-active yeast membranes and non-growing cells of bacteria. The effect of morphology of the surface carbon layer on the adsorption efficiency of the supports and on the biocatalytic properties (stability and activity) of immobilized enzymes has been investigated.

Key words: catalytic filamentous carbon; carbonized ceramics; adsorptive immobilization.

1. INTRODUCTION

The problem of developing efficient supports for adsorptive immobilization of biologically active substances and microorganisms is still relevant because of its importance for various applications including medicine, biotechnology and pharmaceutical industry. Obviously, such *supports-adsorbents* must meet certain criteria. Firstly, they should have sufficient adsorption capacity with respect to such substances and microorganisms and have to strongly bind them to the surface. Secondly, they should retain and stabilize the biological activity of the immobilized substances at a relatively high level. Thirdly, they should possess high mechanical strength and resistance to biological and chemical degradation. Finally, their cost should be relatively low. Apparently, carbon-containing inorganic supports satisfy these requirements [2,3,5].

The present investigation has been carried out to study the granulated catalytic filamentous carbons (CFC) and CFC-coated macrostructured ceramics as supports – adsorbents for immobilization of enzyme glucoamylase, invertase-active yeast membranes and the non-growing bacterial cells. The effect of morphology of surface carbon layer on the adsorption efficiency of the supports and on the biocatalytic properties

S. Guceri et al. (eds.), Nanoengineered Nanofibrous Materials, 265-269.

(stability and activity) of immobilized enzymatic active substances has been investigated.

2. EXPERIMENTAL

Granulated catalytic filamentous carbons (bulk CFC) with diameter up to 5 mm were prepared by catalytic pyrolysis of methane on *Ni* / Al_2O_3 and *Ni, Cu* / Al_2O_3 catalysts [1,2]. The carbon yield (Y) expressed in *g* of carbon deposit per *1 g* of the catalyst was varied from 1 to 180 *g/g*.

Macrostructured ceramics were used as honeycomb monoliths and foams described in [3,4]. These supports were coated by a carbon layer of different morphology, such as *i)* graphite-like (G) layer (Fig.1 a), *ii)* fur-like layer (F) formed by thick and short carbon filaments (Fig.1 b), and *iii)* hair-like layer (H) formed by thin and long carbon filaments (Fig.1 c), which were synthesized by methods described in [3]. Thus, the graphite-like carbon layer was synthesized by thermal decomposition of sucrose impregnated into the support pores at 500°C in H_2 flow for 2 hours. The catalytic filamentous carbon layers were synthesized by catalytic pyrolysis of a propane-butane mixture over *Ni* / Al/Si catalyst at 500°C.

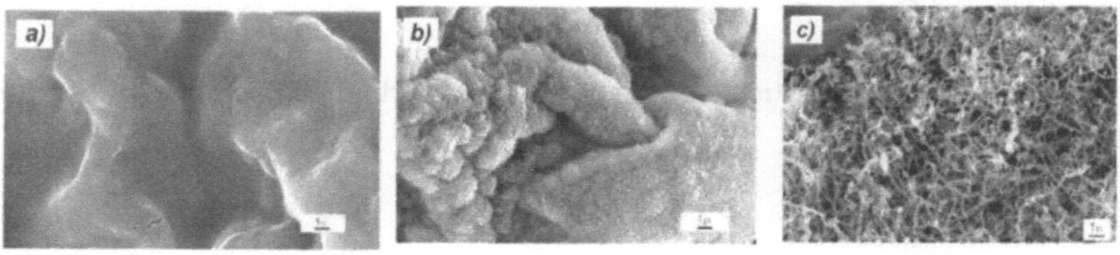

Fig.1. Scanning electron micrographs of carbon layers synthesized on ceramic surface.
a) graphite-like carbon layer (G) on foam-like ceramics,
b) fur-like carbon layer (F) on foam-like ceramics and Al/Si honeycomb monoliths,
c) hair-like carbon layer (H) on cordierite honeycomb monolith.

The adsorptive immobilization of the biologically active substances on the carbon-containing supports was performed under conditions described in [2,5]. The amount of adsorbate was calculated from the difference of its contents in solution before and after adsorption and expressed in *mg/g* of the support.

Scanning electron microscopy (SEM) studies of the support surfaces and the adsorbed bacteria were performed with REM-100Y microscope. The samples for microscopy were prepared as described in [2,5]. The labels in the micrographs correspond to distance in micrometers *(μ.)*

3. RESULTS AND DISCUSSION

Comparative investigation of adsorption properties of granulated bulk CFCs with respect to various biological substances were carried out in [2]. The main conclusion was that the value of *accessible surface area* is of crucial importance in the adsorption of adsorbates of biological origin [2]. As for microorganisms, they were adsorbed exclusively on the external surface of CFC granules; inner surface was not accessible for large bacterial cells (1-2 μ in size) [2,5]. An additional key factor affecting the adsorption of bacterial cells was *roughness* of the surface determined by the carbon yield (Y) [2,5]. It was shown that the adsorption capacity for the non-growing bacterial cells (*E.coli*, *Bacillus* sp. and *Rhodococcus* sp.) was lower for CFCs with dense and smooth surface (at high Y) and higher for CFCs with rough "carbon fur coating" (at low Y) whereas the surface area accessible to bacteria did not depend on Y [2]. The cross-linking of bulk CFC granules by the similar carbon filaments increased significantly both the inner area accessible for bacteria adsorption due to channel formation (Fig.2a) and the roughness of the surface via additional synthesis of "carbon fur coat" (Fig.2b). The adsorptive immobilization of bacteria on the cross-linked CFC granules was found to be highly efficient. In the SEM micrograph the adsorbed bacterial cells of *Rhodococcus ruber* were packed closely together (Fig.3).

As a conclusion, granulated bulk CFC supports with «carbon fur coating» were found to be efficient adsorbents for biologically active substances.

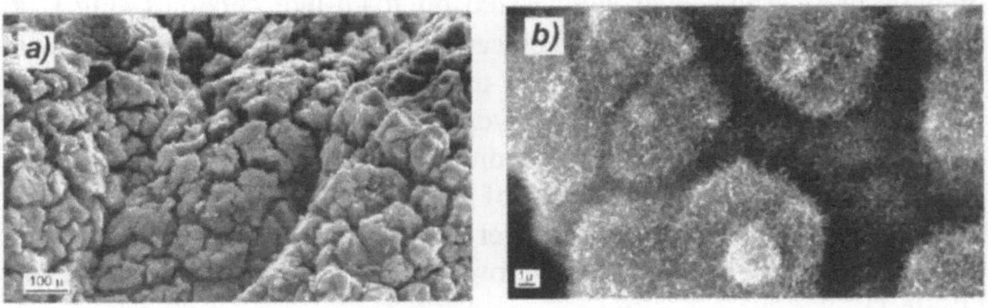

Fig.2. Scanning electron micrographs of the surface of cross-linked CFC granules
a) channels on the external surface of granules;
b) the "carbon fur coating"

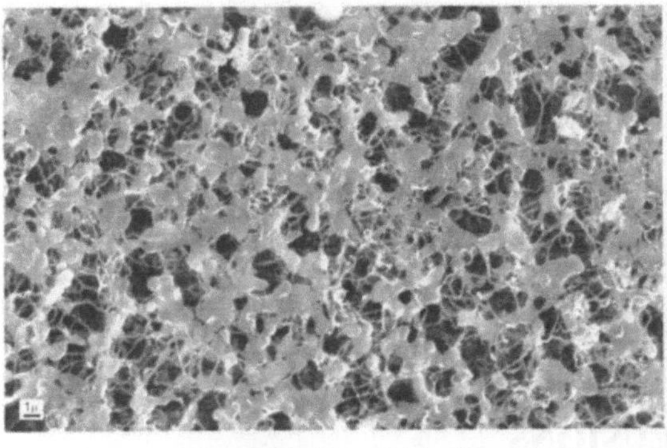

Fig.3. Scanning electron micrograph of the *Rhodococcus ruber* immobilized on cross-linked CFC granules.

To increase the accessible surface area and the mechanical strength of the supports and at the same time to retain the morphology of filamentous carbons, the macrostructured ceramics, in particular foams were used and methods of synthesis of CFC layers on the alumina-silica surfaces were developed. To study the effect of carbon layer morphology on stability and biocatalytic activity of immobilized enzymatic active substances, the ceramics coated by graphite (G)-, fur (F)- and hair-like (H) carbon layers were synthesized. When enzyme *glucoamylase* was adsorbed on the supports with graphite-like carbon layer, the stability of biocatalyst was relatively low [3]. The stability of glucoamylase increased by an order of magnitude via immobilization of enzyme on ceramics coated by filamentous carbon layer [3]. Also, filamentous carbon layer was found to provide the highest operational stability of immobilized glucoamylase. In the continuous hydrolysis of corn dextrinized starch the activity of the biocatalyst remained constant for several hundred hours at 50-55°C.

The effect of carbon layer morphology on the stability and biocatalytic activity of immobilized *invertase*-active cell membranes of autolyzed baker yeast was studied. Supports coated by catalytic filamentous carbon ensured the highest stabilization. The best result on stability was that the biocatalysts retained over 50% of the initial activity for 0,5-year storage at ambient temperature. Unlike glucoamylase, activity of immobilized invertase was found to depend strongly on the morphology of the carbon layer (Table 1). Invertase-active yeast membranes adsorbed on the graphite-like carbon layer synthesized on foam-like ceramics (Fig.1, a) showed a ~10-fold decrease of biocatalytic activity in comparison with fur-like carbon (Fig.1, b) perhaps due to hydrophobicity of graphite. Moreover, invertase adsorbed on hydrophobic carbon supports specially modified to increase the surface hydrophobicity has lost the biocatalytic activity completely. The activity of invertase-active yeast membranes immobilized on hair-like carbon layer synthesized on cordierite monolith was shown to be comparatively low because of diffusion limitations (Table 1).

Table 1.The effect of macrostructure of supports and morphology of carbon layer on activity of biocatalysts prepared via adsorption of yeast cell membrane.

Macrostructure of support	$S_{specific}$, m^2/g of non-carbonized support	Carbon content, w/w%	$S_{specific}$, m^2/g of carbonized support	Morphology of carbon layer*	Activity, *U/g* of biocatalyst
Foam	0.5	0.6	2.7	G	2.2
Foam	0.5	1.0	6.5	F	18.4
Monolith	24	1.4	27	G	12.5
Monolith	24	5.6	60	F	15.6
Monolith	0.15	0.5	0.8	H	7.2

The following designations are used: G – graphite-like carbon lay er (Fig.1a), F – fur-like carbon layer (Fig.1 b), H – hair-like carbon layer (Fig.1c)

4. CONCLUSIONS

Comprehensive investigations of granulated CFC supports and CFC-coated ceramics demonstrate their high adsorption efficiency with respect to biologically active substances (enzymes, yeast membranes and non-growing bacteria) due to filamentous morphology of the surface carbon layer. Cross-linked CFC granules possess the highest adsorption capacity for the non-growing bacterial cells.

Filamentous carbon provides the highest long-term and operational stability of the heterogeneous biocatalysts based on immobilized enzyme glucoamylase. In the continuous hydrolysis of corn dextrinized starch their activity remains constant for several hundred hours at 50-55°C.

The activity of immobilized invertase-active yeast membranes depends strongly on the morphology of carbon layer. Invertase adsorbed on graphite-like carbon layer has low biocatalytic activity. The significant (up to 10-fold) increase of invertase activity is observed when yeast membranes are immobilized on foam-like ceramics with fur-like carbon layer.

The highly efficient CFC-based adsorbents have potential for applications in biotechnology and medicine.

Acknowledgement: The work was partly funded by INTAS grant 01-2151.

References

1. Pat.RU 2064889 (1993).
2. Kovalenko GA, Kuznetsova EV, Mogilnykh YuI, Andreeva IS, Kuvshinov DG, and Rudina NA. (2001) Catalytic filamentous carbons for immobilization of biologically active substances and non-growing bacterial cells, *Carbon* **39**, 1133 – 1143.
3 Kovalenko GA, Komova OV, Simakov AV, Khomov VV, Rudina NA. (2002) Macrostructured carbonized ceramics as adsorbent for immobilization of glucoamylase, *J.Mol.Catal.A: Chemical* **182-183**, 73-80.
4 Pat.RU 2167197 (2001).
5. Kovalenko GA,Semikolenov VA, Kuznetsova EV, Plaksin GV, Rudina NA (1999) Carbonaceous materials as adsorbents for biologically active substances and bacterial cells, *Colloid J.* **61**, 729-736.

HYBRID THREE TERMINAL DEVICES BASED ON MODIFIED DNA BASES AND METALLOPROTEINS

ROSS RINALDI, GIUSEPPE MARUCCIO,
ALESSANDRO BRAMANTI, PAOLO VISCONTI,
PIER PAOLO POMPA, ADRIANA BIASCO,
ROBERTO CINGOLANI
National Nanotechnology Laboratory – University of Lecce , Lecce ITALY

Abstract: The key challenge of current research in nanoelectronics is the realization of biomolecular devices. The use of DNA bases and electron-transfer proteins, such as blue copper protein azurin (Az), is particularly attractive because of their natural self assembly capability. We will present our results concerning the fabrication, characterization and modeling of hybrid three-terminal devices based on deoxiguanosines (modified DNA bases) and azurins. These molecular devices exhibit novel operational functions, thus opening the way to the implementation of a new generation of logic architectures

1. INTRODUCTION

Molecular electronics is one of the potential roadblocks to continue the scaling beyond the 50nm node for the implementation of nanoelectronic devices. Recently key demonstrations of molecular electronics devices have been reported, primarily through the use of carbon nanotubes and polymers [1], suggesting that the obstacles to the accomplishment of molecular electronic devices are more technical than conceptual and mainly concerns issues such as device aging, lifetime and reproducibility. In particular, the demonstration of a three-terminal field effect device (field-effect transistor) consisting of source (s) and drain (d) contact interconnected by a molecular layer, and a third contacts (gate) to modulate the source-drain current (I_{ds}), is a crucial step for the development of a molecular electronics. However, the implementation process of a reliable three terminal device based on a few organic molecules at nanometer scale is still a demanding and difficult step [2]. An alternative route is to exploit conduction of self-assembled biomolecules, in order to achieve devices based on highly ordered engineered layers (bottom-up approach), taking advantage of the specific reactivity of molecules having functional groups with affinity for specific

S. Guceri et al. (eds.), Nanoengineered Nanofibrous Materials, 271-282.

surfaces and/or molecules. While engineers and scientists have been long aspiring to manipulate structures controllably and specifically at the micro- and nanometer scale, nature has been performing these tasks with great accuracy and high efficiency using highly specific biological molecules such as DNA and proteins. Electronic conduction through biological macromolecules and biomolecular monolayers has been demonstrated [3]. Pioneering investigations showed that biomolecules were stable in air and at room temperature and behave like active elements in the devices. In this paper we describe and discuss the basic nanotechnological issues for the implementation of hybrid three terminal devices based on deoxyguanosine derivatives (a DNA base) and metalloproteins. We also present a phenomenological explanation of the operational principles of such devices. Finally the interconnection problem of the hybrid devices is discussed on the basis of a finite element simulation, providing a full electromagnetic characterization of the structure. A thorough analysis of device ageing and degradation is also given.

2. DEVICES BASED ON DNA BASES

Recently, there has been great theoretical and experimental efforts aimed at assessing the problem of electrical transport in DNA strands. Due to the considerable difficulties in connecting single DNA ropes to electrodes separated by few tens of nanometers, we adopted a different strategy choosing a single modified DNA base, the lipophilic deoxyguanosine, as the basic molecular constituent of our nanodevices. The guanine exhibits a low oxidation potential -which favors charge transport- and spontaneous self-assembling properties both in solution and in the solid state [4]. The base was modified, in the lipophilic deoxyguanosine derivative form [5] (Fig.1a), to favor the formation of ordered supramolecolar structures thanks to its peculiar sequence of H-bond donor and acceptor groups and side chains (Fig. 1b).

The nanodevices were fabricated by cast deposition. A 2 μl drop of deoxyguanosine solution in chloroform was prepared (10^{-1} - 10^{-4} M) and deposited in the gap between two metallic contacts. Upon controlled evaporation, Atomic Force Microscopy (AFM) studies of the molecular layer showed that the deoxyguanosines self-assemble in supramolecular ribbon-like structures (Fig. 1c).

To interconnect guanosine-ribbons and implement the hybrid molecular electronic devices, planar metal-insulator-metal nanojunctions (Fig. 2a) were fabricated on the surface of thermally oxidized silicon wafers. The oxide layer was 100 nm thick and constituted the gate oxide, while a silver

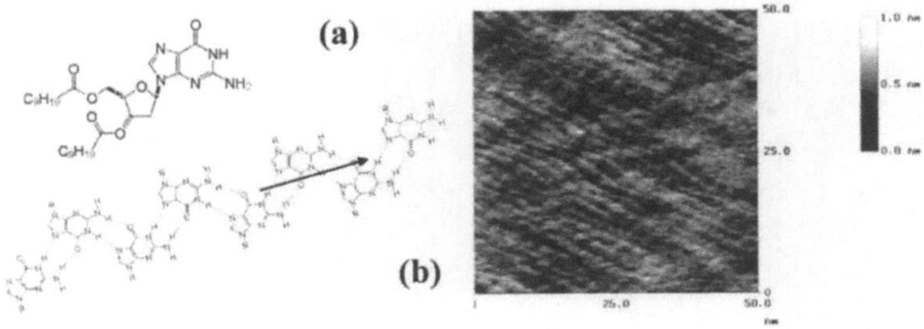

Figure 1 : (a) The molecule used in this study and (b) the self organized ribbons of $dG(C_{10})_2$. (c) AFM image of the guanosine layer morphology. The ribbons form an ordered supramolecular structure, laying parallel to each other with a periodicity of 2.5 nm and a length up to $l \approx 100$ nm. Over such a length scale, the packed lamellar structure of the SAGC, physisorbed on the surface between the electrodes, gives rise to an orthorombic unit cell of size a=1.3±0.1 nm and b=2.5±0.2 nm. For distances longer than 100 nm, that ordering is lost, and the ribbons form randomly oriented SAGCs.

back-electrode –deposited on the degenerated doped silicon substrate- was chosen as the gate electrode. The source and drain Cr/Au (8nm/35nm) electrodes were realized by electron beam lithography followed by lift-off. Typical inter-electrodes distances were in the range between 20 and 40 nm [6]. The fabricated planar electrodes were inspected by plan-view and cross-section scanning electron microscopy (SEM) in order to check the success of the whole technological process and to estimate the separation between the Cr/Au nanotips. All contact pairs were then tested by current-voltage measurements. Open-circuit resistance as high as 100-200GΩ were obtained in all working devices. Typical process yield – i.e. the percentage of good contact pairs where no significant current was observed to flow under open-circuit configuration – was around 90%.

Current-voltage experiments on the guanosine-based devices were focused on the DC behavior and were carried out by using a semiconductor parameter analyzer (HP Agilent 4155B) in the voltage range between –3.5 and 3.5 V at room temperature and in air. The gate and drain voltage were applied with respect to the grounded electrode (source). Drain-source and gate current were recorded simultaneously. Fig. 2b shows a 3D plot of the drain-source current as a function of the drain-source voltage V_{ds} and the gate voltage V_g. In the low drain-source bias region, the device was almost non conductive. All the measured characteristics showed the existence of a well-defined threshold voltage and indicated that conduction can be activated at higher bias. By increasing the gate voltage from 0V to 1V, this threshold voltage gradually increased following an almost linear trend. This behavior can be explained in terms of resonant tunneling.

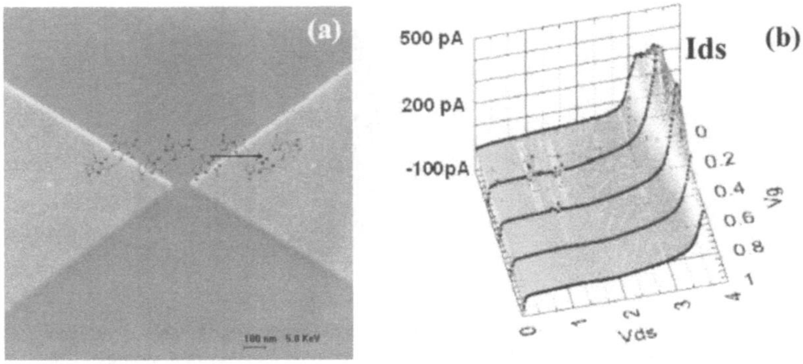

Figure 2 (a) High magnification SEM image of the planar nanoelectrodes. (b)3D plot of the drain-source current as a function of the drain-source voltage V_{ds} and the gate voltageV_g

Due to the difficulty in creating an ideal Ohmic molecule-metal contact, two potential barriers arise at the interfaces between the organic/inorganic layers. Moreover, since the molecules were found to aggregate spontaneously to form ordered self-assembled guanosine crystals (SAGCs), we could model the molecular part as a semiconductor material, characterized by energy bands. At low voltage, the molecular layer behaved as an insulator. On the other hand, transport above threshold can be explained in terms of resonant tunneling. When the applied drain-source bias aligns the electrodes Fermi levels to one molecular miniband, the conduction is activated. The energy shift of the molecular band due to the gate field induces a change in the alignment condition and therefore an increase in the threshold voltage, was required to recover the resonant tunneling process. We found that the threshold voltage increased almost linearly with the gate voltage.

The field effect was found to have a profound effect on the drain-source current at constant drain-source voltage, as manifested by a fivefold decrease of its value as a function of V_g (for drain-source voltage around 3 V, the current I_{ds} changes from 250 pA at V_g=0V to 50 pA at V_g=1V resulting in an on-off ratio around 5). It is worth noting that I_{ds} decreases as in a p-channel MOSFET, but no saturation was observed in the I_{ds}-V_{ds} characteristics. A maximum voltage gain (given by the product of the transconductance and the output resistance) as high as 0.76 was found (small-channel molecular devices usually exhibit gain in the range between 0.3 and 0.5).

Although these nanodevices can be probably further improved, to enhance their performance and reduce their aging, some interesting architecture can already be proposed. The lack of saturation in the source-drain current makes the output current quite sensitive to small changes in the source-drain voltage. The behaviour of our nanodevices is dual with respect to a standard field-effect transistor. In fact, the V_{ds}-I_{ds} is similar to the I_{ds}-V_{ds} curve of a standard FET. Our devices is stable when guided in current rather than in voltage. Moreover, an on-off ratio around 5 could be enough for some logic applications, such as static random access memory (S-RAM).

3. DEVICES BASED ON AZURIN MONOLAYERS

The blue copper protein *Pseudomonas aeruginosa* azurin (14.6 kDa) is one of the best characterized redox metalloproteins [7]. It mediates electron transfer in the denitrifying bacteria; specifically, it transports electrons from cytocrome c_{551} to nitrite reductase in the electron transport chain of respiratory phosphorilation [8]. The central Cu atom, the redox active site, is coordinated to 5 aminoacid ligands in a distorted trigonal bipyramid geometry [8]. The tertiary structure of azurin is strongly asymmetric for the presence of a S-S bridge opposite to copper. Azurin fold into an eight-stranded Greek key β-barrel motif with only a α-helix present. Given to this secondary structure and also to the disulfide bond, it can be regarded as a very stable protein. Due to their physical and chemical properties, Az proteins are good candidates for bio-electronic applications.

Covalent attachment on self assembled monolayers (SAMs), consisting of molecules adsorbed from solution onto solid substrates [9] has been employed to achieve an appropriate and permanent localization of biomolecules on Si/SiO_2. Azurin proteins have been adsorbed on self assembled 3-mercaptopropyltrimethoxysilane (3-MPTS), dissolved in absolute ethanol. A very diluted solution (2%) of 3-MPTS has been used to obtain ordered monolayers and to prevent from the multilayer formation. Sulfide groups are adsorbed from high purity ethanol solution and at the solid-liquid interface a strong chemisorption of the head group (Si) gives rise to highly ordered architectures. The promoted surfaces were exposed to commercial Az solutions and gently dried at room temperature. Adsorption of Az was performed from 10^{-4} M solution with a pH 4.6 (Az isoelectric point) prepared dissolving Az in 50 mM ammonium acetate buffer. In these conditions no charges are present on the biomolecule surface and the adsorption process isn't affected by electrostatic attractive or repulsive forces.

3-MPTS was used as a cross-linking agent because it ensures orientation of the protein deposited film. In fact, the protein adsorption on 3-MPTS SAM occurs by the unique disulfide bridge. Therefore, proteins are well-ordered in the molecular film and the molecular architecture is similar to that of immobilized protein molecules on gold [10]. The obtained tapping-mode AFM image with a scan size of 0.88 μm x 0.88 μm is presented in Fig 3a.

This image is an evidence for azurin adsorption since it reveals features uniformly distributed across the surface, which were not present on SAMs substrates. The presence of these topographic features was reflected in the roughness. The protein adsorption, in fact, is responsible for an rms roughness increase up to ten times. The recognizable granular structures are adsorbed proteins and the surface is totally covered by proteins. The width of these structures is overestimated due to the geometrical effect of the AFM tip. It is known that Az has a dimension of 4.4 nm. Considering the tip convolution effect, the apparent width measured by AFM (19-21 nm) corresponds to d = 4.5-5.5 nm; not very different from values reported in literature.

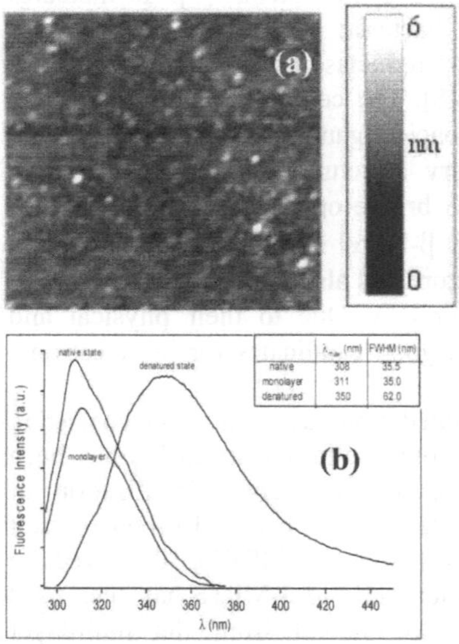

Fig 3: (a) AFM topography image of azurin immobilized on self-assembled 3-MPTS film. (b) Fluorescence spectrum of azurin in solution and adsorbed onto surface.

It is noteworthy that the immobilization procedure is a very demanding aspect because it must preserve the molecular functionality.

Conformational properties of the azurin monolayers were investigated by intrinsic fluorescence spectroscopy. Using this technique, it is possible to analyze the conformational and structural alterations that proteins experience during immobilization on solid films. Since the fluorescence parameters of the aromatic residues are highly sensitive to the microenvironment of fluorophores in protein structures, the use of intrinsic fluorescence spectroscopy allows us to directly monitor the conformational equilibrium induced by the immobilization procedure, thus determining the occurrence of perturbations in protein folding, possibly resulting in physical-chemical changes of the copper active sites. Azurin has only one tryptophan residue (Trp48) which is responsible for protein fluorescence. The photoluminescence spectrum of native azurin is centered at an unusually short wavelength ($\lambda_{max} \approx 308$ nm) owing to the highly hydrophobic microenvironment surrounding Trp48. Despite the holo- and apo-azurins exhibit identical fluorescence spectra, the emission of the holo-protein is strongly quenched by the presence of the copper. Therefore, we utilized the apo-form of azurin to optically investigate the protein monolayers on SiO_2.

With respect to free apo-azurin fluorescence in buffer, the spectrum of the protein immobilized in the film is characterized by a comparable line-shape (Fig. 3b) and a slight red-shift (~3 nm), accounting for the fact that immobilization binding does not interfere with the fold pattern of the native protein. This result is supported by the comparison of the photoluminescence spectrum of protein monolayer with the broadband, red-shifted emission of a denatured sample of azurin (see also the inset of Fig. 3b). According to the experimental evidence and to the photophysics of aromatic amino acids, it seems that the tryptophan residues in native and immobilized proteins are found in similar locations. The small spectral shift is the only minor effect detectable in azurin upon immobilization. This suggests the possibility of a very weak internal rearrangement due to the interaction of the ligand binding, without affecting the actual fold pattern.

Having demonstrated the integrity of the azurin monolayer, we proceeded to the fabrication of the protein transistor prototype. The device structure consisted of a planar metal-insulator-metal nanojunction, comprising of two Cr/Au (6nm/35nm) arrow-shaped metallic electrodes facing each other at the oxide side of a Si/SiO_2 substrate connected by the molecular layer (drain and source electrodes). The fabrication of planar nanojunctions was carried out using the same procedure described in section 2. The disulfide bridge S-S of Azurin was broken to form covalent bond on SiO_2 substrate, properly functionalized by means of 3-MPTS.
Current-voltage experiments were carried out by using a semiconductor parameter analyzer (HP Agilent 4155B) in the voltage range between –6 and 6 volts, at room temperature and ambient pressure. The gate voltage (V_g) was changed in order to investigate its influence on the current (I_{ds}) between the source (s) and drain (d) electrodes. Before the deposition of molecules, all devices were tested to verify that the channel was insulated (resistance =100 GΩ.). The current-voltage characteristic (I_{ds}-V_{ds}) of the protein transistor under forward drain-source bias (V_{ds}) exhibits a low-current plateau at low field and then rises up to hundreds of pA as shown in Fig. 4a.

The room-temperature drain-source current as a function of the gate potential, for V_{ds} = 5.5 V, is displayed in Fig. 4b. The transfer characteristic exhibits a pronounced resonance with a bell shape centred at V=1.25V. In this region, the transconductance changes from positive to negative values. The peak to valley ratio and the FWHM are 2 and 0.3V, respectively. This feature gradually disappears after some cycles of measurement due to the aging of the molecular layer, which is not encapsulated.

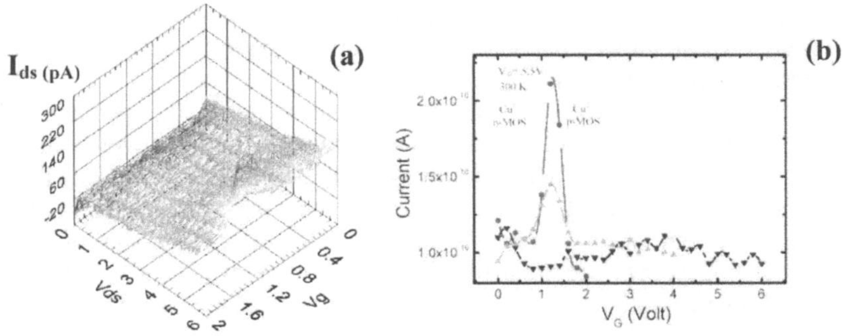

Fig. 4: (a) Characteristic of protein FET. 3D plot of the drain-source current as a function of the drain-source bias (V_{ds}) and of the gate bias (V_g) measured in the dark and at room temperature. (b) Transfer characteristic of protein FET.

These results originate from the unique transport mechanism of our biomolecular devices. Redox proteins devices are very different from standard inorganic semiconductors. Silicon MOSFETs and thin-film transistor (TFTs) are based on a gate field modulating the width and the conductance of a semiconducting channel. In proteins, the long range electron transfer (ET), which represents one of the key processes of living systems involved in photosynthesis and respiration, occurs between a donor (D) and an acceptor (A) site. Two different models for ET have been proposed [12], i.e. a superexchange mechanism (consisting of direct quantum tunneling between the donor and acceptor) or a sequential (incoherent) hopping between adjacent sites.

The transport of electrons through systems containing redox sites occurs via electron hopping from one reduced (Cu(I)) molecule to an adjacent oxidized (Cu(II)) molecule [13]. Therefore, the presence of two adjacent Az molecules in the Cu(I) and Cu(II) redox states is required to have current flux between two planar electrodes. Since the protein is chemisorbed onto the SiO_2 surface with the natural electron transfer route – which joins the copper site to the disulfide-bridge – almost perpendicular to the surface, we believe the vertical field applied along this direction (due to V_g) modifies the oxidation state and induces a change in the equilibrium of the redox reaction, thus modifying the balance between the two populations. As a consequence, the current as a function of V_g is maximum when the populations of protein in the Cu(II) and Cu(I) state become equals, otherwise the current is lower. This phenomenological model explains the presence of the resonance in the transfer characteristics shown in Fig. 4b.

4. DEVICES AGEING AND DEGRADATION ANALYSIS

Finally, the issues of device reproducibility, aging and lifetime must be addressed for molecular electronics [14,15,16]. Among all the fabricated nanodevices, only a limited group exhibited a clear gate effect, whereas in the others significant current fluctuations and failure occurred. Inspection by means of scanning electron microscope demonstrated that the failure was primarily due to the electrode burning and possibly metal electro-migration during operation. Two main phenomena concerning device damage can be distinguished: (1) nanojunctions can blow up due to the local heating during the current flow or (2) the formation of various aggregates can be observed on the contacts . When we apply a bias, a very high electric field arises between the nanoelectrodes, due to the small tip separation and to the drastic change in the dielectric constant at the interface between SiO_2 and the metal. The very intense electric field, combined with the low adherence of the metal layer, favours the diffusive movement of metal mass. We found an experimental relationship, according to which for increasing contact separation the activation voltage for the device damage increases (as expected from a field or current related phenomenon) with a nearly linear trend. Typical values ranged from ~4 V at 50 nm to ~7 V at 300 nm. The extrapolated linear regression, however does not intersect the origin, so predicting a nonzero activation voltage for shot-circuited contact pairs. Clearly, the phenomenon can no longer be treated in terms of macroscopic quantities such as the voltage and gap width – or, which is equivalent, the average electric field. Instead, a more accurate analysis of the local relevant quantities is required, in particular the local electric field, current density and temperature. The simulations where performed solving the Poisson conductive media equation, by the commercial finite element code FEMLAB. In particular, we simulated contact pairs with different separations ranging from 40 to 300 nm. The voltage applied to each configuration was taken according to the linear interpolation of the experimental data; then the electric field in air and current density inside gold were analyzed, to find recurring values or at least orders of magnitude related with breakdown. A thermo-electrical simulation allowed to exclude any significant temperature increase (temperature differences were always less than 1K); therefore we concluded that heating could not be the primary cause for breakdown. The electric field was analyzed in the *highest intensity region* in air, to evaluate its possible effect on the molecular material. We took as *high intensity region* (or hot-spot) the one in which the electric field intensity values were always bigger than the 75% of the maximum of electric field probability distribution. The spatial average for the electric filed values was computed in this region.

Current density too was spatially averaged over its hot-spot (visible in Fig. 5a, inside one contact), defined as for the electric field but this time located inside the gold contact. However, the spot is near the tip's end, where current lines concentrate, while some samples look spoiled also far from the sharp part; thus, we spatially averaged the current density also over the whole contact.

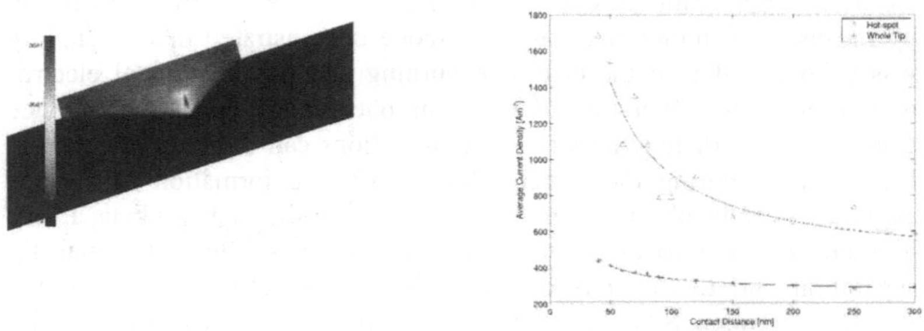

(a) (b)

Fig. 5: (a) Current density hot-spot at the far end of the gold tip (magnitude in Am^{-2}); the other visible region is the SiO_2 layer (air and Si are modeled, but not rendered in this color map). (b)Average current density in the hot-spot (triangles) and over the whole tip (stars) with interpolating curves

In Fig. 5b we report the results for current density, averaged over the hot-spot (triangles) and whole tip (stars). The hot-spot density decreases with increasing contact separation, with a steeper slope in the first part and getting asymptotically closer to a plateau, in the second part; a knee could be located around 100 nm. The reason for this non-uniform descent is that, with small contact separation, the current density distribution around one tip is greatly affected by the distance from the other tip; 'far' contacts, on the other hand, asymptotically resemble a single tip configuration. The electric field-vs.-separation curve (not reported) looks qualitatively similar. The situation is different with current density averaged over the whole tip, where the dependence on the contact distance is less pronounced. This behavior means that the applied voltage-vs.-separation law, obtained by the experiments, is the result of a nearly constant mean current density in the contact. These results provide an indication that breakdown is triggered by a critical level of current density. According to our simulations, such level should be fixed between 300 and 400 A/m^2. These values cannot generate an electromigration proces, since the typical current densities involved are at least 10 orders of magnitude larger [17] than the one we found. The most likely explanation is that current triggers a chemical etching of the contacts, possibly starting from 'weak points' such as bubble defects in the metallic electrodes [18]. This process can cause the contact to blow up and the degradation of the molecular layer.

5. CONCLUSIONS

In conclusion, we demonstrated a transistor based on a deoxyguanosine derivative and a protein FET exploiting the redox and self-assembly properties of the metallo-protein azurin, both operating at room temperature and ambient condition.The novelty of both the active material and the transport mechanism makes such devices very interesting for molecular electronics application.

Acknowledgements: We are thankful for the invaluable support and exciting collaboration with various colleagues. We would like to thank Elisa Molinari, Rosa Di Felice, Francesca De Rienzo, Paolo Facci at S^3-INFM research center in Modena (Italy), Gerard Canters and Martin Verbeet at Leiden University (NL), Salvatore Masiero, Tatiana Giorgi, Gianpiero Spada and Giovanni Gottarelli at University of Bologna (Italy).

Financial support by NNL-INFM, by the Italian Ministry of University and Research (MIUR) through the FIRB project "molecular nanodevices" and by EC through SAMBA project are gratefully acknowledged.

References

[1] Joachim, C., Gimzewski, J.K., and Aviram, A., 2000 Nature 408, p.541;
K.S.Kwok and J. Ellenbogen, Materials Today , Feb. 2002 , p.28;
M.A. Reed, J.Chen, A.M.Rawlett, D.W.Price and J.M.Tour, "Molecular random access memory cell", *Appl.Phys.Lett.*, 78, 3735-3737, 2001

[2]R.Metzger, *Chemical Review*, Oct. 2003

[3]R.Rinaldi, A.Biasco, G.Maruccio, R.Cingolani, D.Alliata, L.Andolfi, P.Facci, F.De Rienzo, R.Di Felice, E.Molinari, 2002, *"Solid-State Molecular Rectifier based on self-assembled metalloproteins"*, Adv. Mater. 20, p.1453, R.Rinaldi, A.Biasco, G.Maruccio, V.Arima, P.Visconti, R.Cingolani, P.Facci, F.De Rienzo, R.DiFelice, E.Molinari, M.Ph Verbeet, G.W.Canters, 2003, *Appl. Phys. Lett.* 82, p.472; D. Porath, A. Bezryadin, S. de Vries and C. Dekker, Nature (London) 403, 635 (2000)

[4] G.Gottarelli, S.Masiero, E.Mezzina, G.P.Spada, P.Mariani and M.Recanatini, *Helv.Chim.Acta,* **81**, 2078, 1998

[5] G.Gottarelli, S.Masiero, E.Mezzina, S.Pieraccini, J.P.Rabe, P.Samorì and G.P.Spada, *Chem.Eur.J.,* **6**, 3242, 2000; R.Rinaldi, G.Maruccio, A.Biasco, V.Arima, R.Cingolani, T.Giorgi, S.Masiero, G.P.Spada and G.Gottarelli, Nanotechnology **13**, 398-403 (2002)

[6] G. Maruccio, P.Visconti, S.D'Amico, P.Calogiuri, E.D'Amone, R.Cingolani, R.Rinaldi, Planar nanotips as probes for transport experiments in molecules. *Microelectronic Engineering* 67-68, 838-844 (2003)

[7] E.T. Adman, R.E. Stenkamp, L.C. Sieker and L.H Jensen, *J. Mol. Biol.* 123 (1978) 35.

[8] M.A. Webb,. C.M. Kwong and G.R. Loppnow, J. Phys. Chem. B101 (1997) 5062.5.

[9] C.H. Kobler and D.K. Scwartz, Curr Opin Colloid Interface Sci 4 (1999) 46.

[10] B. Schnyder, R. Kotz, D. Alliata and P. Facci, Surf. Interface Anal. 34 (2002) 40.7.

[11] S. Heinze, J. Tersoff, R. Martel, V. Derycke, J. Appenzeller, and Ph. Avouris Phys. Rev. Lett. 89 (2002) 106801

[12] O. Kuhn, V. Rupasov and S. Mukamel, J. Chem. Phys. 104 (1996) 5821

[13] D.N. Blauch and J.M. Saveant, J.Am.Chem.Soc, 114 (1992) 3323

[14] J.Park, A.N.Pasupathy,J.I.Goldsmith, C.Chang, Y.Yaish, J.R. Petta, M.Rinkoski, J.P.Sethna, H.D.Abruna, P.L.McEuen and C.Ralph, Nature 417 (2002) 722

[15] J. Lee et al., Nano Letters 3, 113-117 (2003) ; C. R. Kagan et al., Nano Letters 3 (2003) 119
[16] Metzger, R. M.; Chen, B.; Hopfner, U.; Lakshmikantham, M. V.; Vuillaume, D.; Kawai, T.; Wu, X.; Tachibana, H.; Hughes, T. V.; Sakurai, H.; Baldwin, J. W.; Hosch, C.; Cava, M. P.; Brehmer, L.; Ashwell, G. J. J. Am. Chem. Soc 119. (1997) 10455.
[17] Z. Li, G. Wu, and Y. Wang, *Journ. Mat. Sci. Electr.*, vol. 10, pp. 653-659, 1999.
[18] P. E. Burrows, V. Bulovic, S. R. Forrest, L. S. Sapochak, D. M. McCarthy, and M. E. Thompson, *Appl. Phys. Lett.*, vol. 65, no. 23, pp. 2922-2924, December 1994.; V. N. Savvate'ev, A. V. Yakimov, D. Davidov, R. M. Pogreb, R. Neumann, and Y. Avny, *Appl. Phys. Lett.*, vol. 71, no. 23, pp. 3344-3346, December 1997.

POLYPHOSPHAZENE NANOFIBERS FOR BIOMEDICAL APPLICATIONS: PRELIMINARY STUDIES

CATO T. LAURENCIN [1,2,3] AND LAKSHMI S. NAIR[1]

[1]Department of Orthopaedic Surgery, University of Virginia, Charlottesville, VA-22903; [2]Department of Biomedical Engineering, University of Virginia, Charlottesville, VA-22903; [3]Department of Chemical Engineering, University of Virginia, Charlottesville, VA-22903

Abstract: Polyphosphazenes form novel class of inorganic polymers having backbone composed of phosphorous and nitrogen atoms with each phosphorous atom bearing two organic or organometallic groups. The side groups on phosphorous atoms play a crucial role in determining the properties of corresponding polyphosphazenes. Thus, polymers with unusual thermal, electrical, optical, electrolytic conductivity, electroactivity, oxidative stability, biocompatibility and biodegradability have been synthesized by the incorporation of appropriate side groups. Many of these polymers have been identified as excellent candidates for biomedical and high technological applications. Polymeric nanofibers due to their very high surface area, remarkable surface properties and superior mechanical properties are assuming great interest lately. We have demonstrated that nanofibers of polyphosphazenes can be prepared by electrospinning wherein submicron sized fibers are produced from an electrostatically driven jet of polymer solution. The diameter as well as the morphology of the resulting fibers can be efficiently tuned by varying process parameters such as viscosity of the solution and potential gradient. By combining the advantages of nanoscale fibers with the unusual property profile available with polyphosphazenes, novel materials with excellent properties can be developed.

Keywords: Polyphosphazene, nanofibers, biodegradable.

1. INTRODUCTION

Nanotechnology is an emerging field which uses the principles of science and engineering to fabricate materials or structures of dimensions in the nanometer scale [1]. The great interest in nanoscale materials spur from the unusual and unique property profile these materials exhibit compared to macromaterials. Novel physical, chemical and biological

S. Guceri et al. (eds.), Nanoengineered Nanofibrous Materials, 283-302.

properties such as unique shape, orientation, surface chemistry, topology and reactivity exhibited by these materials originate from their small dimensions. These material properties translate into unusual electrical, optical, magnetic, mechanical, thermal and biological properties for these materials. Some of the nanostructures or nanoscale materials currently under investigation include quantum dots and wires, nanoscale self-assemblies and thin films, nanocrystals, nanotubes, nanowires, nanorods, nanofoams, nanospheres and nanofibers. Almost all the classical materials available such as metals, nonmetals, ceramics, organic molecules, organometallics, polymers, biomolecules and composites are actively being investigated for developing these nanoscale materials. Some of the next frontier areas that will benefit from these nano scale materials include nano electronics [2], nano electromechanical devices [2] and nano optical devices [3].

Among the nanostructures discussed above, nanofibers form one of the most extensively investigated areas recently [4]. The word nanofiber refers to fibrous structures usually made of carbon, organic polymers or organometallic polymers with diameter less than one micrometer. This review however, focuses on polymeric nanofibers. Literature shows that polymeric nanofibers can be fabricated using various processing techniques such as drawing [4,5], self assembly [4,6], template synthesis [4,7], phase separation [4,8] and electrospinning [4]. Among these, electrospinning is the most general and simplest technique that can be used to fabricate polymeric nanofibers. The process of electrospinning, initially called electrostatic spinning was first studied by Zeleny in 1914 [9]. The process for developing polymeric fibers via electrostatic spinning was first patented by Formhals in 1934 [10]. Even though there are literature evidences showing work in this area from 1934 onwards, the past few years has seen a renewed interest in polymeric nanofibers and hence the process of electrospinning. This can be attributed to the recent tremendous growth in nanotechnology and also electrospinning forms the simplest and inexpensive way to create ultrafine polymeric fibers.

In electrospinning, a non woven mat of polymeric nanofibers will be created from an electrostatically driven jet of polymer solution. **Scheme 1** shows a general representation of an electrospinning apparatus used to create polymeric nanofibers. The three basic elements of an electrospinning system consist of a high voltage supply (5-40kV), a source electrode and a grounded collector electrode and a capillary tube with a needle of small diameter. Both polymer solution as well as polymer melt can be used to create nanofibers by the process of electrospinning. Briefly, in electrospinning the high voltage electric field is applied to a polymer droplet held by surface tension at the end of the needle. During this process, the polymer droplet gets charged and mutual charge

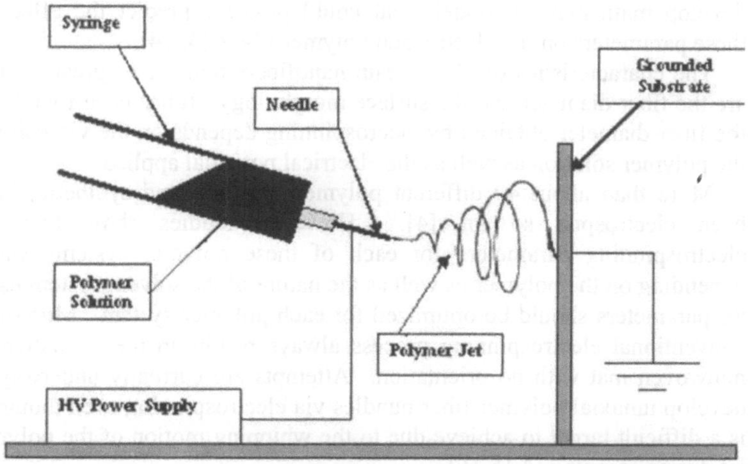

Scheme 1. Schematic representation of electrospinning apparatus

repulsion within the droplet will give rise to a force that opposes the surface tension. At a critical voltage, the droplet elongates to form a cone called Taylor cone at the tip of the needle [11]. When the applied voltage exceeds the critical voltage, the electrical force within the liquid overcomes the surface tension forces and a fine jet of polymer emerges from the cone. The viscosity of the polymer solution/melt is very crucial in producing nanofiber from the emerging jet. In the case of solutions with low viscosity, the emerging jet breaks into droplets due to surface tension and the process is called electrospraying. However, in the case of high viscosity solutions, the emerging jet does not break into droplets. The charged polymer jet will be directed to the grounded electrode and it has been found that the path taken by the jet to the grounded electrode is not straight instead it undergoes a whipping motion resulting in a progressive decrease in the diameter of the jet. As the jet travels in air, the solvent evaporates, leaving behind a charged polymer fiber which is collected as a non woven mat on the collecting screen.

Several parameters affect the process by which polymer solutions are converted into nanofibers. As discussed before the viscosity of the solution plays a very important role in the formation of nanofibers from polymer solutions. The parameters that affect electrospinning can be broadly classified into 1. polymer solution parameters such as viscosity, conductivity/polarity and surface tension of the solution, 2. process parameters such as applied electric potential, the distance between the needle tip and the collecting screen, size of the needle tip, feed rate and the hydrostatic pressure applied to the polymer solution and 3. ambient parameters such as temperature, air velocity and humidity of the electrospinning chamber [4,12,13] . Studies are currently undergoing to develop mathematical models that could possibly predict the effect of these parameters on the electrospun polymer fibers [4, 14].

The characteristics of electrospun nanofibers that are of great interest are the fiber diameter and the surface morphology. It has been found that the fiber diameter obtained by elecrospinning depends on the viscosity of the polymer solution as well as the electrical potential applied.

More than about 40 different polymers (natural and synthetic) have been electrospun so far [4]. However, studies show that the electrospinning parameters for each of these polymer systems varies depending on the polymer as well as the nature of the solvent system used. So parameters should be optimized for each polymer system. Moreover, conventional electrospinning process always results in the formation of nonwoven mat with no orientation. Attempts are currently underway to develop uniaxial polymer fiber bundles via electrospinning even though it is a difficult target to achieve due to the whipping motion of the polymer jet during spinning [4,15,16].

The properties of nanofibers such as fiber diameter, diameter distribution and surface morphology are usually characterized using scanning electron microscopy (SEM) [4]. Other techniques employed for fiber characterization include transmission electron microscopy (TEM) and atomic force microscopy (AFM). The surface chemistry of the electro spun nanofibers can be determined by techniques such as X-ray photoelectron spectroscopy (XPS), contact angle measurements and by Attenuated total reflectance Fourier transform infra red spectroscopy (FTIR-ATR). The structures of the polymer molecules within the nanofibers can be characterized by Fourier transform infra red spectroscopy (FTIR) and Nuclear magnetic resonance (NMR). The formation of thin polymer fibers from polymer solution usually changes the orientation of polymer chains. The configuration of the polymer chains in the nanofiber is usually determined by differential scanning calorimetry (DSC), wide angle X-ray diffraction (WAXD) and optical birefringence.

2. APPLICATIONS OF POLYMERIC NANOFIBERS

Due to the very small diameter of the polymer fibers obtained by electrospinning, the surface area to volume ratio of the fibers is significantly high. The nonwoven mats formed from nanofibers have very small pore sizes, however, the total porosity of the mats will still be very high. This makes them excellent candidates for use in filtration and membrane application. Due to the high filtration efficiency, the polymer nanofiber based filtration devices are now commercially developed for industrial air filtrations, as filters for gas turbine generators and filters for heavy duty engines. Recently it has been demonstrated that the filtration efficiency of these membranes can be increased by using charged fibers which would modify the electrostatic attraction of the particles [17].

Electrospinning has the potential to develop into the future fabrication process for developing protective clothing. Optimal protective clothing should satisfy properties such as it should be permeable to both air and water vapor, should be reactive with toxic gases and chemicals, should be insoluble in solvents and should have light weight. As discussed before due to very small pore size of the nanofiber mats, they provide an impermeable barrier to toxic chemical agents. Also the high surface area to volume ratio of the nanofibers enables them to form light weight fabrics with remarkable breathing properties [18].

Due to the simplicity and versatility of the electrospinning process, conductive fibers can be fabricated from polymers with unusual electrical, electronic, ionic, photoelectric and piezoelectric properties. These fibers could be used in the fabrication of novel nanoelectronic devices, nanoelectronic machines and sensors for high technology applications [19].

Another actively investigated area of application of polymeric nanofiber is the biomedical field [4]. Both non degradable and degradable polymers have been investigated for various applications. As early as 1977, it has been proposed that non degradable polymeric nanofibers can be used for developing prosthetic medical devices such as blood vessels and vascular grafts [20]. Recently it has been shown that electrospinning can be used to create porous thin films with structural gradients and controlled morphology on prosthetic devices in order to facilitate the integration of the devices with the body and thereby enhance their biocompatibility. Another potential medical application of non woven polymer nanofibers is in the area of tissue engineering. The role of biomaterial in tissue engineering is to act as a scaffold for the cells to attach to and organize into tissue. The ideal tissue engineering scaffold should mimic the extracellular matrix, the natural abode of cells. The structure and morphology of non woven nanofiber mats closely match the structure of extracellular matrix of natural tissue. Studies have shown that cells seeded on biodegradable polymeric nanofiber mats can attach, proliferate and maintain their phenotype expression [21,22]. Development of nanofiber based three dimensional scaffolds using various biopolymers and synthetic biodegradable polymers has currently become an active area of research. Polymeric nanofibers have also been investigated as novel wound dressing and as haemostatic devices [4]. The high surface area of the nanofiber matrix allows oxygen permeability as well as prevents fluid accumulation at the wound site. On the other hand, the small pore size of the matrix efficiently prevents bacterial penetration making them ideal candidates for wound dressings. Further, flexibility of the electrospinning process allows co-spinning polymers with drugs or proteins thereby making the non woven nanofiber mat as a drug delivery matrix which could enhance wound healing.

3. POLYPHOSPHAZENES

The wide range of organic polymers developed from petrochemical sources has undoubtedly contributed to the advancement of modern science and technology. However, materials with better physical, chemical, mechanical and biological properties are constantly in demand to tune with the requirements of modern technology. The advancement in synthetic organic chemistry and development of novel polymer synthetic routes provide us with a wide array of organic polymer with wide spectrum of properties. However, the organic polymers are associated with some innate drawbacks [23]. Most of the organic hydrocarbon based polymers have low oxidative stability, low resistance to temperature and organic solvents, low fire and light resistance and most of them except polyanilines have low electrical conductivity. In the biomedical field, even though organic polymers are extensively investigated as candidates for various applications, the performance of most of them are far less than optimal. The low thermal, optical and oxidative stability of organic polymers can be attributed to the presence of aliphatic C-C and C-H bonds present in these polymers. On the other hand, inorganic materials such as ceramics combine the properties of high thermal and oxidative stability, high solvent resistance and exhibit high mechanical strength and biocompatibility. However inorganic systems are often associated with high brittleness and low processability. Logically, the development of inorganic polymers with organic substituents is one way to solve most of the problems posed by organic polymers and inorganic materials. Main group elements of group III, IV, V and VI such as aluminum, silicon, phosphorous, sulfur, germanium, arsenic and selenium are ideal candidates for hetero atom polymer formation and these elements can form bonds with other elements which have longer bond lengths than that formed by carbon [24]. The longer bond length in these polymers could impart high flexibility to these polymers compared to organic polymers. The search led to the development of the first synthetic inorganic polymer poly(organosiloxane) by the ring opening polymerization of organocyclosiloxanes in 1940's. The polymer exhibits high flexibility, elasticity, high resistance to temperature and organic solvents, has ease of fabrication and biocompatibility. The initial success led to the development of other inorganic polymer systems such as polysilazenes, polysilanes, polygermanes, polystannanes and polyphosphazenes. Among these, polyphosphazenes form an interesting class of polymer which could develop as promising candidates for high technology and biomedical applications due to their synthetic flexibility and versatile adaptability for applications.

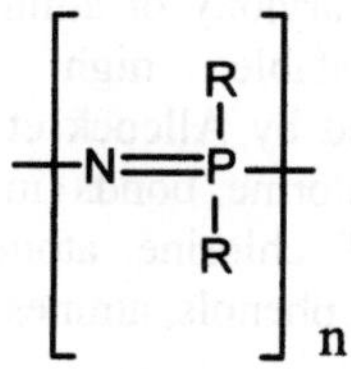

R = alkoxy, aryloxy or amino groups

Scheme 2. General structure of polyphosphazene

Polyphosphazenes are inorganic polymers having a backbone of alternating phosphorous and nitrogen atom and each phosphorous atom is attached to two organic / organometallic side groups [25]. The general structure of polyphosphazene is shown in **Scheme 2**. The synthesis of polyphosphazene involves a two step process: The first step involves the thermal ring opening polymerization of a cyclic trimer, hexachlorocyclotriphosphazene (**Scheme 3**; **1**) in the melt at 250°C, to form the linear high molecular weight polymer poly(dichlorophosphazene) (**Scheme 3**; **2**). This route was perfected by Allcock and Kugel in 1965 by carefully controlling the time and temperature of the ring opening polymerization [26]. The intermediate, poly(dichlorophosphazene) is

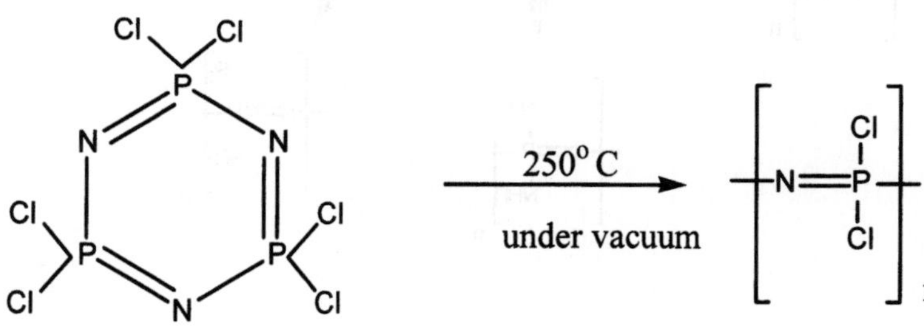

1 **2**

Scheme 3. Scheme showing the ring opening polymerization of hexachlorocyclotriphosphazene

unstable due to the highly reactive phosphorous-chlorine bonds present in it. Consequently, the polymer rapidly hydrolyzes upon exposure to atmospheric moisture to form phosphoric acid, ammonia and hydrochloric acid. The second step in polyphosphazene synthesis is the macromolecular substitution reaction which involves the replacement of reactive chlorine atoms of poly(dichlorophosphazene) with hydrolytically stable nucleophiles such as alkoxy, aryloxy or amino groups. The synthesis of the first hydrolytically stable, high molecular weight linear polyphosphazene was reported by Allcock et al in 1966 [27]. The high reactivity of phosphorous-chlorine bonds in poly(dichlorophosphazene) enables the replacement of chlorine atoms with a wide range of nucleophiles such as alcohols, phenols, amines or organo-metallic reagents.

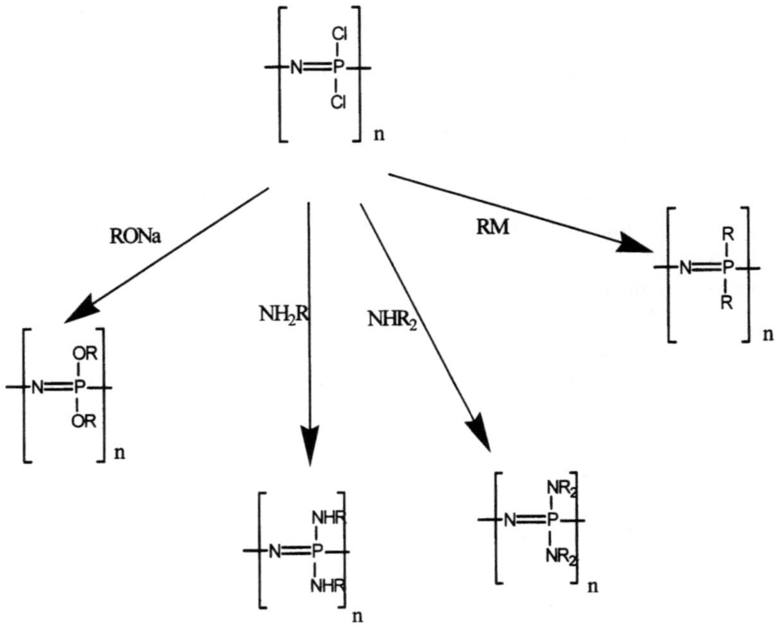

Scheme 4. Scheme showing the general route for synthesizing substituted polyphosphazenes from poly(dichlorophosphazene)

This forms the basis of the unprecedented diversity of polyphosphazenes, wherein different polymers can be synthesized from a single macromolecular precursor. Different classes of polyphosphazenes have been synthesized so far by replacing the chlorine atoms of poly(dichlorophosphazene) by alkoxides/aryoxides, primary/secondary amines and organometallic reagents [25]. **Scheme 4** shows the general route for synthesizing substituted polyphosphazenes.

Ring opening polymerization of hexachlorocyclotriphosphazene results in high molecular weight poly(dichlorophosphazene) with broad polydispersities. Attempts were therefore made to develop poly(dichlorophosphazene) with controlled molecular weight and narrow polydispersity since it is the intermediate from which almost all the polyphosphazenes are currently being made. Thus, an ambient temperature method to synthesize poly(dichlorophosphazene) intermediate with precise molecular weight control and narrow polydispersities by living cationic polymerization of phospharanimines was developed by Allcock et al. [28]. One advantage of the living cationic polymerization route is that the living sites at the ends of the polymer chain offer the possibility for the formation of block co-polymers either with other phosphazene monomers or with organic monomers [29]. In addition to the two step synthetic route described above a direct synthetic route for polyphosphazenes has also been attempted by Allcock et al. It involves the ambient temperature polymerization of mono- and di-substituted organophosphoranimines such as $PhCl_2P{=}NSiMe_3$ induced by PCl_5 [30].

As in the case of other inorganic polymers discussed before, one versatile property of polyphosphazenes is their unusual backbone flexibility which arises from the torsional and angular freedom within the –P-N- skeletal system. Unlike organic systems, the conjugated double bonds in polyphosphazenes are made up of $d\pi$-$p\pi$ bonds which do not impose severe conformational restrictions upon the molecule since any one of the five phosphorous 3d orbitals can π bond to the nitrogen p-orbitals [31].

Due to the innate skeletal flexibility, the steric and polar interactions of the side groups determine the polymer flexibility and the resulting physico-chemical properties [32]. This offers the opportunity for synthesizing polymers with wide spectrum of properties in terms of crystallinity, solubility, processability, hydrophobicity/ hydrophilicity and biodegradability depending on the nature of the side groups from poly(dichlorophosphazene). Another unique feature of polyphosphazene chemistry is the ability to incorporate two or more different side groups, which could add another dimension, to fine tune the properties of the polymer. Thus, two or more different side groups can be introduced along the polymer chain by simultaneous, sequential or metathetical ligand exchange reaction [33]. More than about 700 different polyphosphazenes have been synthesized so far, by varying the nature and ratios of the side group from a single macromolecular precursor poly (dichlorophosphazene).

The effect of side groups on the nature and properties of resulting polyphosphazenes can be illustrated using examples as below. Poly[bis(trifluoroethoxy)phosphazene] (**Scheme 5:** polymer **3**), which in fact is the first stable poly(organophosphazene) made, is a flexible film or fiber forming polymer, soluble in organic solvents and highly water repellent due to the presence of fluorinated side groups. It has a Tg of -66°C and remains flexible from its Tg to its melting point (242°C). The regular packing of the trifluoroethoxy side groups along the polymer chain imparts microcrystallinity and the polymer is thermoplastic. However, the polymer can be easily converted into a useful elastomer by the random introduction of other fluoro-alcohols along with trifluoroethoxy side groups by simultaneous substitution or using metathetical ligand exchange reaction. These elastomers are potential candidates for developing gaskets, O-rings and low-temperature fuel lines.

Similarly the Tg of polyphosphazenes can be increased by incorporating rigid, bulky side groups which will restrict the main chain mobility, the extent of which depends on the nature and size of the side groups. The highest Tg polyphosphazene synthesized so far is the polymer (**Scheme 5**: polymer **4**) with a phenyl phenoxy side groups (Tg 93°C). Incorporation of phenoxy side groups along with phenyl phenoxy side groups (**Scheme 5**: polymer **5**) by sequential macromolecular substitution significantly decreases the Tg of the polymer (Tg 43°C) [34].

The alkoxy and aryloxy polyphosphazenes discussed above are soluble in organic solvents. However, the hydrophilicity of polyphosphazenes can be significantly improved by incorporating hydrophilic side groups such as amino groups or alkoxy groups such as polyethylene glycols/glyceryl groups. All the amino phosphazene polymers are film forming thermoplastic polymers with Tg usually higher than alkoxy and aryloxy derivatives. This can be attributed to inter and intra molecular hydrogen bonding possible with these groups which consequently reduce the torsional mobility of the phosphazene backbone. Some of the amino phosphazenes are soluble in aqueous media rather than organic solvents.

The development of amino acid ester polyphosphazenes paved the way for a new class of polyphosphazenes called biodegradable polyphosphazenes [35]. Biodegradable polymers are those which degrade *in vitro* and *in vivo* either into products that are normal metabolites of the body or into products that could be completely eliminated from the body with or without further metabolic transformations. The biodegradable polyphosphazenes undergo hydrolytic degradation resulting in neutral and

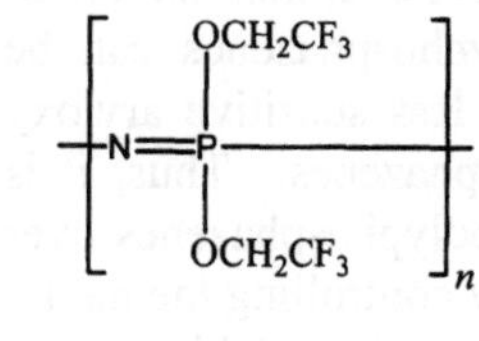

3

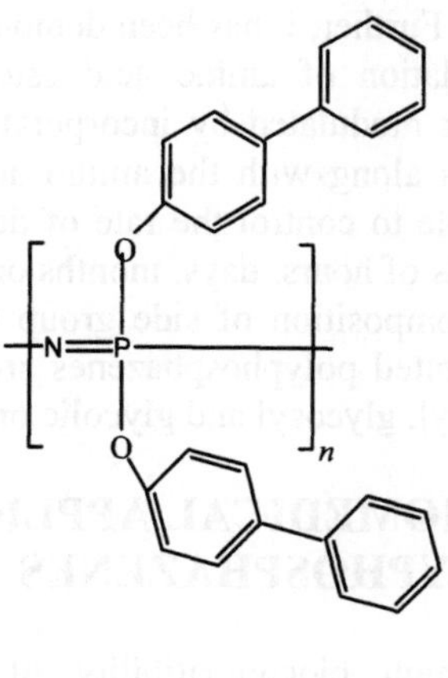

4

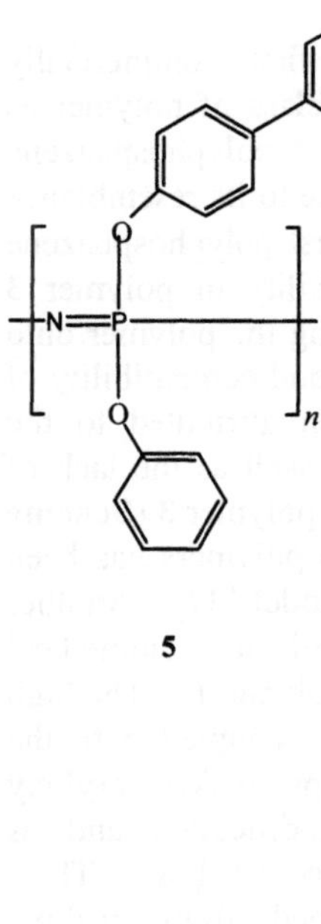

5

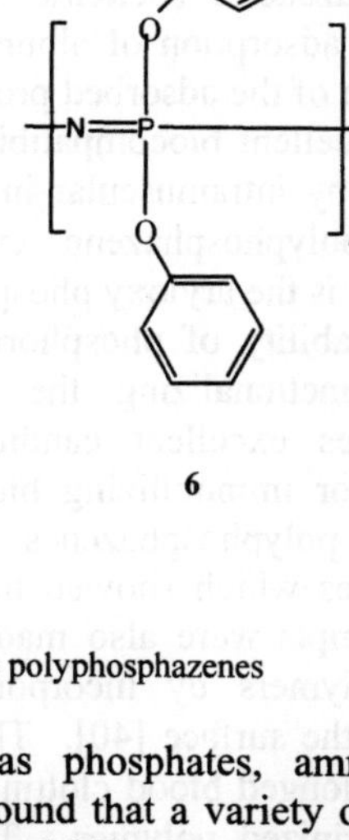

6

Scheme 5. Structures of various polyphosphazenes

non toxic degradation products such as phosphates, ammonia and corresponding side groups. It has been found that a variety of aminated and alkoxy side groups can impart biodegradability to polyphosphazenes. Biodegradable aminated polyphosphazenes include polyphosphazenes with side groups such as amino, amino acid ester and imidazole. The rate of degradation of these polymers depends on the nature of the side groups with imidazole substituted polyphosphazenes being the most hydrolytically sensitive phosphazene synthesized. In the case of amino acid ester substituted polyphosphazene the rate of degradation depends on the nature of the ester groups and the α-substituents on the amino acid ester. Further, it has been demonstrated by Laurencin et al that the rate of degradation of amino acid ester substituted polyphosphazenes can be further modulated by incorporating hydrolytically less sensitive aryloxy groups along with the amino acid ester polyphosphazenes. Thus, it is possible to control the rate of degradation of the polyphosphazenes over periods of hours, days, months or years by carefully controlling the nature and composition of side group substituents. The biodegradable alkoxy substituted polyphosphazenes are those which have side groups such as glyceryl, glycosyl and glycolic or lactic acid esters.

4. BIOMEDICAL APPLICATIONS OF POLYPHOSPHAZENES

The high biocompatibility of polysiloxane, the first commercially developed inorganic polymer raised interest in this class of polymer as biomaterials. The fluorinated polyphosphazene poly[bis(trifluoroethoxy)]phosphazene (polymer **3**) due to its resemblance to Teflon® a biocompatible organic polymer is the first polyphosphazene investigated as a biomaterial. The blood compatibility of polymer **3** (**Scheme 5**) was investigated by Welle et al by coating the polymer onto various surfaces [36]. The study showed the high blood compatibility of polyphosphazene **3** (**Scheme 5**), and this has been attributed to the preferential adsorption of albumin on the surface as well as the lack of denaturation of the adsorbed protein on the surface of polymer **3** (**Scheme 5**). The excellent biocompatibility of many of these polymers has been established by intramuscular implantation in a rat model [37]. Another class of polyphosphazene extensively investigated for biomedical applications is the aryloxy phosphazene (**Scheme 5**: polymer **6**). The high chemical stability of phosphorous-nitrogen backbone coupled with the ease of functionalizing the aromatic side groups makes aryloxy phosphazenes excellent candidates for surface modification and as substrates for immobilizing biologically active molecules [38]. Thus, sulfonated polyphosphazenes have been developed from aryloxy phosphazenes which showed high blood and cytocompatibility *in vitro* [39]. Attempts were also made to develop the blood compatibility of aryloxy polymers by incorporating the biologically active molecule heparin on the surface [40]. The heparin immobilized polyphosphazene showed prolonged blood clotting time which is several-folds higher than the unheparinized polymer. The ease of functionalization of aryloxy phosphazenes makes them preferred candidates for developing carrier matrices for chemotherapeutic drugs as well as other biologically active agents. Several prototypes for chemotherapeutic agents were developed by Allcock et al. based on polyphosphazene carriers [41].

Biodegradable polyphosphazenes such as amino acid ester polyphosphazenes are perhaps the most extensively investigated polyphosphazenes for biomedical application. One of the active area of research in biomedical field where biodegradable polymers find application is in developing controlled drug delivery systems. The controlled drug delivery systems are aimed to achieve effective therapies using a drug by partially or completely eliminating the potential for under or overdosing of drugs commonly associated with traditional drug administration. The synthetic flexibility and non toxicity of degradation products makes polyphosphazenes potential candidates for the development of polymer drug conjugates [35]. Several biodegradable polyphosphazene drug conjugates were developed which include steroids such as dexoestrone, estrone and 17β-estradiol, anesthetics such as procaine and chloroprocaine as well as anti-inflammatory drugs such as naproxen have been immobilized and released in a controlled manner by the hydrolytic degradation of the backbone.

Polymeric monolithic drug delivery systems where in drug is dispersed though out the matrix polymer is the most extensively investigated controlled drug delivery system so far. The rate of drug release from such as system depends on a multitude of factors such as nature of the polymer matrix, matrix geometry, nature of drug, initial drug loading and drug–matrix interaction if any [42]. Most of the initially developed monolithic devices were two dimensional (2-D) matrices where, drug is dispersed in a thin film of the polymer. However, the 2-D matrices suffer from the disadvantage that they need to be implanted in the body. Therefore, various other dosage forms such as microspheres and nanospheres were later developed which could be administered via parenteral routes. The microsphere based drug delivery system has been extensively investigated for the controlled delivery of various drugs such as anti-inflammatory agents, anticancer agents, antibiotics, hormones, steroids and vaccines. Many of these delivery systems are currently available in the market. In order to form an ideal candidate for developing a drug delivery system, the polymer should be biodegradable, degradation products should be non-toxic, have appropriate solubility, ease of fabrication and have appropriate degradation rate. Biodegradable polyphosphazene due to its synthetic flexibility and neutral and non-toxic nature of the degradation products has the potential to develop as an ideal candidate for fabricating drug delivery systems [35]. The initial studies using monolithic polyphosphazene drug delivery systems were carried out by Laurencin et al.[43]. The copolymer poly[(imidazolyl)(p-methyl phenoxy)phosphazene] was used to develop a delivery system for progesterone and the efficacy was evaluated *in vitro* and *in vivo* using a rat model. The rate of release of the drug from these matrices was modulated by varying the ratio of the imidazolyl and p-methyl phenoxy groups in the polymer. It has been found that increasing the amount of imidazolyl side groups in the polymer significantly increases the rate of degradation of the polymer and consequently the release of drug from the matrix. Followed by this several monolithic drug delivery systems for anti-inflammatory agents and antibiotics have been developed using various amino acid ester polyphosphazenes. The high solubility and processability allows the fabrication of these polymers into microsphere monolithic devices using various processing techniques such as emulsion/solvent evaporation or solvent emulsion/extraction-evaporation and spray drying [44].

Due to its high matrix permeability, biodegradable polyphosphazenes can form potential candidates as protein delivery vehicles [35]. The feasibility of using polyphosphazenes as delivery vehicle for macromolecules was first demonstrated by Laurencin et al using poly[(imidazolyl)(p-methyl phenoxy)phosphazene] and bovine serum albumin (BSA) as the model drug [43]. Controlled delivery of various macromolecules such as insulin, inulin and dextran has been investigated so far using different biodegradable polyphosphazenes [35].

Over the last decade considerable efforts are going on to engineer/develop tissues *in vitro* which can specifically meet the needs of individual patients to circumvent the problems associated with autografts and allografts. Tissue engineering is now considered as a novel therapeutic strategy to repair or reconstruct damaged tissues or organs. Tissue engineering has been defined as the application of biological, chemical and engineering principles towards the repair, restoration or regeneration of living tissues using biomaterials, cells and factors alone or in combination [45]. Biodegradable polymers are usually employed as biomaterials in tissue engineering. The role of the biomaterial is to provide a three dimensional scaffold onto which cells can grow and organize into tissue. The biomaterial should satisfy certain criteria to be considered as appropriate candidate for tissue engineering applications. The material should be biocompatible and have appropriate mechanical properties to provide structural support to cells, depending on the corresponding tissues to be developed. The rate of degradation of the polymer is also very crucial and should match the rate of matrix synthesis in vitro/in vivo by the cells. The processability of the material in order to form a 3-D matrix is also important, since the volume of tissue that can be developed using the scaffold depends on the pore size and pore tortuosity of the matrix developed. The wide range of degradation rate, wide spectrum of properties, ease of processability and the neutral and non-toxic degradation products of polyphosphazenes make them potential candidate as scaffolds for tissue engineering.

The first attempt towards using biodegradable polyphosphazenes for tissue engineering was carried out by Laurencin et al [46]. They investigated various amino acid ester polyphosphazenes as candidate materials for bone tissue engineering. It has been found that the rate of cell adhesion and proliferation significantly depends on the nature of the side groups present and rate of degradation of the polymer. Thus the rate of adhesion of osteoblasts on the homopolymer poly[bis(ethyl glycinato)phosphazene] was found to be low even though the polymer as well as the degradation products was found to be non-toxic to cells. This has been attributed to the high rate of degradation of the polymer.

However, by modulating the degradation rate of ethyl glycinato polymer by incorporating hydrolytically less sensitive groups such as p-methyl groups in polymer poly[(ethyl glycinato)(p methyl phenoxy)phosphazene], the rate of cell adhesion was found to be significantly improved [47]. In many formulations, the number of cells present on polyphosphazenes was found to be higher than tissue culture polystyrene (TCPS). The ease of processability of these polymers was demonstrated by developing a three dimensional (3-D) matrices of poly[(50% ethyl glycinato)(50%p-methyl phenoxy)phosphazene] via salt leaching. After 21 days of incubation of osteoblast like cells (MC3T3-E1) on 3-D and 2-D matrices as well as TCPS, the number of cells on the 3-D matrices were found to be significantly higher than that found in 2-D polyphosphazene matrices as well as TCPS [48]. The high *in vivo* osteocompatibility of the above mentioned polyphosphazenes were demonstrated using a 5 mm defect model in rabbit [49]. Other biodegradable polyphosphazenes investigated for tissue engineering application includes poly[bis(ethyl alanato)phosphazene] and poly[(ethyl alanato)(imidazole)phosphazene] as conduits for nerve regeneration and poly[(80% phenyl alanine ethyl ester)(20% imidazole)phosphazene] for guided tissue regeneration in the periodontal cavity [50].

5. NANOFIBER MATRICES FOR TISSUE ENGINEERING AND DRUG DELIVERY

As discussed earlier, the architecture of non woven nanofiber matrices, due to their structural similarity to natural extracellular matrix could present an ideal substrate on which cells can adhere, proliferate and organize into tissue. Further, the high surface area to volume ratio of the fibers can lead to high rates of delivery of drugs or bioactive molecules from these matrices making them potential candidates for developing controlled drug delivery systems. In our initial studies we have used a commercially available FDA approved biodegradable polymer poly(lactide-co-glycolide) (PLAGA) to develop nanofiber based devices for tissue engineering and drug delivery. Nanofibers of PLAGA were prepared by the process of electrospinning discussed earlier, using a solvent system consisting of 1:1 tetrahydrofuran and dimethylformamide [21]. Ultrafine fibers of diameters in the range 500-800 nm were obtained by the process. **Figure 1** shows the SEM of electrospun non woven mat of PLAGA.

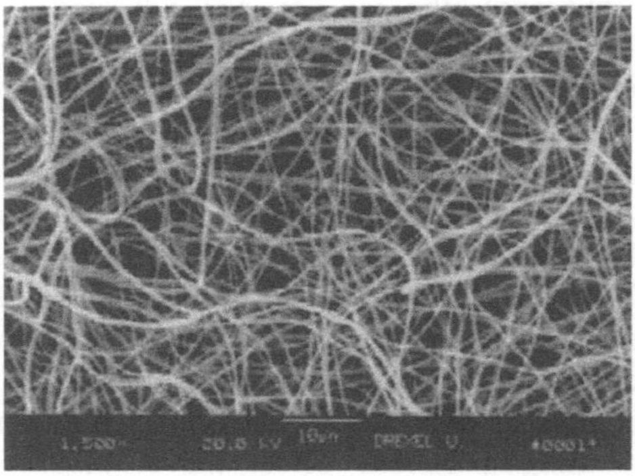

Figure 1. SEM showing the nanofibers of PLAGA
[Reprinted from Li et al (21)]

The mechanical properties of the electrospun non woven mats of PLAGA were evaluated and have been found to be comparable to those of human cartilage and bone. The ability of cells to attach and proliferate on the scaffold was evaluated using mouse fibroblast cells. It has been found that the macroporous nanofibrous structure, due to the high surface area, allows for high cell adhesion and proliferation. **Figure 2** shows the SEM of mouse fibroblast cells on PLAGA nanofiber mat after 3 days of culture. However, it has been found that by 7 days of culture, the cell number plateau on the PLAGA nanofiber scaffolds. According to the authors, one possible reason for this observation is that the acidic degradation products of PLAGA may be hindering the cell proliferation. The nanofiber structure, due to the high surface area to volume ratio, would significantly increase the rate of degradation of the polymer. This would decrease the local pH of the environment due to the acidic nature of PLAGA degradation products.

This shows the pressing need to develop novel biodegradable polymer systems which degrade to non toxic degradation products, for the development of nanofiber based matrices for biomedical application. Biodegradable polyphosphazenes due to the non-toxic and neutral degradation products and other versatile properties discussed earlier could form a potential candidate for developing nanofiber based scaffolds for tissue engineering. We have recently demonstrated (US Patent pending)

Figure 2. SEM showing the mouse fibroblast cells on PLAGA nanofiber mat developed by electrospinning after 3 days of culture **[Reprinted from Li et al (21)]**

the ability to develop nanofibers from a wide range of non degradable and degradable polyphosphazenes via electrospinning. It has been found that diameter and morphology of the resulting fibers depends on various process and solution parameters that effect electrospinning technique as discussed earlier, however the optimal parameters vary with each polymer system. **Figure 3** shows the electrospun nanofibers obtained from a non degradable polyphosphazene poly[bis(methyl phenoxy)phosphazene]. It can be seen that defect free nanofibers with smooth surface morphology having diameters varying from 0.5-1μm can be obtained from electrospinning the polymer under optimized conditions (**Figure 3**). It has been found that the same electrospinning process can be extended to other non degradable polyphosphazenes of biomedical interest such as poly[bis(carboxylato phenoxy)phosphazene] (PCPP) and poly[bis(phenoxy)phosphazenes]. Some of the biodegradable polyphosphazenes electrospun so far include various amino acid ester polyphosphazenes and the copolymers of amino acid ester phosphazenes where, the degradation rate of the polymers has been modulated by the incorporation of various aryloxy side groups. The electrospinning process also enabled us to co-spin polyphosphazenes with other polymers as well as low molecular weight drugs or other nanoparticulates. These composite nanofiber matrices will allow us to develop novel drug delivery systems, wound dressings or tissue engineering scaffolds based on polyphosphazenes for biomedical applications.

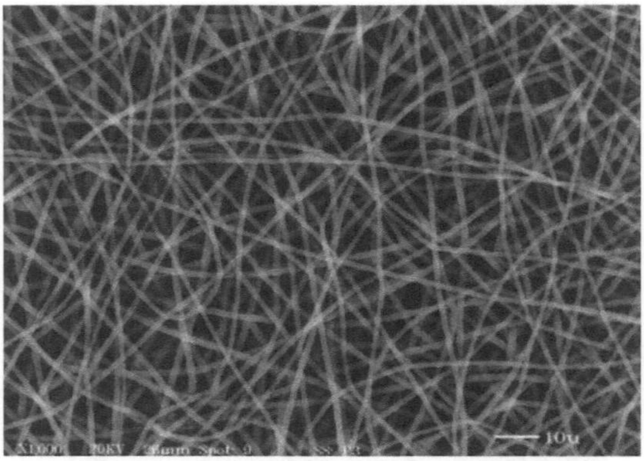

Figure 3. SEM showing the nanofibers of poly[bis(methyl phenoxy)phosphazene].

6. CONCLUSIONS

Polyphosphazene nanofibers which combine the versatile property profile and biocompatibility of the base polymer and the amazing characteristics of the nanostructure could form potential candidates for various biomedical applications. Current work is in progress in our laboratory to develop nanofiber matrices from polyphosphazenes of appropriate physical, chemical, mechanical and biological properties and evaluate their performance as scaffolds for regenerating a broad range of tissues *in vitro* and *in vivo*.

Acknowledgement: Authors acknowledge NIH#46560 for financial assistance.

References

1. Nalwa, H.S., Handbook of nanostructured materials and nanotechnology, Academic Press, San Diego, 2000.
2. Krupke R., Hennrich F., Lohneysen H.V., Kappes M.M., Separation of Metallic from semiconducting single walled carbon tubes. Science 2003; 301:344-47.
3. Joannopoulos J.D., Villeneuve P.R., Fan S. Nature 1997; 386:143-
4. Huang Z.M., Zhang Y.Z, Kotaki M., Ramakrishna S. A review on polymer nanofibers by electrospinning and their applications in nanocomposites. Composites Science and Technology 2003:63:2223-53.
5. Ondarcuhu T., Joachim C. Drawing a single nanofibre over hundreds of microns. Europhys Lett 1998; 42:215-20.
6. Whitesides G.M., Grzybowski B. Self–assembly at all scales. Science 2002; 295:2418-21.
7. Martin C.R. Membrane-based synthesis of nanomaterials. Chem Mater 1996;8:1739-46.
8. Ma P.X, Zhang R. Synthetic nano-scale fibrous extracellular matrix. J Biomed Mater Res 1999;46:60-72.
9. Zeleny J. Phys Rev 1914;3:69-91.
10. Formhals A. US patent 1,975,504, 1934.

11. Taylor G.I. Disintegration of water drops in an electric field. Proc R. Soc London, Ser A 1964; 280:383-397.
12. Zong X., Kim K., Fang D., Ran S., Hsiao B.S., Chu B. Structure and process relationship of electrospun bioabsorbable nanofiber membranes. Polymer 2002;43:4403-12.
13. Deitzel J.M., Kleinmeyer J., Harris D., Beck Tan N.C. The effect of processing variables on the morphology of electrospun nanofibers and textiles. Polymer 2001;42:261-72.
14. Abdel-Ghani M.S., Davies G.A. Simulation of non-woven fiber mats and the application to coalescers. Chemical Engineering Science 1985;40:117-29.
15. Matthews JA, Wnek GE, Simpson DG, Bowlin GL. Electrospinning of collagen nanofibers. Biomacromolecules 2002; 3:232-38.
16. Theron A., Zussman E., Yarin A.L. Electrostatic field-assisted alignment of electrospun nanofibers. Nanotechnology 2001;12:384-90.
17. Tsaia P.P., Schreuder-Gibson H., Gibson P. Different electrostatic methods for making electret filters. Journal of Electrostatics 2002; 54:333-41.
18. Smith D., Reneker D.H. PCT/US00/27737, 2001.
19. MacDiarmid A.G., Jones W.E., Noris I.D etal. Electrostatically generated nanofibers of electronic polymers. Synthetic Met 2001; 119:27-30.
20. Martin G.E., Cockshott I.D., Fildes F.J.T. US patent 4044404, 1977.
21. Li W.J., Laurencin C.T., Caterson E.J., Tuan R.S., Ko F.K. Electrospun nanofibrous structure: A novel scaffold for tissue engineering. J Biomed Mater Res 2002;60:613-21.
22. Yoshimoto H., Shin Y.M., Terai H., Vacanti J.P. A biodegradable nanofiber scaffold by electrospinning and its potential for bone tissue engineering. Biomaterials 2003; 24: 2077-82.
23. Allcock H.R. Polyphosphazenes, and the inorganic approach to polymer chemistry. Sci Prog Oxf 1980;66:355-69.
24. Allcock H.R. "Inorganic macromolecules and the search for new electroactive and structural materials" In Design of New Materials, Cocke D.L and Clearfield A, eds, Plenum Publishing Corporation, 1987.
25. Allcock H.R. Chemistry and Applications of Polyphosphazenes. Wiley Interscience, New Jersey, 2002
26. Allcock H.R., Kugel R.L. Synthesis of high polymeric alkoxy and aryloxy phosphonitriles. J Am Chem Soc 1965;87:4216-17.
27. Allcock H.R., Kugel R.L., Valan K.J. Phosphonitrilic compounds. VI. High molecular weight polyalkoxy-and aryloxy phosphazenes. Inorg Chem 1966;5:1709-15.
28. Allcock H.R., Crane C.A., Morrissey C.T., Nelson J.M., Reeves S.D., Honeyman C.H., Manners I. Living cationic polymerization of phosphoranimines as an ambient temperature route to polyphosphazenes with controlled molecular weights. Macromolecules 1996; 29:7740-47.
29. Allcock H.R., Reeves S.D., Nelson J.M., Crane C.A., Manners I. Polyphosphazene block copolymers via the controlled cationic, ambient temperature polymerization of phosphoranimines. Macromolecules 1997;30:2213-15.
30. Allcock H.R., Nelson J.M., Reeves S.D., Honeyman C.H., Manners I. Ambient temperature direct synthesis of poly(organophosphazenes) via the living cationic polymerization of organo-substituted phosphoranimines. Macromolecules 1997;30:50-53.
31. Allcock H.R. Inorganic macromolecules: Developments at the interface of inorganic, organic and polymer chemistry. Chem Eng News 1985;63:22-36.
32. Allcock H.R., Connolly M.S., Sisko J.T., Al-Shali S. Effects of organic side group structures on the properties of poly(organophosphazenes). Macromolecules 1988; 21:323-34.
33. Allcock H.R. Poly(organophosphazenes)-Unusual new high polymers Angew Chem Int Ed Engl 1977;16:147-56.

34. Allcock H.R., Mang M.N., Dembek A.A., Wynne K.J., Poly[(aryloxy)phosphazenes] with phenylphenoxy and related bulky side groups. Synthesis, thermal transition behavior and optical properties. Macromolecules 1989;22:4179-90.
35. Lakshmi S., Katti D.S., Laurencin C.T. Biodegradable polyphosphazenes for drug delivery applications. Adv Drug Delivery Rev 2002;55:467-82
36. Welle A., Grunze M., Tur D. Blood compatibility of poly[bis(trifluoroethoxy)phosphazene]. J Appl Medical Polym 2000;4:6-10.
37. Wade C.W.R., Gourlay S., Rice R., Hegyeli A., Singler R., White J. Biocompatibility of eight poly(organophosphazenes) In: *Organometallic polymers*, Carraher C.E., Sheats J.E., Pittman C.U., Editors., Academic Press, New York, 1978.
38. Allcock H.R., Morrissey C.T., Way W. K., Winograd N. Controlled formation of carboxylic acid groups at polyphosphazene surfaces: oxidative and hydrolytic routes. Chem Mater 1996;8:2730-38.
39. Allcock H.R., Fitzpatrick R.J. Sulfonation of (aryloxy) and (arylamino) phosphazene: Small molecule compounds, polymers and surfaces. Chem Mater 1991;3:1120-32.
40. Neenan T.X, Allcock H.R. Synthesis of a heparinized poly(organophosphazene). Biomaterials 1982;3:78-80.
41. Allcock H.R., Neenan T.X., Kossa W.C. Coupling of cyclic and high polymeric (aminoaryloxy)phosphazenes to carboxylic acids: Prototypes for bioactive polymers. Macromolecules 1982;15:693-96.
42. Mathiowitz E, *Encyclopedia of Controlled Drug Delivery*. New York, John-Wiley, 1999. .
43. Laurencin C.T., Koh H.J., Neenan T.X., Allcock H.R., Langer R. Controlled release using a new bioerodible polyphosphazene matrix system. J Biomed Mater Res 1987;21:1231-46.
44. Veronese F.M., Marsilio F., Caliceti P., De Filippis P., Giunchedi P., Lora S., Polyorganophosphazene microspheres for drug release: polymer synthesis, microsphere preparation, in vitro and in vivo naproxen release. J Controlled Rel 1998; 52:227-37.
45. Laurencin C.T., Ambrosio A.A., Borden M.D., Cooper J.A. Tissue Engineering: Orthopaedic Application: In : Annual Review of Biomedical Engineering Yarmuch M.L., Palo Alto, editors., Annual Reviews Inc. 1999.
46. Laurencin C.T., Morris C.D., Jacques H.P., Schwartz E.R., Keaton A.R., Zou L. Osteoblast culture on bioerodible polymers: Studies of initial cell adhesion and spread. Polym Adv tech 1992;3:359-64.
47. Laurencin C.T., Norman M.E., Elgendy H.M., El-Amin S.F., Allcock H.R., Pucher S.R., Ambrosio A.A. Use of polyphosphazenes for skeletal tissue regeneration. J. Biomed Mater Res. 1993;27:963-73.
48. Laurencin C.T., El-Amin S.F., Ibim S.E., Willoughby D.A., Attawia M.A., Allcock H.R., Amborsio A.A. A highly porous 3-dimensional polyphosphazene polymer matrix for skeletal tissue regeneration. J. Biomed Mater Res. 1996;30:133-38.
49. Laurencin C.T., Ambrosio A.M.A., Bauer T.W., Allcock H.R., Attawia M.A., Borden M.D, Gorum W.J., Frank D. The biocompatibility of polyphosphazenes. Evaluation in Bone, Soc for Biomat. 24th annual meeting in conjuction with 30th international symposium, 1998 April 22-26, San Diego, CA.
50. Aldini N.N., Fini M., Rocca M., Martini L., Giardino R., Caliceti P., Veronese F.M., Lora S., Maltarello M.C. Peripheral nerve reconstruction with Bioabsorbable polyphosphazene conduits. J Bioact Comp Polym 1997;12:3-13.

BIONANOCOMPOSITES BASED ON NANO-CARBON MATERIALS FOR CULTURE CELLS MEDIA

LUMINITA STAMATIN*, IOAN STAMATIN**

* *"Longhin Scarlat" Dermato- Venereology Clinical Hospital, Biocomposites & Sensors Lab, Serban Voda, 214, Bucharest*

** *University of Bucharest, Faculty of Physic Diamond films, Plasma Polymerization &Novel Carbon lab, Bucharest- Magurele MG-11, Romania*

Abstract: This study focused on an appropriate bionanocomposite, containing nano-carbon, to replace the cells culture media with aim to find a rapid answer to staphylococcal infections this study enlarges implications, which the nanoparticles may have in our daily life. The interrelations nanoparticles - environment- microbes development are taken in account.

Keywords: Bionanocomposites, nanocarbons, cell biology, environment

1. INTRODUCTION

The diversity of micro-particles, existent in dust or as aerosols often influences the microbe's development and infectious diseases spreading. Micro-particles and aerosols from the industrial activity induce a large disease class: mines (silicoses), paper printing (the lead disease), diamond micro/ nano-powder (lung caverns) etc. Once with the advances in materials nanotechnology and electronic at nanometric scale have appeared a large class of nanopowders and nanometric devices dimensional compatible with biological structures. Accidentally or not, they escape to environment direct from industry, cars, or human activity. If they have or do not a major influence to the people healthy is the question, to which the scientific world is trying to find answers. It is worthy of note that nanoparticles may have an unimaginable impact on cells/microbes and viruses, owing to their reactivity and the photo-catalytic influence in environment. Nanoparticles can modify the growth rate of the microbes like staphylococcus and streptococci revealed by this contribution. The aim is to find useful applications with carbon nanoparticles to get timed information about microbes in a disease. In particular, a nanocarbon-based bionanocomposite (BNC) for the culture cell media has been designed. Any microbial specie involved in cutaneous pathology (a particular case) need to be identified, after they grow in a culture media for 24h in a thermostat at 37^{0}C. Further, biochemical tests and the antibiotic response assessment give a medical treatment. Unfortunately takes more than 2 days and clinical analysis claim a short evaluation. Also in any place, it is need to have a rapid evaluation of the microbial attack and their virulence. The actual

S. Guceri et al. (eds.), Nanoengineered Nanofibrous Materials, 303-312.

active media, used for the bacteria growing and identification, are complex gels with specific nutritive elements and indicators, which define if a microbial class is present. The species into a class need more biochemical investigations. The need to have a rapid response in clinical analysis impose the designing of new materials where bacteria should be seeded and identified in same day if not possible shorter. The bionanocomposites based on nanocarbons (gel-nanocarbons mixed in a proper way at nanometric level) speed up the microbes growing, especially aerobic bacteria. They give shorter response if a bacteria colony is in development and specifying, by pigmentation, which kind of infection took place (in this case Staphylococcus Aureus). The method proved to be useful in bioanalytical and clinical laboratory methods for staphylococcal infection. The first preliminary observations when was started with nanocarbon insertion into a cell culture media have been: microbes grow faster, the colonies are much more developed, different type of microbes have more ore less affinity to nanocarbon (carbon black and active- coal are not so specific in effects). At a glance seems to be nothing new than a simple add-on of growth centers. The carbon in forms of active –coal and carbon black has been used long time in drug delivery and curative aim. The nanocarbon resulted from laser or CVD pyrolysis is quite different as carbon –black. With an apparently similarity in turbostratic organization the LPNC (laser pyrolysis nanocarbon) is much more rich in free bonds and hydrogen adsorbed on its surface. It can be defined oxygen-free NC. In contact with oxygen, nanocarbons absorb large quantities becoming a good supplier for aerobic bacteria. Nanocarbons exist in environment from amorphous to fullerene –like structures and many common features are with the synthesizable one. Three directions arise: 1. The carbon nanoparticles have a major role in the interaction with living cells and with unpleasant microbes or viruses. 2. The carbon particles from micrometric to nanometric level, dispersed in atmosphere, are compatible vectors to transport microbes and interact with the skin (an excellent natural culture media for microbe growing and then attack to human body). 3. The environment is "full" of micro and nanoparticles of different species compatible or not with viruses and bacteria. Each particle is a possible vector to transport them. Are nanoparticles benign or actives? Bionanocomposites (BNC), special designed for culture cell media, are usefully instruments to analyze these kinds of nanoparticles and their role giving good results in much more shorter time. In our particular case BNC were used to clinical diagnostic in dermatology. Many pathogen agents (bacteria, fungi, parasites) are involved and a fast and clear diagnostic is sometimes impossible without a precise laboratory analysis. The status in this field consists of a gel media (culture media) where bacteria are grown at all with no specificity in identification. In next step the biochemical analysis are approached for identifications. *The study one of the most aggressive pathogens, Staphylococcus Aureus (Gram-positive pathogen agent), given us a perspective to interrelate data with bionanocomposites based on different species of nanocarbons and to make a projection to environment implications.* The results with BNC based on nanocarbon species for microbe's growing, implied in a large class of infectious diseases are presented. In addition, a collection of nanocarbon typology, with features close to what is finding atmosphere is given.

2. EXPERIMENTAL DETAILS

2.1 Nanocarbons with Compatible Features as Those in the Environment

Carbons resulted from pyrolysis of organic precursors at moderate temperature (<1700^0 C) are usually called carbon blacks. All are under 100 nm. They are often described as amorphous or graphitic (turbostratic structures). As a primary definition, they are forms with irregular arrangement of its carbon atoms and a randomized arrangement of graphene can be considered. Sometimes they contain polynuclear hydrocarbon (fused aromatic rings or interconnected rings with sp3 and sp1 carbon chains). Carbon blacks may be produced by different techniques, low-tech [5] (incomplete burning of hydrocarbons like methane, acetylene, oil) and high-tech (laser pyrolysis and thermal assisted-CVD) [5-7]. The variety of the synthesis techniques give results in many morphological forms of carbon particles with diameters in a typical range of 10-100 nm. The multi-scale organization (structure, micro-texture, texture) of carbons resulting from pyrolysis of organic precursors at moderate temperature (<1400^0 C) is not yet fully understood, despite numerous works carried out for half a century. These carbons are frequently met in nature or synthesized for industrial applications (blast furnace cokes, active carbon). The nanocarbons are very hard graphitizing like any amorphous carbon (over 2500^{0}C), good criteria for appreciation the level of organization in its structure. Studies based on X-rays techniques and transmission electron microscopy (TEM), have introduced the concept of basic structural units (BSU) [18,19-21]. Such carbons described as elemental bricks were constituted by stacking a few aromatic layers of nanometric size. The BSU size was usually considered as quasi-constant, in the entire carbonization temperature range. It must be noticed that only BSU (intuitively, sufficiently organized as stacks of graphene layers) give the 002 diffracted beams used in X-ray diffraction (XRD) or in the dark-field and lattice fringe TEM modes. The multi-scale organization, of nanocarbons [20], is studied over more than six orders of magnitude by using successively different techniques. The structure (organizational the atomic scale) and micro-texture (spatial distribution of the aromatic layers) can be directly observed by HRTEM, due to its high-resolution mode allowing imaging the profile of the aromatic sheets. Table 1, sketches the most representatives nanocarbons considered for study with designed bionanocomposites. For convenience, there is assigned carbon black (CB) that nanocarbon produced with low-tech (including that of engine gas exhaust or furnace) and nanocarbon that is produced with high-tech methods (NC). Some of the representative features with TEM for four classes, sketched in figure 1, are as follows ; regular carbon black, turbostratic of laser pyrolysis with alkenes/ alkynes precursors, with polyaromatics and fullerence fragments content [6] of benzene precursors produced by LP and of toluene produced by CVD. Consequently, these techniques are blind for a more or less important fraction of possible single layers, pore activity, biological activity. Transmission IR spectra recorded using FTIR spectrometers (Digilab model 2002) at a resolution of 4 cm^{-1} is a good tool to appreciate the composition. The determination allows direct comparison between IR spectra of deposits obtained with different conditions of reaction and if any changes appear with oxidation or oxygen absorption.

Table 1- Nanocarbons compatible with dust and aerosols of environment, used for NBC designing

Type	Source/method	Main feature	Principal components
Regular CB –R	Incomplete burning: oil and gases, acetylenes	The surface area of blacks used for rubber reinforcement ranges (10-150 m2/g) when determined by nitrogen adsorption that the ultimate black particle have an average diameter of 20 to 300 nanometers.	Oxygen, hydrogen, and sulfur are main impurities. Oxygen and hydrogen are combined chemically with carbon on the surface of the particle. The ratio of hydrogen to oxygen varies with the manufacturing process. In contact (channel) blacks, oxygen content is 2-5% while hydrogen amounts to about 0.5%; in furnace blacks, the oxygen content of the black depends on that of the hydrocarbon feed.
Pigment CB- P	Incomplete burning in furnace with high cyclone	Pigment-grade blacks have surface areas of 300-500 m2/g.	
NC-LP1 NC- CVD1	Precursors ethylene, butadiene	Turbostratic structures,	Hydrogen adsorbed on surface
NC- LP2	Acetylene with sensitizer	Turbostratic structures	Sulfur and fluorine are in large quantity
NC-LP3 NC-CVD2	Benzene Benzene, toluene,	Fullerene fragments in very small quantity.	Multiple Polyaromatics (PAH)

CB-R & CB-P, mixtures supplied by Research Inst for Electrical Engineering- Advanced research (ICPE-CA). In experiments all samples were sterilized; NC-LP, supplied by National Institute, Laser Plasma and Radiation (NILPRP), Bucharest; NC-CVD, soot synthesized in Diamond films, Plasma Polymerization &Novel Carbon lab.

Nanocarbon samples were deposited on Si-100 using electro-spray technique. The IR spectra, after subtraction of the silicon substrate contribution on samples used in NBC have had typical features: The bands can be assigned to CH and CC bonds, on positions well established for hydrogenated carbon compounds [22, 23]. Three groups are representatives. In the range 2700-3400 cm^{-1} corresponds to the C-H stretching vibrations. The band at 3040-3050 cm-1 attributed to aromatic compounds is always the most intense; the C-H stretching modes (2970 and 2915cm-1) of aliphatic compounds are much weaker. A second group of bands, between 1000 and 2000cm-1 comprises essentially the in plane vibration modes of aromatic CC and CH bonds. The emergent bands at 1600 and 1440 cm^{-1} are due to aromatic CC stretching, symmetric and anti-symmetric ring modes, respectively. Both are infrared-allowed here, due to a high level of non-symmetric substitution in the aromatic units as expected from an irregular network of aromatic units. The in-plane bending CH modes give a large bump around of 1150 cm^{-1}. A third group of bands is found, between 600 and 1000 cm^{-1}, characteristic of the aromatic C-H out of plane bending modes. The large intensity of these bands shows the strong aromatic character of the deposits. These modes are very sensitive to the ring substitution and give rise to three main components according to the number of adjacent hydrogen atoms on a ring.

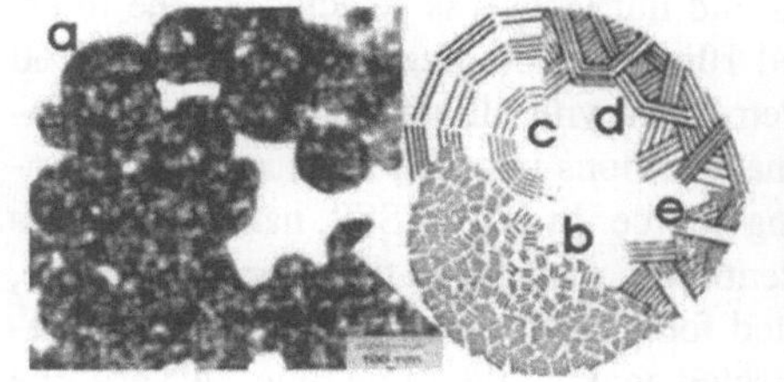

A

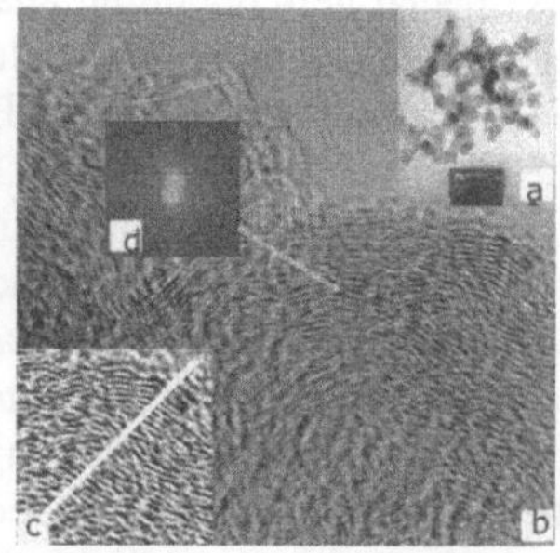

B

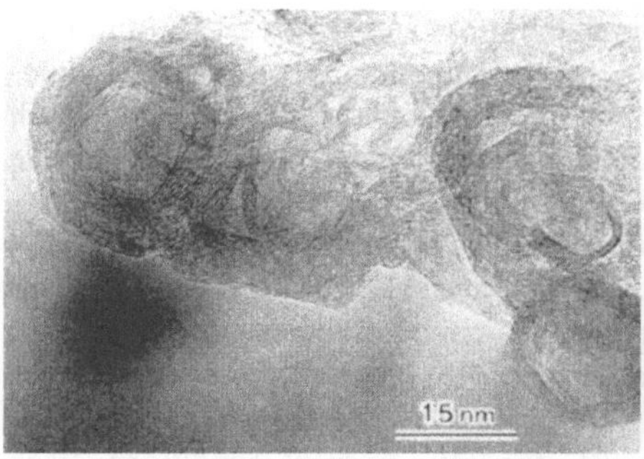

C

Figure 1- Nanocarbons from different synthesis sources, TEM features

D

A. a) CB-R, scale 100nm; b) model for CB,randomized distribution of BSU a CB sphere c) CB heated at 3000K, BSU- circumferential alignment, d)the graphene stacking in CB e) CB partial etched [11,14, 16]; **B**. NC-LP1-ethylene/acethylene precursors, typical turbostratic structures a) scale 100nm b) 10 nm c), morphological details. d) TEM-phase contrast; **C**. LP-NC3 of benzene precursor, structure with fullerene like multishell and large quantity of polyaromatics **D.** NC-CVD2, of toluene precursors, chained structures of nanocarbon spherules

The band at 752 cm^{-1} corresponds to three and four adjacent H, the feature exhibiting two maxima at 839cm-1 and 815 cm^{-1} is due to two adjacent H, and the band at 885 cm^{-1} is attributed to lone H. The corresponding summation bands can be found in the 2000-1650 cm^{-1} region. Finally, minor bands can be observed in the spectra, due to alkynes groups at 3290 and 2103 cm-1, and to mono and di-substituted aromatic rings as well as olefin groups in the 1200-400 cm^{-1} region. The samples NC-LP3 and NC-CVD2 did not present the specific lines for any type of fullerene (C60, 1424 and 1183cm^{-1}) but the figure 1 C and D show some similar fullerene-like multi-shell structures of 20-30 nm. The third band group is not well defined for these structures. When NCs are dispersed and sonicated in PSS9 (see experimental) C-H bands decreased and bands corresponding for three, two and mono adjacent hydrogen vanished with appearance of the bands centered on 1650cm^{-1} (carbonyls group) and –OH (band centered at 3370 cm^{-1}). The effect was prominent in case of turbostatic structures NC-LP1 and NC-CVD1 weak for NC-LP3 and NC-CVD-2 not relevant for CB-R.

2.2 Culture Media with Nanocarbon

A special gel-containing agar and nutrients was selected. In the reference sample (R) there was added 10ml physiological saline solution 9%0 (PSS9), boiled and poured in Petri dish with all sterile conditions maintained. For a fine dispersion of nanocarbons into gel, an ultrasonically instrument was adapted to a heating source. In 10ml PSS9, nanocarbon was added for a specific weight concentration reported to the reference sample, R. The solution was was sonicated for 15 min at room temperature, ultrafiltered on G1-G3 glass and sonicated again at the same time. A very fine nanocarbon dispersed solution can be obtained and particles keep this state without coalescence or sedimentation for long time. In the reference sample, heated in sonic field, they poured the nanocarbon-dispersed solution by using a milliliter syringe with continuous stirring. This stage took 20 min. Finally, the sample is cooled at room temperature and kept in the same condition as the reference sample. A series of 0.01, 0.1 and 1% weight concentrations with nano carbons has been prepared. The nanobiocomposites with agar-gel and nanocarbon have all the properties for a culture cell medium. In addition to that, the nano carbon is very finely dispersed in lattice without coalescence. The samples inseminated with Staphylococcus aureus bacteria were processed according to the bacteriological protocol: thermostat at 37^0C for 24h, ambient conditions. Each hour, optical image microscopy inspected the samples watching the virulence of bacteria growth.

3. RESULTS AND DISCUSSIONS

To reduce the enormous experimental data a trial series was done in a concentrated formula. The interest is to set the minimum concentration level, with any effect, and to set a maximum level where there is any saturation or the inhibition is present. It is typical for Staphylococcus aureus (pathogen) in proper conditions to develop colonies with yellow citrine color (pigmentation genesis) and it is more virulent and pathogen when the pigmentation gene-sis appears in early growth stages. Table2 presents the virulence induced by the three types of nanocarbons with the specific features: carbons of incomplete com-bustion, turbostratic of alkenes, and aromatics (benzene, toluene), which have in content PAH and fullerene like fragments. Figure 2 presents some exemplars of bacteria colonies grown with BNC containing nanocarbon species with detailed comments of table 2.

Table 2 Qualitative comparison of the virulence in BNC by comparison with reference sample

BNC	0.01%	0.1%	1%
CB-R	N. O	M	as in R
CB-P	N.O	M to WD	as in R
NC-LP1	W D	**Explosive**	Atypical Pigment genesis
NC-LP2	weak	no	no
NC-LP3 & NC-CVD2	atypical shapes	W. D with atypical pigment genesis	Not distinguishable

N.O- sample has same aspect as the reference; M- moderate development but colonies are not well developed ; WD- well developed as in R;

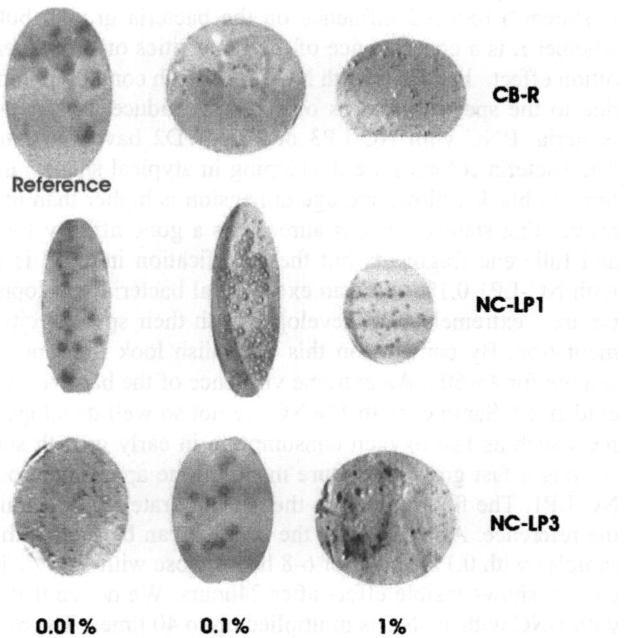

Figure 2. Bacteria colonies grown with BNC containing nanocarbon species

All BNC samples, being seeded in the same conditions according to standard protocol, are compared to the sample R. BNC with CB-R and CB-P did not show any spectacular improvement in the development of fast growing process and no any virulence has been noticed. The analysis of the structure in CB-R (fig.1A) gives details of randomized graphene stacking in small units of crystallites. In graphitized samples, the crystallites are ordered with graphene perpen-dicular on radius. The oxidizing process in-duces etching of the side part and active centers at nanopore level are created. Some impurities originated from raw materials, neutralize the specific activity and quantity of absorbed oxygen is reduced. In all studied cases, it

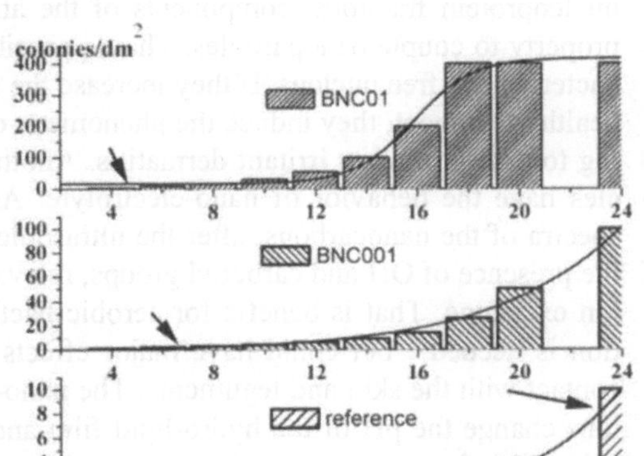

Figure. 3 The growing rate for two concentrations and the reference. The arrow indicate the time when colonies are eye visible. BNC with 0.1% nanocarbon LP-1, develops in 24 h, 40 times more colonies than the reference.

is shown a reduced influence on the bacteria growth but it is not clear whether it is a consequence of the impu-rities or of its inertness. An inhibition effect had BNC with NC-LP2. High content of sulfur and fluorine due to the specific process of synthesis induce a clear poisoning of the bacteria. BNC with NC-LP3 or NC-CVD2 have very strange behavior. The bacteria colonies are developing in atypical shapes; the pigmentation turns to black yellow, average dimension is higher than in reference. This proves that staphylococcus aureus has a good affinity for poly-aromatics and fullerene fragments but the modification induced is not clear. BNC with NC-LP1 0.1% show an exceptional bacteria development. All bacteria are extremely well developed with their specific citrine yellow pigmentation. By comparison this Petri dish look like one in the reference sample for 48-60h. An extreme virulence of the bacteria colonies has been evidenced. Samples with 1% NC are not so well developed. Inhibition affects such as fast oxygen consumption in early growth stage takes place. Aiming a fast growing culture medium, the appropriate one is NBC with NC-LP1. The figure 3 shows the growing rate for two concentrations and the reference. After 4 hours, the bacteria can be seen with the naked eye, samples with 0.1% and after 6-8 hours those with 0.01%. The sample, reference, shows visible effect after 24hours. We notice that the growth rate with BNC with 1%NC is multiplied up to 40 times. The bioche-mical tests and specific coloration shows the bacteria colonies did not have any modification. When BNC is used with NC-LP3, some atypical forms appeared but it was not clear which kind of modifications had been induced (figure3, NC-LP3, 0.1% shows a yellowish center surrounded by a black cage). It can be appreciated that nanoparticles dispersed in the gel lattice, work like oxygen absorbent and as oxygen pump for aerobic bacteria. It is interesting to prove, for Staphylococcus aureus, whether polypeptides and nucleoprotein fractions, components of the antigenic structure, have the property to couple nanoparticles. That is possible as the above-mentioned bacteria have free nucleus. If they increase the adherence in contact with a healthy tegument, they induce the phenomena of pathogenesis or necrotizing toxins in primary irritant dermatitis. Another aspect is that nanoparticles have the behavior of nano-electrolyte. As mentioned above, FT-IR spectra of the nanocarbons, after the ultrasonic processing in PSS9, show the presence of OH and carbonyl groups, relevant to consider it has had an ion exchange. That is benefic for aerobic bacteria where oxygen absorption is needed but could have major effects when nanoparticles are in contact with the skin and teguments. The nano-electrolytes with hydrogen ions change the pH of the hydro-lipid film and the acidity of the healthy skin. Therefore many channels are opened to a microbial infection.

4. CONCLUSIONS

BNC designed, as a gel, specific nutrients and turbostratic nanocarbons are good culture cells medium that accelerate the aerobic bacteria growth, particularly Staphylococcus.

Not all nanocarbons are good suppliers of oxygen, the turbostratic structures proved to have this property. Nanocarbons with polyaromatic structures and with fullerene fragments induce some modifications, which need a much deeper investigation. At the superior limit, carbon black has, apparently, no major influence, but many aspects have not been considered yet, such as a study on the activity of different absorbed gases. BNCs designed in this study have two aspects. Firstly, with minimum conditions a diagnosis on an infection can be predictable in less than 6h. This is very important to establish the main diagnosis and the suitable antibiotic class for a specific treatment. Secondly the existent nanoparticles in atmosphere can be tested using specially designed nanobiocomposites. Micro and nanoparticles are in direct contact with microbes. They interfere either by attachment or by other uncontrolled ways. It is not excluded an inductive effect such as increased resistance to antibiotics. In contact with teguments, atypical behavior and diseases are induced.

Acknowledgement: The present research developed as an extension of the NATO-SfP 974214 project with all the support assured by its infrastructure.

References

1. Hamburg, Margaret A. (2002) Bioterrorism: responding to an emerging threat, *Trends in Biotechnology*, **20**(7):296–298.
2. Demidov, V.V. (2002) Bioterrorism: how well are we protected? *Trends in Biotechnology*, **20**(5):192.
3. Leong, John M., Parveen N., Goguen Jon D, (2002) New approaches and old prob lems in the shadow of bioterrorism, *Trends in Microbiology*, **10**(3):112–114.
4. Venter, A. (2001) Defending against bioterrorism. *Trends in Microbiology*, **9(11)**:52
5. Lahaye J, Prado G.(1978) vol. 14, In: Walker PL, Ed, *Chemistry and physics of carbon*, New York: Marcel Dekker, Inc, p. 167.
6. Galvez A, Herlin-Boime Nathalie, Reynaud Cecile , Clinard Ch. Rouzaud Jean-Noel, (2002) Carbon nanoparticles from laser pyrolysis, *Carbon* **40**, 2775–2789
7. Haggery J. S. et. al,(1981) Chap. 3, in. Steinfeld J. I Ed, *Laser – Induced Chemical Processes*, Plenum Press, New York
8. G. Benedek, P. Milani and V.G. Ralchenko (2001) *Structure, Properties and Applications of Nanostructured Carbon Architectures* in *Nanostructured Carbon for Advanced Applications,* Kluwer Academic Publishers,
9. Kroto H.W., Heath J.R., O'Brien S.C., Curl R.F. and Smalley R.E., (1985) *Nature* **318**, 162
10. Kaae J.L., (1985) Carbon **23**, 665
11. Bockhorn H., in Bockhorn H. Ed, (1994) *Soot Formation in Combustion,* Springer Verlag: New York, p. 4
12. Robertson D.H., Brenner D.W. and White C.T. (1992) *J. Phys. Chem.* **96** 6133
13. Heckman T.A., Harling D.F., (1966) *Rubber Chem. Technol.* **39**, 1
14. Siegla D.C. and Smith G.W. Editors (1981) *Particulate Carbon,* New York: Ple num Press, p. 33-55
15. Grieco W., Howard J., Rainey L.C.,Van Sande J.B.(2000) *Carbon* **38**, 4 635
16. Symons G.H., Ed. (1953) *Fluorine Chemistry, Vol. 1,* Inostr. Lit., Moscow, p. 81 (translated into Russian)
17. Alexandrescu R., Armand X., Cauchetier Herlin M., Petcu N., S., Voicu I., (1998) *Carbon* **36**, 1285.
18. Franklin R E. (1950), Crystallites size in graphitizing and non-graphitizing carbons. *Proc R Soc*, **209**:196–218.
19. Thrower PA, Editor, (1989) *Chemistry and Physics of Carbon*, vol. **22**, New York: Marcel Dekker, pp. 1–143.
20. Lahaye J, Ehrburger P, Editors,(1991) *Fundamental issues in control of carbon gasification reactivity*, Dordrecht: Kluwer Academic Publishers, pp. 257–68.

21. Rouzaud JN, Duber S, Clinard C, Pusz S. (2000) Optical properties of porous car bons, *Extended Abstracts, International Conference, Carbon 2000*, 9–14 July, pp. 187–188.
22. Dischler B, Bubenzer A, Koidl P.(1983); Solid State Commun **48:**105–8.
23. Lin-Vien D, Colthup NB, Fateley WG, Grasselli JG. (1991) *The handbook of infra red and Raman characteristic frequencies*, New York: Academic Press.

CONTACT-INDUCED PROPERTIES OF SEMICONDUCTING NANOWIRES AND THEIR LOCAL GATING

Eric Gallo, Amro Anwar and Bahram Nabet
Electrical and Computer Engineering Department
Drexel University – Philadelphia, PA 19104, USA

Abstract: Contacts and interfaces between three-dimensional and quasi-one dimensional structures may dominate charge transfer behavior in the nano-scale regime. Properties arise as a result of such contacts that reside in neither system alone. Of particular interest is when a Schottky barrier is formed between the metallic contact and a semiconducting one-dimensional (1D) nanowire. We show how current conduction is modified as a result of reduced dimensionality and present a model that examines the effect of Coulombic force on the Schottky barrier formed between a 1D semiconductor and a metallic electrode. Using existing theories of Schottky contacts, we introduce a capacitively coupled charge that further modifies the potential profile of the interface and changes both thermionic and tunneling currents through such a metal-1D junction. We model changes in the Schottky barrier height due to a superposition of the electric field, the image force and this Coulombic potential.

Keywords: Schottky Contacts, Thermionic Emission, Heterodimensional Contacts

1. INTRODUCTION

As new materials and techniques are discovered for use in nanoscale electronics, a deeper understanding of the physics at this scale becomes necessary. In conventional solid state devices, transitions from p to n and metal to semiconductor often dominate device behavior; examples include PN junctions, metal-semiconductor junctions and heterojunctions that can be used to create transistors, detectors, light emitters and many other useful devices[1]. The nanoscale regime introduces a new type of contact, a heterodimensional junction between materials of differing dimensions. By contacting a quasi-one-dimensional (1D) semiconductor to a three dimensional metal contact new behaviors arise due to the confined states of the semiconductor wire[2, 3]. One can expect these contacts to dominate charge transport, as contacts between materials dominate such behavior in bulk devices. In this paper we examine nanoscale Schottky contact between a metallic electrode and a 1D semiconductor and investigate some of the expected behavior changes. In particular we demonstrate the changes

S. Guceri et al. (eds.), Nanoengineered Nanofibrous Materials, 313-322.

in the model of thermionic emission from the metallic contact to the wire which is quite applicable to room-temperature operation of such a device. We also describe a new mechanism of affecting current through this junction, which is not observed in bulk semiconductor contacts. By using an electronically isolated local gate, the height and shape of the Schottky barrier can be modified, thereby affecting current through the junction.

2. HETERODIMENSIONAL SCHOTTKY CONTACTS

Contacts between heterogeneous materials generally play a dominating role in most conventional semiconductor devices. One would expect that at the nanoscale the same to be true. The characterization of a contact on the nanoscale can show significantly different properties than that of a bulk contact. We will address the specific case of a 3D metallic contact to a 1D semiconducting wire. The properties of this junction, such as the shape of the depletion region and the current through the junctions can be changed due to the heterodimensional nature of the contact. Contacts to nanowires can be rectifying, ohmic or tunneling barriers, our interest is on rectifying contacts, hence what follows applies to room temperature operation of a diode more so than transport at Fermi level in low temperatures.

2.1 Schottky or Blocking Contacts

Formation of an energy barrier between metal and semiconductor upon contact has been known since the 1930's.[4,5] Current transport between metal and semiconductor has also been intensely studied[6,7] although proper treatment of current conduction mechanisms, and effects such as the role of interface states are still actively debated.[8,9] Schottky contacts to 1D systems have also been studied especially as they apply to the carbon nanotube.[10,11,12,13] Here we outline a semi-classical treatment that can demonstrate expected differences between current conduction across a metal-1D junction compared to a metal-3D. Rectifying contacts result in systems of different electron affinities with current transport being based on kinetic energy requirements for motion from one system to another. Hence, in Schottky diodes thermionic emission over the barrier, rather than tunneling through it, is the main transport mechanism for semiconducting wires although it is possible that the latter dominates in other 1D devices.[14] In a 3D-1D contact, calculation of thermionic emission current proceeds similar to its classic derivation[7] or derivation of 3D-2D by summing the flux of electrons with proper velocity that overcome the barrier[15]

(1)
$$J_{th} = \int_{v_x(E_F+q\phi_B)}^{\infty} q v_x \, dn$$

Here dn is the incremental change in carrier concentration given by the product of the density of states (DOS) and Fermi function $dn = N_{1D}(E)f(E)dE$, and v_x is carrier velocity perpendicular to the contact along the axis of the wire; the lower limit of the integral is the required minimum kinetic energy. Using Boltzmann distribution and the 1D density of states (DOS) function, thermionic emission current is derived as[16]

$$(2) \qquad J_{th} = \frac{qk_B}{\pi\hbar} T \exp(-\frac{E_{oy}+E_{oz}}{k_B T}) \exp(-\frac{q\phi_{Bn}}{k_B T}) [\exp((qV)/k_B T) - 1]$$

where E_{oy} and E_{oz} are the first discrete energy levels due to confinement in y and z, respectively. This equation then describes thermionic emission current for a metal-1D Schottky diode. For the case of a cylindrical wire $E_{oy} = E_{oz} = E_o$, and the reverse saturation current becomes

$$(3) \qquad J_{th} = \frac{qk_B}{\pi\hbar} T \exp(-\frac{q\phi_{Bn}}{k_B T}) \exp(-\frac{2E_o}{k_B T})$$

Equation (3) can now be compared to the 3D-2D case[15]

$$(4) \qquad I_{2D} = WA_{2D}^{*} T^{3/2} \exp\left(-\frac{q\phi_B}{kT}\right) \exp\left(-\frac{E_0}{kT}\right),$$

or to the conventional (3D-3D) contact[7,1]

$$(5) \qquad I_{3D} = A_{eff} A^{*} T^2 \exp\left(-\frac{q\phi_B}{kT}\right) \exp\left(\frac{\Delta\phi_{im}}{kT}\right).$$

Important differences exist between these cases. A barrier height increase by $2E_0$ is observed that for a typical GaAs system causes a current reduction at room temperature by about a factor of 150. The pre-factor also shows a linear dependence on temperature as opposed to T^2 and $T^{3/2}$ in 3D and 2D cases, respectively. Hence, there exist critical thicknesses below which injected current to a nanowire significantly changes in magnitude, turn-on behavior, saturation current and temperature dependence.

In addition to room temperature operation of a Schottky (nano)diode, formation of a Schottky barrier strongly affects current transport, with carbon nanotubes being probably the best understood family of molecular conductors and closest to an ideal 1D system. Recent experimental progress has made it clear that contacts between a nanotube and metallic electrodes define the properties of the resulting circuit. This is particularly true for single wall carbon nanotubes (SWNTs), where contact to a metal electrode forms a Schottky barrier.[17,18] This barrier is the critical factor that determines the operating characteristics of a carbon nanotube field effect transistor (CNFET). When the transistor is "off", transport may be reduced to the level of thermionic emission (3), while in "on" state tunneling can draw large currents ultimately limited by quantum conductance. In above derivation, one dimensionality has been incorporated through inclusion of a

1D DOS, and quantized confined energies in two dimensions, hence we expect the treatment to be applicable to other wires, including polymer nanowires.

3. LOCAL GATING

The nanoscale regime also allows for new opportunities in controlling the behavior of junctions. It has been shown that gating in nanotube devices affects band bending at the contacts to control the behavior of a particular device[14,17,19]. In what follows, we examine how the presence of charge in the vicinity of a wire can modify the electrostatic potential profile at the junction thus affecting the barrier height and its shape. This leads to significantly altering current conduction across the barrier, be it over the barrier as is the case of thermionic emission or through the barrier via tunneling processes of field emission and thermionic field emission. One point, also worth stressing, is that it is the small scale of nanowires that allows for electrostatic forces outside the wire to directly affect carriers within the wire. The small scale allows the electric field to penetrate and influence the carriers involved in conduction, unlike in bulk material where the electric field would be screened near the surface and not affect the behavior of most carriers.

The most straightforward mechanism of introducing isolated charge in the vicinity of a metal-1D wire is by means of the voltage biased conductive tip of an AFM, as schematically shown in Figure 1a[17,20,21]. If no current flows between the tip and the surface below, it acts as an isolated charge suspended over the sample. Another possibility is the introduction of a thin static gate. This gate could be another nanowire positioned perpendicular to the contacted wire and separated by an insulating layer, as shown schematically in Figure 1b. For simplicity we consider a point charge and analyze its effect on conduction.

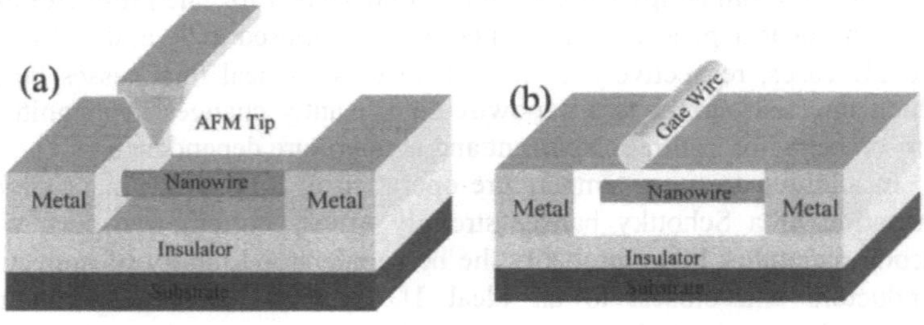

Figure 1: Schematic of two practical systems with Coulombic interaction at the Schottky Barrier interface. (a) is a voltage biased AFM tip. (b) is a perpendicular gating wire separated from the semiconductor by an insulating layer. The AFM tip is charge contained at a point in space, while the wire is charge distributed over a region.

3.1 Barrier Lowering due to Coulombic Potential

Consider a point charge, Q, held over the semiconducting wire at height h and located a distance r from the carrier as shown in Figure 2. The charge is free to move laterally and vertically near the contact. The potential due to the isolated charge is modeled by considering the electric field generated by its presence. The electric field generated by Q is:

$$E = \frac{Q}{4\pi\varepsilon r^2} \tag{6}$$

where ε is the permittivity of free space and Q is charge in Coulombs. The force exerted on a carrier in the semiconducting wire is found by multiplying this electric field by the charge on a single carrier, q. This force can be decomposed into two parts, the force in the x direction (along the wire) and the force in the y direction (perpendicular to the wire), neglecting the height of the wire. Since the carrier is confined to a quasi-one dimensional wire as it moves through the interface we need only consider the lateral force in x, which can be found by multiplying the force due to Q by the cosine of the angle between r and h. This force generates a potential energy experienced by a carrier moving through the wire, which can be found by integrating the force from infinity to a position x:

$$PE = -\frac{Qq}{4\pi\varepsilon\sqrt{X_Q^{\ 2} - 2X_Q x + x^2 + h^2}} \tag{7}$$

where X_Q is the position of Q along the x axis and $r = \sqrt{(X_Q - x)^2 + h^2}$. Assuming Q is a charge equal and opposite to the carrier, the potential seen by the carrier, as defined by (7), is shown in Figure 3.

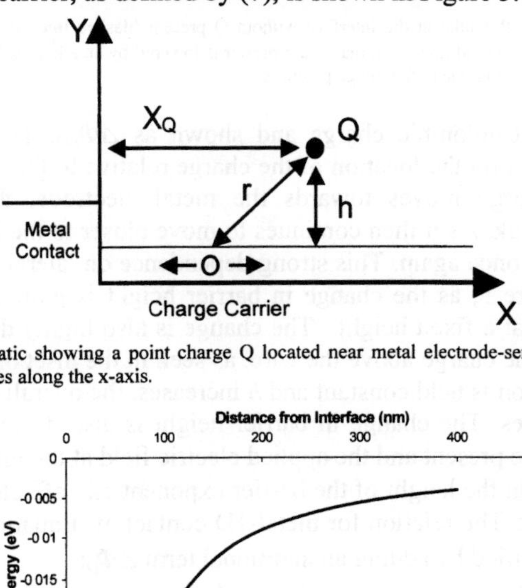

Figure 2 Schematic showing a point charge Q located near metal electrode-semiconductor interface. The charge moves along the x-axis.

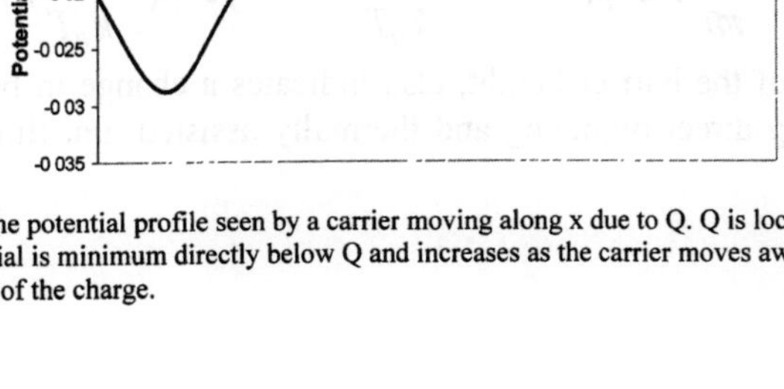

Figure 3 The potential profile seen by a carrier moving along x due to Q. Q is located at x=50nm. The potential is minimum directly below Q and increases as the carrier moves away from the x position of the charge.

It is known that at the interface of a metal-bulk semiconductor, the total barrier seen by a carrier is the result of a superposition of the image force potential and the electric field potential[1]. Adding the effect of the charge Q on the potential landscape can further modify this barrier. By locating the charge at a fixed distance X_Q from the interface and a fixed height h, the potential landscape of the overall interface is altered as seen in Figure 4. The barrier height is now further reduced by the attractive force

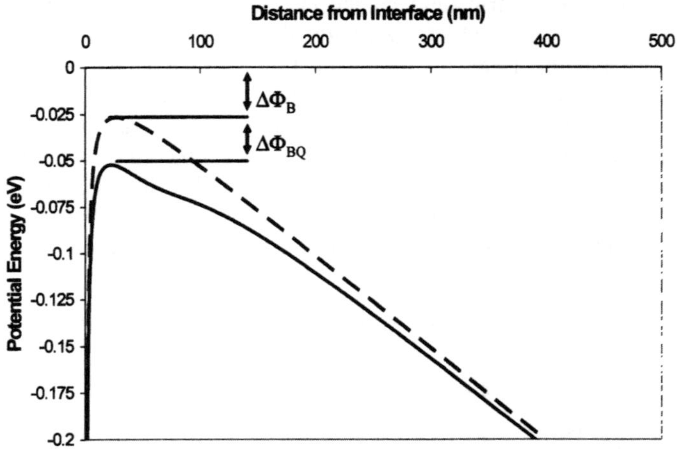

Figure 4 – Plot of the Potential at the interface without Q present (dashed line) and with Q located 50nm from the interface (solid line). The maximum barrier is lowered by an additional value $\Delta\Phi_\Theta$ from the original barrier lowering due to the image potential.

of the isolated Coulombic charge and shown as $\Delta\Phi_{BQ}$. This change is greatly dependent on the location of the charge relative to the interface. As the isolated charge moves towards the metal electrode, the lowering increases to a peak. As it then continues to move closer to the interface the effect decreases once again. This strong dependence on lateral position can be seen in Figure 5, as the change in barrier height is plotted versus the charge location at a fixed height. The change is also highly dependent on the position of the charge above the wire, as seen in the inset of Figure 5. If the lateral position is held constant and h increases, the overall effect of the charge diminishes. The change in barrier height is also dependent on the amount of charge present and the applied electric field at the interface.

This change in the height of the barrier exponentially affects thermionic emission current. The relation for metal-1D contact of Equation (2) above can now be modified by adding an additional term $\Delta\Phi_Q$:

$$(8) \qquad J_{th} = \frac{qk_B}{\pi\hbar} T \exp\left(-\frac{q\phi_B - \Delta\varphi_B - \Delta\varphi_Q}{k_B T}\right) \exp\left(-\frac{2E_o}{k_B T}\right) (A/cm^2)$$

The change of the barrier height, also indicates a change in barrier shape thus affecting direct tunneling and thermally assisted tunneling currents[1].

As seen in Figure 4, the barrier can be thinned at the top due to the charge Q. This thinning and reduction of barrier height will increase the observed

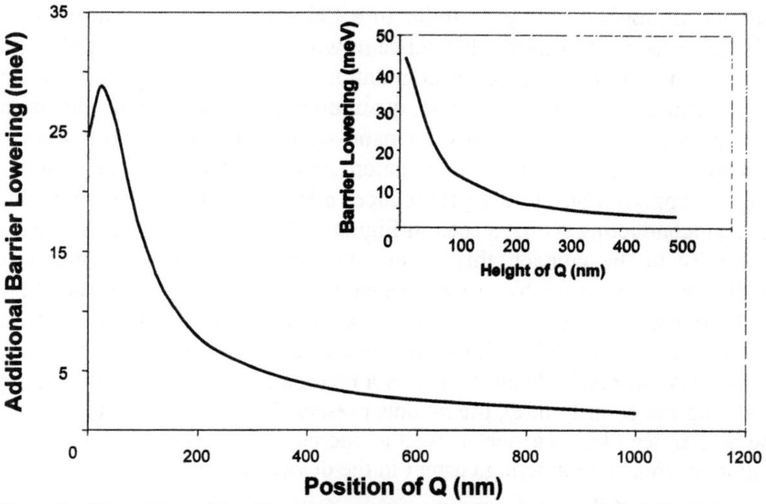

Figure 5 – Plot of the additional barrier lowering as a function of the position of Q. As Q moves towards the interface the barrier lowering increases, until it reaches a peak and then begins to decrease as it further approaches the metallic electrode. The position of the peak barrier lowering is dependent on the electric field present in the interface. The inset compares barrier lowering to height of Q above the nanowire in a fixed lateral position

tunneling currents. This can be seen if we consider the equation for tunneling current from a metal to a semiconductor:

$$J_{tun} = \frac{A^*T}{k}\int_0^{q(V_o-\Delta\Phi_B-\Delta\Phi_Q)} F_M T(\eta)(1-F_S)d\eta \tag{9}$$

where F_M and F_S are occupancy factors in metal and semiconductor, respectively. The change in barrier height modifies the limits of integration, and the change in height and shape will modify the transmission coefficient, $T(\eta)$. As expected, thinning the barrier will increase transmission probability and therefore the tunneling current[1].

3.2 Barrier Width Change due to Coulombic Potential

Further change in the potential profile at the interface can also be found by considering the perpendicular component of the force due the isolated charge. If we assume the charge Q is held on a metallic gate we can estimate the effects of the capacitive coupling between the charged material and the quantum wire. As with metal-oxide-semiconductor (MOS) capacitors, it is expected that the capacitive coupling should affect the carrier concentration within the wire in a region near the gate. The issue of modeling the capacitance at this small scale becomes complex due to the close proximity to the wire and its small size and length[22].

In traditional MOS models, the difference between the work functions of the metal gate and the semiconducting wire would create band bending at the interface. This bending creates accumulation or depletion of charge without an applied voltage, similar to a Schottky contact separated by a thin oxide layer[1]. Given the difference in work functions would lead to the conclusion that this charge accumulation, with no voltage difference, would also occur in the metal gate, quantum wire configuration. This issue has not yet been explored fully for nano-scale systems and more work is needed to understand the accumulation or depletion effect in quantum wires. Applying bias to the gate induces additional carriers locally within the semiconducting wire, as seen in Figure 6. If the gate is located in close proximity to the contact, this change in carrier concentration will also modify the barrier seen by carriers. Using the Schottky model, the height of the barrier will change, by changing the internal electric field of the region, and the width will be altered, by varying the carrier density.

We have primarily dealt with Q as a method to increase conduction. By changing the polarity of Q, the models presented will also serve to increase the barrier seen by a carrier, as well as increase the width of the depletion region, decreasing tunneling current in the device. The model of barrier

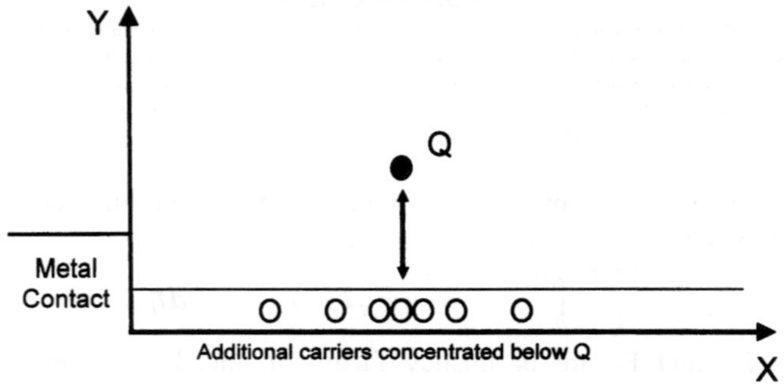

Figure 6 The presence of Q above the semiconductor creates additional carriers in the semiconductor below Q. This increase in carrier alters the potential landscape of the interface by altering the electrical field in the barrier region.

lowering using a point charge gives an understanding of what a Coulombic potential might do to a barrier at a Schottky interface. The point charge model can be extended to create a practical model of an actual device.

4. DISCUSSION AND CONCLUSIONS

Research has primarily concentrated on tunneling and thermally assisted tunneling currents as the only current mechanism in nano-scale contacts. With respect to carbon nanotube circuits, the Schottky Barrier Transistor Model theorizes that gating modifies the Fermi level within a carbon nanotube and in doing so thins the tunneling barriers seen by a carrier entering or leaving the tube[14]. Other proposals suggest that in nanoscale contacts, tunneling dominates over thermionic due to the dimensions of the materials involved[23].

One basis for this tunneling only theory has been the observed temperature dependence of devices. For instance, tunneling currents into a Luttinger liquid should follow a $T^{0.7\sim0.9}$, which has been observed in single-walled carbon nanotube devices[24]. We have shown here, however, that Thermionic emission between a metal contact and a 1D semiconducting wire should have a T dependence, which is much closer to this observed dependence than the bulk value of T^2. Further experimental evidence that thermionic emission should not be ruled out on the nanoscale is found in results from large diameter carbon nanotube circuits. Larger diameter tubes are expected to have a smaller bandgap than narrower tubes. These devices clearly show two separate temperature dependencies in their current-voltage characteristic[25,26], further evidence that tunneling may not be the only mechanism present in the device. We expect that different mechanisms should dominate in particular devices, however thermionic emission must be incorporated into an accurate model of conduction in nanoscale devices, particularly for room temperature operation and for calculation of reverse saturation currents.

In summary, we have derived the change in thermionic emission in a heterodimensional metal-semiconductor contact. We have also shown how an electronically isolated local Coulombic potential can alter the behavior of a nanoscale Schottky contact. The Coulombic force can penetrate into the device to directly affect the height and shape of the barrier region at the junction. These changes to the barrier in turn modify the current flow through the contact. Reduction in the height of the barrier increases thermionic emission, while a reduction in height and width can increase tunneling currents. This behavior presents a new technique for controlling conduction in the nanoscale that is not applicable for large scale devices. Heterodimensional contacts and their properties present new possibilities for devices and behaviors in electronics.

Acknowledgements: This work was supported in part by Ben Franklin Technology Partnership of Pennsylvania through a Nanotechnology Institute funding; EG was supported by NSF graduate fellowship.

References

1. S.M. Sze, Physics of Semiconductor Devices, 2nd Ed., J Wiley and Sons, 1981.

2. *Electronic Transport in Mesoscopic Systems*, S. Datta, Cambridge University Press (1995).

3. "Conductance Viewed as Transmission" Y. Imry, R. Landauer Reviews of Modern Physics 71, 2 (1999)

4. W. Schottky, "Halbleitertheorie der sperrschicht," Naturwissenschaften, vol.26, no.52, p.843 (1938).

5. J. Bardeen, "Surface states and rectification at a metal semi-conductor contact", Phys. Rev., vol. 71, no. 10, pp. 717-727 (1947)

6. H. A. Bethe, "Theory of boundary layer of crystal rectifiers," MIT Radiation Lab Report, pp. 43-52, (1942).

7. C. R. Crowell and S. M. Sze, "Current Transport in Metal-Semiconductor Barrier Diode," Solid State Electron, 9, 1035. (1966)
8. E.H.Rhoderick and R.H. Williams, Metal-Semiconductor Contacts, 2nd ed, Oxford Press (1988)
9. R.B.Darling, "Current-voltage characteristics of Schottky barrier diodes with dynamic interfacial defect state occupancy", IEEE Trans on Elect. Devices, Vol. 43, no. 7, 1153-1160 (1996).
10. Z. Yao, H.W.Postma, L. Balents, and C. Dekker, "Carbon nanotube intramolecular junctions" Nature, Vol. 402, pp. 273-275 (1999)
11. M.P.Anantram, S.Datta, Y. Xue, "Coupling of Carbon Nanotubes to Metallic Contacts", Phys. Rev. B, vol. 61, no, 20,. pp.14219-14224 (2000)
12. A.A. Odintsov and Y. Tokura, "Contact phenomena and Mott transition in carbon nanotubes" Journal of Low Temp. Phys. 118 pp. 509-518, (2000)
13. F. Leonard and J. Tersoff, "Role of Fermi-Level Pinning in Nanotube Schottky Diodes", Phys. Rev. Lett., vol. 84, no. 20 pp. 4693-4696 (2000)
14. S. Heinze, J. Tersoff, R. Martel, V. Derycke, J. Appenzeller, and Ph. Avouris, "Carbon Nanotubes as Schottky Barrier Transistors", Phys. Rev. Lett. 89, 106801. pp1-4 (2002)
15. A. Anwar and B. Nabet , "Barrier Enhancement mechanisms in heterodimensional contacts and their effect on current transport", IEEE Trans. On Microwave Theory and Techniques, Vol. 50, No. 1,. pp68-71(2002)
16. A. Anwar, "Heterodimensional Schottky Contacts to Modulation-Doped Heterojunction with Application to Photodetection," PhD Thesis, Drexel University, Philadelphia, PA, 2001.
17 "Controlled Creation of a Carbon Nanotube Diode by a Scanned Gate", M.Freitag,,M.Radosavljevic, Y.Zhou, A.T.Johnson, and W.F.Smith; App.Phy.Lett., 79 (2001).
18. M. Radosavljevic, M. Freitag, K.V. Thadani and A.T. Johnson, "Nonvolatile molecular memory elements based on ambipolar nanotube field effect transistors" NanoLetters 2, 761 – 764 (2002).
19. "Ambipolar Electrical Transport in Semiconducting Single-Wall Carbon Nanotubes", R.Martel, V.Derycke, C.Lavoie, J.Appenzeller, K.K.Chan, J.Tersoff, and Ph.Avouris; Phys.Rev.Lett., 87 (2001).
20. "Scanned Probe Microscopy of Electronic Transport in Carbon Nanotubes", A.Bachtold, M.S.Fuhrer, S.Plyasunov, M.Forero, E.H.Anderson, A.Zettl, and P.L.McEuen; Phys.Rev.Lett., 84 (2000).
21. "Local Electronic Properties of Single-Wall Nanotube Circuits Measured by Conduction-Tip AFM", M.Freitag, M.Radosavljevic, W.Clauss, and A.T.Johnson; Phys.Rev.B, 62 (2000).
22. *Classical Electrodynamics*, J.D. Jackson, John Wiley and Sons (1962).
23. "Scaling of nano-Schottky-diodes", G.D. Smit, S. Rogge, T.M. Klapwijk; Appl. Phys. Lett. 81, 20 (2002).
24. "Luttinger Liquid Behaviour in Carbon Nanotubes", M.Bockrath, D.H.Cobden, J.Lu,,A.G.Rinzler, R.E.Smalley, L.Balents, and P.L.McEuen; Nature, 397 (1999).
25. "Electrical Measurement of Individual Semiconducting Single-Walled Carbon Nanotubes of Various Diameters", C.Zhou, J.Kong, and H.Dai; Appl.Phy.Lett., 76 (2000).
26. "Electrical Properties and Devices of Large-Diameter Single-Walled Carbon Nanotubes", A.Javey, M.Shim, and H.Dai; App.Phy.Lett., 80 (2002).

BESPOKE CARBON NANOTUBE DEVICES AND STRUCTURES

D.C. COX*, R.D. FORREST, P.R. SMITH and S.R.P. SILVA
Advanced Technology Institute, School of Electronics and Physical Sciences, University of Surrey, Guildford-Surrey, GU2 7XH, UK

Abstract: For carbon nanotubes to be used in large-scale electronic applications a suitable method of interconnection between individual nanotubes must be available. Recently it has been shown that it is possible to manipulate and 'solder' individual nanotubes in a scanning electron microscope (SEM), using carbon-based contaminating material present in differential pumped vacuum systems, or metal inorganic vapours, by focusing the electron beam on the region to be connected for several minutes.

By comparison, in this paper we report the welding of individual carbon nanotubes to other nanotubes and metal substrates, on a selective basis, to produce joints of both good electrical conductivity and mechanical integrity, without the need for a 'soldering' material. Carbon nanotubes have been manipulated *in-situ* in a SEM and have subsequently been cut to specific, pre-determined lengths, joined to substrates or to each other using only carefully applied voltages with limited currents.

Our studies of the connections formed between carbon nanotubes and substrates have indicated that only certain metals appear to form mechanically strong, low resistance, connections. Results on connections to metals indicate preliminarily that welds form in carbide forming metals, whereas non-carbide formers behave poorly.

This procedure may be utilised as a proof of concept technique to construct carbon nanotube-based devices and connections in electronic applications.

Keywords: Nanotubes, Manipulation, Devices, Connections

1. INTRODUCTION

The extreme electronic and microstructural properties of carbon nanotubes have been well documented[1]. However, to realise applications in the area of electronics, such as quantum wires[2], ballistic conductors[3], microchip interconnects[4] and transistors[5] the need to reproducibly fabricate connections between individual nanotubes has been identified as a major impediment[6]. The fabrication of all but the most

S. Guceri et al. (eds.), Nanoengineered Nanofibrous Materials, 323-328.

simple electronic devices and structures using carbon nanotubes, reliably, with high yield and accuracy is being hampered by the unavailability of a technique that allows in-situ manipulation coupled with direct observation. Connecting and "soldering" of individual nanotubes using an AFM[7] and an SEM[8] has been demonstrated, but in both cases nanotube connections can take upwards of tens of minutes per joint, and hence fabrication of even simple structures is at best very slow, and often not possible at all.

In this paper we introduce a method that allows for the design and fabrication of tailor-made freestanding carbon nanotube structures with unrivalled accuracy and nano-construction capability on a 3-D field. We show the nanotube constructions to be both mechanically robust and electronically active. The technique shows the means to select and manipulate carbon nanotubes, and produce joints between nanotubes and substrates, and other nanotubes, without the need for a solder material, and with significantly reduced time taken per connected joint.

2. EXPERIMENTAL DETAILS

Two types of nanotubes were used in this study: Catalytically grown multiwall CVD nanotubes on silicon substrates, and arc-discharge grown multiwall nanotubes embedded in polystyrene, but exposed via cleavage of the polymer[9]. The nanotubes, on their substrates, were mounted on an electrically isolated specimen stage in a Cambridge Instruments SEM that includes electrical feedthroughs connected to a pair of sharp tungsten tips, and the specimen substrates on the stage. The tungsten tips are sharpened via electrochemical etching using a method used to prepare tips for STM. It is not necessary to achieve tips as sharp as that required for STM, but we routinely use tips with tip radii of less than 1μm. Two computer-controlled Keithley 238 source/measurement units, running under a Labview program, allow us to electrically characterise the nanotubes, and provide the means for welding and cutting of the nanotubes.

The tips can be moved within the SEM, independently of the stage, and each other, by six piezo-sliders (Omicron Nanotechnology, MS5), with a minimum step size of 40 nm, and up to 5 mm total travel. The piezo-sliders are grouped into two sets of three providing three-axis movement for both tips. Control of the piezo-sliders is provided by a simple keypad control box (Omicron) that allows real-time movement of the tips, and hence nanotube manipulation, while the SEM is in operation.

Also attached to the stage, and electrically connected, is a range of substrates to which the nanotube constructions can be attached. For this study a number of suitable substrates such as 50 μm wires, thin metal foils and deposited films on insulating substrates were used.

3. RESULTS AND DISCUSSION

The first stage in producing nanotube structures by our *in-situ* manipulation method is

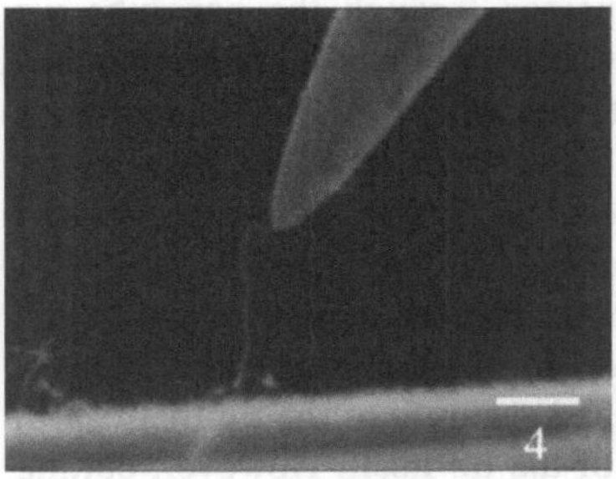

Figure 1: Nanotube connected to tungsten tip

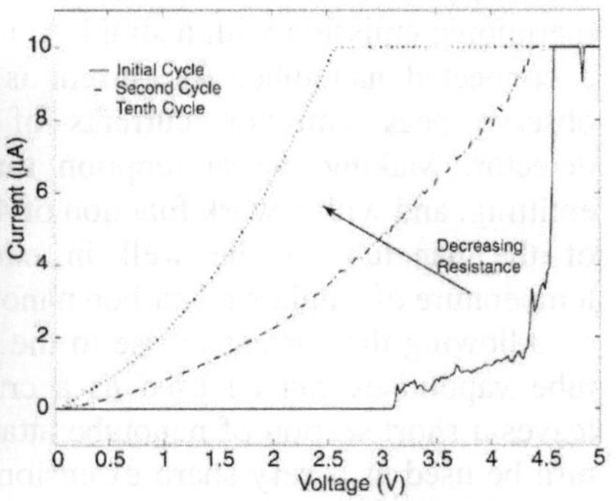

Figure 2: I-V curves for the nanotube-tip-substrate connection showing conditioning

to choose and then connect one of the tips to a suitable nanotube. Examination of the nanotubes on their substrates in the SEM reveals nanotubes extending from the substrate, and when one is chosen, one of the tips is brought close to the free end of the nanotube. The gap between tip and nanotube end is then reduced until contact is made (Figure 1). Often electrostatic attraction pulls the nanotube onto the end of the tip relieving the need for accurate positioning. A voltage is then applied across the nanotube to establish an electrical contact. Typically, a poor contact is initially achieved with almost no current flow detected until a threshold voltage (~3V) is reached. At this point, the current then increases by several orders of magnitude up to a pre-determined limit of 10^{-5}A, as the voltage is raised. On reducing the voltage we observe that the connection has improved and the resistance lowered. Subsequent cycling of the voltage to the current limit results in a further lowering of the resistance with each successive cycle down to approximately 150 kΩ for the connection shown (Figure 2). We have however, seen values as low as 12 kΩ depending on CNT and substrate. The CNT to substrate or nanotube contact is now mechanically robust and electrically active. We believe this conditioning behaviour is due to the intense ohmic heating, caused by the large current densities, facilitating alloying at the connection.

Typically the conditioning process can be completed in under a minute, and despite the high current densities observed, the total power dissipated is under 5 mW, demonstrating that creating a stable connection is inherently a low power process. We observe that the resistance also decreases with increasing power, and hence temperature, in agreement with Peng *et al* [*10*]. Initially, the current limit of 10^{-5}A is observed at around 4.5V, but following cycling the same current can be achieved at lower than 2.5V. After typically ten cycles stable I-V behaviour is obtained, with no further conditioning. At this point it is possible to increase the voltage further until the current is high enough to evaporate the central region of the tube (>40x10^{-5}A). As these current levels are reached thermionic emission from the nanotube is detected by the secondary electron detector in the SEM. We further monitored the thermionic emission with a simple Cu-wire detector placed few mm from a connected nanotube. As current is passed through the nanotube we observe peak emission currents of 6.2×10^{-8}A in the non-optimised detector. Making the assumption that the whole of the nanotube is emitting, and with a work function of 4.5 eV, we estimate the temperature of the nanotube to be well in excess of 2700K, the vaporisation temperature of single wall carbon nanotubes [*11*].

Allowing the current to rise to the point that the central region of the tube vapourises can be used as a crude form of cutting. This method leaves a short section of nanotube attached securely to the tip that can in turn be used as a very sharp extension of the tip itself. However, cutting nanotubes to a pre-determined length is made possible by utilising the second tip. A nanotube is held between the first tip and the substrate with a voltage of 2-3 V applied. On bringing the second tip, with a similar voltage, into contact with the nanotube, at the point cutting is required, and then raising the voltage on both tips simultaneously, a small region of around 100-150 nm vaporises near to the point of contact of the second tip, leaving the desired length of nanotube attached to the first tip. This process is somewhat easier with a section of nanotube attached to the second tip (Figure 3). It is now possible to bring a new substrate to the tip

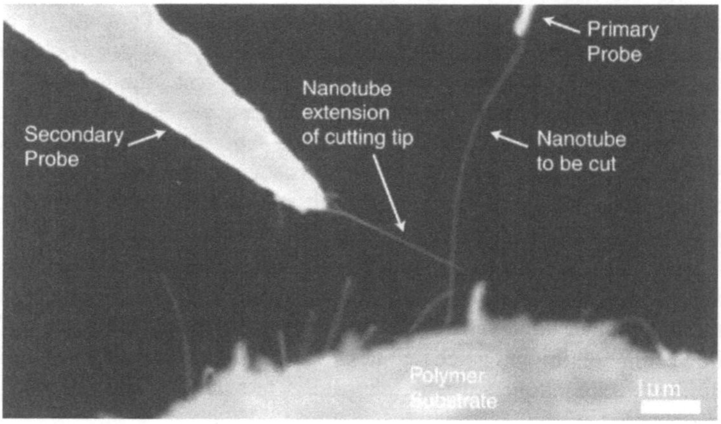

Figure 3: Micrograph showing selected nanotube about to be cut

and attach the nanotube to it by moving it into contact and applying a voltage in the same manner as the initial nanotube to tip connection. Removing the nanotube from the tip is achieved by cutting the nanotube near to the tip connection, in the same manner as before, leaving the tube attached securely to the substrate.

Our studies of the connections formed between carbon nanotubes, metal tips and substrates have indicated that only certain metals appear to form mechanically strong, low resistance connections. The tungsten tips are effective, as is iron wire, chromium foils, and 7 μm diameter carbon fibre. However, with gold, platinum and nickel, after many voltage cycles, and using higher voltages, we were unable to form connections reliably.

This observation infers that welds form easily in carbide forming metals, whereas non-carbide formers behave poorly. A study is currently underway to elucidate this further. In addition to attaching nanotubes to new substrates, it is further possible to connect additional nanotubes to the previously deposited ones. As with attaching to substrates, the nanotubes are brought into contact at the desired point on the previously deposited nanotube and cut off from the tip as before. However, when nanotubes are connected together no conditioning behaviour is observed and the I-V behaviour of the connection is stable and reproducible despite repeated cycling.

As a demonstration of the versatility of this new technique, Figure 4 shows several aspects that would be useful in the construction of bespoke electronic devices incorporating carbon nanotubes. This figure shows the letters A, T and I constructed from arc discharge grown carbon nanotubes and deposited onto the edge of a chromium foil. The height of the three letters is approximately 1.8 μm, occupying a length of less than 4 μm. The figure demonstrates the most important requirements necessary for bespoke devices; the ability to pick and place a single nanotube onto a substrate with a specific length (letter I), the ability to attach a second nanotube to the first, (letter T) and the ability to connect nanotubes together that have previously been attached to a chosen substrate, mechanically as well as electrically (letter A). The combination of these procedures realises the desire to produce nanotube based electronic components and provides the tools necessary to achieve this. Another important point is that we have routinely removed these structures from the vacuum of our lower-resolution 'manipulation SEM' and placed them in other higher resolution instruments, clearly demonstrating the robustness of the structures formed.

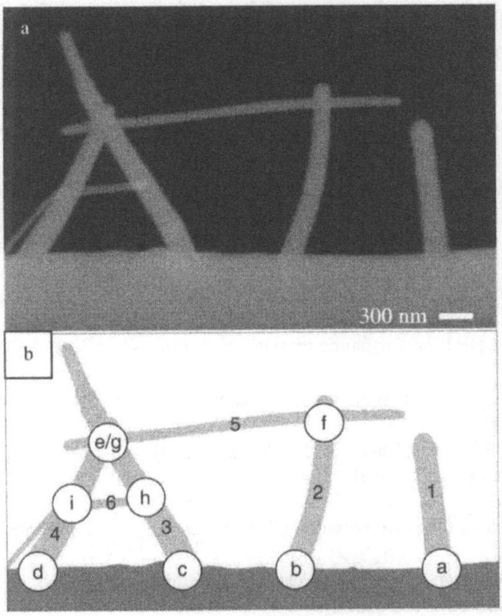

Figure 4: Secondary electron micrograph of nanotube construction of letters A,T and I (a), and schematic diagram showing order in which the nanotubes were attached and the connections formed (b). The nanotubes were placed in the order shown numerically (1-6) and the connections made in the alphabetical order also shown in the schematic diagram.

4. CONCLUSIONS

In conclusion, we demonstrate that it is possible to weld nanotubes to metal contacts and to each other with both good electrical and mechanical quality. Providing the manipulation tips and substrates are of a suitable 'carbide-forming' metal it is possible to construct robust and relatively complex freestanding structures *in-situ*, at the rate of one connection every few minutes. Additionally no connecting medium is required and both the cutting and connecting of nanotubes is carried out entirely by carefully applied voltages and regulated electrical currents. This procedure may be utilised as a proof of concept technique to construct carbon nanotube-based devices and connections in electronic applications

Acknowledgements: We acknowledge the support received from the EPSRC, UK in the form of a Portfolio Partnership grant and the Carbon Based Electronics Programme.

References

1. Dresselhaus, MS, Dresselhaus G and Avouris P, *(eds), Carbon Nanotubes: Synthesis, Structure, Properties and Application*, Springer, (2000).
2. Tans, SJ, *et al.*, Individual single-wall carbon nanotubes as quantum wires. *Nature*, 386, 474 (1997).
3. White, CT and Todorov, TN. Carbon nanotubes as long ballistic conductors. *Nature*, 393, 240 (1998).
4. Li, J, *et al.*, Bottom-up approach for carbon nanotube interconnects. *Appl. Phys. Lett.*, 82, 2491 (2003).
5. Martel, R. Nanotube electronics: High-performance transistors. *Nature Materials*, 1, 203 (2003).
6. Baughman, RH, Zakhidov AA, de Heer WA. Carbon nanotubes - the route toward applications. *Science*, 297, 787 (2002).
7. Henk, W, Postma C, de Jonge, M, Yao, Z and Dekker, C. Electrical transport through carbon nanotube junctions created by mechanical manipulation. *Phys. Rev. B*, 62, 10653 (2000).
8. Barnhardt, F. The Formation of a Connection between Carbon Nanotubes in an Electron Beam. *Nano Lett.* 1, 329 (2001).
9. Poa, CH *et al.*, Field emission from non-aligned carbon nanotubes embedded in a polystyrene matrix. *Appl. Phys. Lett.* 80, 3189 (2002).
10. Peng, L, *et al.*, Polarized incandescent light emission from carbon nanotubes. *Appl. Phys. Lett.* 82, 1763 (2003).
11. Collins, PC Arnold MS, Avouris P. Engineering carbon nanotubes and nanotube circuits using electrical breakdown. *Science*, 292, 706 (2001).

VACUUM ELECTRONIC APPLICATIONS OF NANO-CARBON MATERIALS

A.N. OBRAZTSOV
Physics Department of Moscow State University
Moscow 119992, Russia

Abstract: Carbon materials attract great attention for use in vacuum electronic devices as cold cathodes. High aspect ratios and strong interatomic bonds provide stable and intensive electron emission at moderate electric fields from nanotubes, fibres as well as from other forms of essentially graphitic carbon. The evaluation of device applicability of carbon cathodes requires a better understanding of the process used for their synthesis and the mechanisms of electron emission. Chemical vapor deposition of carbons was studied in our work to find conditions providing better electron emission properties of cathodes. Raman spectroscopy, cathodoluminescence, various microscopy techniques, XPS, UPS and other methods were used in our experimental studies for a comparative study of various carbon materials. We propose an original model of emission sites in nano-carbon emitters having heterogeneous sp^2-sp^3 structure. This model is in good agreement with theoretical analyses and numerical estimations. Applicability of the nano-carbon emitters are demonstrated by a few examples of vacuum light emitting devices including flat panel display prototype, triode- and diode-type lamps.

1. INTRODUCTION

Electron field emission (FE) from surface of conductive materials induced by strong electric field has great importance for fundamental and applied science [1,2]. Various carbon related materials show low-field electron emission and are therefore highly attractive as replacement for metals and semiconductors for FE cold cathodes. One of the main advantages of carbon materials for FE applications is the extremely strong interatomic σ–σ covalent bonding force providing high mechanical strength and chemical inertness appropriate for cathode materials operating in vacuum devices under intense electric field of the order of 10^7-10^8 V/m and in the harsh conditions of residual gas ion bombardment. However, despite the strongest interatomic bonds in graphite, the usual graphitic carbons with sp^2 atomic coordination are not suitable for FE because of their layered structure with rather weak interlayer bonding. So, while each layer itself is very stable, the entire material combined of a number of parallel layers can be destroyed easily by applying a strong electric field, and this instability is known to take place for carbon fibers [3] and graphite particles [4].

Carbon nanotubes (CNT) consisting of sp^2 coordinated atomic layers are much more stable under strong electric field action because of their cylindrical shapes preventing layers chipping off. An additional advantage is the high aspect ratio of CNT emitters, providing electric field focusing and corresponding significant reduction of applied voltages (see, for example, [5]). Similar increase of FE stability and voltage reduction can be obtained in materials composed of nano-graphite crystallites (NGC), which have bent atomic layers at the crystallite edges [6]. Layer bending at graphite crystallite edges is rather usual phenomena

S. Guceri et al. (eds.), Nanoengineered Nanofibrous Materials, 329-339.

(see, for example, [7-9]) that may explain the efficient and stable electron emission from various carbon materials containing NGC: polycrystalline diamonds [10], diamond-like films [11], carbon fibers [3], and graphite powders [4].

Other carbon materials with sp^3 coordination (diamond and diamond-like) have been believed to be prospective for FE applications also due to their unique property of negative electron affinity (NEA). NEA decreases up to zero the potential barrier for electrons escaping into vacuum from the diamond conduction band [12]. However, undoped diamond (as well as other diamond-like materials) has a wide band gap (WBG) with a negligible density of free electrons in the conduction band and it is therefore unlikely that NEA and the required high electron conductivity can be combined in the doped forms of any diamond-like material. This consideration allows the attribution of the numerous experimental observations of low-field electron emission from diamond materials to non-diamond carbon inclusions existing in polycrystalline diamond films obtained by chemical vapor deposition (CVD) [4,10], in single crystal diamonds [13] and in diamond composites and powders [14,15].

Thus two main approaches for the explanation of field electron emission from carbon materials exist. One of them considers FE on the bas of Fowler-Nordheim (FN) theory, assuming that emission occurs from conductive carbon material under a strong electric field enhanced due to morphological features on the emitter surface. Another approach assumes that FE originates from the NEA property of diamond and diamond-like species. The various mechanisms explaining electron injection into the conduction band of WBG diamond and diamond-like material and electron transport to the emitter surface have been proposed for this second approach. These two approaches consider separately either conductive (graphite-like) or non-conductive (diamond-like) carbons. At the same time, experimentally observations show that graphite-like (conductive) emitters may exhibit properties similar to dielectric one, including electroluminescence accompanying FE [16,17], charging of the carbon nanotube surface during FE [18,19], and non-linear I-V dependence behavior in FN coordinates [20]. Other distinctions of carbon nanotube FE from those known for conductive tip emitters consists in the controversial facts that single-wall carbon nanotubes having higher aspect ratios than multiwall carbon nanotubes show a surprisingly higher emission threshold for electric field [21] and that better FE occurs from CNT aligned across the electric field direction [22].

In our studies we performed comparative characterization of various properties of carbon materials with different compositions obtained using the CVD method. Different features of the CVD process were investigated to evaluate the optimal conditions for fabrication of thin film field emitters. The obtained carbon cold cathodes were used for the manufacturing of the vacuum electron device prototypes.

2. CARBON THIN FILM MATERIALS FABRICATION BY CVD METHOD

The remarkable flexibility of the chemical vapor deposition process allows one to employ similar technologies for the production of materials with dramatically different properties (e.g. diamond [23], carbon nanotubes [24]) by changing the geometry of the reactor chamber and the discharge type used in the plasma enhanced CVD process. A fully controllable growth of various carbon films can be achieved in the same dc discharge CVD system by changing the pressure and composition of the hydrogen-methane mixture along with the substrate temperature, discharge current and voltage [25-29]. Standard Si wafers,

Ni, W, Mo and some other metal sheets and films were used in our experiments as substrates with sizes up to 50 mm in diameter. The carbon films depositions were carried out using a hydrogen-methane gas mixture. Our experimental CVD setup was described in detail elsewhere [29]. In brief, the distance between the cathode and tungsten anode in the reactor chamber is 50 mm. The substrate for carbon film deposition is located on the water-cooled anode. The internal diameter and height of the cylindrical reactor chamber is 400 mm. The chamber has water-cooled walls with quartz windows, which allows visual observation and optical measurements.

TABLE 1 Experimental parameters for carbon films deposition in dc discharge plasma.

Type of CVD film material	Substrate temperature, °C	Methane concentration, %	Total gas pressure, Torr
diamond	850-900	0.5-2	60-90
nano-diamond	900-1000	2-5	60-100
graphite-like	1050-1100	5-10	60-100
soot	1100-1250	over 15	50-100

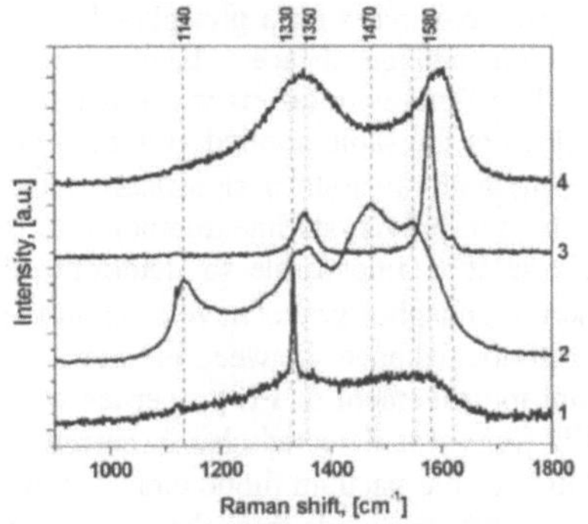

Figure 1. Raman spectra of polycrystalline diamond (1), nanocrystalline diamond (2), nanostructured graphitic materials (CNT and/or NGC) (3), and carbon soot (4).

The CVD chamber is equipped with optical systems allowing in-situ measurements of Raman spectra for deposited carbon materials [27] and optical emission spectra (OES) of the plasma in the CVD reactor [28]. Also dc discharge current and applied voltage were measured to characterize the deposition process [28]. The typical sets of parameters of the CVD process providing deposition of diamond, CNT and disordered graphite (soot) films are listed in TABLE 1. Raman spectra of these materials are shown in Fig. 1. In these spectra, Raman lines at 1140 and 1470 cm^{-1} are specific for diamond particles with sizes smaller than 2 nm, while 1330 cm^{-1} corresponds to "usual" diamond particles, which have larger diameters [30]. Raman lines at 1350 cm^{-1} and in the vicinity of 1580 cm^{-1} (between 1550 to 1620 cm^{-1}) are attributed to different forms of disordered graphite. It should be noted that a similar 1580 cm^{-1} line is typical for multiwall carbon nanotubes [31,32].

The most important CVD films composed of nanosized graphite-like species seen as multiwall CNT and graphite nanocrystals (NGC) were obtained at relatively high concentrations of methane in the gas mixture. OES data show the presence of C_2 dimers in the plasma containing a high concentration of methane [28]. These carbon dimers play a dominant role in the synthesis of the nanocrystalline diamond (see e.g. [33]). In particular, insertion of the C_2 into acetylene-like C-C bond to produce a carbene structure is the most energetically favorable scenario. Empirical calculations have shown [33] that the addition of C_2 may result in the evolution of critical nuclei into two-dimensional graphite-like sheets, which become mechanically unstable and roll up spontaneously to form CNT in a non-catalytical process [26]. The experimental confirmations of this model were found by using scanning electron microscopy (SEM) observations [28]. Another important circumstance of the CNT and NGC species formation is a substrate temperature of about 1000°C (see TABLE 1).

3. COMPARATIVE STUDY OF FE FROM CARBON MATERIALS

The FE tests were performed in our experiments using a vacuum diode configuration with a flat parallel anode and cathode. The anode was a glass plate coated with a transparent, conducting indium and tin oxide (ITO) film. This schema provided a very easy yet adequate method of determining the macroscopic values of electric field strength and FE current density in contrast to any tip-like or ball-like anode probe configurations (see [5,15,34] e.g.). The conducting anode film was covered with a phosphor layer allowing an image of FE site distribution over the cathode surface to be obtained. Typical images of FE site distributions for carbon films with different composition are shown in Fig. 2. The images were obtained at the same applied voltage (1600 V) and cathode-to-anode distance (200 μm) and illustrate a significant increase of emission site density from about 10^2 for polycrystalline diamond to 10^5-10^6 for graphitic material. In the last case it is impossible to determine the actual density of emission sites because the phosphor grains have sizes in the range of 3 to 7 μm, which is comparable with the distances between FE sites.

The significant improvement of FE properties with decreasing diamond component in the CVD films was detected also in current-voltage dependencies. Fig. 3 shows typical plots of the vacuum diode emission current (I) as a function of applied voltage (V) for cathodes made from those three type of CVD films. It is seen from the plots that threshold macroscopic electric field (E), determined as $E=V/d$ (where d is the interelectrode distance), at which electron emission can be detected is 4 V/μm for polycrystalline diamond, 2 V/μm for nanocrystalline diamond, and less than 1.5 V/μm for nano-graphitic material. With the reduction of the threshold field we detect an increase of the maximal values of emission current.

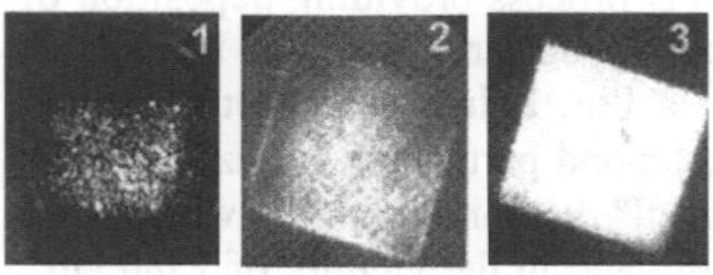

Figure 2. The images of anode screen obtained for polycrystalline diamond (1), nanoscrystalline diamond (2), and nano-graphite (3) cathodes.

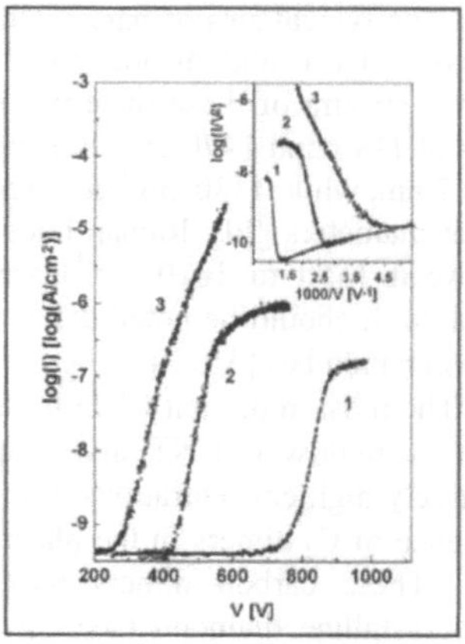

Figure 3. Current-voltage dependencies for field emission from CVD films having various composition (polycrystalline diamond – 1, nano-diamond – 2, and nano-graphite –3). Inset shows Fowler-Nordheim plots for the same curves.

Our comparative experimental studies performed for various CVD film materials as well as for bulk and powdered graphite, carbon fibres and other carbon materials clearly show similarity in the field emission properties for all of them. The most efficient electron emission including, the lowest threshold field, and the highest current and emission site densities was obtained for well structured nano-sized graphite type materials. The most distinctive features of the FE characteristics of these materials are: (i) a linear *I-V* curve in FN coordinates over a wide range of voltages and currents; (ii) substantial curvature of the FN *I-V* plot at low voltages and currents; (iii) homogeneity of the FE site distribution over the film surface. An evaluation of the FE mechanism requires a corresponding analysis of these features.

4. STATISTICAL ANALYSIS OF FIELD EMISSION CHARACTERISTICS

The linear character of the FN plot corresponds to classical electron tunneling mechanism of FE [1] while its substantial curvature may be explained by the statistical distribution of the emitting sites: geometrical, structural and electronic characteristics [35]. Although FE site characteristics have complex physical origins, a normal (Gaussian) distribution can be a good first approximation to the actual distribution of their heights and/or diameters. The FE image homogeneity at moderate electric field (see Fig.2) is evidence of a narrow width of the normal distribution with a small standard deviation σ. To simplify further consideration we assume that the subject for the statistical distribution has an effective size (radius *r*) of emission sites only. The normal distribution of the FE site radius is given by $n(r) = \frac{N}{\sqrt{2\pi\sigma^2}} \exp[-(r-r_0)^2/2\sigma^2]$, where r_0 is an average size of radius. By having the total number of emission sites on the cathode surface equal to *N*, it may be written that $N = \int n(r)dr$. Thus the total FE current is $I = \int JdS = \int J\pi r^2 n(r)dr$, by assuming that emission occurs from the entire surface area ($S = \pi r^2$) of each emission site.

In FN theory the density of FE current is expressed by the equation

$$J = \frac{A}{\varphi} F^2 \exp[1.03Be^3\varphi^{-1/2}]\exp[-0.95B\varphi^{3/2}F^{-1}], \quad (1)$$

where *F* is the local electric field at the emission site surface, φ is the work function of the cathode material; constants *A* and *B* are combinations of electron charge (*e*), mass (*m*) and Planck's constant *h*: $A = \frac{e^3}{8\pi h}$, $B = \frac{8\pi\sqrt{2m}}{3he}$. The constants values for the units of current density (*J*) A/m^2, electric field (*F*) – V/m, and work function (φ) – eV are: A=1.5414×10^{-6} A eV V^{-2}, B=6.8309×10^9 eV$^{-3/2}$ V m^{-1}, and 1.03Be^3=10.1 eV.

The most common method to interpret local field value *F* relationships with applied voltage *V* is to believe their linear conversion: $F = \frac{H}{r}E = \frac{H}{r}\frac{V}{d}$, where *E* is the macroscopic field in planar electrode configuration with an interelectrode distance *d* and aspect ratio for a separate FE site equal to the ratio of its height (*H*) over the substrate surface to its radius which may be believed to be the same order as the radius of the emission site area - *r*. By assuming a value unity for the height *H* for all of the emission sites we obtained (see [36]) a formula describing the total emission current dependence on the macroscopic field (or on the applied voltage $V=Ed$) as

$$I = CNE^2 \exp\left[-\frac{Dr_0}{E}+\frac{\sigma^2}{2}\left(\frac{D}{E}\right)^2\right];\ C=\frac{A\pi H^2}{\varphi}\exp[10.1/\sqrt{\varphi}]\ ;D=0.95B\frac{\varphi^{3/2}}{H}. \qquad (2)$$

The last formula (2) shows that the *I-V* dependence differs from the classical expression deduced from FN theory for FE from metals in the second term of exponent. Fig. 4 illustrates the corresponding agreement of the experimental results with the calculation, derived by using equation (2) for $\sigma=0.1r_0$. The second term causes excess current and upward curvature in the FN plot. This term vanishes when all emission sites have the same geometrical characteristics since $\sigma=0$. But the term increases at lower values of macroscopic field (or applied voltage) if $\sigma\neq0$. The inputs of both terms in (2) are comparable at σ values in the range of 0.5Å to 2Å when we assume that r=5nm, H=1μm, φ=5eV and with E in the range of 1V/μm to 10V/μm, correspondingly. This numerical estimation is in agreement with our initial assumption about the narrow width of the normal distribution with a small standard deviation σ. The FN plot curvature increases with σ, whereas for the linear part of the plot we assume that σ and the corresponding second term in (2) are equal to zero.

Comparison of the theoretical results (formula (2)) and the experimental *I-V* curve (see Fig.4) allows us to obtain empirical relationships for two parameters - work function φ and field enhancement factor β: $\beta=H/r$: $\frac{\varphi^{3/2}}{\beta}=7.4\times10^{-3}eV^{3/2}$.

This relationship is shown in Fig. 5 graphically. One can observe that the expected range for work function is in the range of 4 to 5 eV corresponds to the field enhancement factor of about 10^3. This value of β contradicts to our SEM and TEM observations of the geometrical shapes of CNT and GNC species on the carbon cathode surface. Since a typical size of the sharp edge is about 10nm, these carbon structures have heights up to 5 μm and the corresponding geometrical field enhancement factor may be in the range of 50 to 500 (see [25,26,28,29].

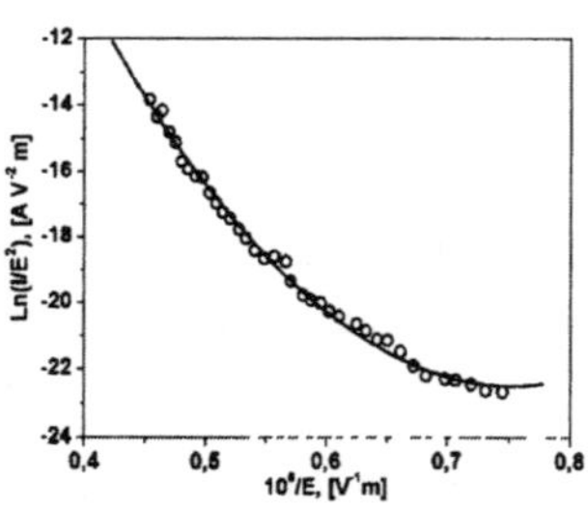

Figure 4. Typical I-V experimental dependence plotted in FN coordinates (dots) and fitted curve calculated by using Eq. 3 with $\sigma=0.1r_0$ (line).

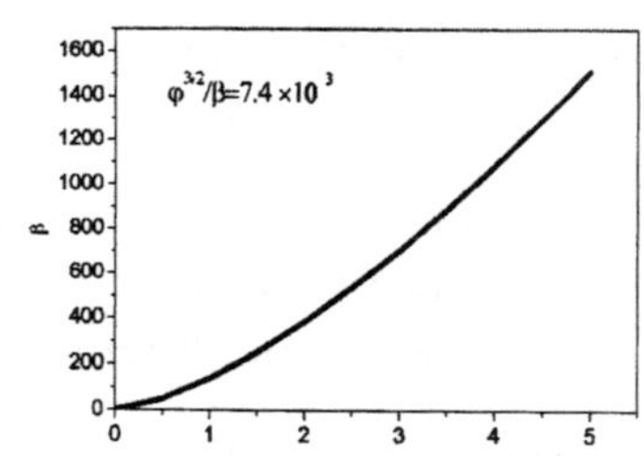

Figure 5. Empirical relationships for work function φ and field enhancement factor β obtained from comparison of theoretical I-V dependence and experimental curve.

Another contradiction relates to the estimate of the number of emission sites N. This estimation may be made for the linear part of the FN plot from an intercept by assuming that all emission sites have the same surface area $S=\pi r^2$. Thus, for the particular case of the FN plot in Fig. 6 we have $N\approx10^6$ cm^{-2}, that is in well agreement with our observations made using the phosphor screen (see Fig. 2 and [25,26,28,29]). However, for the low-voltage range curved part of the plot our estimation shows a value of $N\approx10^{13}$ cm^{-2}. This estimation was made by the fitting theoretical curve (3) to the experimental dependence by variation of N as a parameter. The emission site density decrease with voltage increase may be explained by a screening effect leading to the suppression of emission from the sites having smaller field enhancement factors. But even in the case of screening, the number $N \approx 10^{13}$ cm^{-2} is possible only if the emission sites have atomically small sizes.

The alternative explanation consists in the suggestion that emission occurs from an area beyond the top edge of the nano-carbon species and including some part of their lateral surface also. In this case, the estimated number of emission sites will be decreased on two or three orders without an extreme decrease of the characteristic sizes of the emitting carbon species. This explanation agrees with the model of emission sites proposed in our publications (see, e.g. [25,26]) as well as with other observations showing the origination of field emission from CNT lateral walls [37,38]. A mechanism providing such lateral emission is nonmetallic behavior of nano-carbon emitters and electron tunneling through a potential barrier, which is reduced in the vicinity of defects of sp^2 coordinated graphite-like carbon atom network [25,26].

5. TWO-BARRIER MODEL FOR FIELD EMISSION FROM NANO CARBON MATERIALS

The model of FE sites on the surface of nano-carbon material is comprised of graphitic material covered by an atomically thin layer of carbon with diamond-like electronic structure. The energy-band diagram of such a heterogeneous surface includes two potential barriers on the interfaces between the different carbon phases and between the cathode surface and vacuum (see Fig.6). In the band diagram analysis one should consider the atomistic structure of the emitter material, emphasizing the importance of the outer atomically thin layer. Thus a bulk graphitic part of the cathode is represented schematically in Fig. 6 by a periodic array of quantum wells each corresponding to carbon atoms with the Fermi level lying above the wells. The outer layer of carbon atoms with diamond-like properties is represented by a separate quantum well with the Fermi level inside a gap between the lowest unoccupied and highest occupied bands of electronic states (LUS and HOS correspondingly). The LUS-HOS gap width is assumed to be similar to that in defective diamonds and according to our cathodoluminescent studies it can be taken slightly larger than 4 eV [25].

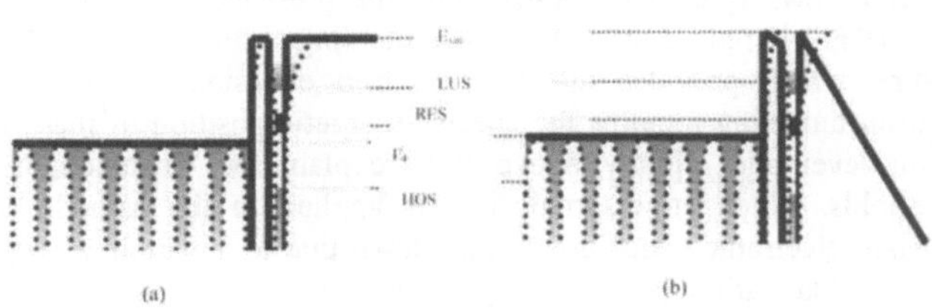

Figure 6 Schematic energy band diagram presentation of vacuum-cathode interface without (a) and with (b) external field. The straight lines show simplified two-barrier structure including rectangular barrier between two carbon phases and outer potential barrier. The dashed lines show quantum wells corresponding to separate carbon atoms in the presurface layer with graphite properties and surface cluster with diamond-like properties similar to WBG semiconductor having gap in density of electron states between highest occupied state (HOS) and lowest unoccupied states (LUS). The electron states in the quantum wells are shown by gray and resonance electron states (RES) in HOS-LUS gap are shown by black. The height of rectangular barrier and work function of surface cluster (E_{vac}-E_F) is about 4.5 eV.

The application of an external field to the cathode surface leads to the deformation of this two-barrier structure. In the simplest approximation the first barrier remains to have a rectangular shape and the second one becomes triangular as sketched in Fig. 6. The main properties of FE are determined by the probability of electrons to tunnel through these two barriers at various values of the external field F. By considering the probabilities for electron tunneling through this two-barrier structure we obtain an expression for emission current density [39]:

$$J = \frac{e^3}{8\pi h} \frac{1}{\left[\frac{\varphi^{1/2}}{eF} - \frac{w}{\sqrt{\chi}} \right]^2} \exp\left(-\frac{4\pi\sqrt{2m}}{h} \left(\frac{2(\varphi)^{3/2}}{3eF} - \sqrt{\chi} w \right) \right) \quad (3)$$

where w is the width of the rectangular barrier, χ – the energy difference between the Fermi level and the top of the rectangular barrier, and φ is the work function of the cathode surface.

This formula is different from the usual Fowler-Nordheim law by the second additional term in the exponent. This new term ($\chi^{1/2}w$) may lead to a significant increase in the current density. Using for estimation φ=4.5 eV, χ=4.5eV and w =4 Å, we obtain a current density increase of 4 orders for the same field in comparison with the value predicted by the FN model for metal emitters as is evident from Fig. 7, which clearly show a reduction of the threshold field and a dramatic increase of the current density for the two-barrier emitters with resonant electron states.

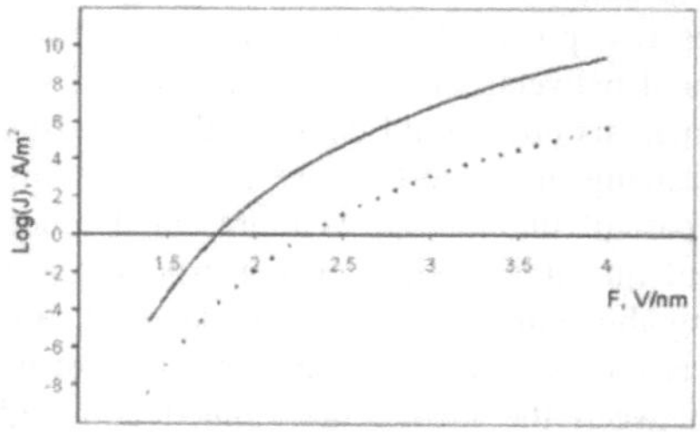

Figure 7. Current-vs-field dependencies for electron emission calculated in accord with FN low (dots line) and with respect of two-barrier model using formula (4) (straight line).

It follows from our analysis, that the presence of the resonant electronic states (RES) in the quantum well corresponding to the WBG layer is an important feature, which provides intensive electron emission at low fields. Efficient electron tunneling requires the special energetic position of these RES near the Fermi level and slightly above it to explain the existence of nonzero FE thresholds. When an external field is applied to the cathode surface, these resonant electronic states are shifted down due to a partial voltage drop in the diamond-like carbon layer (see Fig. 6). Various carbon materials including CNT, GNC, nanodiamonds etc. show very similar threshold values in the range of 1 to 1.5 V/μm (see e.g., [2-6]) that may be considered as evidence of the similarity their emission mechanisms.

6. LIGHTING ELEMENT PROTOTYPES

Our studies show that the field emission characteristics of carbon cathodes result from the unique structure and physical properties of the nanosized graphitic species. These species may be found in various carbon-related materials including diamonds (single crystal, polycrystalline and nanocrystalline films and powders), diamond-like and tetrahedral amorphous films, carbon fibres and carbon nanotubes, graphite powders etc. The most important parameters for application: low-field emission threshold, high density of emission sites and emission current, emission stability of the carbon emitters, are determined by the arrangement of these nano-carbon species in the cathode material. To evaluate device applicability of our nano-carbon (nC) cathodes made by the CVD method, a few vacuum light emission lamps were designed and manufactured [40].

The simplest prototype of the light emitting device with a nC cold cathode was made as a flat vacuum diode lamp (see Fig. 8). The glass envelope of the lamp cells was assembled using two glass slides 40×50 mm in size coated with ITO and a phosphor layer. The nC cold cathode (25×25 mm) was fixed to the bottom slide using conductive epoxy. The glass slides were fixed to each other using vacuum cement allowing spacing between the phosphor layer and the cathode within the 0.1-1 mm range. The total cell thickness is about 3 mm. The residual pressure in the flat glass cell is about 10^{-6} Torr after pumping and sealing. The operation voltage for this type of lamp ranges from 200V to 2 kV depending on the type of phosphor and the vacuum gap distance. A very uniform, stable and high-brightness light emission was obtained from these lamps and they can be considered as a prototype of the flat-panel display. However, the intensity of the light emitted by the diode lamps is limited to the value of about 1000 cd/m^2 and the total power efficiency is about 10% due to the low efficiency of the phosphors operated at low anode-to-cathode voltages. This low voltage value is, in fact, one of the main advantages of the nC cold cathode, so, an improvement of phosphor properties is necessary to use the simplest flat diode scheme.

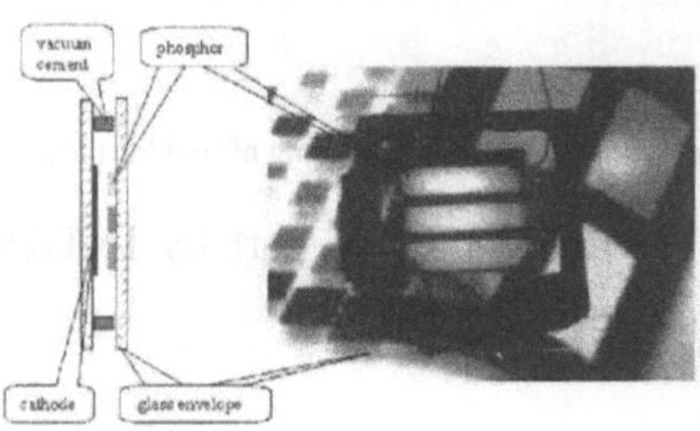

Figure 8. Schema and photograph of flat diode lamp taken with CRT screen image as background.

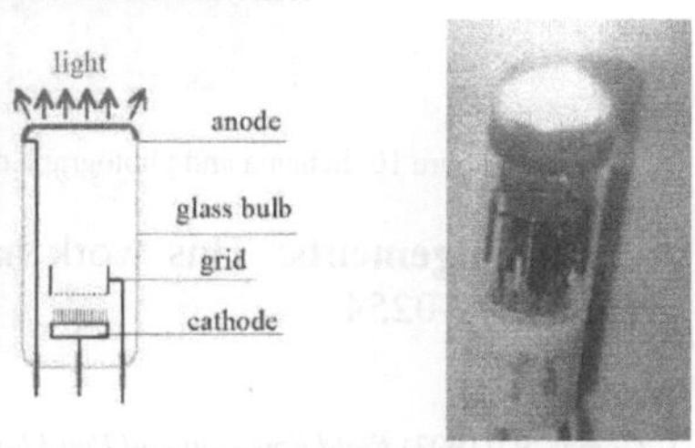

Figure 9. Schema and photograph of triode lamp with flat FE cathode of nC film material.

A significant increase in light intensity and power efficiency was obtained for the lamps with a triode configuration (see Fig. 9). This type of vacuum device has a more complicated structure including a grid electrode located near the flat nC cathode with approximately the same distance between them as between the cathode and anode in the flat diode lamps. This allows electron emission at the same low voltages applied between the cathode and the grid. The anode-to-grid distance and voltage can be much higher, providing high-energy electrons and high-efficiency radiation from the phosphor layer with the use of standard TV phosphors. The intensity of emitted light for such lamps exceeded 20000 cd/m^2 when an accelerating voltage of 5 kV was applied between the anode and the grid. The estimated power efficiency of the triode lamps is about 15%. The triode lamp seems to be the most convenient in such applications where low-voltage switching is necessary.

An alternative scheme providing high electron energy and corresponding high efficiency of light emission exploits a cylindrical diode configuration. An important advantage of the cylindrical diode configuration in comparison with the planar one is that for the same applied voltage V, the macroscopic electric field

$F=V/[r\ ln(R/r)]$, as compared to $F=V/d$ for two planar electrodes with an interelectrode distance d. Practically, this means that the macroscopic field at the cathode surface $F(r)$ will be much higher in the cylindrical case than for the identical $d=R-r$ and V. This makes possible to realize a field emission diode working at suitable voltages applied between electrodes separated by a rather large distance. An essential improvement of the lamp parameters was achieved by using a scheme with the reflecting Al layer deposited on the glass tube instead of ITO. This configuration and the lamp photograph are shown in Fig. 10. The lamp brightness is achieved up to 2×10^5 cd/m^2. The power efficiency of the lamps with a reflecting Al anode exceeds 30%. Such power efficiency is an absolute record for light emission devices and is obtained due to the combination of an excellent nC cold cathode and a lamp configuration providing more optimal conditions for excitation of the phosphor by electrons in comparison with other similar devices. The other advantage of our lamp is the high efficiency obtained from the direct output of light generated in the phosphor layer in contrast to lamps using transparent anodes where CL light must penetrate through the phosphor layer.

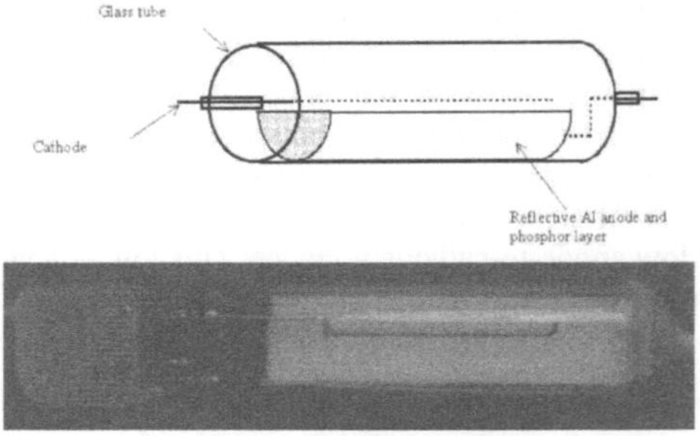

Figure 10. Schema and photograph of the cylindrical diode lamp using nC cold cathode.

Acknowledgements: This work has been supported in part by INTAS grant no. 01-0254.

References

1. Gomer, R. (1993) *Field Emission and Field Ionization*, AIP, NY.
2. Temple, D. (1999) Recent progress in field emitter array development for high performance and applications, *Mat. Sci. and Eng.* **R24**, 185-239.
3. Latham, R.V., Wilson, D.A. (1982) The development of high-definition cathode-ray tube using a carbon-fibre field-emission electron source, *J. Phys. E: Instrum.* **15**, 1083-1093.
4. Xu, N.S., Tzeng, Y., Latham, R.V. (1993) Similarities in the 'cold' electron emission characteristics of diamond coated molybdenium electrodes and polished bulk graphite, *J. Phys. D: Appl. Phys.* **26**,1776-1780.
5. Gröning O., Küttel O.M., Emmenegger Ch., Gröning P., Schlapbach L. (2000) Field emission properties of carbon nanotubes, *J. Vac. Sci. Techol. B* **12**, 665-678.
6. Obraztsov A.N., Volkov A.P., Pavlovsky I.Yu., Chuvilin A.L., Rudina N.A., Kuznetsov V.L. (1999) Role of the curvature of atomic layers in electron field emission from graphitic nanostructured carbon, *JETP Lett.* **69**, 411-417.
7. Hiura H., Ebbesen T.W., Fujita J., Tanigaki K., Takada T. (1994) Role of sp^3 defect structures in graphite and carbon nanotubes, *Nature* **367**, 148-151.
8. Kuznetsov V.L., Chuvilin A.L., Butenko Yu.V., Stankus S.V., Khairulin R.A., Gutakovskii A.K. (1998) Closed curved graphite-like structures formation on micron-size diamond, *Chem. Phys. Lett.* **289**, 353-360.
9. Huang J.Y., Yasuda H., Mori H. (1999) Highly curved carbon nanostructures produced by ball-milling, *Chem. Phys. Lett.* **303**, 130-134..
10. Zhu W., Bower C., Kochanski G.P., Jin S. (2001) Field emission properties of diamond and carbon nanotubes, *Diamond and Related Mat.* **10**, 1709-1713.
11. Cui J.B., Robertson J., Miln W.I. (2001) Field emission site densities of nanostructured carbon films deposited by cathodic arc, *J. Appl. Phys.* **89**, 5707-5711.

12. Himpsel F.J., Knapp J.A., VanVechten J.A., Eastman D.E. (1979) Quantum photoyield of diamond(111) – A stable negative-affinity emitter, *Phys. Rev. B* **20**, 624-627.
13. Steeds J.W., Gilmore A., Bussmann K.M., Butler J.E., Koidl P. (1999) On the nature of grain boundary defects in high quality CVD diamond films and their influence on physical properties, *Diamond and Related Mat.* **8**, 996-1005.
14. Cui J.B., Stammler M., Ristein J., Ley L. (2000) Role of hydrogen on field emission from chemical vapor deposited diamond and nanocrystalline diamond powder, *J. Appl. Phys.* **88**, 3667-3673.
15. Karabutov A.V., Frolov V.D., Konov V.I., Ralchenko V.G., Gordeev S.K., Belobrov P.I. (2001) Low-field electron emission of diamon/pyrocarbon composites, *J. Vac. Sci. Technol. B* **13**, 965-970.
16. Latham R.V., Wilson D.A. (1981) Electroluminescence effects associated with the field emission of electrons from a carbon fibre micropoint emitter, *J. Phys. D: Appl. Phys.* **14**, 2139-2145.
17. Bonard J.-M., Stoeckli Th., Maier F., de Heer W.A., Chatelain A., Salvelat J.-P., Forro L. (1998) Field-Emission-Induced Luminescence from Carbon Nanotubes, *Phys. Rev. Lett.* **81**, 1441-1444.
18. Wang Z.L., Gao R.P., de Heer W.A., Poncharal P. (2002) In-situ imaging of field emission from individual carbon nanotubes and their structural damage, *Appl. Phys. Lett.* **80**, 856-858.
19. Zhou G., Duan W., Gu B., Kawazoe Y. (2002) Emission probabilities of π electrons in carbon nanotubes, *Appl. Phys. Lett.* **80**, 1999-2001.
20. Dean K.A., von Allmen P., Chalamala V.R. (1999) Three behavioral states observed in field emission from single-walled carbon nanotubes, *J. Vac. Sci. Technol. B* **17**, 1959-1969.
21. Choi W.B., Chung D.S., Kang J.H., Kim H.Y., Lee Y.H., Jung J.E., Lee N.S., Park G.S., Kim J.M. (1999) Fully sealed, high-brightness carbon-nanotube field-emission display, *Appl. Phys. Lett.* **75**, 3129-3131.
22. Chen Y., Shaw D.T., Guo L. (2000) Field emission of different oriented carbon nanotubes, *Appl. Phys. Lett.* **76**, 2469-2471.
23. Spitsyn B.V. (1994) Growth of Diamond Films from the Vapour Phase in D.T.J. Hurtle *Handbook of Diamond of Crystal Growth, vol.3*, Elsevier Science, pp.403-456.
24. Saito R., Dresselhaus G., Dresselhaus M.S. (1998) *Physical Properties of Carbon Nanotubes*, Imp. College Press.
25. Obraztsov A.N., Pavlovsky I.Yu., Volkov A.P., Petrov A.S., Petrov V.I., Rakova E.V., Roddatis V.V. (1999) Electron field emission and structural properties of carbon vhemically vapor-deposited films, *Diamond and Related Materials* **8**, 814-819.
26. Obraztsov A.N., Volkov A.P., Nagovitsyn K.S., Nishimura K., Morisawa K., Nakano Y., Hiraki A. (2002) CVD growth and field emission properties of nanostructured carbon films *J. Phys. D: Appl. Phys.* **35**, 357-362.
27. Obraztsov A.N., Pavlovsky I. (1997) Raman spectrometer for in-situ study of diamond growth in DC discharge plasma CVD reactor *SPIE Proc.* **3484**, 53-58.
28. Obraztsov A.N., Zolotukhin A.A., Ustinov A.O., Volkov A.P. Svirko Yu.P. (2003) Chemical vapor deposition of carbon films: in-situ plasma diagnostics *Carbon* 41, 836-839.
29. Obraztsov A.N., Pavlovsky I.Yu., Volkov A.P. (1998) Field emitter and method for producing the same, *PCT Application* WO 99/44215.
30. Lopez-Rios, T., Sandre, E., Leclercq, S., Sauvain, E. (1996) Polyacetylene in Diamond Films Evidenced by Surfac Enhanced Raman Scattering *Phys. Rev. Lett.* **76**, 4935-4938.
31. Lespade, P., Al-Jishi, R., Dresselhaus. M.S. (1982) Model for Raman scattering from incompletely graphitized carbons *Carbon* **20**, 427-431.
32. Saito, R., Dresselhaus, G., Dresselhaus, M.S. (1998) *Physical Properties of Carbon Nanotubes*, Imp. College Press, Singapore.
33. Krauss, A.R., Auciello, O., Gruen, D.M., Jayatissa, A., Sumant, A., et. al. (2001) Ultrananocrystalline diamond thin films for MEMS and moving mechanical assembly devices *Diamond and Related Mat.* **10**, 1952-1961.
34. Zhirnov, V.V., Lizzul-Rine, C., Wojak, G.J., Sanwald, R.C., Hren, J.J. (2001) "Standartization" of field emission measurements *J. Vac. Sci. Technol. B* **19**, 87-93.
35. Levine, J.D. (1995) Statistical analysis of field emitter emissivity: Application to flat displays *J. Vac. Sci. Technol. B* **13**, 553-557.
36. Obraztsov, A.N., Zakhidov, Al.A., Volkov, A.P., Lyashenko, D.A. (2003) Non-classical electron field emission from carbon materials *Diamond and Related Mat.* **12**, 446-449.
37. Wang, Z.L., Gao, R.P., de Heer, W.A., Poncharal, P. (2002) In situ imaging of field emission from individual carbon nanotubes and their structure damage *Appl. Phys. Lett.* **80**, 856-858.
38. Chen, Y., Shaw, D.T., Guo, L. (2000) Field emission of different oriented carbon nanotubes *Appl. Phys. Lett.* **76**, 2469-2471.
39. Obraztsov, A.N., Zakhidov, Al.A. (2003) Low-field electron emission from nanocarbons, Rep. On Diamond 2003 Conf. (Abs. DM 15 PR), Salzburg (Austria), Sept. 2003, to be published in *Diamond and Related Mat.*
40. Obraztsov, A.N., Volkov A.P., Zakhidov, Al.A., Lyashenko D.a., Petrushenko, Yu.V., (2003) Satanovskaya, O.P., Field emission characteristics of nanostructured thin film carbon materials *Appl. Surf. Sci.* **215**, 214-221.

$MoS_{(2-x)}I_y$ NANOTUBES AS PROMISING FIELD EMITTER MATERIAL

A. MRZEL[1], V. NEMANIČ[1], M. ŽUMER[1], B. ZAJEC[1], J. PAHOR[1], M.REMŠKAR[1], E. KLEIN[2], D. MIHAILOVIČ[1]
[1] *Jozef Stefan Institute, Jamova 39, 1000 Ljubljana, Slovenia*
[2] *Weizmann Institute of Science, Rehovot 76100, Israel*

Abstract: We have investigated the field emission (FE) properties of $MoS_{(2-x)}I_y$ (y~1\3) nanotubes. The $MoS_{(2-x)}I_y$ nanotubes are all sub-1-nm diameter and grow in sharp bundles, each typically consisting of up to million identical nanotubes. As-grown material was dispersed in ethanol and the individual bundles were then glued with the silver conducting paste to an aluminized glass fiber. The FE measurements of single bundles were performed under continuous bias conditions at room temperature with voltages from 0.1 to 2.4 kV. The emission current at cycling experiments from individual sites reached up to 80 nA at a macroscopic field 0.9 $V\mu m^{-1}$. On a time scale of some ten hours the current was stable. Short term FE currents up to 10 microamperes were obtained from an individual bundle without a noticeable degradation of the FE performance. All single bundle measurements have shown virtually identical FE properties. They appear to be inert and unsusceptible to vacuum-related problems, including intermediate exposure to air. The ease of processing and good dispersion characteristics in combination with their stable reproducible FE performance suggest that $MoS_{(2-x)}I_y$ NT may offer specific advantages in some applications.

1. INTRODUCTION

Nanowires and nanotubes have attracted research interest in recent years as their functional properties hold promise for incorporation into nanoscale devices and systems [1]. Field-emission properties of nanotubes are one of the forefront of worldwide nanotechnology research. The unique geometry of nanotubes suggests that they are likely to be ideal field emitters for some applications like flat panel displays [2].

So far, mostly carbon nanotubes (CNT) were studied extensively in several laboratories, which had also led to many improvements of their FE performance. Carbon nanotubes have been investigated as field electron emission sources for a number of applications including field-emission displays [3], microwave power amplifiers, gas discharge tubes [4], and nanolithography systems [5]. These applications require that nanotubes

S. Guceri et al. (eds.), Nanoengineered Nanofibrous Materials, 341-347.

provide a stable field-emission current at required current densities without compromising device lifetime. It was shown that CNTs do exhibit serious advantage compare to technology using tungsten or molybdenum metal tip arrays, such as operating under moderate vacuum conditions of 10^{-6} mbar. They also appear to be self-healing to some extent [6], which is thought to be the reason for their stable operation in excess of 10^4 h at average current densities of up to 0.1 A/cm^2 [7]. Bright displays were fabricated in a few years [8] but the point-like character of the light emission from these displays was probably the main disappointment. The reason is uncontrolled emission site density arising also from the difficulty in processing and dispersing CNTs [9].

In addition to carbon nanotubes, other materials were also recently tested as field emitters. Though promising results were achieved with hydrogenated diamond like carbon [10] and silicon nanowires [11], none of them displayed enough feasibility for technological applications. A hybrid emitter is a relatively a new trend. The examples are carbon nanotubes grown on the top of a tungsten–wire tip [12], diamond coated silicon tips [13] and sharp molybdenum needles coated with BN nanoparticles [14]. Also several other nanostructured materials have been reported to exhibit FE properties (BCN [15], TiN [16], Nb_2N [17], GaN [18], Mo nanowires [19]....).

It is important to stress that a quantitative comparison of their FE properties with CNTs can not be easily made, since they have not been characterized by identical methods [20]. Although comparisons are difficult to make at this stage, the search for the most suitable nanotube-shaped material for use in different FE applications is clearly just beginning.

2. EXPERIMENTAL WORK

Here we report on the FE properties of $MoS_{(2-x)}I_y$ (y~1/3) transition metal chalcogenide nanotubes which were grown by a catalyzed transport method [21]. The iodine in the formula signifies the presence of interstitial iodine between the MoS_{2-x} nanotobes and could be largely removed by heating in vacuum. The nanotubes are all sub-1-nm in diameter and grown in form of bundles, each typically consisting of up to 10^6 identical nanotubes. Bundles range from 20 nm up to 0.5 μm in diameter and can be up to several 100 μm long. The bundles end in a sharp tip with a single or few protruding NTs at the end (Figure 1).

Figure 1 SEM image of as-grown material in the form of bundles of $MoS_{(2-x)}I_y$ The bundles end in a sharp tip with a single or few protruding nanotubes at the end.

The bundles can be dispersed in some polar solvents (e.g., ethanol; N,N-dimethylformamide, water) into smaller bundles and eventually into individual nanotubes. Transport measurements on individual small diameter bundles of $MoS_{(2-x)}I_y$ nanotubes suggest that they are metallic for transport along the tube axis [22].

2.1 Experimental Setup

The measurements were performed in a FE microscope (FEM), where the emission pattern was observed on a 22-mm-diameter aluminized luminescent screen. The screen material (designated as P-43) is composed of ultrafine particles of gadolinium oxysulfide, activated with terbium. After deposition of the luminescent material, the screen was aluminized, forming a point-to-plane geometry with the cathode pin. The distance from the pin to the screen was set to 5 mm. The base pressure below 10^{-7} mbar was obtained readily by a turbomolecular and ion-getter pump within a few hours. The FE measurements were performed under continuous bias conditions at room temperature at high voltage supplied by a stabilized voltage supply (EMI PM 28B) ranging from 0.1 to 2.8 kV or a custom designed supply for potentials up to 5 kV. The emission current was recorded by an ammeter (Keithley K 485). A 33 MΩ series resistor was used to limit the current. This combination allowed us to measure the time-averaged emission current versus voltage and the current stability at the set voltage. The photo of the setup is represented in Figure 2a and the schematic presentation in Figure 2b.

a) b)

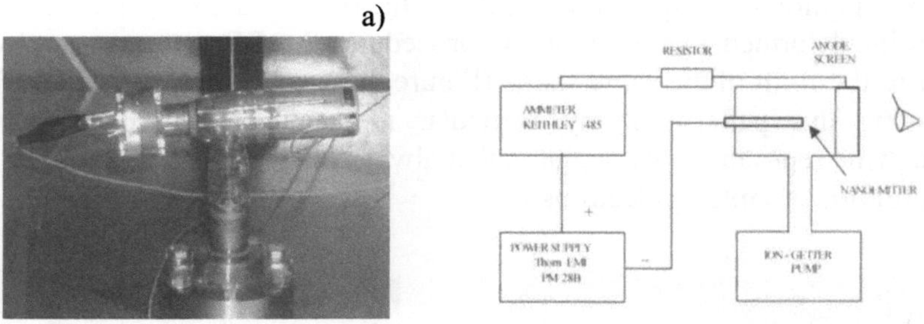

Figure 2 (a) The photo of the setup is represented (b) The schematic presentation of the setup.

3.RESULTS

In order to compare FE data with published results for other materials, we can either: (a) measure the field emission from an individual site or (b) measure the average current from a well-defined and sufficiently large emitting area. The latter requires a sizeable amount of material and a suitably reproducible deposition method, which is quite difficult to achieve.

This was the reason that in the first step of measurements only the observation and preliminary FE measurements from a large surface but with unknown number of sharp emitters were performed. First results in this initial step when the emission was observed from several bundles (global monitoring) showed that the current was some tens of μA at 2.7 kV and

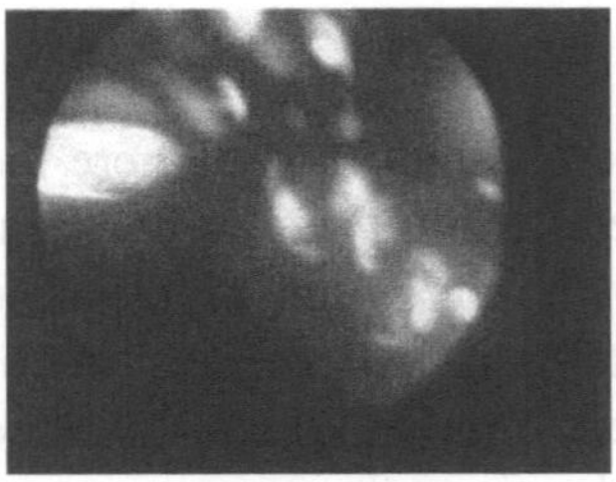

Figure 3 First results in this initial step when the emission was observed from several bundles (global monitoring) caused an interesting image on the FEM screen

caused an interesting image on the FEM screen (Figure 3).

All further measurements were made with single bundles, which were supposed to end into a few FE tips. This allowed more quantitative results to be obtained. As-grown $MoS_{(2-x)}I_y$ material was first dispersed in absolute ethanol using an ultrasonic bath, and the individual bundles were then glued with the silver conducting paste (Flash-Dry^{TM} Paint, SPI) to an aluminized 5–10 μm diameter glass fiber. The assembly was then glued with the same silver paste to the metal pin. In all our experiments, the field-emission current, originating from a few sites, prevailed, which enabled us to reproducibly measure its emission current characteristics. A scanning electron microscope (SEM) image of a typical FE tip assembly, prior to mounting into the FEM, is shown in Figure 4a. In spite of the fact that the bundles are very malleable and are easily deformed in the mounting procedure, the FE pattern was always found in the form of the same shape (Figure 4b). At present, we refrain from attributing this pattern to any particular molecular orbital configuration, although the reproducibility suggests that always the same nanotube-end configuration is emitting electrons.

a) b)

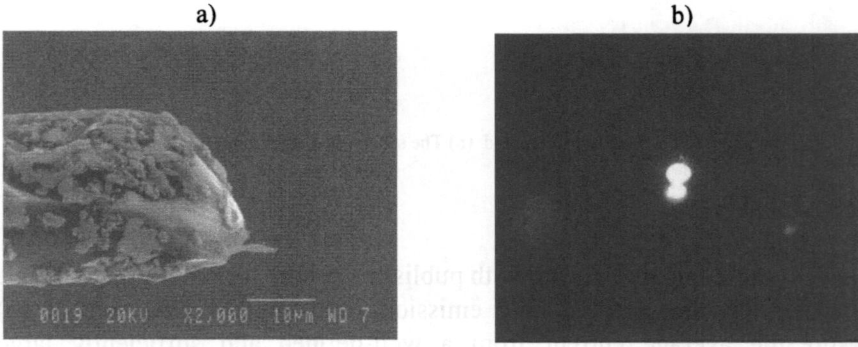

Figure 4 (a) A SEM image of a bundle of $MoS_{(2-x)}I_y$ on the fiber prior to mounting into the FEM. (b) Typical sequence of field emission patterns from a single bundle on the FEM screen.

The field emission characteristics (*I–U*) of the same sites from Figure 4b are shown in Figure 5a. The corresponding Fowler–Nordheim (F–N) plots are shown in Figure 5b. Each point in the diagram represents the current averaged over 15 s. We omitted the low-field region where the FE current cannot be clearly distinguished from the leakage current. The leakage current, which may limit the accuracy of the measurements at low FE currents, was determined by reversing the polarity of the FEM terminals, and was found to be between 10 – 20 pA at 4.5 kV. A small hysteresis was observed when comparing measurements in increasing and decreasing current modes at start and after the initial stability measurement at the fixed voltage U = 2800V. The improvement after 17 hours is in accordance with the graph given in Figure 6.

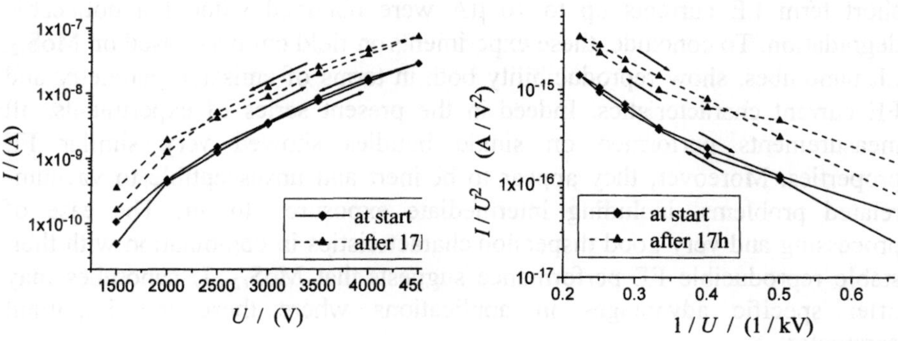

Figure 5 (a) FE characteristics (I-U) of the same sites as shown in Figure 5(b) at start and after the initial 17 hours stability measurement. The corresponding Fowler-Nordheim plot shows that the FE current fits to the standard model, usually used to describe field emission.

The low-field region is frequently misinterpreted in terms of a "low emission threshold," whereas in fact the leakage current is being measured. For viability in commercial applications, the high current region is important, with the requirement for *stable* emission under appropriate vacuum conditions. We present a typical initial stability measurement of the emission current from a single bundle in Figure 6. The voltage was fixed to 2800 V. The current fluctuations were calculated to lie within 6 % during 15 h after the burn-in phase (2 h).

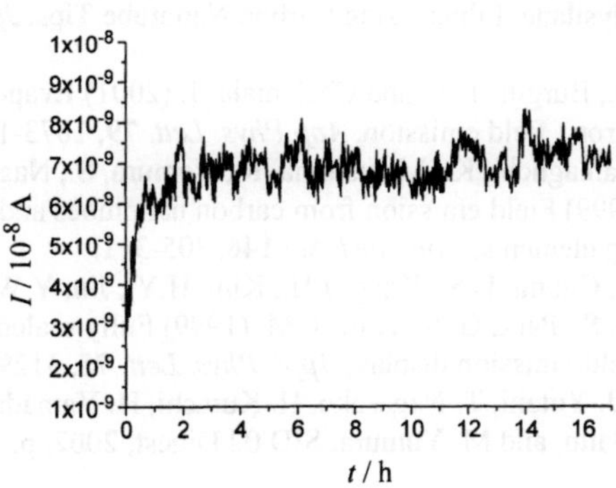

Figure 6 The initial stability of the emission current from a single bundle.

Although the current cycling experiments were performed up to 80 nA, short term FE currents up to 10 μA were obtained from individual bundles *without a noticeable degradation* of the FE performance. Because of the risk of damage to our measuring instruments and the P43/aluminized anode above 10 μA, we did not test the FE tips for burn-out.

4. CONCLUSIONS

The initial investigations of FE properties of $MoS_{(2-x)}I_y$ nanotubes show that they have very promising FE properties. Although the emission current at cycling experiments from individual sites reached up to 80 nA at a macroscopic field 0.9 $V\mu m^{-1}$ and the emitter-to-anode distance equal to 5 mm, short term FE currents up to 10 μA were obtained without a noticeable degradation. To conclude, these experiments on field emitters based on $MoS_{(2-x)}I_y$ nanotubes, show reproducibility both in terms of emission geometry and FE current characteristics. Indeed in the present series of experiments, all measurements performed on single bundles showed very similar FE properties. Moreover, they appear to be inert and unsusceptible to vacuum-related problems, including intermediate exposure to air. The ease of processing and very good dispersion characteristics in combination with their stable reproducible FE performance suggests that $MoS_{(2-x)}I_y$ nanotubes may offer specific advantages in applications where these are important parameters.

References

1. Baughman, R., Zakhidov, A. A. and de Heer, W. A. (2002) Carbon Nanotubes-the Route Toward Applications, Science **297**, 788-792.
2 Bonard, J.-M., Kind, H., T. Stockli T. and Nilsson L.A. (2001) Field emission from carbon nanotubes: the first five years, *Solid State Electron.* **45**, 893-914.
3. N. S. Lee, D. S. Chung, J. H. Kang, H. Y. Kim, S. H. Park, Y. W. Jin, Y. S. Choi, I. T. Han, N. S. Park, M. J. Yun, J. E. Jung, C. J. Lee, J. H. You, S. H. Jo, C. G. Lee, and J. M. Kim, Jpn. (2000) *J. Appl. Phys.,* Part **1 12B**, 7154.
4. Rosen, R., Simendinger, W., Debbault, C., Shimoda, H., Fleming, L., Stoner, B. and O. Zhou (2000) Application of carbon nanotubes as electrodes in gas discharge tubes, *App.Phys. Lett.* **76**, 1668-1670.
5. Okazaki, A., Akita, S., Nishijima, H. and Y. Nakayama, (2000) Nanolithography of Organic Polysilane Films Using Carbon Nanotube Tips, *Jpn. J. App.Phys.Part 1* **39B**, 3744-3746.
6. Dean, K. A., Burgin, T. P. and Chalamala B. (2001) Evaporation of carbon nanotubes during electron field emission, *App.Phys. Lett.* **79**, 1873-1875.
7. Saito, Y., Hamaguchi, K., Mizushima, R., Uemura, S., Nagasako, T., Yotani, J. and T. Shimojo (1999) Field emission from carbon nanotubes and its application to cathode ray tube lighting elements, *App. Surf. Sci.***146**, 305-311.
8. Choi, W. B., Chung, D.S., Kang, J.H., Kim, H.Y., Jin, Y.W., Han, I. T., Lee, Y. H., Yung, J.E., Lee, N. S., Park, G. S., Kim, J. M. (1999) Fully sealed, high-brightness carbon-nanotube field-emission display, *Appl. Phys. Lett.* **75**, 3129-3131.
9 S. Uemura, J. Yotani, T. Nagasako, H. Kurachi, H. Yamada, T. Ezaki, T. Maesoba, T. Nakao, Y. Saito, and M. Yumura, SID 02 Digest, 2002, p. 1135.

10. Bandis, C., Pate, B. B. (1996) Simultaneous field emission and photoemission from diamond, *App.Phys. Lett.* **69**, 366-368.
11. Zhang, Y. F., Tang, Y. H., Wang, N., Yu, D. P., Lee, C. S., Bello, I. and Lee, S. T. (1998) Silicon nanowires prepared by laser ablation at high temperature, *App.Phys. Lett.* 72 , 1835-1837.
12. Sharma, R. B., Tondare, V. N., Joag, D. S., Govindaraj A., and. Rao, C. N. R (2001) Field emission from carbon nanotubes grown on a tungsten tip, *Chem. Phys. Lett.* **344**, 283-286.
13. Rangelow, I.W., Shi, F., Hudek, P., Grabiec, P., Volland, B., Givargizov, E.I., Stepanova, A.N., Obolenskaya, L.N., Mashkova, E.S. and Molchanov V.A. (1998) Micromachined ultrasharp silicon and diamond-coated silicon tip as a stable field-emission electron source and a scanning probe microscopy sensor with atomic sharpness , *J. Vac. Sci.Technol.* **B 16** , 3185-3191.
14. McCarson, B. L., Schlesser, R., McClure, M. T., and Sitar Z.(1998) Electron emission mechanism from cubic boron nitride-coated molybdenum emitters, *App.Phys. Lett.* **72**, 2909-2911.
15. Bai, X. D., Guo, J. D., Yu, J., Wang, E. G., Yuan, J. and Zhou, W. Z. (2000) Synthesis and field-emission
behavior of highly oriented boron carbonitride nanofibers, *App.Phys. Lett.* **76**, 2624-2626.
16. Lee, D. G., Baik, D. K., Kang, N. S., Cho, W. K., Yoon, S. J., Kim, T. Y., Hwang, H. D., Ahn, D. H. and Park, M. H. (2000) Fabrication of volcano-type TiN field emitter arrays, *J. Vac. Sci. Technol. B* 18., 1085-1088.
17. M. Endo, H. Nakane, and H. Adachi, (2001) *J. Vac. Sci. Technol. B* **14**, 2114-2118.
18. Hwa-Mok Kim, Kang, T. W.,Chung, K. S., Hong J. P., and Choi W. B.(2003) Field emission displays of wide-bandgap gallium nitride nanorod arrays grown by hydride vapor phase epitaxy, *Chem. Phys. Lett.* **377**, 491-494.
19. Zach, M. P., Kwok H. Ng. Penner R. M. (2000) Molybdenum Nanowires by Electrodeposition, *Science* **290**, 2120-2123.
20. Zhirnov, V., Lizzul-Rinne, C., Wojak, G. J.,. Sanwald, R. C. and Hren, J. J. (2001) "Standardization" of field emission measurements, *J. Vac. Sci. Technol. B* **19**, 87-93.
21. Remskar, M., Mrzel A., Skraba, Z., Jesih, A., Ceh, M., Demsar, J., Stadelmann, P., Levy F. and Mihailovic D (2001) Self-Assembly of Subnanometer-Diameter Single-Wall MoS_2 Nanotubes, *Science* **292**, 479-481.
22. Remskar, M., Mrzel, A., Sanjines, R., Cohen, H. and Levy F. (2003) Metallic Sub-Nanometer $MoS_{2-x}I$ Nanotubes, *Adv. Mate.* **15**, 237-240.

X-RAY SPECTROSCOPY CHARACTERIZATION OF CARBON NANOTUBES AND RELATED STRUCTURES

A.V. OKOTRUB, A.V. GUSEL'NIKOV AND L.G. BULUSHEVA
Nikolaev Institute of Inorganic Chemistry SB RAS, pr. Ak. Lavrentieva 3, Novosibirsk 630090, Russia

Abstract: X-ray emission spectroscopy was applied for revealing the peculiarities in the electronic structure of different carbon nanotubes. The CKα-spectra of graphite and arc-produced multiwall carbon nanotubes (MWNTs) were shown to be similar in appearance indicating essentially no difference in the electronic state of carbon atoms constituting both materials. Slight changes detected in the spectrum of arc-produced single-wall carbon nanotubes (SWNTs) were attributed to the π-system perturbation due to curvature of small diameter graphite shells. The CKα-spectrum of catalytic MWNTs showed the enhanced density of high-energy states that by the results of quantum-chemical calculations on various tube models could be caused by development of vacancies in the tube walls. The angular dependence of X-ray fluorescence from the samples of the oriented MWNTs was compared with that measured for graphite. The dependence can be used for qualitative characterization of tube alignment and definition of the width of occupied π-band and defect structure of graphitic materials.

Key words: Carbon nanotubes; graphite; X-ray fluorescent spectroscopy; electronic structure; anisotropy of chemical bonding.

1. INTRODUCTION

Morphology and atomic arrangement of carbon nanotubes are mainly determined by synthetic conditions [1]. The most perfect multiwall carbon nanotubes (MWNTs) are produced using the arc-discharge evaporation of graphite in an inert atmosphere [2]. The nanotubes formed in the cathode deposit have a diameter of 5–15 nm and free from catalyst. Inserting the different catalysts into electric arc makes it

S. Guceri et al. (eds.), Nanoengineered Nanofibrous Materials, 349-362.

possible to prepare single-wall carbon nanotubes (SWNTs) with varied diameter [3]. Chemical vapor deposition (CVD) of hydrocarbons over metallic catalysts yields the carbon nanotubes and filaments with diameter changed from 1 to 100 nm depending on the synthetic technique, temperature, kind and size of catalyst [4]. Relatively mild conditions of CVD method can result in imperfections of tube arrangement. Really, the catalytic carbon nanotubes are often bent and curved, have compartments (bamboo-shaped tubes) and layers oriented at an angle to the tube axis (fishbone packing).

Defect structure of carbon nanotubes is revealed using various structural (electron microscopy, X-ray diffraction, etc.) and spectroscopic (Raman and X-ray spectroscopy) methods. X-ray spectroscopy is one of the direct methods for probing the electronic structure of solids [5]. X-ray emission spectroscopy indicates features of the occupied electronic states while X-ray absorption spectroscopy gives information about the unoccupied states. Carbon atoms formed a perfect hexagonal network and those constituted a defect should be different in the electronic state and increase of the defect portion in a material will have stronger effect on its electronic structure. The measured energy and density of electronic states can be directly related with the results of quantum-chemical calculations. The comparison between experimental and theoretical data can be considered as the test on the applicability of the concrete quantum-chemical method to the investigated objects. From the other side, the quantum-chemical modeling itself can explain the many peculiarities existing in the X-ray spectra of the real samples. The plotting of the theoretical spectra for proposing models and the comparison of them with experiment allow choosing the most appropriate type of defects and estimating defect concentration.

With investigation of the electronic structure of carbon nanotubes, graphite could be a reference compound. Chemical bonding of perfect graphite layer has clearly defined anisotropic character caused by two-dimensional structure of solid. The measurement of angle dependence of X-ray fluorescence from high quality graphite single-crystals allows separation of σ- and π-bands [6]. Namely the distribution of π-electrons should be more sensitive to the imperfections in the graphite structure. It is very interesting the experimental definition of σ- and π-bands in carbon nanotubes that could be possible only for the samples having a certain orientation of tubes. However the recently attempted measurement of angle dependence of near-edge X-ray absorption of catalytic MWNTs had no advances [7].

The aim of the present work is demonstration of possibilities of ultra-soft X-ray spectroscopy for structural characterization of carbon nanotubes and graphite and revealing the electronic state of defects.

2. EXPERIMENTAL AND CALCULATION DETAILS

2.1. Synthesis of the Samples

The apparatus for arc-discharge graphite evaporation is described elsewhere [8]. The cathode deposit was produced by a simultaneous evaporation of seven graphite rods in helium gas of 800 Torr. The d.c. arc current was typically 1000 A at 35–40 V. The inner part of the deposit had a laminated texture and as was estimated by transmission electron microscopy (TEM) to consist of 80% MWNTs [9]. The soot containing SWNTs was prepared by arc-discharge evaporation of a rod filled by a mixture of graphite with 2.5 wt% Co and 2.5 wt% Ni [10]. To purify SWNTs the soot was heated in nitric acid.

The catalytic MWNTs were produced by decomposition of acetylene at 700°C over 2.5 wt% Co - 2.5wt% Fe / NaY zeolite catalyst. After reaction, the catalyst was removed by repeated dissolution in hydrofluoric acid. Since the nanotubes contained some pyrolytic carbon, it was eliminated by oxidation using $KMnO_4$ in diluted sulphuric acid solution at room temperature [11]. The tube annealing was carried out in argon atmosphere at 1000°C.

The films of vertically aligned nanotubes on silicon and quartz substrates were prepared by thermal catalytic chemical vapour decomposition (CCVD) of ferrocene/ fullerene mixture [12]. The horizontal CCVD apparatus consists of a stainless steel gas flow reactor of 1 m length and 3.4 cm diameter and a tubular furnace with a heating length of 30 cm. A ceramic boat with a mixture of fullerene C_{60} and ferrocene $Fe(C_5H_5)_2$ taken in the ratio of (1:1) was placed inside the quartz tube under the silicon (quartz) plate of size 10×10 mm. The pyrolysis was performed at 950°C and atmospheric pressure in an argon flow (3 l/min).

The micrographs of the products were obtained with a JSM-T200 scanning electron microscope (SEM) and a JEM-2010 transmission electron microscope.

2.2. Measurement of X-Ray Spectra

CKα-spectra of the carbon nanotube and graphite samples were recorded with a laboratory X-ray spectrometer «Stearat» using ammonium biphthalate (NH_4AP) single crystal as a crystal-analyzer. This crystal has nonlinear reflection efficiency, which is corrected by the procedure described elsewhere [13]. Angle dependence of X-ray emission spectra of graphite and MWNTs was measured at different orientations of sample surface to the take-off θ direction of the emitted radiation. An angle between exiting beam and emitting one was constant and equal to 70°. The sample of arc MWNTs was evenly abraded on the polished copper support. The film of aligned MWNTs was forced against the polished copper support by a nickel mask. The samples were cooled down to liquid nitrogen temperature in the vacuum chamber of the X-ray tube with copper anode (U=6 kV, I=0.5 A). Determination accuracy of X-ray band energy was ±0.15 eV with spectral resolution of 0.5 eV. The spectra were normalized at the maximal intensity.

2.3. Computation

The quantum-chemical calculations of carbon models were carried out using the semi-empirical AM1 method within the GAMESS package [14]. The geometry of ideal (8,8) tube and bent tube was optimized by standard BFGS procedure to the gradient value of 10^{-4} Ha/Bohr. X-ray transition intensity was calculated by summing the squared coefficients with which carbon 2p atomic orbitals (AOs) involved in the concrete occupied molecular orbital (MO). The energy location of intensity corresponded to the MO eigenvalue. Calculated intensities were normalized by maximal value and broadened by convolution of Lorenzian functions with half width at half maximum (HWHM) of 0.5 eV.

3. RESULTS AND DISCUSSION

3.1. Electronic Structure of Diverse Carbon Nanotubes

TEM pictures of the measured carbon materials are shown in Fig. 1. The arc-produced SWNTs having diameter of 1.3 nm form long ropes (Fig. 1(a)). The sample heated in acid still contains some amount of amorphous carbon and metal particles. The material from the inner part of cathode deposit consists of MWNTs and multiwall polyhedral particles (Fig. 1(b)). The outer tubes diameter is in the range of 10–15 nm. The arc-produced MWNTs have well graphitized shells. The catalytic MWNTs with diameter of 15–20 nm are characterized by bent and wavy walls (Fig. 1(c)). The tube tips were opened during the oxidation step.

Figure 2 compares the CKα-spectra measured for polycrystalline natural graphite and materials containing diverse carbon nanotubes. The spectra show three main features. The maxima A and C in the graphite spectrum originate from the occupied π- and σ-states respectively. Both types of states are involved in the formation of shoulder B [15]. The spectra of graphite and multiwall carbon nanoparticles from the cathode

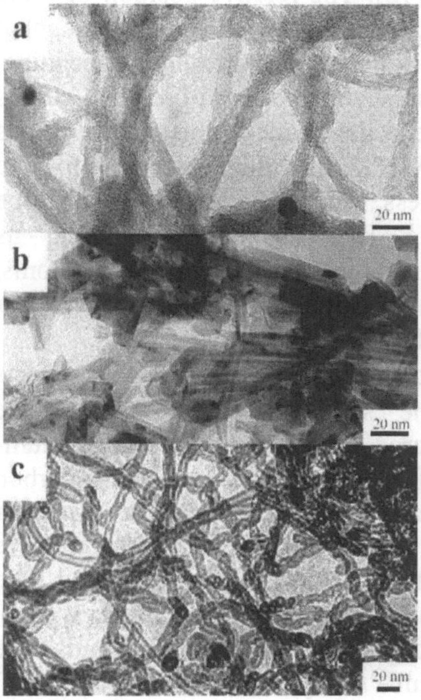

Figure 1. TEM pictures of SWNTs (a) and MWNTs (b) prepared in electric arc and catalytic MWNTs (c).

deposit are in good agreement (Fig. 2(a, b)). The typical arc-produced MWNTs are straight with the defects concentrated close to the tips mainly. Hence, the portion of defects in such tubes may be assumed to be about the same as that in natural graphite, the boundary grain atoms in which are different from the bulk atoms in the electronic state.

The CKα-spectrum of arc-produced SWNTs (Fig. 2(c)) is somewhat distinguished from the top spectra by intensity and structure of main features, namely, the maxima C and A are broadened and the latter maximum is slightly increased. The broadening of the maximum C is likely to be due to the residual contaminations in the sample. The changes in the maximum A can be attributed to the redistribution of π-electrons with rolling of graphene sheet into a small diameter cylinder [16]. The recently performed measurements on the SWNTs produced by laser vaporization method support this conclusion. The comparison between CKα-spectra of graphite and purified SWNTs in 'bucky paper' had detected the high-energy maximum in the latter spectrum was slightly broadened and increased in relative intensity by about 10% [17].

CKα-spectrum of catalytic MWNTs is significantly different from other spectra by increase of intensities of the features B and A (Fig. 2(d)). As π-electrons participate in origin of both these features the valence band of catalytic tubes is characterized by rising density of π-states that may be connected with high portion of defects breaking the uniformity of hexagonal carbon network.

Three kinds of defects can be generated in the graphitic shells of carbon nanotubes: topological defects, sp^3-hydridized carbon atoms, and incomplete bonding [1]. Topological defect in the graphite layer is a ring, which is different of hexagon. Occurrence of pentagons and heptagons gives a clearly visible effect on tube structure, namely, tube

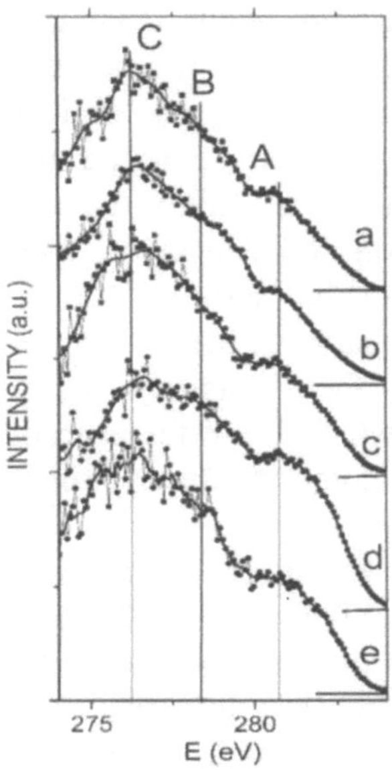

Figure 2. CKα-spectra measured for polycrystalline graphite (a), arc-produced MWNTs (b), arc-produced SWNTs (c), catalytic MWNTs (d), and catalytic MWNTs annealed at 1000°C (e).

bend [18]. The relative position of these rings in the otherwise perfect hexagonal network determines the angle of bending. The sp^3-hybridization of carbon atoms can be realized due to intershell linkages in MWNTs that might account for the waviness and segmentation of tube layers. The incomplete bonding is attributed to the edge carbon atoms, which constitute the open tube ends or boundary of vacancies.

To check the effect of various defects on the shape of CKα-spectrum we compared the results of quantum-chemical calculations of ideal carbon tube and defective ones. The armchair (8,8) tube with 1.1 nm diameter was chosen as a model. The calculated tube structures are shown in Fig. 3 (right part). The ideal (8,8) tube has closed ends. The bent tube is a junction of armchair (8,8) and zigzag (14,0) tubes realized by insertion of pentagon and heptagon in diametrically opposite position. The sp^3-hybridized carbon atoms link the inner and outer shells of double (4,4)@(8,8) tube [19]. Geometry of the double tube fragment was relaxed by the molecular mechanic MM+ force field. The optimized length of the bonds between sp^3-hybridized atoms is equal to 1.61 Å, the maximal intershell spacing is about 2.92 Å, that is somewhat different from that in MWNTs. The dangling bonds at the ends of bent tube and double tube fragments were saturated by hydrogen atoms. The vacancies in the (8,8) carbon nanotube were generated by removing of two neighboring atoms (two-atomic vacancy). Theoretical CKα-spectra calculated for the tube models are compared in Fig. 3 (left part). The spectrum of ideal (8,8) tube shows three features A, B, C (Fig. 3(a)) with relative intensity and separation being in well agreement with those of the arc-produced MWNTs spectrum (Fig. 2(b)). Slightly enhanced intensities of the features A and B in the theoretical spectrum could be attributed to rather small diameter of the calculated tube.

The density of 2p-electron states for carbon atoms composing the pentagon-heptagon defect in the bent (8,8)–(14,0) tube is presented in Fig. 3(b). Compared with the spectrum of ideal tube, the spectrum of the topological defect is characterized by increase of intensity of the

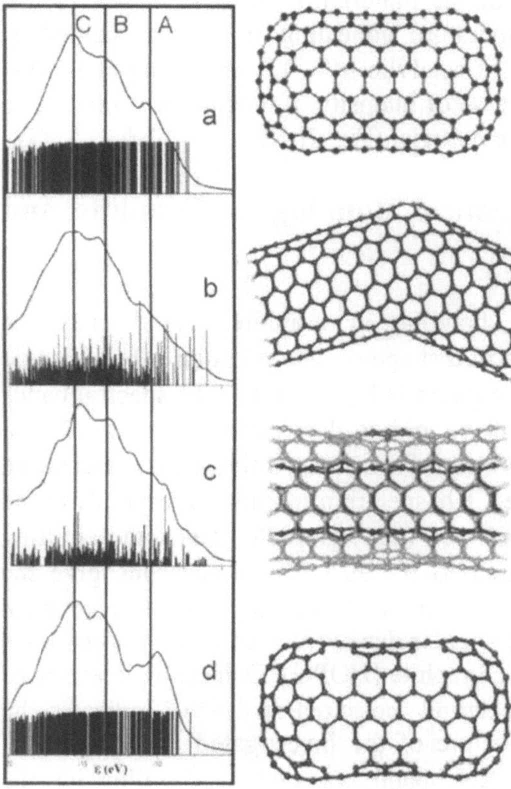

Figure 3. Theoretical CKα-spectra calculated for ideal (8,8) tube (a), pentagonal and heptagonal rings in bent (8,8)-(14,0) tube (b), marked hexagon from the outer shell of double (4,4)@(8,8) tube (c), and tube with two-atomic vacancies (d). The dangling edge bonds of bent and double tube fragments are saturated by hydrogen atoms.

shoulder B and almost the same relative intensity of the maximum A. It is believed that occurrence of pentagons and heptagons in the carbon nanotube structure cannot provide high density of weekly bonding π-states. The sp^3-hybridized atoms have a perturbing action on the π-system of graphite shell that causes a pronounce enhancement of both high-energy features, A and B, in the density of states calculated for the atoms composing a convexity of the outer shell of double (4,4)@(8,8) tube (Fig. 3(c)). Nevertheless, incorporation of such kind of defects into the tube construction cannot satisfactory explain the features in the spectrum of catalytic MWNTs. Theoretical spectrum plotted for the (8,8) tube with vacancies shows a considerable rise of the maximum A (Fig. 3(d)). The changes in the spectrum of ideal (8,8) tube with producing of vacancies have a tendency similar to that revealed at the comparison between CKα-spectra of arc-produced and catalytic MWNTs. Increase of density of high energy states is caused by contribution of electrons of dangling bonds at the vacancy boundaries. The vacancy can be developed during the synthesis in the case of deficient in carbon atoms for perfect network formation or with the purification of nanotube containing material. Healing of vacancies can be achieved with high temperature annealing of carbon nanotubes. Actually, the CKα-spectrum of catalytic MWNTs annealing at 1000°C (Fig. 2(e)) demonstrates decrease of intensity of maximum A relative to that in the spectrum of pristine material (Fig. 2(d)).

3.2. Anisotropy of Chemical Bonding in Graphite and Carbon Nanotubes

Anisotropy of chemical bonding in graphite crystal causes a dependence of the CKα-spectrum shape on the radiation take-off angle θ measured with respect to the graphite layer planes. The spectral bands originated from transition of π- and σ-electrons exhibit a different angular dependence that provides information on the spatial orientation of C2p-orbitals. The graphite CKα-spectrum mainly corresponds to the σ-subband when the angle between (001) graphite planes and emitting fluorescence direction is equal to θ = 90°. The π-electrons give the greatest contribution to the spectrum with the angles θ close to zero. Figure 4 (a) demonstrates the angular dependence of X-ray fluorescence from highly oriented pyrolytic graphite (HOPG). Detectable intensity of the feature A in the CKα-spectrum measured at θ = 90° indicates the imperfection and mosaic structure of the investigated HOPG sample. Separation of the graphite π-subband was achieved following the procedure described in [20]. The spectral intensity measured at θ = 90°

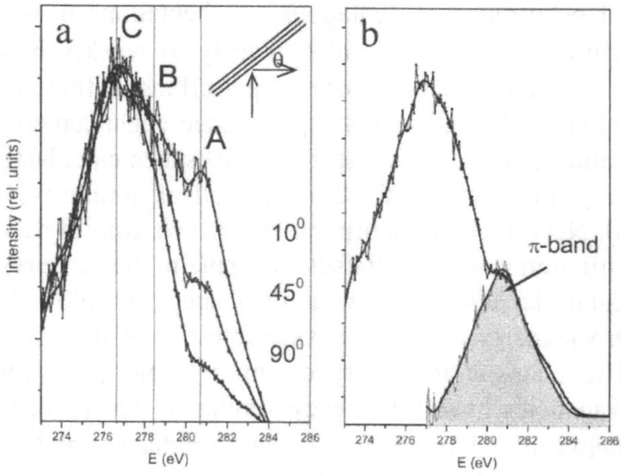

Figure 4. CKα-spectra of HOPG measured for the radiation take-off angles θ = 10°, 45°, and 90° (a). The scheme shows relation between θ and graphite layers plane. Selection of π-band in the HOPG spectrum measured at θ = 45° (b).

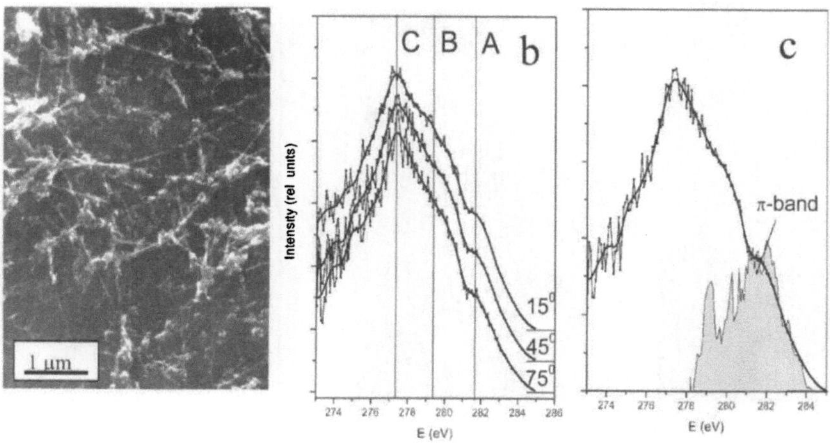

Figure 5. STM picture of inner part of the cathode deposit (a) shows predominant orientation of carbon nanotubes in a layer. CKα-spectra of arc-produced MWNTs measured for the take-off angles θ = 15°, 45°, and 75°. Selection of π-band in the carbon nanotube spectrum measured at θ = 45° (c).

was subtracted from that obtained for θ = 10° and than the resultant difference was fitted by intensity to the short-wave side of the CKα-spectrum measured at θ = 45° (Fig. 4(b)). The distribution of occupied π-electrons in HOPG has almost symmetrical shape and width about 7 eV.

Carbon nanotubes, the walls of which are represented by the rolled graphene sheets, have the chemical bonding anisotropy also. Radial directed carbon AOs are analogous to the π-orbitals in graphite. The AOs, which are tangential to a tube surface, are σ-like orbitals. Because of a cylindrical, nonplanar, surface of tube shells, the observed angular dependence of X-ray emission for the carbon nanotubes sample should be smaller than that for the graphite. The reason is the angular independence of X-ray emission intensity when the radiation is perpendicular to the graphitic cylinder surface.

Optimization of conditions of arc-discharge graphite evaporation, particularly configuration of anode and value of helium pressure, had lead to the predominant orientation of carbon nanotubes in the layers of material deposited onto cathode [9]. SEM image of the material presented in Fig. 5(a) shows the bundles of MWNTs. The sample for the angular measurements was mounted by such way that MWNTs were mainly adhesive parallel to the support surface with even azimuthal distribution. The CKα-spectra were obtained for three orientations of the sample surface θ = 15°, 45°, and 75° with respect to the take-off direction of the emitted radiation (Fig. 5(b)). The principal distinguish between the spectra is a change of relative intensity of the features A

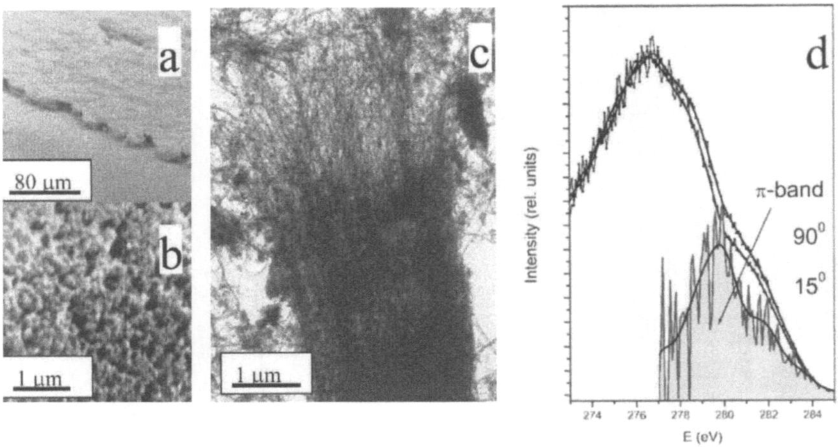

Figure 6. SEM images of carbon nanotube forest grown on a Si substrate: side view (a) and top view (b). TEM picture of a bundle splitting into individual carbon nanotubes (c). CKα-spectra of the film measured at two take-off angles 15° and 90° of X-ray fluorescence (d). The π-band was obtained as difference of the spectra.

and B. Both these features are formed with the participation of π-electrons, as the maximum C corresponds to the σ-electrons only [21]. The angular dependence of X-ray fluorescence for arc-produced MWNTs is considerably less than that obtained for HOPG. In addition to the cylindrical symmetry effect, the other reason of reduced variation in the high energy spectral intensity is poor alignment of MWNTs. Distribution of π-electrons in the valence band of arc-produced MWNTs was obtained by subtraction of the CKα-spectrum taken at $\theta = 75°$ from the spectrum obtained for $\theta = 15°$. The resultant intensity was normalized to the 281 eV feature of the CKα-spectrum measured at $\theta = 45°$ (Fig. 5(c)). The obtained π-electrons distribution is characterized by large fluctuations that could be due to week interactions between carbon nanotubes in the samples. Compared to HOPG, the π-band of arc-produced MWNTs is slightly wider and has an asymmetric shape. The asymmetry of π-band in the low-energy spectral region arises from high intensity of the feature B indicating rather perfect arrangement of graphitic shells of arc-MWNTs.

One of the simplest methods for preparation of MWNTs with good alignment is CCVD technique [22]. Figure 6(a) shows a carbon nanotubes forest grown on a Si substrate. The tubes are closely packed and vertically aligned in the sample. The SEM image of the film surface demonstrates rather uniform distribution of the tube tops (Fig. 6(b)). The TEM picture reveals the MWNTs being about 100 Å in diameter form ropes (Fig. 6(c)). The CKα-spectra of the sample produced measured at two take-off angles θ = 15° and 90° of X-ray fluorescence are shown in Fig. 6(d). The spectra have the dominant maximum around 277 eV, the shoulder at 278.5 eV, and the less intense maximum at 281 eV, which are close in energy to main features in the CKα-spectra of graphite and arc-produced MWNTs. The enhanced intensity of the 278.5 eV shoulder in the measured spectra compared to the spectrum of arc-produced MWNTs might be caused by an additional contribution from the iron carbide Fe_3C [23]. The principal distinguish between two measured spectra is a change of relative intensity of the high-energy features. In contrast to the angular dependence obtained for the CKα-spectra of HOPG and arc-produced MWNTs the spectrum taken at θ = 90° exhibits the higher intensity of the features. The inverse angular dependence of X-ray emission from CCVD carbon nanotubes is due to their perpendicular deposition to the support. The smaller variations in the CKα-spectra of vertically aligned MWNTs compared to those observed for other graphitic samples could be due to larger defectness of the CCVD tubes. Really, the π-band being a difference between CKα-spectra taken for θ = 90° and 15° is considerably different from the π-bands of HOPG and arc-produced MWNTs. The π-band of the CCVD tubes exhibits a new feature around 281.5 eV indicating lowering of density of weekly bonding π-states. The reason is most likely to be disruption of the hexagonal carbon network uniformity.

The comparison of π-systems selected from the angular dependence of X-ray fluorescence measured for graphitic-like materials showed that the arc-produced MWNTs are characterized by most perfect atomic arrangement in the considered series. The density of π-occupied states of arc-MWNTs has maximum around 282 eV while this value is equal to 280.8 and 279.8 eV for HOPG and CCVD MWNTs respectively. The detected shift toward the low-energy region indicates increase of π-electrons bonding that could be caused by decrease of the π-system size. Really, the defects in arc-produced MWNTs are mainly concentrated at the tubes ends and, hence, the perfect shells areas should be compared with the average tubes length (~1000 nm). The investigated HOPG sample was estimated from X-ray diffraction measurements to consist of the graphitic crystallites with about 100 nm dimension. Finally, the layers of CCVD MWNTs often have the fishbone packing that can reduce the size of graphitic fragments up to 10 nm.

4. CONCLUSION

X-ray fluorescent spectroscopy is one of the direct methods for probing the electronic structure of solids. Furthermore, the measurements of angular dependence of X-ray fluorescence from anisotropic solids allow experimental definition of π-system. We applied the ultra-soft X-ray emission spectroscopy for investigation of carbon nanotubes prepared in different synthetic conditions. CKα-spectra of arc-produced MWNTs and SWNTs were shown to be similar to the spectrum of natural graphite. High temperatures of the electric arc provide generation of the perfect graphitic shells and the portion of topological defects needed for tube closing is comparable with that in polycrystalline graphite, which boundary grain atoms are differed by electronic states from the central atoms. Slight change in intensity detected for the high-energy region of the SWNTs CKα-spectrum was attributed to the π-system perturbation with formation of small diameter graphitic cylinder. The processes of decomposition of hydrocarbons over metallic nanoparticles proceed at rather mild conditions that often results in tube bending and coiling, formation of compartments and poor layer graphitization. Imperfection in the tubes arrangement led to considerable enhancement of density of high-energy occupied states in the CKα-spectrum of catalytic MWNTs. Three tubular models involving pentagon-heptagon pair, sp^3-hybridizated carbon atoms, and incomplete bonds were calculated using a quantum-chemical method for revealing the effect of different kinds of defects on the density of occupied states. Comparison of the CKα-spectra plotted by the results of defective structures calculations with the spectrum of ideal tube showed only vacancies can provide the spectral changes observed in the experiment. Electrons of dangling bonds at the vacancy boundaries have almost the same energy as the localized π-states that gives essential addition to the intensity of high-energy maximum of the theoretical spectrum. The sp^3-defects, which could be realized as the linkages between tube shells or as the hydrogenated carbon atoms, increase the π-electrons localization that however is insufficient for the experiment interpretation. The topological defects were found to produce no gain in the intensity of high-energy occupied states.

The angular dependence of CKα-spectra was compared for HOPG and MWNTs produced by arc-discharge and CCVD methods. The predominant orientation of MWNTs in a certain direction was achieved directly in the synthesis. The measurements of X-ray fluorescence for different take-off angles were shown to provide information about the sample texture, namely, orientation of graphitic layers respect to the substrate surface and layer ordering. The smallest variations in the CKα-spectra observed for the catalytic MWNTs are associated not only with poor alignment of tubes in the sample but with high defect structure of tube shells as well. The systematic low-energy shift of the maximum of density of π-occupied states was detected for arc-produced

MWNTs, HOPG, and catalytic MWNTs and correlated with the size of graphitic fragments.

Acknowledgments: We thank Dr. A. Fonseca for the sample of catalytically produced nanotubes, Dr. V.S. Danilovich, Dr. A.L. Chuvilin and O.G. Abrosimov for SEM and TEM characterization of samples. The work was financially supported by the INTAS (projects 01-254, 01-822) and the RFBR (grants 03-03-32336, 03-03-32286).

References

1. Ebbesen, T.W., Takada, T. (1995) Topolodgical and sp^3 defect structures in nanotubes, *Carbon* **33**, 973-978.
2. Ebbesen, T.W., Ajayan, P.M. (1992) Large-scale synthesis of carbon nanotubes, *Nature*, **358**, 220-222.
3. Journer, C., Maser, W.K., Bernler, P., Loiseau, A., Lamy de la Chapelle, M., Lefrant, S., Denlard, P., Lee, R., Fisher, J.E. (1997) Large-scale production of single-walled carbon nanotubes by the electric-arc technique, *Nature*, **338**, 756-758.
4. Ivanov, V., Nagy, J.B., Lambin, Ph., Lucas, A., Zhang, X.B., Zhang, X.F., Bernaerts, D., Van Tendeloo, G., Amelinckx, S., Van Landuyt, J. (1994) The study of carbon nanotubules produced by catalytic method, *Chem. Phys. Lett.* **223**, 329-335.
5. Maizel, A., Leonhardt, G., Sargan, R. (1981) *X-ray Spectra and Chemical Binding*, Naukova Dumka Publishers, Kiev.
6. Beyreuther, Chr., Hierl, R., Wiech, G. (1975) Zur elektronenstruktur von graphite and bormitrid, *Berichte der Bunsen-Gesellschaft fur Physikalische Chemie* **79**, 1081-1085.
7. Tang, Y.H., Sham, T.K., Hu, Y.F., Lee, C.S., Lee, S.T. (2002) Near-edge X-ray absorption fine structure study of helicity and defects in carbon nanotubes, *Chem. Phys. Lett.* **366**, 636-641.
8. Okotrub, A.V., Romanov, D.A., Chuvilin, A.L., Shevtsov, Yu.V., Gutakovskii, A.K., Bulusheva, L.G., Mazalov, L.N. (1995) Frame carbon nanoparticles: synthesis, structure and properties, *Phys. Low-Dim. Struct.* **8/9**, 139-158.
9. Okotrub, A.V., Bulusheva, L.G., Romanenko, A.I., Chuvilin, A.L., Rudina, A.L., Shubin, A.L., Yudanov, N.F., Gusel'nikov, A.V. Anisotropic properties of carbonaceous material produced in arc discharge (2001) *Appl Phys A* **71**, 481-486.
10. Okotrub, A.V., Romanenko, A.I., Yudanov, N.F., Bulusheva, L.G., Asanov, I.P., Anikeeva, O.B., Chuvilin, A.L., Boronin, A.I., and Bonard J.-M. (2002) Fluorine effect on the structure and electrical transport of arc-produced carbon nanotubes, in H. Kuzmany, J. Fink, M. Mehring, and S. Roth (eds.), *Structural and Electronic Properties of Molecular Nanostructures,* AIP Conference Proceedings **633**, pp.263-266.
11. Bulusheva, L.G., Okotrub, A.V., Asanov, I.P., Fonseca, A., Nagy, J.B. (2001) Comparative Study on the electronic structure of arc-discharge and catalytic carbon nanotubes, *J. Phys. Chem. B* **105**, 4853-4859.
12. Grobert, N., Hsu, W.K., Zhu, Y.Q., Hare, J.P., Kroto, H.W., Walton, D.R.M., Terrones, M., Terrones, H., Redlich, Ph., Ruhle, M., Escudero, R., Morales, F. (1999) Enhanced magnetic coercivities in Fe nanowires, *Appl. Phys. Lett.* **75**, 3363-3365.

13. Okotrub, A.V., Bulusheva, L.G. (1998) CKα- spectra and investigation of electronic structure of fullerene compounds, *Fullerene Science and Technology* **6**, 405-432.
14. Schmidt, M.W., Baldridge, K.K., Boatz, J.A., Elbert, S.T., Gordon, M.S., Jensen, J.H., Kosoki, S., Matsunaga, N., Nguyen, K.A., Su, S.J., Windus, T.L., Dupluis, M., and Montgomery, J.A. (1993) *J. Comput. Chem.* **14**, 1347-1363.
15. Bulusheva, L.G., Okotrub, A.V., Romanov, D.A., Tomanek D. (1998) Electronic structure of carbon nanotubes, *Phys. Low-Dim. Struct.* **3/4**, 107-134.
16. Okotrub, A.V., Bulusheva, L.G., Gusel'nikov, A.V., Chuvilin, A.L., Fonseca A. (2000) Comparative study of electronic structure of diverse carbon nanotubes, in P.V. Kamat, D.M. Gidi, K.M. Kadish (eds.), *Recent Advances in the Chemistry and Physics of Fullerenes and Related Materials* **10**, 229-235.
17. Eisebitt, S., Karl, A., Eberhardt, W., Fischer, J.E., Sathe, C., Agui, A., Nordgren, J. (1998) Electronic structure of single-wall carbon nanotubes studied by resonant inelastic X-ray scattering, *Appl. Phys. A* **67**, 89-93.
18. Han, J., Anantram, M.P., Jaffe, R.L., Kong, J., Dai, H. (1998) Observation and modeling of single-wall carbon nanotube bend junction, *Phys. Rev. B* **57**, 14983-14989.
19. Bulusheva, L.G., Okotrub, A.V., Fonseca, A., Nagy, J.B. (2001) Electronic structure of multiwall carbon nanotubes, *Synthetic Metals* **121**, 1207-1208.
20. Simunek, A., Wiech, G. (1987) σ- and π-bands of potassium-intercalated graphite C_8K studied by angle-resolved X-ray emission spectroscopy (ARXES), *Solid State Commun.* **64**, 1375-1378.
21. Okotrub, A.V., Bulusheva, L.G., Tomanek, D. (1998) X-ray spectroscopic and quantum-chemical study of carbon tubes produced in arc-discharge, *Chem. Phys. Lett.* **289** 341-349.
22. Huczko, A. (2002) Synthesis of aligned carbon nanotubes, *Appl. Phys. A* **74**, 617-638.
23. Fabian, D.G. (ed.) (1968) *Soft X-ray Band Spectra,* Academic Press, London and New York, 117.

HIGHLY FUNCTIONAL MAGNETIC NANOWIRES THROUGH ELECTRODEPOSITION

N. D. Sulitanu
Department of Solid State Physics, Faculty of Physics, "Al.I.Cuza" University, Iasi 700506, Romania

Abstract: There has been increasing interest in the fabrication of nanostructured magnetic materials because of their unusual properties compared to bulk materials and for their potential applications to ultra-high-density magnetic recording, sensors and other devices. Many different techniques have been employed to produce nanoscaled materials. Among them, electrodeposition is a simple, low-cost and high throughput technique. In this paper, we focus on a particular category of magnetic nanowires and multilayered nanowires fabricated by electrodeposition into templates. The template synthesis has recently proved to be an elegant chemical approach for the fabrication of nanoscale materials and an alternative to sophisticated methods such as molecular beam epitaxy and microlithography. In this method, thin fibers are electrochemically synthesized into nanometer-wide cylindrical pores of a porous material. Interestingly, depending on the nature of the material and on other factors, such as the chemistry of the pore wall, these deposited cylinders may be solid (a nanowire) or hollow (a nanotube). This work is referring to thin fibers of magnetic materials that were electrochemically synthesized within the nanosized pores of nuclear track-etched porous membranes (PET films typically of 5-50 μm thick). Obtained nanowires have a diameter in the range 10-50 nm for a length of the order of 10 μm, and can be composed of a stack of layers of different metals with thickness in the nanometer range (multilayered nanowires). The paper describes the preparation methods, presents structural characterization and magnetic properties of nanowires. Finally, are suggested some possible applications.

1. NANOSCIENCE AND NANOTECHNOLOGY – TOWARDS AN ULTIMATE LIMIT

Nanometer scale structure and nanostructured materials play an increasing pervasive role in a wide range of scientific disciplines ranging from solid state physics through to molecular biology. In response to the ever-increasing demands for high performance data storage media and sensing devices, great effort is currently put into the development and studies of innovative magnetic materials [1-3]. As the sizes of devices are reduced, and as more and more information is packed into smaller and

S. Guceri et al. (eds.), Nanoengineered Nanofibrous Materials, 363-376.

smaller volumes, nanotechnology tends *towards an ultimate limit*, which is set by the *quantum properties* of matter. Just where does that limit lie? How small are the smallest devices which can be made? Should nanoscale devices be built up atom by atom or engineered by refining conventional fabrication techniques? What will be the applications of nanotechnology in computing? in medicine? Will nanotechnology revolutionize our future? Does the nanoscale hold new and fundamental scientific and social challenges? Many of these questions will found its answer in the near future as a result of large contribution of the scientists' cooperative from different preoccupation field: physics, chemistry, materials science and engineering, biology, medicine etc.

Nanofabrication offers the technical ability to fabricate objects with nanometric scale precision having unique optical and electronically properties, and which may be tailored by manipulating the size, the shape, or the composition of the nanostructures. Many of the unusual magnetic properties of the nanomagnets come about by imposing a geometrical restriction on the magnetization. As the size of the nanomagnets becomes comparable to key magnetic length scales, such as the exchange length (in the nm range), or the domain wall width (nm – μm range), the magnetization configuration may indeed be strongly affected. The understanding of how finite lateral size may affect the magnetic behavior of nanostructures is of primary importance from practical and theoretical viewpoints. For example, necessary to achieve ultra high density in magnetic storage media is that the lateral size of the information bits be as small as possible, without though the bit volume reaches the super-paramagnetic limit beyond energy needed to switch the magnetization of a bit becomes less than the thermal energy [1]. Also required is that the transmission region between two bits of opposite magnetization be as narrow as possible, which in continuous thin film media is clearly hindered by the finite width of the domain walls. A promising way to achieve small lateral size, with comparatively large volume and hence high magnetic anisotropy energy of the bits is to realize nanowire arrays.

Magnetic nanowires, fabricated by various methods, represent an important family of magnetic nanostructures. Many different techniques have been employed to produce magnetic nanowires [2, 4, 5]. Among them, electrodeposition is a simple, low-cost and high throughput technique.

In this article, we review the recent investigations as well as some open issues in magnetic nanowires, and arrays of nanowires, in terms of the fabrication technique (electrodeposition) as well as their physical properties. This subject has been briefly reviewed earlier [1-12], and here we present a more exhaustive description of the state-of-the-art. Note that electrodeposition techniques discussed here are also common to other types of nanostructures, magnetic or not, ordered or not.

2. FABRICATION

Nanowires and arrays of nanowires are usually fabricated by filling a porous template that contains a large number of straight cylindrical holes with a narrow size distribution. Filling proceeds in solution by electrochemical deposition. Useful templates for electrochemical deposition of nanowire arrays include porous alumina films formed by anodic oxidation of aluminium (alumite membranes or anodized aluminium films), nuclear track-etched porous membranes, nanochannel array-glass and mesoporous channel hosts. The first two are commercially available. The properties of nanowire arrays are directly related to properties of the nanoporous templates such as the relative pore orientations in the assembly, the pore size distribution, and the surface roughness of the pores. Most works have focused on nanowire arrays grown in polymer (polycarbonate, PET, PVDF etc) membranes. These membranes are made using the nuclear track-etched technology in which a film (typically 5-50 μm thick) of polymer is bombarded by heavy ions of high-energy accelerated in a cyclotron. The irradiated film is then etched in an adequate media where tracks are revealed, leading to the formation of pores with a diameter depending on the etching time. Recently, the technology has been improved in order to produce reliable and reproducible nanoporous polymer membranes. These membranes exhibit improved properties in terms of pore shape, minimal size, size control and parallel arrangement of the pores [1, 12].

Co, Ni and Fe nanowire arrays can be fabricated using commercially available polycarbonate membranes (Nucleopore Inc.). These membranes are approximately 6 μm thick and have relatively high pore densities of about 6×10^8 cm^{-2}. Prior to electrodeposition, a metallic (Cu, Au) film serving as cathode is evaporated on one side of the membrane. The membrane sample is then placed in an electrodepositing cell (made by glass or teflon). A schematic illustration of the experimental arrangement is shown in Fig. 1. In Fig. 2 is schematically shown an array of nanowires in a nanoporous membrane.

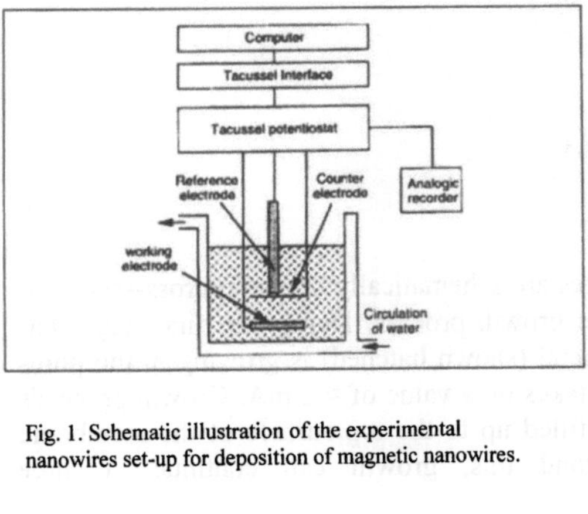

Fig. 1. Schematic illustration of the experimental nanowires set-up for deposition of magnetic nanowires.

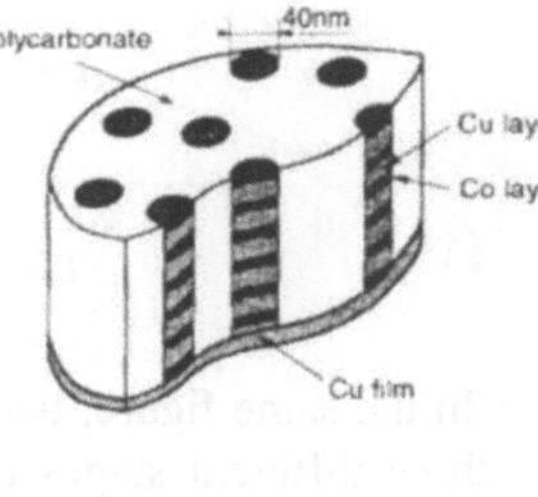

Fig. 2 Schematic of an array of in nanoporous polymer membrane.

Cobalt nanowires can be electrodeposited from a solution of 40 g/l $CoSO_4 \cdot 7H_2O$, 40 g/l H_3BO_3, 20 g/l ammonium citrate buffered to pH 3.5 [5]. The solution for nickel nanowires deposition can be 270 g/l $NiSO_4 \cdot 7H_2O$, 40 g/l H_3BO_3, 40 g/l $NiCl_2 \cdot H_2O$ buffered to pH 3.6 [5]. Iron nanowires can be electrodeposited using a sulfate bath containing 200 g/l $FeSO_4 \cdot 7H_2O$, 0.50 g/l $H_2C_2O_4$, 8.50 g/l glycine buffered to pH 4.5 [6]. Similarly, NiFe (14 at % Fe, 86 at % Ni) alloy nanowires can be electrodeposited from an electrolyte containing 6 g/l $FeSO_4 \cdot 7H_2O$, 218 g/l $NiSO_4 \cdot 7H_2O$, 25 g/l H_3BO_3, 10 g/l NaCl buffered to pH 3.6 [5]. The electrodeposition process is controlled by a computer which continuously integrates the charge during nanowires deposition. The deposition process is stopped when the wires emerge from the surface (as evidenced by a sudden increase of the plating current). Figure 3a shows a potentiostatic electric current I versus time t characteristic (I-t characteristic) for the reduction of Ni^{2+} on track-etched polycarbonate membrane [12]. The schematic displays three different stages of the growth process.

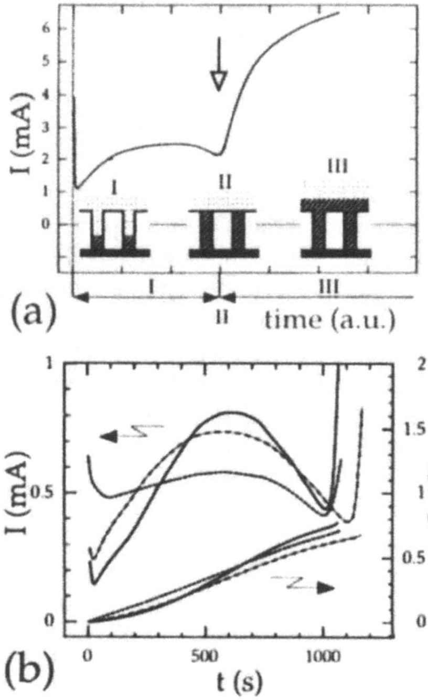

Fig. 3. (a) Electrochemical reduction current I(t) as a function of time t for the potentiostatic plating of Ni in pores of polycarbonate membrane with pore-diameter d = 80 nm. (b) Three detailed characteristics of I(t) and the charge Q(t) for Ni growth at -1.0 V_{SCE} (SCE – saturated-calomel electrode).

In the same figure, two pores are schematically drawn in cross-section at three different stages of the growth process. During the first stage (left schematics labeled I) the metal (shown hatched) is growing in the pores while the reduction current takes on a value of ≈ 2 mA. Growth proceeds in the pores until they are filled up to the top surface of the membrane (middle schematics). Beyond this, growth can continue in three dimensions. This is the transition region (labeled II), which starts at the position of the arrow in Fig. 3a when current rapidly increases since the effective electrode area increases (a right schematics labeled III).

Electrodeposited multilayered nanowires are usually obtained from a single bath using a pulsed deposition technique. Typical cathode current transients resulting from potentiostatic pulses of pulse-plated Co/Cu multilayered nanowires are represented in Fig. 4 [8]. A computer continuously integrates the charge during each layer deposition.

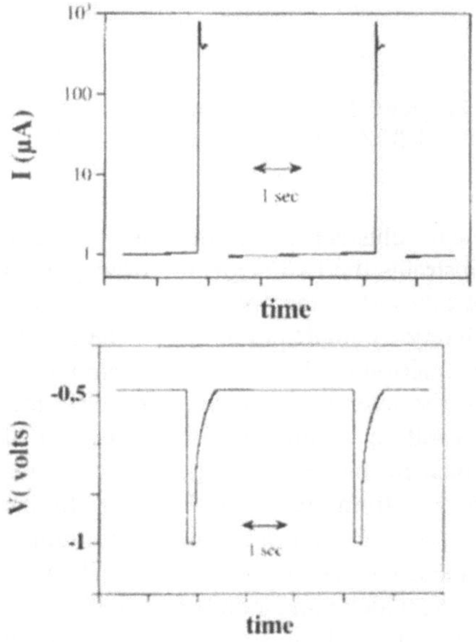

Fig. 4 Time traces of the potential and plating current of pulse-plated Co/Cu multilayered nanowires.

The potential is switched when the deposition charges for the nonmagnetic (Cu) and the magnetic layers reach the set value. Such a procedure is required to give uniform layer thicknesses all along the multilayered filament (nanowire). Dividing the membrane thickness by the number of cycles gives the average period for the wires involved in the nanowire arrays.

The control of the wire structure can be achieved via the overpotential used to initiate growth. Therefore, nanotubes, whose formation could result from a complexation between metal ions and the polymeric membrane, have been also synthesized [6]. By changing the pulse cycles of the two pulse steps electrodeposition current, metallic hallow nanotubes can be electrochemically synthesized. In Fig. 5a, a TEM micrograph clearly shows interlaced iron nanotubes [6].

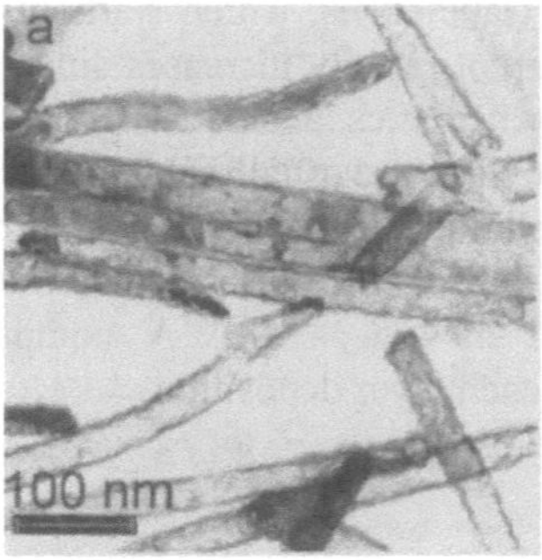

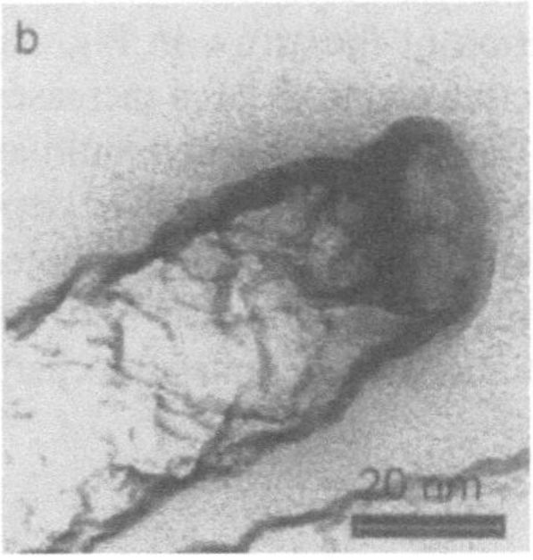

Fig. 5. (a) TEM image of Fe nanotubes growth in a polycarbonate membrane by using two pulses steps electrodeposition current. (b) An aperture is clearly visible at the last extremity of a Fe tube.

The observation of the tubes reveals clusters coalescing inside a tube (dark spot). This image clearly illustrates the two steps proposed growth mechanism with the filling of the tubes after the formation of the outer wall. The early stage of growth is likely a complexation of Fe^{2+} metal ions together with the CO_3^{2-} carbonate function of the polymeric membranes, which is highly favorable because the interaction between the chemical species is strong. This first stage results in a coating of the inner part of the membrane pores with metal ions. In the second stage, reduction via electrochemistry of the metal species from Fe^{2+} into Fe leads to the formation of the tubes (deposition time ~ 6-8 min). At higher resolution, an aperture at one extremity of a tube is clearly visible (Fig. 5b). The wall was evaluated to be ~ 1 to 2 nm thick, which corresponds to a few atomic planes. The polymer and alumite membranes have successfully been used to prepare colloidal suspensions in which the nanorods are dispersed and stabilized in water. For example, the polycarbonate membranes can be dissolved in 40^0 C dichlormethane (Cl_2CH_2), rinsed in fresh, dichlormethane, chloroform and ethanol. In case of the aluminium oxide model system, the oxide can be dissolved in a 1.25 M NaOH solution containing 0.1 mM polyvinylpyrrolidone (PVP, M = 40, 000). Figure 6 shows SEM image of Ni nanorods from a colloidal suspension obtained after completely dissolving the polycarbonate membrane in dichlormethane (CH_2Cl_2) and collecting the wires on holey carbon grides [12].

It has been found that there is a significant influence of deposition conditions on the deposit properties. It was found that pH of the electrolyte shows a strong effect on the phase structures obtained. The increasing the bath acidity favors formation of more phase structures. It also decreases the amount of deposited alloys, which is explained by the lower Faraday efficiency in the solutions of lower pH. The temperature of solution exerts a profound influence on the deposit and it favors the formation of new phases. It was also found that the rate of deposition depends strongly on electrolyte temperature. Among all parameters investigated the deposition current density exerts the greatest effects, both on the appearance of the deposit and its phase structure. Figure 7 shows anodic linear sweep voltammetry (ALSV) diagrams of Fe nanowires deposited at different current densities (4, 7, 9 and 12 A dm^{-2}) at t = 40^0 C for t = 8 min [1, 12].

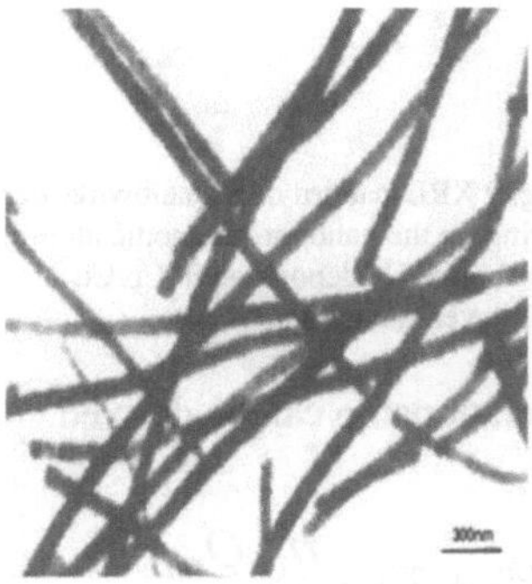

Fig. 6. SEM image of Ni nanorods nanowires from a colloidal suspension.

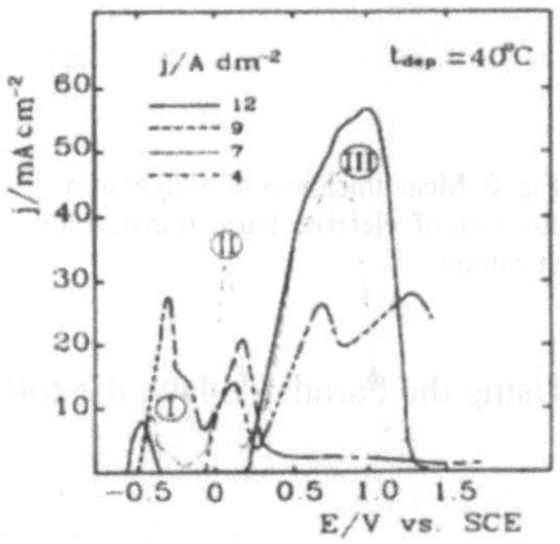

Fig. 7. ALSV voltammograms of Fe in a sulfate solution of pH 2.5 at 23^0 C.

Alloys deposited at low current densities (4 A dm^{-2}) contain only one phase structure, since only one dissolution peak is observed. Alloys deposited at higher current densities (7 and 9 A dm^{-2}) show more dissolution peaks, indicating the presence of more different phase structures. However, at current densities larger than 12 A dm^{-2}, peak II is not observed, indicating that the phase structure which corresponds to peak II and which is the only one present in the deposits obtained at low current densities, is absent in the deposits obtained at very high current densities. It can be seen from Fig. 7 that the deposition current density also influences the total Faraday efficiency of the deposit and increases with current density irrespective of the temperature.

3. CHARACTERIZATION

Microanalysis of nanowires and multilayered nanowires can be performed in TEM and scanning TEM (STEM) by X-rays energy dispersive spectroscopy (XEDS) and electron energy loss spectroscopy (EELS), the former being more quantitative and the latter having a higher spatial resolution [1, 12]. In the case of multilayered nanowires, we can alternatively use another method based on the relationship between the layer thickness d_{NM} and d_M and the electric charges Q_{NM} and Q_M transferred at the cathode during the corresponding pulses. Figure 7 shows $d_{bilayer}$ as a function of Q_{Cu} for the Co/Cu system [8].

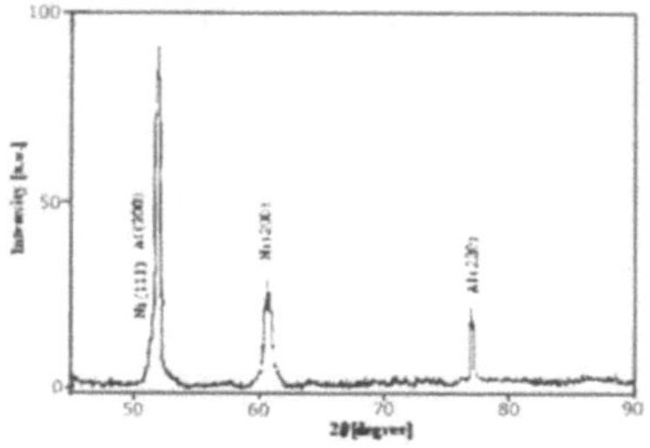

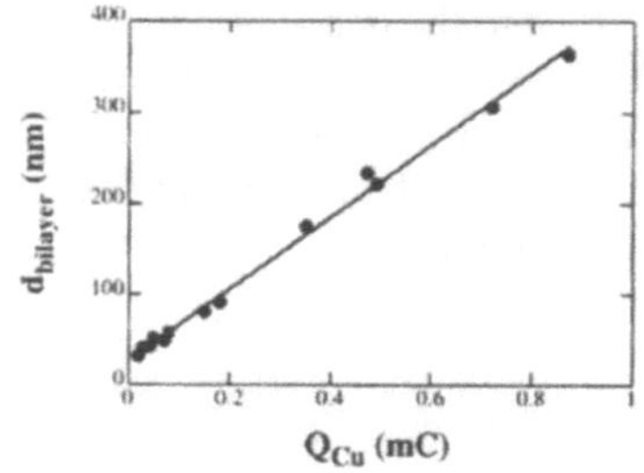

Fig. 8. Mean thickness of bilayer as a function of electric charge transferring at cathode

Fig. 9 XRD pattern of Ni nanowires deposited during in the nanoporous anodic alumina membrane. Cu deposition in Co/Cu multilayered nanowires.

Using the Faraday's law, the following relationship can be derived:

$$d_{bilayer}(nm) = 36.7\frac{\eta_{Cu}Q_{Cu}}{A} + 34.3\frac{\eta_{Co}Q_{Co}}{A} \tag{1}$$

where A is the cathode surface area (in mm^2) which depends on the pore density diameter; η_{Cu} and η_{Co} are the cathode current efficiencies for pulsed electrodeposition of Cu and Co, respectively; the two charges Q_{Cu} and Q_{Co} in Eq. (1) are expressed in mC. From the slope and intercept of the straight line of Fig. 7, we can obtain $\eta_{Co}/\eta_{Cu} = 0.72$ and d_{Co} 24 nm.

Detailed structural characterization of nanowires and multilayered nanowires can be done by X-ray diffraction (XRD), scanning electron microscopy (SEM) analytical transmission electron microscopy (TEM) and atomic force microscopy (AFM) [1]. Figure 8 shows XRD pattern of Ni nanowires deposited in the nanoporous anodic alumina membrane [1]. Figure 9 shows SEM images of Ni wires grown in 80 nm diameter pores of a polycarbonate membrane [1]. Hemispherical caps form on top of the wires if the growth will proceed after the pores will be filled up to the top surface of the membrane. The total length of a wire is ≈ 6 μm. The right most image shows the top a bottom 2 μm of one and the same wire. Figure 10 shows TEM images of a short-period Py/Cu multilayer grown on a pure permalloy ($Ni_{80}Fe_{20}$-Py). The structure is FCC and the <110> axis is parallel to the growth direction [1, 8]. The period of the super lattice shown in Fig. 10 is about 5 nm. It is noted that for a 20 μm-thick membrane, multilayered nanowires composed of more than 3000 bilayers can be prepared.

Fig. 10. SEM images of Ni wires grown in a polycarbonate membrane.

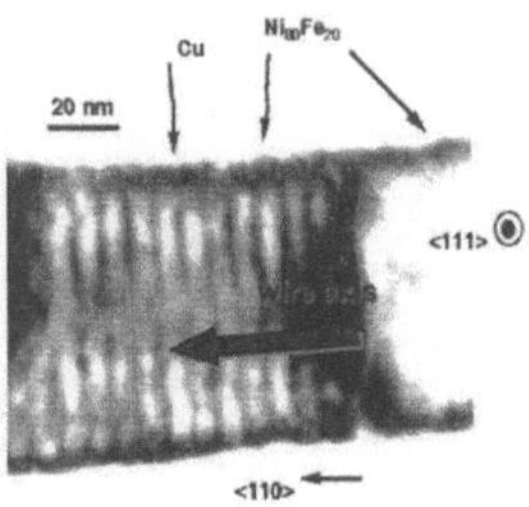

Fig. 11. TEM observations in <111> zone axis of a Py/Cu multilayered nanowire.

4. PROPERTIES OF ELECTRODEPOSITED NANOWIRES

Magnetization of the nanowires can be determined at room temperature using a vibrating sample magnetometer (VSM). Low temperature magnetization measurements can be carried out using a superconducting quantum interface device (SQUID). Magnetic domain structure can be evidenced by magnetic force microscopy (MFM) [4, 9, 12]. Generally, the enhancement of coercivities of ferromagnetic nanowires increases for decreasing wire diameter due to the decreasing probability of multiple domains in the nanowires. The high anisotropy observed for the nanowire arrays has possible application to high density perpendicular magnetic recording media. By having their preferred magnetization perpendicular to the sample plane (parallel to nanowire axis), magnetic media with perpendicular anisotropy can allow smaller bit size and thus increase the recording density. As has been shown, the magnetic nanowire arrays studied exhibit high perpendicular anisotropy due to their unique artificial structure. For example, highly ordered $Fe_{14}Ni_{86}$ alloy nanowire arrays (with diameter 25-43 nm, distance between the neighboring pore centers 60 nm) exhibit enhanced coercivity (about 80 kA/m) and remanent magnetization up to 70 % [10]. Therefore, the coercivity is comparable to that of a 1.44 Mb, 3.5 inch floppy disk (64 kA/m). The easy magnetization axis of the magnetic nanowire arrays is perpendicular to the membrane. Thus, it has potential application to be used as rewritable ultra-high density vertical magnetic storage media. Its storage density would be more than 2 x 10^5 times than that of 3.5 inch floppy disk.

The transport properties of nanowire have been mainly studied by measurements on arrays of nanowires because the high pore density and the random arrangement of the pores make difficult the resistance measurement of a simple metal nanowire. However, it has been recently developed strategies that allow performing measurements on single metal nanowires and multilayers. Till now, three different approaches have been proposed [1, 8, 12]. A single nanowire has been connected by evaporating a thin metallic layer on the face of the porous membrane exposed to the electrolyte. The film is thin enough in order to ensure that it does not cover the pores (Fig. 11a).

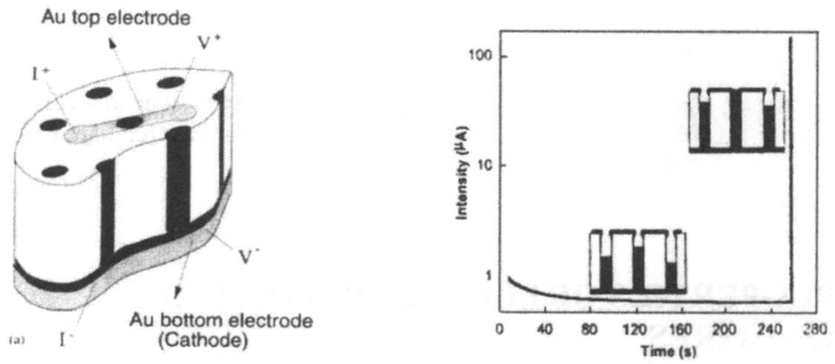

Fig. 12. (a) Schema of the polymer membrane used for the electrodeposition. The bottom Au thin film is used as cathode; the top Au thin film strip used for contacting a single nanowire. (b) Short increase of the plating current when the wire is contacted to the metal strip.

In the second method, combination of the chemical methods and e-beam lithography is used for contacting a single wire after electroplating and formation of metallic caps at the surface [1]. Another method, also combining e-beam lithography and electrodeposition was recently reported [1]. In this method, submicron cylindrical cavities of typical height 0.3 μm and diameter 0.1 μm were fabricated and precisely located in a PMMA layer deposited on a pattered substrate. In a second step, electroplating of nanowires was performed into these cavities and caps are formed at the surface. The final step of the process consists in sputtering a metal layer at the top of the device allowing contacting electrically single pillars. Figure 11b shows the short increase of the plating current when the wire is contacted to the metal strip.

The reversal of magnetization in individual magnetic nanowires was studied using transport measurements, i.e., anisotropic magnetoresistance (AMR) measurements. Multilayered nanowires are ideal structures to study conveniently the giant magnetoresistance (GMR) and its temperature dependence in the current perpendicular to the layer plane

(CPP) geometry. Up to now, most measurements have been performed on Co/Cu, Fe/Cu and Py/Cu systems. Magnetoresistance and magnetization curves obtained with magnetic fields parallel to the layers at 4.2 K on $Ni_{80}Fe_{20}$ (12 nm)/Cu (14 nm) multilayered nanowires are shown in Fig. 12a and Fig. 12b [8]. In order to reduce saturation field, can be used another structure composed of Py (3 nm)/Cu (10 nm)/Py (3 nm) trilayers magnetically isolated by thick layers of Cu (100 nm or more) [8]. A typical magnetoresistance curve for in-plane field is shown in Fig. 3 c. The multilayered nanowires exhibit high MR ratio with a fairly large resistance (between a few Ω and a few kΩ depending on the number of wires which are connected in parallel). It has also been shown that the field required to saturate the magnetoresistance can be relatively small, of the order of 8-10 kA/m, in nanowires composed of Py/Cu/Py trilayers spaced by thick Cu layers.

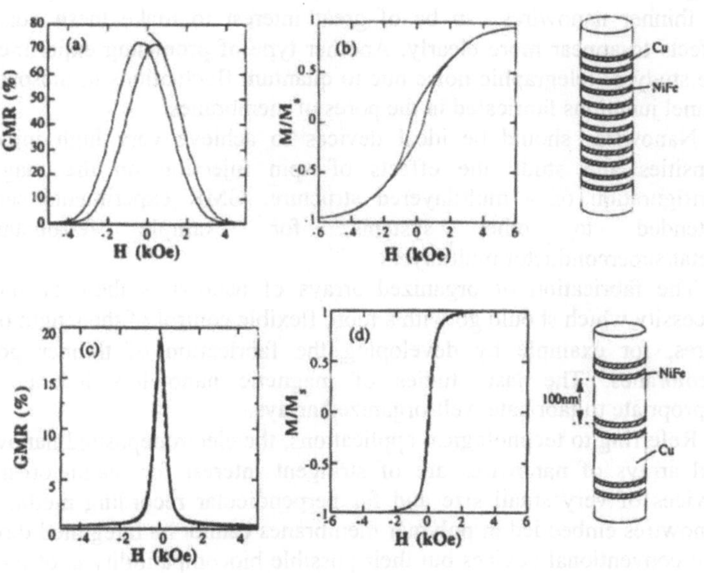

Fig. 13 CPP – GMR and magnetization curves versus applied field parallel to the layers at 4,2 K for Py/Cu multilayered nanowires (a, b) and for multilayered nanowires composed of Py/Cu/Py trilayers separated by a thick (90 nm) long Cu rods (c, d).

In hybrid structures, (e.g., Ni/NiO/Co system), the most striking result is the observation of two-level telegraphic noise (RTSN) which is is ascribed to the trapping and untrapping of a single electron by a defect in ultra-small NiO barrier [11].

5. PERSPECTIVES

Research on magnetic nanowires are driven towards the progress of the fabrication techniques and to novel properties what implies to obtain wires of smaller diameter, down to 10 nm or smaller than certain characteristic length scales, such as spin diffusion length, carrier mean free path, magnetic domain wall width etc. The effect of confinement, proximity and order govern the interplay between the relevant physical length scales and the sizes of the nanowires. For example, ballistic transport appears in material confined to the appropriates length, smaller than the electron mean free path, or magnetization reversal processes can be drastically modified in nanowires confined to sizes that preclude the domain wall formation. The recent microsquid (μ - SQUID) experiments have suggested the existence of quantum tunnelling effects in the magnetization reversal magnetic wires (45 nm thick Ni) and experiments of thinner nanowires can be of great interest to make these quantum effects to appear more clearly. Another type of promising experiment is the study of telegraphic noise due to quantum fluctuations in mesoscopic tunnel junctions fabricated in the pores of membranes.

Nanowires should be ideal devices to achieve very high injection densities and study the effects of spin injection on the magnetic configuration of a multilayered structure. GMR experiments can be extended to other systems, for example ferromagnetic metal/superconductor multilayers.

The fabrication of organized arrays of nanowires there is another necessity which should go with a more flexible control of the length of the wires, for example by developing the fabrication of thinner porous membranes. The last studies of magnetic nanowires indicate less appropriate to fabricate well-organized arrays.

Referring to technological applications, the electrodeposited nanowires and arrays of nanowires are of stringent interest for magnetoresistive devices of very small size and for perpendicular recording media. The nanowires embedded in polymer membranes cannot be integrated directly into conventional devices but their possible biocompatibility is of interest to medical devices. For application to conventional devices, it will be necessary to develop techniques of electroplating into other types of templates, supported membranes on silicon or patterned insulator layers.

6. CONCLUSIONS

Electrodeposition method is capable of producing high-quality magnetic nanowires and ordered arrays of nanowires with desirable features, including nanowires with reproducible sizes in the range of 10 nm, and arrays which can be extended over large areas (~ 100 x 100 nm^2). Usually there are many variable parameters that can be used to tune the properties of nanowires, such as material, crystalinity (polycrystalline, single-crystal, amorphous), structure single layer, multilayered, array geometry etc.

The structure of the nanowires and arrays of nanowires can be investigated by XRD, nuclear magnetic resonance (NMR), SEM, TEM, AFM, etc. The magnetic properties of the nanowires and arrays of nanowires can be characterized by techniques such as magnetization measurements (VSM, SQUID, MFM, AGM – alternating gradient magnetometry, MOKE – magneto-optical Kerr effect) or transport measurements.

The main magnetic properties of the nanowires present important differences with respect to continuous films: the remanent state is usually metastable shape anisotropy usually plays an important role in magnetization reversal process; coercive field are often larger than the values found in the unpatterned samples; the spin wave spectrum may be quantized due to the small size of the nanowires. The shape anisotropy of individual wires dominates the magnetic properties. Dipolar interactions between nanowires are fundamental key to understand the experimental behavior of the array. The arrays of the magnetic nanowires are not only interesting for their intrinsic magnetic properties, but also due to their interaction with other systems. Arrays of magnetic nanowires can constitute effective ordered spinning centers for the vortex lattice when they interact with type II superconducting films. Also, ordered magnetic nanowire arrays can be used in hybrid semiconductor/ferromagnetic systems to produce a periodic magnetic field that modulates, in a controlled fashion, the transport properties of two-dimensional electron gases (2 DEG). Technologically, applications of nanowire arrays are becoming increasingly important, especially in fields like high density magnetic recording (HDMR), magnetic random access memory (MRAM), sensors, switching and magneto electronics.

The fabrication of large arrays of ordered, ultra fine (~ 10 nm or smaller), uni-disperse nanoelements, remains challenging. More systematic studies of the magnetization reversal processes are necessary, including the associated dynamic effects in the short time scales. It also crucial to achieve a good understanding of interaction effects in arrays of nanowires or between magnetic nanowire arrays and other systems.

References

1. Wang, L.W., Liu, Y. and Zhang, Z. (2002) *Handbook of Nanophase and Nanostructured Materials*, Kluwers Academic Publishers, Dordrecht.

2. Sulitanu, N. (2002) *Electrochemical deposition of novel nanostructured magnetic thin films for advanced applications*, Mater. Sci. Eng. **B95**, 230-235.

3. Osaka, T. (2000) *Electrodeposition of highly functional thin films for magnetic recording devices of the next century*, Electrochimica Acta **45**, 3311-3321.

4. Raposo, V., Garcia, J.M., Gonzalez, J.M. and Vasquez, M. (2000) Long-range magnetostatic interactions in arrays of nanowires, *J. Magn. Magn. Mater.* **222**, 227-232.

5. Chiriac, H., Moga, A.E., Urse, M. and Ovari, T. – A. (2003) Preparation and magnetic properties of electrodeposited magnetic nanowires, *Sensors and Actuators* A **206**, 348-351.

6. Tourillon, G., Pontonnier, L., Levy, J.P. and Langlais, V. (2000) Electrochemically synthesized Co and Fe nanowires and nanotubes, *Electrochem. Solid-State Lett.* **3**, 20-23.
7. Yin, A.J., Li, J., Jian, W., Bennett, A.J. and Xu, J.M., (2001) Fabrication of highly ordered metallic nanowire arrays by electrodeposition, *Appl. Phys. Lett.* **79**, 1039-1041.
8. Fert, A. and Piraux, L. (1999) Magnetic nanowires, *J. Magn. Magn. Mater.* **200**, 338-358.
9. Henry, Y., Ounadjela, K., Piraux, L., Dubois, S., George, J.-M. and Duvail, J.-L. (2001) Magnetic anisotropy and domain patterns in electrodeposited cobalt nanowires, *Eur. Phys. J. B* **20**, 35-54.
10. Zhu, H., Yang, S., Ni, G., Yu, D. and Du, Y. (2001) Fabrication and magnetic properties of $Co_{67}Ni_{33}$ alloy nanowire array, *Scripta Mater.* **44**, 2291-2295.
11. Doudin, B., Redmond, G., Gilbert, S.E. and Ansermet, J. P. (1997) Magnetoresistance governed by fluctuations in ultrasmall Ni/NiO/Co junctions, *Phys. Rev. Lett.* **79**, 933-936.
12. Cohen, M.L. (2001) Nanotubes, Nanoscience and Nanotechnology, *Mat. Sci. Eng.* C **15**, 1-11.

AFM AS A MOLECULAR NUCLEIC ACID SENSOR

Ş. DOĞUTAN ÜLGEN, CENGIZ KOÇUM, EMRECAN ÇUBUKÇU, ERHAN PIŞKIN

Hacettepe University, Department of Chemical Engineering and Bioengineering Division 06532 Beytepe, Ankara, Turkey

Abstract: Model single strand DNA (ssDNA) was covalently immobilized onto Atomic Force Microscope (AFM) tips (cantilevers) as specific ligand. Then these tips were interacted with the buffer solutions with or without free ssDNA molecules as strand were used for comparison. Immobilization and hybridization onto the cantilever surfaces were observed by optical and fluorescence microscopies. Interactions between the AFM cantilever (tip) and the aqueous medium (therefore with the target ssDNAs) were quantified by obtaining the "percent descenting distance" ("PDD") as the main variable. The PDD values obtained for the buffer solutions were between -2.07 and +4.91%. There were slight increases in the negative values when a non-complementary ssDNA molecules were introduce into the buffer. However, after hybridization with its complemenray ssDNA, the PDD values were significantly increased (between -32.24 and -43.47 %). There was a correlation between the concentration of the complementary target ssDNA in the medium and the PDD value. However, there were significant changes from one AFM cantilever to another, which was considered as a problem to be solved. However, it was concluded that this approach may be further develop to create AFM based molecular sensors.

Keywords: AFM cantilevers/tip; nucleic acid molecular sensors; percent descenting distance

1. INTRODUCTION

The ability for AFM to measure forces of 10 pN or less will make it an essential tool for measuring biological interaction. A number of studies have been performed to investigate serum albumin and its associated antibody using AFM (Ducker *et al.*, 1991; Stuart and Hlady, 1995; Hinterdorfer *et al.*, 1996; Williams and Blanc, 1994; Chowdhury and Lunckham, 1998). Streptavidin-biotin interactions have been studied with AFM (Lee *et al.*, 1994). Recently, the concept of cantilever-based biosensor techniques driven from AFM was proposed, in which the mechanical responses of the microcantilever due to changes surfaces stresses upon the interaction of the

S. Guceri et al. (eds.), Nanoengineered Nanofibrous Materials, 377-384.

ligands immobilized onto cantilever and the target molecules in the medium (Moulin *et al.*, 2000; Battiston *et al.*, 2001).

Sequence-specific hybridizations between nucleic acids, either in solution or immobilized on fixed support, are widely used for the detection and analysis of genetic material for diverse applications including identification of genetic diseases and disorders, detection and characterization of viruses, bacteria, and parasites, etc. (Symons, 1989; Downs, 1991).

Also interaction forces between the strands of DNA were measured with the AFM by a procedure in which DNA strands were covalently attached on probe and surface (Lee *et al.*, 1994). Recently, Holmberg et al. investigated the formation and modulation of a mixed molecular layer, containing single strand DNA (ssDNA) and a spacer arm (mercaptohexanol, MCH), immobilized on gold using in-situ AFM (Holmberg *et al.*, 2003). The introduction of the complementary DNA to the mixed monolayer of ssDNA and MCH resulted an increase in smoothness and thickness of the monolayer, which, they suggested in, is due to hybridization.

In this study we propose to use ssDNA, covalently bound (immobilized) onto AFM tips as a molecular nucleic acid sensor for the detection of the complementary strand by a very simple measurement using AFM.

2. MATERIALS AND METHODS

2.1 Materials

Three model DNA single strands given in Scheme 1 were studied. The first one was the "ligand" ssDNA (MWG Biotech AG, USA) it has one extra amino group at the 5'th-end for immobilization onto the AFM tips. The second one was the "target", complementary ssDNA (MWG Biotech AG, USA). The third one was an arbitrarily selected non complementary ssDNA (Xeragon, USA), for compare and to show the nonspecific interactions between the non-complementary DNA chains. Nitric acid, aminopropyltrimethoxysilane, gluteraldehyde, fluorescent dyes, Hoechst, Orange G were purchased from Sigma Chemicals (USA). Standard silicone nitride cantilevers (tips) were purchased from Digital Instrument (USA).

2.2. Imobilization of the Ligand ssDNA onto AFM Tips

A two step procedure was applied to covalently attach the ligand ssDNA to the silicon nitride AFM tips for immobilization of antibody molecules onto AFM tips (Chowdhury and Lunckham, 1998) (Scheme 2). The conditions applied in these steps were optimized in the preliminary studies (Ülgen, 2003). Briefly, in the first step ("the silanization step"), the silicon nitride tips were treated with 10% nitric acid solution which was left in a silicone bath for 20 min at 80°C. This causes the formation of surface hydroxyl groups on the

Si_3N_4 tips. The tips were then thoroughly rinsed with distilled water and placed into 10% aminopropyltrimethoxysilane (APTS) solution with a pH of 7.0 adjusted with acetic acid for 4 h at 80°C. This treatment provides reactive primary amine groups on the nitride surface, and is known as the linker group. The surfaces were thoroughly rinsed with water. The tips were then reacted in 10% gluteraldehyde solution at room temperature for 1 h. Gluteraldehyde acts as a bridging agent between the solid support and the ssDNA. At the last step, the tips were left in 2 pmol/μL of the ligand ssDNA for 2.5 hr for immobilization, then the unbound DNA molecules were removed by rinsing with the buffer solution. A schematic view of the AFM tip as a molecular sensor is shown in Fig. 1.

(A) The "Ligand" DNA

5'-NH_2-TTT TTT TTT TTT TTT TTT TT-3'

(B) The "Target" Complementary DNA

3'-AAA AAA AAA AAA AAA AAA AA-5'

(C) The "Non-complementary DNA

5'-(AGT) CT AGA CCT GTC ATC AGT TAG GGT TAG-3'

Scheme 1. Oligonucleotides used in this study: (A) Ligand ssDNA which is immobilized onto AFM tips; (B) the complementary strand (target DNA) which was dissolved in the medium; (C) the non-complementary DNA (used for comparison).

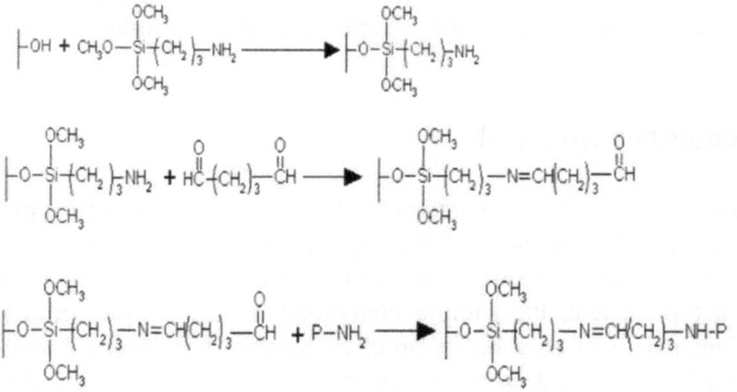

Scheme 2. Immobilization of the ligand ssDNA onto AFM tips: (A) Silanization; (B) attachment of glutaraldehyde; and (C) Immobilization of the ligand ssDNA, where "P" denotes the ssDNA strand.

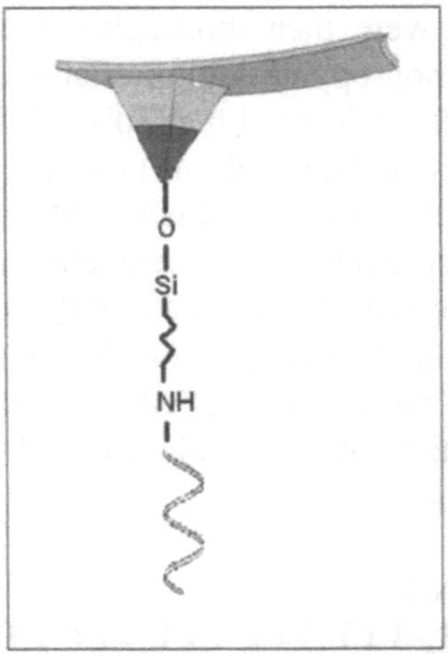

Fig.1. Schematic description of the AFM tip prepared as a molecular nucleic acid sensor carrying the ligand ssDNA immobilized through gluteraldehyde molecules on the tip.

2.3. Microscopic Observations

In order to follow changes on the AFM cantilever surfaces after treatments, images of the surfaces were taken both with an optical microscope (Axiosckop 20, Zeiss, Germany) and fluorescence inverted microscope (Olympus, Model IX70, Japan). Hoechst and Orange G (Sigma, USA) were used as fluorescence dyes for single strand and double strand DNA, respectively.

2.4. Measurements with AFM

The apparatus used for the measurements was a homemade atomic force microscope. To control the approach of the target solution to the AFM tips, a Burleigh Aris 11 Approach Module was used, which was controlled by a computer. In a typical test, the sample containing a pure buffer solution (phosphate buffer, pH: 7.4) or a buffer solution having the target (or non-target) ssDNA molecules was driven (+z direction) to the AFM tip (untreated or carrying the ligand ssDNA) until the tip touches the solution. It is waited at this position for about 10 min to complete hybridization (if there is any), and then the sample was driven to the opposite direction (+z direction) with a predetermined rate, 900 nm/sec (Ülgen 2003). The "separation distance", or the so-called "the discenting distance" (in nm), at which the AFM tip gets rid of the solution that sticks on its surface, was measured. The "percent discenting distance" was calculated from the following equation and used as the main variable in this study.

"Percent Discenting Distance"(%) $= \{(L_1 - L_2) / L_1\} \times 100$ **(1)**

Where, L_1: discenting distance for the reference solution (i.e., phosphate buffer) (nm); and L_2: discenting distance for the same buffer solution containing the target ssDNA (nm). Note that the negative value shows the decrease in the interaction between the AFM tip and liquid phase.

For each AFM tip, phosphate buffer (pH: 7.4) was used as the reference. Then, separation distance was measured, without replacing the tip from tip holder or changing the position. The target DNA solution diluted with reference buffer solution, was added to the buffer and the same procedure was applied. To determine nonspecific bindings to the unmodified surface, unmodified AFM tips were also used. Measurements were repeated with the solutions containing the target complementary ssDNA at different concentrations (46.8, 117, 234, and 468 pmol ssDNA in 10 μL buffer). To determine the non-specific interaction between the non-complementary DNA strands, the AFM tips carrying the ligand ssDNA were dipped in buffers containing only the non-complementery ssDNA (the third ss DNA given in Scheme 1).

3. RESULTS AND DISCUSSION

3.1. Microscopic Observations

Representative micrographs of the AFM cantilever taken with an optical microscope before and after chemical modifications are given in Fig. 2. There were no changes after the silanization step, while structures with different sizes on different parts of the cantilever surface, including at the top corner where the tip is located, were very much observable. It is not possible to see the individual glutaraldehyde (GA) molecules by optical microscopy, therefore, most probably these structures most probably some GA polymeric chains or aggregates that may occur during the treatment of the surfaces with GA. From these images we assumed that the other parts of the surface of both the cantilever and the tip were cover with GA surface groups (maybe evenly) that are available for further modifications.

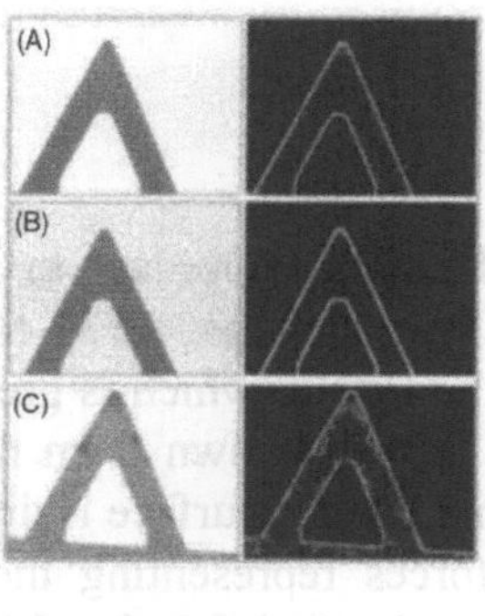

Fig.2. Representative optical micrographs of the AFM cantilever: (A) untreated; (B) after sialanization step; and (C) after glutaraldehyde attachment. Magnifications: x100. Pictures at the left colufiltering.

Fluoresence microscopy images of the untreated and carrying the ligand ssDNA (before and after hybridization) were also taken. Note that two different fluroscent dyes, Hoechst and Orange G were used. The first one was utilized to observe the immobilized ligand ssDNA, while the second one was applied after the hybridization of the ligand DNA with its complementary, again on the AFM cantilever surfaces.

Figs. 3A and 3B give the representative fluorescent microscopy images of the untreated AFM cantilever, which show that there was no significant nonspecific adsorption of the dye molecules on the surfaces before ssDNA immobilization. Figs. 3C and 3D clearly exhibit that a high percentage of the cantilever surface (including the tip) was covered with the fluorescence dye Hoechst (means with the ligand ssDNA). In order to examine the specificity of the second fluorescence dye, namely Orange G, which should be specific for only double strand DNA (rather than for single strands), the fluorescence microscopy images of the cantilevers carrying only the immobilized ligand ssDNA. As seen in Figs. 3F and 3H, there was very low absorption of Orange G onto the cantiver carrying ssDNA. While Figs. 3E and G demonstrated that very significant orange color on a large part of the the cantilever surface carrying the hybridized (double strand) DNA molecules.

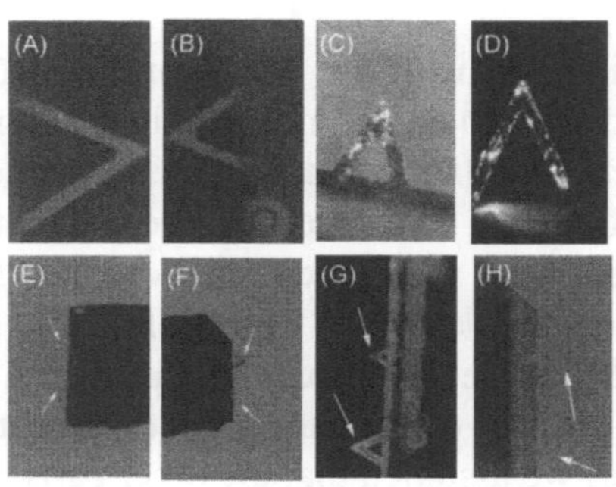

Fig.3. Representative fluorescence microscopy images of the AFM cantilever: (A, B) before ssDNA immobilization; (C, D) carrying the immobilized ligand ssDNA (dyed with Hoechst); and (E, G) carrying the immobilized ligand ssDNA hybridized with its complementary (dyed with Orange G). (F, H) carrying the immobilized ligand ssDNA (dyed with Orange G)

3.2. Studies with AFM

In this study we applied a simple approach to detect a model target single strand ssDNA in solution by using an AFM tip (cantilever) carrying its complementary ssDNA as the ligand, which is prepared by the method given in the previous sections. It is well-known from the surface tension (contact angle) measuring systems that when a surface is dipped in ("descent") and out ("ascent") from a liquid, forces representing the interactions between the surface and liquid are measured, which is related to the surface tension of the liquid (Hiemenz, 1986) and depend on not only the liquid phase properties but also on the surface properties (hydrphobicity, ionic charges, etc.)

In this study we used the "percent descenting distance" ("PDD") as the main parameter measured in the AFM sensor, which is calculated from Eq.1 as described above. The PDD values for five different nonmodified AFM cantilevers at room temperature are as follows: -3.88, +4.91, -3.59, -2.07 and +4.77 %. As seen here there are both positive and negative values, but not very significantly different from each other. This is expected, because the cantilevers, even if they are the producted by the same company, have different surface characteristics. Also there is no specific interaction between SiN tips and DNA solution.

These AFM tips were then dipped into buffer solution containing the target (the complementary) ssDNA with a concentration of 468 pmol/10 μL. The PDD values determined are as follows: -43.47, -35.12, -32.24, -32.45 and -38.31 %. These significant changes (about ten times compared to those obtained with the buffer solution alone, containing no DNA) are indication of the hybridization (double strand formation onto the surfaces, which caused changes of the surface properties, and therefore the PDD values. Interestingly, the values are quite close to each other, the differences are most probably due to differences in the surface properties of the original AFM cantilevers.

Table 1 shows the PDD values obtained for the experiments done with different concentration of the target ssDNA in the buffer solutions. Note that five AFM cantilevers were used in each group.

Table 1. Effects of concentration of the target DNA in phosphate buffer on the PDD values.

Concentration of the target ssDNA (pmol/10 μL)	PDD Values (%) For the AFM Tip 1	2	3	4	5
468	-43.47	-35.12	-32.24	-32.45	-38.31
234	-30.69	-31.64	-28.98	-34.40	-33.42
117	-28.85	-27.67	-32.33	-27.77	-29.81
46.8	-29.72	-13.84	-12.85	-16.64	-10.41

As a general trend, can be seen here the PDD values decrease with the concentration of the target ssDNA in the medium. There is an non-linear relationship between PDD values and concentration. However, there are also strong fluctuations in the PDD values (changing from one AFM cantilever to another) which, we believe, may be improved by special design of the cantilevers that will be more suitable for these kinds of molecular sensor.

In order to exhibit the specificity of these AFM sensors, we have performed a final group of study, in which buffer solutions containing 468 pmol/10 μL the non-complementary ssDNA (the third ssDNA given in

Scheme 1). The PDD values obtained with five different AFM cantilevers were as follows: +5.45, -6.26, -12.30, +6.07 and -6.98 %. Note that these values were slightly higher than those observed with pure buffer solutions, but clearly much lower that were observed for the hybridization with the complementary ssDNA (see the first line Table 1).

In conculison these preliminary results show very strong promising application of AFM nucleic acid sensors not only for qualitative but also quantitative characterisation of the target complementary DNA in aqueous medium.

References

1. Battiston, F.M., Ramseyer, J.-P., Lang, H.P., Baller, M.K., Gerber, Ch., Gimzewski, J.K., Meyer, E., Güntherodt, 2001, A Chemical sensor based on a multifabricated cantilever array with simultaneous resonance-frequency and bending readout, *Sensors and Actuators B*, 77, 122.
2. Chowdhury, P.B., Luckham, P.F., 1998. Probing recognition process between an antibody and antigen using atomic force microscopy, Colloids and Surfaces A, *Physicochemical and Engineering Aspects*, 143, 53.
3. Downs, M.E.A., 1991. Prospects for nucleic acid biosensors. Biochemical Society Transactions 19, 39.
4. Ducker, W.A., Senden, T.J., Pashley, R.M., 1991. Direct measurements of colloidal forces using an atomic force microscope, *Nature* 353, 239.
5. Hiemenz, P.C., Ed., 1986. Principles of colloid and surface chemistry. *Marcel Dekker, Inc., New York*, 1986.
6. Hinterdorfer, P., Baumgartner, W., Gruber, H.J., Schindler, K., Schindler, H., 1995. Detection and localization of individual antibody-antigen recognition events by atomic force microscopy. *Proceedings of National Academy of Science USA* 93, 3477.
7. Holmberg, M., Kühle, A., Garnaes, J., Boisen, A., 2003. Hybridization of short DNA molecules investigated with in situ atomic force microscopy. *Ultramicroscopy* 97, 257.
8. Lee, G.U., Kiddwell, D.A., Colton, R.J., 1994. Sensing discrete streptavidin-biotin Interactions with atomic force microscopy, *Langmuir* 10, 354.
9. Lee, G.U., Chrisey, L.A., Colton, R.J., 1994. Direct measurement of the forces between complementary strands of DNA, *Science*, 266, 771.
10. Moulin, A.M., O'Shea, S.J., Welland, M.E., 2000, Microcanvantilever-based biosensors, *Ultramicroscopy,* 82, 23.
11. Stuart, J.K., Hlady, V., 1995. Effects of Discrete Protein-Surface Interactions in Scanning Force Microscopy Adhesion Force Measurements, *Langmuir* 11, 368.
12. Symons, R.H., 1989. Nucleic Acid Probes. *CRC Press, Boca Raton,* FL.
13. Ülgen, D., 2003. AFM based molecular biosensors. M.Sc. thesis, Hacettepe University, Ankara.
14. Williams R.A., Blanch H.W., 1994. Covalent immobilisation of protein monolayers for biosensor applications, *Biosensor and Bioelectronics* 9,159.

ELECTROPHORETIC DEPOSITION OF ZnO THIN FILM IN AQUEOUS MEDIA

A. DOGAN, E. SUVACI, G. GUNKAYA, E. UZGUR
Anadolu University, Department of Materials Science and Engineering, Eskisehir/Turkey

Abstract: In thin films, controlled porous structures are desired for certain ceramic gas sensors. The aim of this study was to fabricate ZnO coated substrates starting with aqueous suspensions with electrophoretic deposition technique. Sub-micron sized ZnO particles were synthesized by using homogenous precipitation method. A 5 wt% ZnO aqueous suspension prepared for Electrophoretic Deposition (EPD) and an anionic dispersant was used as stabilizer. Electroded surface of the alumina substrates were coated with EPD technique at various voltages. Coated samples were sintered at 1200°C for two hours. Characteristics of the sintered ZnO coated substrates were investigated with SEM and XRD techniques.

1. INTRODUCTION

Chemical sensor research is focused on new studies to develop specific sensors. Producing smaller, more efficient devices is very important in the electronic industry and is pursued with increasing demand [1]. Some of the ceramic semiconductors such as tin oxide and zinc oxide with controlled porous bodies show an electrical conductivity as the response of trace amount of hydrocarbon based gases, hydrogen, and ammonia etc. at elevated temperature in air [2,3]. ZnO is one of the first and most widely used ceramic materials for sensor applications. Gas sensing in n-type semiconductor gas sensors like ZnO is achieved by the decrease in resistivity as a response of increasing amount of reducing gases in the air. The reason for a decrease in resistivity is explained by desorption of oxygen absorbed on the surface and grain boundaries of metal-oxides at high temperatures in air [3]. Furthermore, thin film type sensor devices are rising because of their low power consumption and easy integration into other devices. Film type devices can be fabricated with various techniques such as spray pyrolysis, sputtering, laser ablation, sol-gel method, MOCVD, CVD, PECVD, LPCVD, and LPE [4]. There is also another deposition technique, called electrophoretic (EPD), that involves migration of charged particles driven by an electric field in a liquid media by electric field to the surface of a substrate. Although, the above-mentioned techniques produce films with a limited thickness, the EPD

S. Guceri et al. (eds.), Nanoengineered Nanofibrous Materials, 385-389.

process can easily achieve thicknesses from 10nm to 1m, as seen on Fig.1. Because of its simplicity and versatility electrophoretic deposition might have potential for being the preferable technique for ceramic sensor applications. Accordingly there are some studies reporting the use of deposited ZnO films for sensor application. In this investigation, EPD techniques were used for the formation of ZnO thin films with controlled porosity on alumina substrates for sensor application.

2. EXPERIMENTAL PROCEDURE

Sub-micron sized ZnO particles were synthesized by using homogenous precipitation method [5]. The synthesized powders were characterized by using a particle size analyzer. It was observed that the particles were highly agglomerated even in 4μm size. To disperse the agglomerated particles a grinding process was employed. In this study, two types of milling, conventional and chemical added milling (CAM), were used to see the effect of dispersants on particle de-agglomeration during the milling of ZnO particles. The powder was milled for 22 hours.

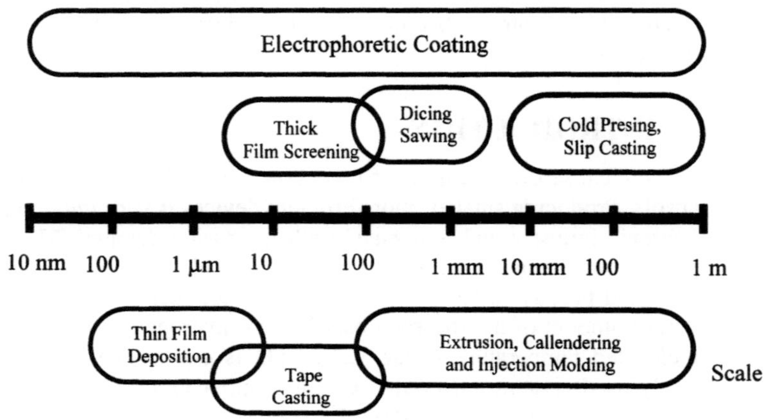

Figure. 1. Comparison of different techniques by EPD [1].

An anionic dispersant was used for the chemical aided milling to stabilize the 5 wt% ZnO aqueous suspension prepared for Electrophoretic Deposition (EPD). After milling sedimentation experiments were performed to test the stability of suspensions from each milling process.

After obtaining a well dispersed medium EPD works were carried out under constant DC voltage. High-grade alumina (Coors Ceramics) with 1 in.^{2} dimension was used as the substrate. The surfaces of the substrates were electroded with gold-palladium by sputtering process. A Pyrex glass deposition chamber was used for EPD studies due to its chemical and electrical inertness. By using stable suspension of ZnO, coating was done altering voltage and time. Coating was carried out in the range of 5-240 V. Coated samples were dried at room temperature to eliminate surface tension originated cracks [6]. Samples were sintered at 1200°C for two hours. Characteristics of the sintered ZnO coated substrates were investigated by scanning electron microscope (SEM) and X-ray difractometer (XRD).

3. RESULTS AND DISCUSSION

XRD pattern of synthesized powder shows that the formed phase is pure ZnO Fig.2.

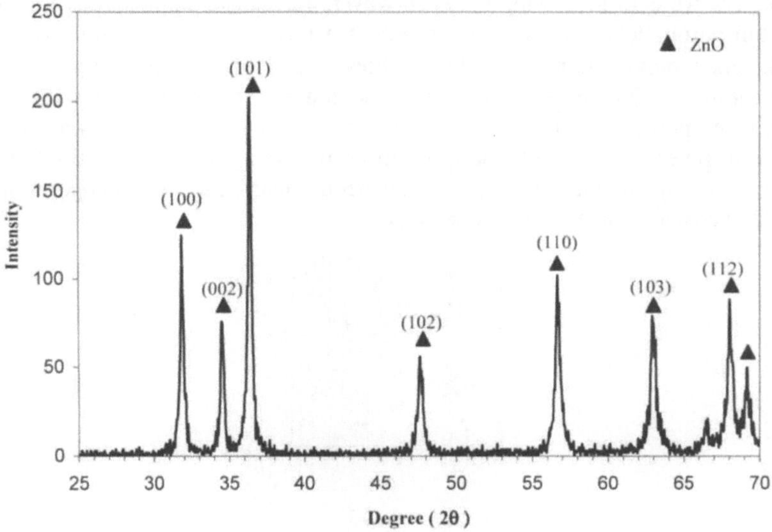

Figure. 2. XRD chart of the synthesized ZnO powder.

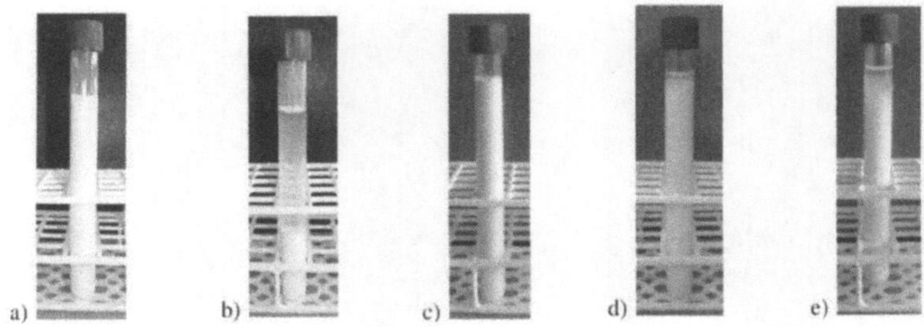

Figure. 3. Effect of dispersant to the stability of ZnO particles in aqueous media. a) Beginning of sedimentation test of ZnO particles without dispersant b) Sedimentation of ZnO particles without dispersant after 10 minutes c) Starting point of sedimentation test of ZnO particles with dispersant d) Sedimentation of ZnO particles with dispersant after 6 hours e) Sedimentation of ZnO particles with dispersant after 24 hours.

ZnO powder in aqueous medium is stabilized using anionic polyelectrolyte by negatively charging the surface, Fig.3. shows the well dispersed suspension. The reason of choosing anionic polyelectrolyte was the dissolution of ZnO in acid.

The particles in chemical aided milling form a sediment within about 10 minutes and have a diameter of approximately 120nm. The prepared suspension had a conductivity of 0.87mS/cm and a pH of 8.4 so it was stable enough for EPD. This study shows that chemical milling is required to achieve stable suspension of ZnO particles Fig.3. Coating layers were formed with EPD. However, for sensor applications rather thin porous layers are preferable. For that reason we focused on coating layers around 10μm Fig.4. Fig.5 shows SEM photographs of sintered coatings for 2 hours at 1200°C. As it is seen on the figure it is possible to create porous structure. However, our pore distribution was not homogeneous and for sensor applications uniform porosity is required. As a next step of this research, homogeneous distribution of nanopores in EPD coated films will be investigated.

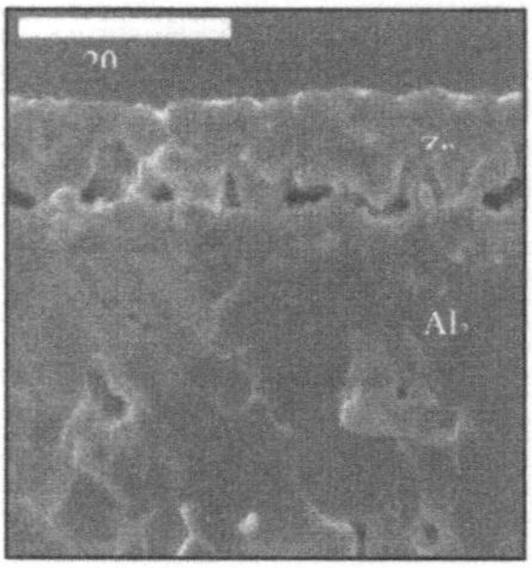

Figure. 4. SEM image of porous Al_2O_3 substrate and ZnO coating.

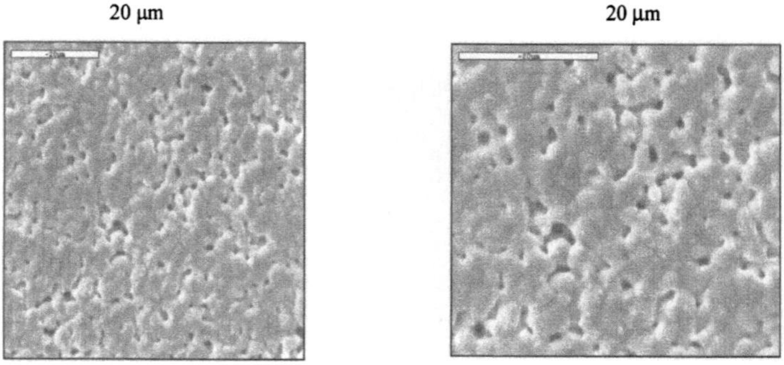

Figure. 5. SEM images of porous body by EPD after sintering 1200°C for 2 hours.

Density of the sintered body should be changed by using different powders, which have different grain sizes and sintering temperatures. In this experiment nearly 5μm powders and 2 □m pores were get after sintering as seen on Fig.5.

4. SUMMARY

It has been demonstrated that the EPD technique can be utilized to form ZnO films with ~10μm thickness for sensor applications. A stable suspension is necessary for a successful EPD process. The stable suspension of submicron ZnO particles can be achieved by chemically-aided milling approach.

Acknowledgements: The authors gratefully thank to C. A. Randall and J. Van Tassel of The Pennsylvania State University, MRI for their support. They also would like to thank Cem Caglar for the ZnO synthesis. This study is sponsored by Turkish State Planning Organization DPT (2003K120170).

References

1. OGATA, N., TASSEL, J. V. and RANDALL, C.A. (2001) Electrode formation by electrophoretic deposition of nanopowders, Mater. Let., 49, 7-14.
2. TRIVIKRAMA RAO, G.S. and TARAKARAMA RAO, D. (1999) Gas sensitivity of ZnO based thick film sensor to NH_3 at room temparature, Sensors and Actuators, B 55, 166-169.
3. MOON,W. J., YU, J. H. and CHOI, G. M. (2002) The CO and H_2 gas selectivity of CuO-doped SnO_2-ZnO composite sensor, Sensors and Actuators, B 87, 464-470.
4. CHOI. W. K., SONG, S.K., CHO, J. S., YOON, Y. S., CHOI, D., JUNG, H.J. and KOH, S. K., (1997) H_2 gas-sensing characteristics of SnO_x sensors fabricated by a reactive ion-assisted deposition with/without an activator layer, Sensors and Actuators, B 40, 21-27.
5. SAKKA, Y., HALADA, K., and OZAWA, E. (1988) *Ceramic Transactions-Ceramic Powder Science II A, Synthesis of ZnO Particles by the Homogeneous Precipitation Method,* The American Ceramic Society, Inc, Westerville, OH.
6. TANG, F., UCHIKOSHI, T. and SAKKA, Y. (2002) Electrophoretic Deposition Behaviour of Aqueous Nanosized Zinc Oxide Suspensions, J. Am. Ceram. Soc., 85, 2161-2165.

SYNTHESIS AND OPTICAL SPECTROSCOPY OF SINGLE-WALL CARBON NANOTUBES

E.D. OBRAZTSOVA[1], M. FUJII[2], S. HAYASHI[2], A.S. LOBACH[3], I.I. VLASOV[1], A.V.KHOMICH[4], V.Yu.TIMOSHENKO[5], W.WENSELEERS[6], E.GOOVAERTS[6]

[1] *Natural Sciences Center of A.M. Prokhorov General Physics Institute, RAS, 38 Vavilov str., 119991, Moscow, Russia,* *elobr@kapella.gpi.ru*
[2] *Department of Electrical and Electronics Engineering, Kobe University, 1-1 Rokkodai, Nada, Kobe 657-8501, Japan*
[3] *Institute of Chemical Physics Problems, RAS, 142432, Chernogolovka, Russia*
[4] *Institute of Radio Engineering & Electronics, RAS, 11 Mokhovaya str., 103907, Moscow, Russia*
[5] *Physics Department of M.V. Lomonosov Moscow State University, 119899, Vorob'evy gory, Moscow, Russia*
[6] *Department of Physics, University of Antwerp, 2610 Antwerp, Belgium*

Abstract: The abilities of three optical techniques (Raman scattering, UV-VIS-IR optical absorption and photoluminescence spectroscopy) for diagnostics of single-wall carbon nanotubes have been overviewed on example of nanotubes synthesized by two different methods - arc discharge and catalytical CO decomposition under high pressure.

Key words: Single-wall carbon nanotubes, Raman scattering, UV-VIS IR optical absorption, photoluminescence spectroscopy

1. INTRODUCTION

Discovery of carbon fullerenes and nanotubes has opened a new era in investigations of nanostructured materials. The existence of closed carbon shells has shown a possibility of self-organization of materials at nano-level leading to appearance of unique physical properties, often very different from those of bulk materials. As soon as a mass-production of carbon nanotubes has been developed [1], the diagnostics techniques being informative at typical distances 0.4-2 nm have been demanded. The optical methods appeared to be as effective as well-known high resolution structural methods (transmission electron microscopy, tunneling electron microscopy, atomic force microscopy). Data of optical measurements allowed to estimate a variety of single-wall carbon nanotube (SWNT) parameters: their diameters and chiralities, the distribution over diameters, rope diameters, conductivity type, positions of van Hove singularities in

S. Guceri et al. (eds.), Nanoengineered Nanofibrous Materials, 391-400.

one-electron density of states (DOS) for nanotubes of different geometry [2,3]. This work overviews possibilities of three optical methods – Raman scattering, optical absorption and photoluminescence spectroscopy- for diagnostics of single-wall carbon nanotubes, synthesized by two different techniques – direct dc-arc discharge and catalytical CO-gas decomposition under high pressure (HipCO).

2. EXPERIMENTAL

Nowadays more than fifteen different techniques are used for production of single-wall carbon nanotubes [4]. In this work we dealt with two of them: dc-arc discharge [5] and HipCO [6]. They have been chosen due to a big difference in geometrical parameters of tubes synthesized. For arc-produced material an average tube diameter is about 1.3-1.4 nm. Tube distribution over diameter is quite narrow (0.2-0.3 nm). For HipCO nanotubes the diameters vary in much wider range - from 0.7 nm to 2.0 nm.

The arc nanotubes have been produced with a home-made apparatus (Fig.1). The synthesis has been performed in helium atmosphere with

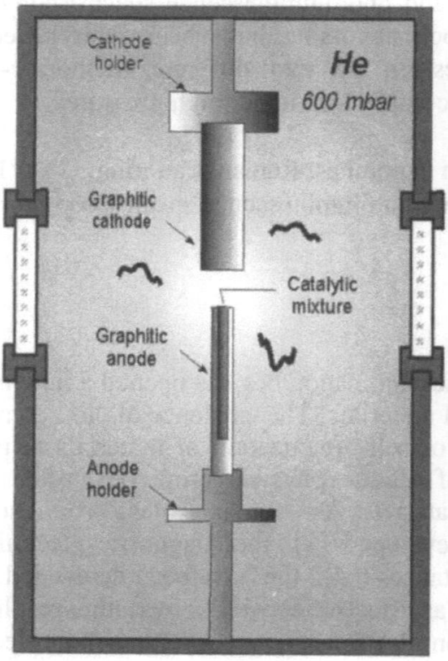

FIGURE 1. A scheme of dc-arc apparatus for SWNT synthesis.

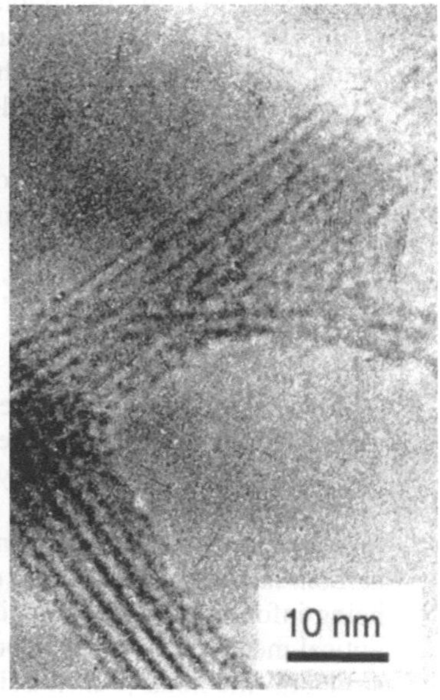

FIGURE 2. HRTEM image of arc-produced SWNT rope .

$Ni:Y_2O_3$ catalyst [7]. The raw material, containing about 20 weight % of SWNT, aggregated into ropes (Fig.2) [8], has been purified partially by chemical method, including the successive steps of acid treatment and oxidative annealing [9].

A commercially available raw HipCO material (Carbon Nanotechnologies, Inc., Houston, TX) has been used. A nanotube content was 80% already in the raw material. A high pressure (30 atm) of a buffer gas during synthesis (Fig.3) [6] provided a super-saturation of the catalyst particle with carbon atoms, resulting in a high yield of nanotubes and a presence of small-diameter nanotubes in the soot. HRTEM confirms a wide distribution of nanotubes over diameter in HipCO material [10].

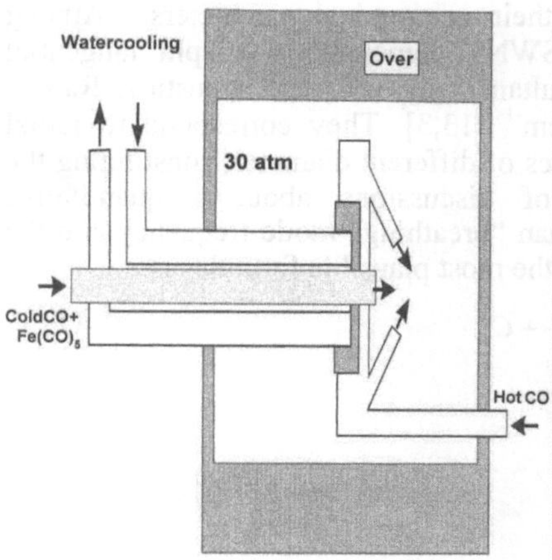

FIGURE 3. A scheme of apparatus for HipCO SWNT synthesis [6].

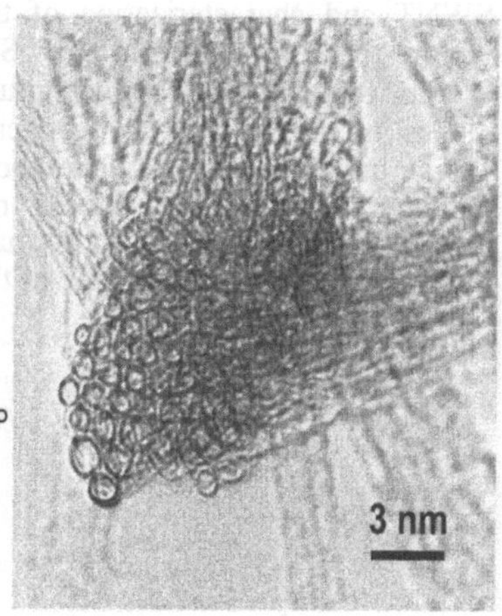

FIGURE 4. A cross-sectional HRTEM image of HiPCO nanotube rope.

The UV-VIS-NIR absorption spectra of SWNT dispersed in 2%-solution of SDS (sodium dodecyl sulfate) in D_2O were taken with a spectrophotometer (*Shimadzu UV-3101PC*), equipped with a double monochromator. It covered a wavelength range between 200 and 1800 nm. The SDS/D_20 solution without nanotubes was used as a reference sample in a second channel of the spectrometer. D_2O (instead of H_2O) was used to avoid the strong absorption bands of water in IR spectral range. For the rope disintegration the suspension was treated with *Tomy ultrasonic disruptor UD-201* (1 hour, power 40 W), followed by ultracentrifugation (*Himac CS150GX, Hitachi*) during 1 hour with acceleration 265 000 g. The upper fraction of the solution has been taken for measurements.

The Raman spectrometer *Jobin Yvon S-3000* was used. The spectra were excited by irradiation of Ar+-ion laser with different wavelengths. In some cases the radiation of *Ti: sapphire laser* was used also.

For photoluminescence *(PL)* measurements a 90°- geometry was used: the SWNT suspension in a quartz cell was excited from the side and the photoluminescence signal was collected from the front surface of the cell. The emitted light was focused on the slit of a single monochromator (30 cm focal length) and detected by a liquid nitrogen cooled InGaAs CCD (Princeton instruments). The excitation was made by a continuous wave *Ti: sapphire laser* (Spectra Physics Model 3900S) pumped with 532-nm diode-pumped solid state laser. The wavelength range covered by *Ti: sapphire* laser was between 700 and 1000 nm.

3. RAMAN SPECTROSCOPY OF SINGLE-WALL CARBON NANOTUBES

Raman technique appeared to be very efficient for identification of SWNT and characterization of their geometrical parameters. Among various carbon materials only SWNT demonstrates a split tangential Raman mode at 1592 cm^{-1} simultaneously with the acoustical Raman modes in the range 100- 400 cm^{-1} [13,3]. They correspond to radial "breathing" vibrations of nanotubes of different diameter, constituting the material. There were a lot of discussions about a quantitative correspondence between the Raman "breathing" mode frequency and the nanotube diameter value. One of the most plausible formulas is

$$\omega = \frac{C_1}{d} + C_2 \ , \qquad (1)$$

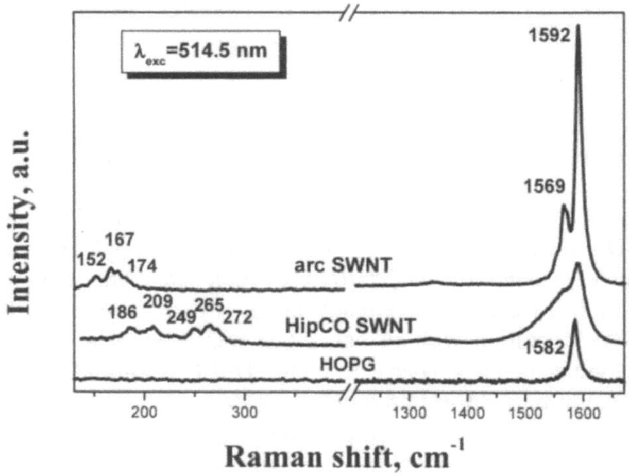

FIGURE 5. The typical Raman spectra for arc- and HiPCO-produced single-wall carbon nanotubes and for HOPG (as a reference).

where ω is measured in cm^{-1}, d is measured in *nm*, C_1=234, C_2=10 cm^{-1} [11]. The tube diameter estimation may be improved after comparison of

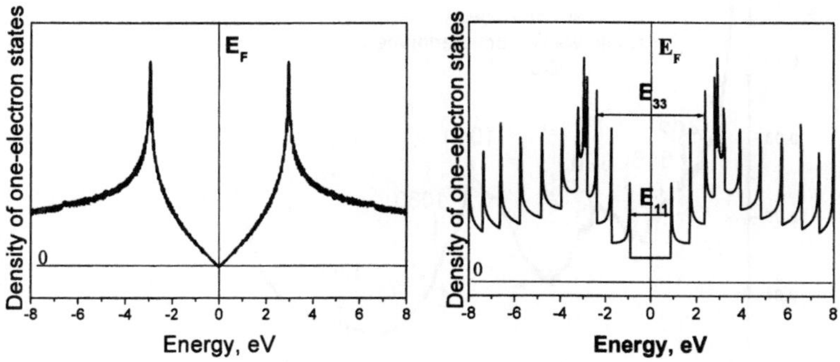

FIGURE 6. A schematic view of one-electron DOS for a two-dimensional graphite sheet (left), and for a single-wall carbon nanotube (right) .

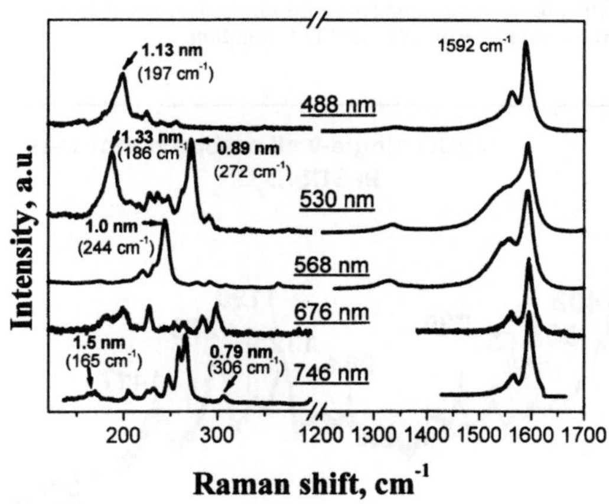

FIGURE 7. The Raman spectra of HiPCO single-wall carbon nanotubes registered with different laser excitation wavelengths.

the Raman and photoluminescence *(PL)* data for the same tubes [12].The Raman spectra of two types of SWNT materials (arc and HipCO) and for HOPG (highly oriented pyrolytic graphite), are shown in Fig.5. The number and positions of the "breathing" Raman modes indicate that a tube diameter distribution is much wider for HiPCO material than for the arc one. A different shape of the tangential modes [3] points to a preferable contribution to the Raman signal of semiconducting nanotubes in arc material and metallic nanotubes in HiPCO material (for λ_{exc}=514.5 nm).

Due to a quantum-scale circumference of a nanotube its electronic structure is different from that of graphite (Fig.6). Two-dimensional graphite is a zero-gap semiconductor (Fig.6, left). A nanotube may have semiconducting or metallic properties depending on the diameter value

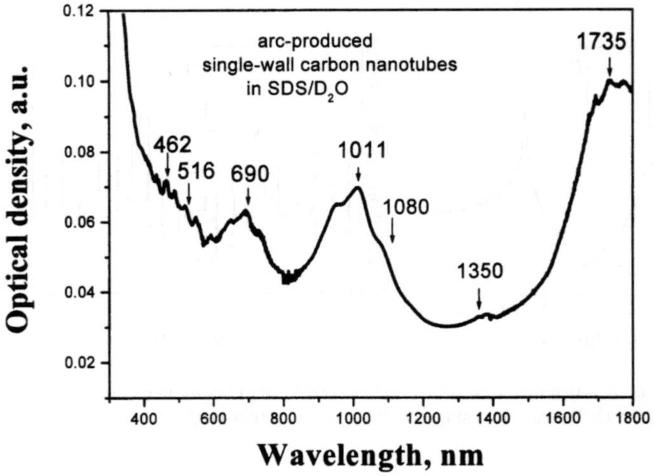

FIGURE 8. The optical absorption spectrum for suspension of arc-produces single wall carbon nanotubes in 2% SDS/D_2O solution.

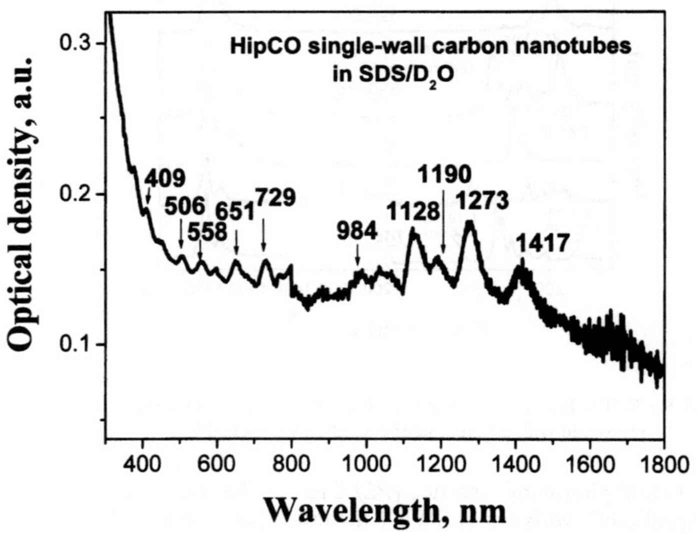

FIGURE 9. The optical absorption spectrum for suspension of HipCO single-wall carbon nanotubes in 2% SDS/D_2O solution.

[2]. One electron density of states (DOS) for nanotube looks as a set of the discrete maxima symmetric relative to Fermi level (E_F) (Fig.6, right). A separation of the maxima depends on the tube diameter. Optical transitions are allowed between the symmetrical maxima: E_{11}, E_{22}, E_{33}, etc. If the laser photon energy coincides with one of the transitions a resonance Raman scattering condition is realized for the tube with a fixed diameter. A laser tuning provides a step-by-step selective Raman

excitation of all tubes constituting the material. This is an efficient way to estimate a tube diameter distribution. Fig. 7 shows Raman spectra of HipCO material registered with the different excitation energies ranging from 1.66 eV (746 nm) to 2.54 eV (488 nm) [10]. The contributions of nanotubes with diameters from 0.8 nm to 1.5 nm are evident.

4. OPTICAL ABSORPTION SPECTROSCOPY OF SINGLE-WALL CARBON NANOTUBES

The efficiency of UV-VIS-IR (ultraviolet-visible-infrared) optical absorption spectroscopy for SWNT diagnostics has been demonstrated in the first time by H. Kataura et al. [14]. The absorption demonstrates features corresponding to E_{11}, E_{22}, E_{33}... optical transitions. Each nanotube material contains an ensemble of different nanotubes rather than a mono-diameter fraction. Due to this the features in the absorption spectrum should be broadened in accordance with the nanotube distribution over diameter. A raw nanotube material usually includes a lot of contaminations (polyhedral graphitic particles, amorphous carbon, catalyst particles, etc.). Their optical absorption contributes to the spectrum background. To reduce the background a suspension of SWNT in 2% solution of SDS in D_2O, treated by ultrasonication and ultracentrifugation, was used for the measurements.

The optical absorption spectrum of arc-produced nanotubes is shown in Fig.8. The absorption bands corresponding to $\mathbf{E_{11}}$ (centered at 1735 nm), $\mathbf{E_{22}}$ (1011 nm) and $\mathbf{E_{33}}$ (516 nm) transitions for semiconducting nanotubes and $\mathbf{E_{11}}$ (690 nm) transition for metallic nanotubes are well pronounced. The absorption band widths (about 0.1 eV) confirm a narrow distribution of arc nanotubes over diameter: 1.4 ± 0.15 nm. There are two local maxima in absorption at 1350 nm and 1080 nm. They may be ascribed to the smallest arc nanotubes. The photoluminescence measurements have shown that these maxima actually correspond to E_{11} transition.

The spectrum of HiPCO nanotubes is shown in Fig.9. Due to a wide distribution of nanotubes over diameter the bands corresponding to E_{11}, E_{22}, E_{33}... optical transitions are not separated in this case. The resolved peaks observed may be considered as the absorption of the individual tubes of fixed geometry. The peak widths (about 0.03 eV) are very close to a thermal broadening of the absorption bands.

5. PHOTOLUMINESCENT SPECTROSCOPY OF SINGLE-WALL CARBON NANOTUBES

Recently an efficiency of *photoluminescence (PL) spectroscopy* for characterization of single-wall carbon nanotubes has been demonstrated [15-17]. From general principles the semiconducting tubes should demonstrate the photoluminescence in any case of optical excitation with the photon energy exceeding the gap value. In practice, during several years all attempts to observe PL signal from carbon nanotubes failed. The problem has been solved only when experimentalists have succeeded in disintegration of ropes into individual nanotubes. The rope is not a chaotic agglomeration of individual nanotubes (Fig.2, Fig.4). It is a one-dimensional crystal consisting of aligned nanotubes with van der Waals interaction between them. The tubes in the same rope may have different geometry [18]. Depending on geometry they are metals or semiconductors. Usually there are few metallic nanotubes in each rope. They serve as a non-radiative channel for relaxation of excitation occurred in the neighboring semiconducting tubes. Due to this relaxation no photoluminescence signal is observed.

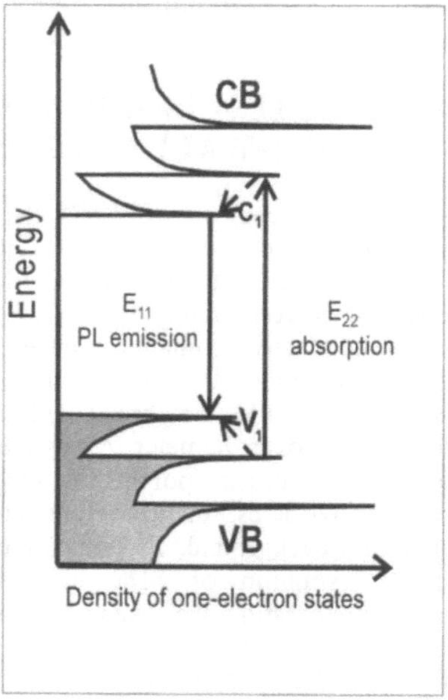

Figure 10. A scheme of PL excitation in individual semiconducting carbon SWNT. **CB**- conductance band. **VB**- valence band. DOS has shown as a set of van Hove singularities. The dashed lines show non-radiative transitions for electrons and holes [16].

This obstacle may be overcome by disintegration of nanotube ropes via interaction with the surfactant molecules under ultrasonic treatment in the water suspensions. This process leads to formation of surfactant-coated micelles. There are three types of micelle nucleus: an individual tube, a rope fragment (containing a few tubes) or a carbon nanoparticle. The single-tube micelles constitute a lightest fraction of the suspension. Namely this fraction should demonstrate PL. For the lightest micelle separating the suspension is subjected to ultracentrifugation followed by decanting of the upper fraction.

An electronic DOS for SWNT has only one real gap ($\mathbf{E_{11}}$). All others are called "pseudo"-gaps due to a presence of a very small, but non-zero DOS between the symmetric van Hove singularities (Fig.10). This means that PL emission process is possible only for $\mathbf{E_{11}}$ electronic transition, while for PL excitation any wavelength in a wide spectral range may be used. As a result, a spectral structure of PL peak for a given nanotube material informs about a set of energies of the first electronic transitions and, as a consequence, about the diameters and geometries of nanotubes, constituting the material.

The photoluminescence spectra of *arc* (Fig.11) and *HiPCO* (Fig.12) nanotubes have demonstrated different sets of E_{11} peaks. A comparison of PL peak positions with the Raman data (Fig.7) taken with the same excitation energy (E_{exc}=2.33 eV, λ=532 nm) has allowed to estimate the individual tube geometries [17]. The indices *(n,m)* are indicated above the PL peaks in Fig.11 and Fig.12. PL data have confirmed the estimations of the tube diameter distributions for *arc* and *HipCO* materials made by Raman and UV-VIS-NIR techniques.

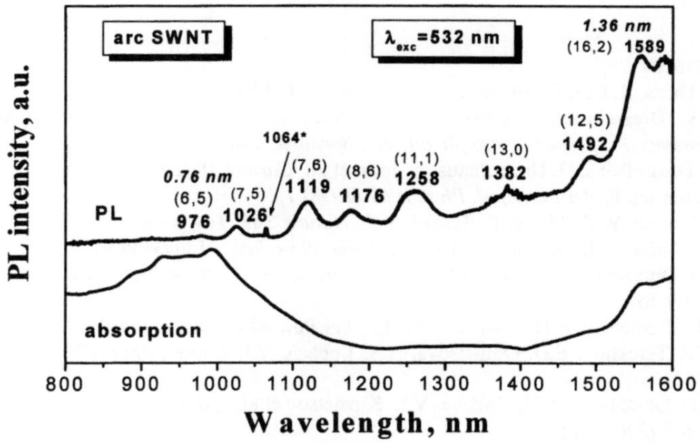

FIGURE 11. The photoluminescence (upper) and the optical absorption (lower) spectra for the suspension of *arc-produced* SWNT in 2% - SDS/D_2O solution.

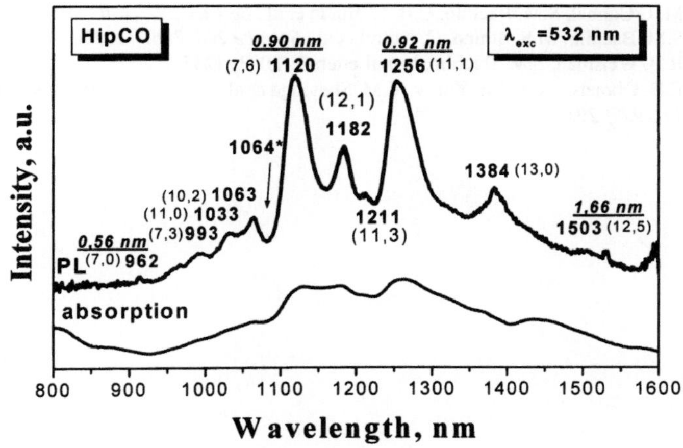

FIGURE 12. The photoluminescence (upper) and the optical absorption (lower) spectra for the suspension of HipCO SWNT in 2% -SDS/D_2O solution.

5. CONCLUSION

A complex optical diagnostics based on Raman, UV-VIS-NIR absorption and photoluminescence techniques, has been demonstrated as an efficient tool for structural characterization of single-wall carbon nanotubes synthesized by different methods.

Acknowledgements: The work is supported by INTAS project 01-254 and by programs of Russian Academy of Sciences "Low-dimensional Quantum structures" and "Optical Spectroscopy and Frequency Standarts".

References

1. A. Thess, R. Lee, P. Nikolaev et al., *Science* ***273*** *(1996) 483.*
2. M.S. Dresselhaus, G.Dresselhaus, P. Avouris, *"Carbon Nanotubes: Synthesis, Structure, Properties and Applications", Springer, 2001.*
3. M. Dresselhaus, G. Dresselhaus, A. Jorio et al., *Carbon* ***40*** *(2002) 2043.*
4. C. Journet, P. Bernier, *Appl. Phys.* ***A 67*** *(1998) 1.*
5. C. Journet, W.K. Maser, P. Bernier et al., *Nature* ***388*** *(1997) 756.*
6. P. Nikolaev, M.J. Bronikovski, et al., *Chem. Phys. Lett.* ***313*** *(1999) 91.*
7. E.D. Obraztsova, J.-M. Bonard, V.L. Kuznetsov et al., *NanoStructured Materials 12 (1999) 567.*
8. V.L. Kuznetsov, A.N. Usoltseva, et al., *Phys.Rev.* ***B64*** (2001) 235401.
9. S.V. Terekhov, E.D. Obraztsova, A.S. Lobach, V.I. Konov, *Appl. Phys. A* **74,** 393 (2002).
10. E.D. Obraztsova, S.N. Bokova, V.L. Kuznetsov et al., *AIP Conferences Proceedings. v. 685 (2003) 215.*
11. H. Kuzmany, W. Plank, M. Hulman et al., *Eur. Phys. J.* ***B22*** *(2001) 307.*
12. M.S. Strano, *J. Am. Chem. Soc. 125 (2003) 16148.*
13. A. M. Rao, E. Richter, S. Bandow et. al., *Science* ***275*** *(1997) 187.*
14. H. Kataura, Y. Kumazawa, Y. Maniwa et al., *Synth. Met. 103 (1999) 2555.*
15. M. O'Connel, S.M. Bachilo, C.B. Huffman et al., *Science 297 (2002) 593.*
16. S.M. Bachilo, M.S. Strano, C. Kittrel et al., *Science 298, 2361 (2002).*
17. R.B. Weisman, S.M. Bachilo, NanoLetters 3 (2003) 1235.
18. E.D. Obraztsova, V.Yu. Yurov, V.M. Shevluga et al., *NanoStructured Materials 11(1999) 295.*

AQUEOUS FLUIDS IN CARBON NANOTUBES: ASSISTING THE UNDERSTANDING OF FLUID BEHAVIOR AT THE NANOSCALE

Almila G. Yazicioglu, Constantine M. Megaridis
Department of Mechanical and Industrial Engineering, University of Illinois at Chicago, Chicago, IL, 60607, USA

Nevin Naguib, Haihui Ye, Yury Gogotsi
Department of Materials Engineering, Drexel University, Philadelphia, PA, 19104, USA

Abstract: Transmission electron microscopy (TEM) experiments are performed on carbon nanotubes filled with a multiphase aqueous fluid. It is shown through high resolution TEM that the dense aqueous fluid wet the inner surface of the carbon nanotube, possibly due to the presence of oxygen-based groups, which modify the carbon surface such that it becomes hydrophilic. Electron energy loss spectroscopy (EELS) analysis confirms the presence of oxygen inside the fluid-filled nanotubes, as well as on the inner and outer surfaces of the nanotube walls. Fluid experiments involving controlled liquid evaporation and condensation, as driven by external heating, demonstrate that liquid transport in nanochannels of less than 100 nm in diameter is achievable. The present results offer promise for the development and functional performance of future micro- and nanofluidic devices.

Keywords: Carbon nanotubes, liquid transport, wetting, hydrophilic

1. INTRODUCTION

The study of carbon nanotubes has been a rapidly progressing research area for the past decade. Most of the ongoing research is focussed on nanotube structure and their electronic, transport and optical properties [1]. However, multiwall carbon nanotubes also show potential for use in various micro- and nanofluidic devices, since they resemble cylindrical channels used in the macroscopic world. They provide a test platform orders of magnitude smaller than capillaries used in previous fluid experiments that allowed in-situ observations of fluid phenomena. It is therefore essential to investigate their thermal and fluidic properties.

S. Guceri et al. (eds.), Nanoengineered Nanofibrous Materials, 401-408.

Understanding fluid behavior at the nanoscale would facilitate the effective design and operation of future micro- and nanofluidic devices, such as drug delivery systems, biosensors and micro heat pipes. Fluid behavior in carbon nanotubes may be significantly different from that in conventional pipes, even microcapillaries. However, very few studies have been reported on the dynamic behavior of fluids at the nanoscale [2,3]. Some reports have shown that nanotubes can be filled partially with salts, molten metals, and metal oxides, such as lead [4,5], silver [5,6], chromium oxide [7], and cobalt [8] through capillarity if the liquid shows good wettability of the inner carbon walls. The procedure in these studies was either a chemical method, such as wet chemistry, or a physical method, such as capillarity induced filling. In either case, little has been reported on the *dynamic* aspects of fluid filling and fluid transport in the nanotubes. Molecular dynamics simulations of water in cylindrical nanopores showed that confined water remains fluid and bulk-like [9]. Hummer et al. [10] performed molecular dynamics simulations and observed pulse-like transmission of a one-dimensionally ordered chain of water molecules through a nanotube, due to the tight hydrogen-bonding network inside the tube. Noon et al. [11] reported that water molecules inside nanotube segments tend to organize themselves in a solid-like network, due to the hydrogen-bonding, in forms of wrapped-around ice sheets. In all of the abovementioned cases, the diameter of the channel was no more than 2 nm, which may be too small for nanofluidic applications. Furthermore, most of these theoretical studies lack experimental verification.

The objectives of this study are to investigate high quality multi walled carbon nanotubes with aqueous fluid inclusions, and to perform high resolution in-situ thermal experiments using electron irradiation as a means of heating the contents of individual nanotubes in the high vacuum of a transmission electron microscope (TEM) column. Wetting of these nanotubes by the aqueous liquid is observed with high resolution microscopy as well. As the fluids investigated herein are contained at very high pressures (hundreds of atmospheres), the present work offers a unique opportunity for studying the dynamic behavior of multicomponent, multiphase fluids in nanosize channels at conditions corresponding to sub-, near- or even super-critical regions of the thermodynamic diagram, which may help create an experimental database for the parallel development of theoretical models describing fluid behavior in nanosize channels.

2. EXPERIMENTAL

The nanotubes are synthesized hydrothermally from a mixture of ethylene glycol, de-ionized water, and Ni powder (catalyst) sealed into gold capsules, using a two-step process, see procedure in Ref. 12. This procedure produces nanotubes, which have high aspect ratios, high degree of graphitization, and smooth internal walls, with wall thickness ~10-25 nm (30-70 fringes per wall), outer diameter ~100 nm, and lengths 1-10 μm. They are rarely open-ended, while most of the closed-end nanotubes encapsulate a high-pressure multiphase aqueous fluid, displaying segregated liquid and gas inclusions with clearly defined interfaces [13,14]. A schematic of this system can be seen in Fig. 1. The composition of the multicomponent fluid at room temperature is predicted to be 85.2% H_2O, 7.4% CO_2, and 7.4% CH_4, with traces of H_2 and CO, as based on thermodynamic calculations performed at the synthesis conditions [15]. It is also estimated that this composition is almost independent of pressure at room temperature [16].

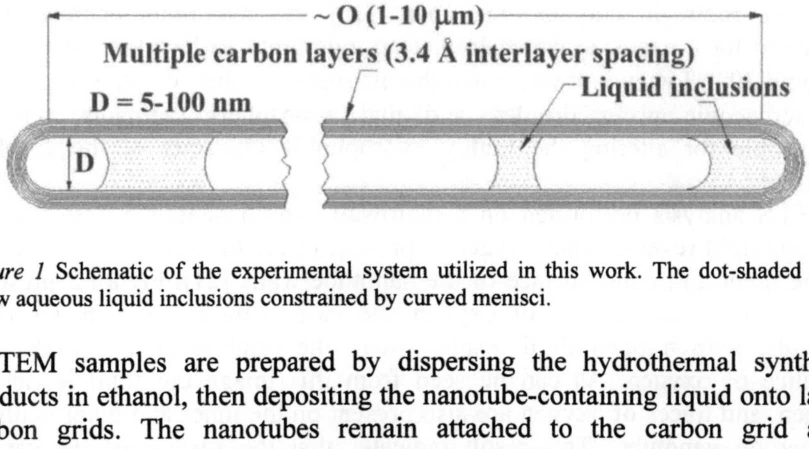

Figure 1 Schematic of the experimental system utilized in this work. The dot-shaded areas show aqueous liquid inclusions constrained by curved menisci.

TEM samples are prepared by dispersing the hydrothermal synthesis products in ethanol, then depositing the nanotube-containing liquid onto lacey carbon grids. The nanotubes remain attached to the carbon grid after evaporation of the liquid carrier. Thermal experiments are performed using electron irradiation as a means of heating the contents of individual nanotubes in the high vacuum of the TEM (JEOL JEM-3010, 300 kV, 0.14 nm lattice resolution), where the multiphase fluid is excited by expanding/contracting the low-intensity electron beam. The dynamics of the fluid phenomena in the nanotubes are visualized and recorded using a CCD camera and a Betacam video recorder.

The nanotubes in the present study are leak-tight, often for days, and they can maintain the pressurized fluid in the vacuum environment of a TEM column (~10^{-6} Pa) even at temperatures above 500°C, thus demonstrating a high strength and wall perfection. This makes them very suitable for studying dynamic fluid behavior at length scales possibly approaching the continuum limit. Moreover, the superior smoothness of their inner surfaces and the excellent definition of fluid interfaces make these nanotubes a perfect candidate for contact-line and wettability studies at the nanoscale.

3. RESULTS AND DISCUSSION

High-resolution TEM experiments performed on liquid-carbon interfaces, such as the ones presented in Fig. 1, revealed that the aqueous fluid wet the inner surface of the carbon nanotube. Previously there have been contradictory reports on the hydrophobic/philic behavior of carbon. Dujardin *et al.* [17] suggested that materials with surface tensions lower than 100-200 mN/m should be drawn into the nanotube via capillarity. Also, Ebbesen [18] reported experiments of placing a droplet of water (surface tension 73 mN/m) on a bundle of carbon nanotubes; the water was sucked into the nanotube channels, leaving a dry surface. On the other hand, Li *et al.* [19] studied experimentally the wettability of aligned carbon nanotubes, and reported the contact angle for water droplets on untreated aligned nanotube films to be ~160°, which confirmed super-hydrophobic behavior of the aligned carbon nanotube films. In another experiment [20], the maximum contact angle measured for a nanosize water droplet on a graphite surface was reported to be about 30°. However, it was noted therein that the value would be higher for a macroscopic water droplet, and that atmospheric aerosols may be responsible for altering the highly hydrophobic character of the graphite substrate.

EELS analysis performed on a multiwall carbon nanotube with trapped aqueous fluid revealed that oxygen is present inside the nanotube, as well as on the inner and outer surface of the nanotube walls [21]. Figure 2 presents the composite EELS map of oxygen and carbon distribution in a carbon nanotube with trapped fluid inside, where the brighter areas indicate the presence of oxygen. As can be seen from this image the fluid is rich in oxygen, and traces of oxygen are also present on the inner and outer walls of the carbon nanotube. This result indicates that the surface of the carbon nanotube can adsorb oxygen, leading to the formation of functional groups, such as hydroxyl, which have been shown to transform the carbon nanotube surface to a hydrophilic one [19].

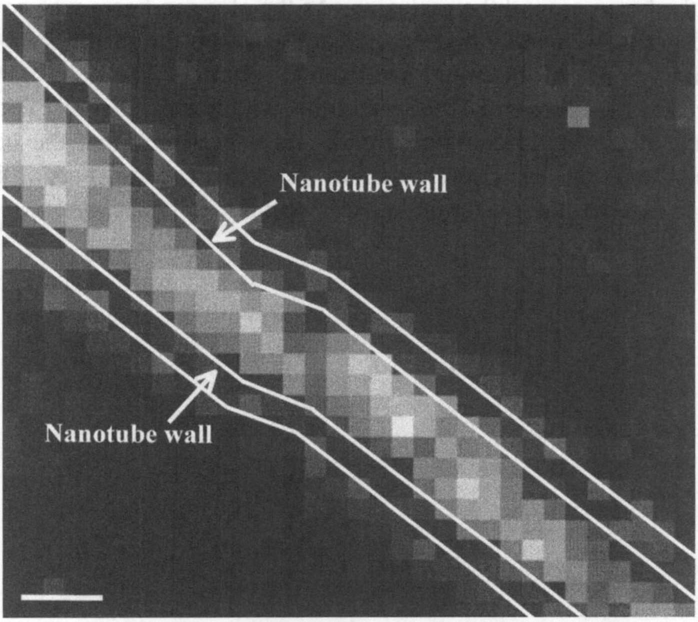

Figure 2 Composite EELS map of oxygen and carbon distribution in a carbon nanotube with trapped fluid inside. Brighter areas indicate more intense presence of oxygen. The nanotube wall is marked by thick white lines. Traces of oxygen are present on the inner and outer walls of the nanotube. Scale bar 50 nm.

Structural changes have been seen in the graphite layers in contact with the liquid inside the carbon nanotube. This concept is presented in Fig. 3. Layers of graphite sheets terminating at the inner surface of the nanotube often bend away from the wall, towards the axis of the nanotube, in the presence of a thin aqueous liquid layer. It should be mentioned that similar changes occur on the outer nanotube wall due to the presence of the fluid on both sides of the wall during the hydrothermal synthesis process, which clarifies the adsorption of oxygen on the outer wall, as shown on the EELS map. Three inner sheet endings are shown, marked by arrows, in Fig. 3(a), with the corresponding schematic for possible attachments of functional groups presented in Fig. 3(b). When graphite sheets are terminated with oxygen based groups, as shown in Fig. 3(b), this allows water bonding to the inner wall of the nanotube, thus rendering the surface hydrophilic.

As an example of the dynamic phase change experiments performed in this work, an amount of the aqueous liquid was transported from one end of a closed carbon nanotube compartment, 100 nm in diameter, to the other end via evaporation and condensation. Initially, the liquid inclusion was constrained between an internal closure and a long gas bubble. Adjacent to the gas bubble was another internal closure, encapsulating the liquid and the gas in a closed compartment of the carbon nanotube, about 1 μm long. The experiment consisted of gently heating the liquid via electron irradiation, by focusing the beam around the inclusion. As the electron beam was contracted, the liquid gradually evaporated upon heating. When the heating was ceased, approximately 65% of the liquid had been evaporated. The field of view was then moved to the other end of the nanotube, which was relatively cool during the heating of the liquid-containing end, and it was seen that the liquid had condensed there. This experiment demonstrates that the transport of liquids in response to external thermal stimulation is possible in nanochannels.

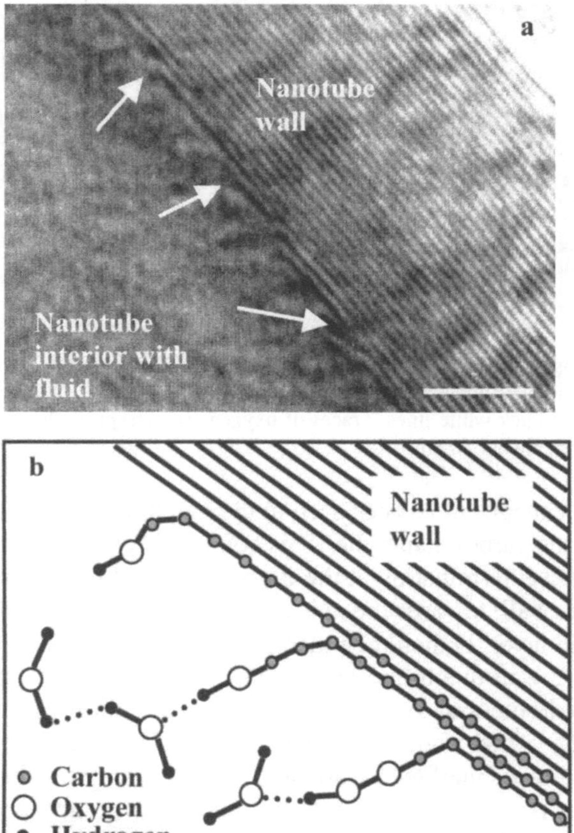

Figure 3 Structural changes occur in the graphite layers that come in contact with the liquid inside the carbon nanotube. (a) Arrows show graphite sheets terminating at the inner surface of the nanotube, bending away from the wall in the presence of the liquid. Scale bar 2 nm. (b) Oxygen based groups attached to carbon allow water bonding to the inner wall of the nanotube.

4. CONCLUSIONS

Carbon nanotubes with aqueous fluid inclusions have been used as a test platform for unique in-situ nanofluidic experiments in the TEM. It has been shown that the aqueous fluid wet the inner surface of the carbon nanotube, possibly due to the formation of functional groups, such as hydroxyl, which render the surface hydrophilic. The results demonstrate the feasibility of liquid transport in response to external thermal stimuli in pipes of <100 nm in diameters, which could allow for future development of various micro- and nanofluidic devices.

References

1. R. Saito, G. Dresselhaus and M. S. Dresselhaus, Physical Properties of Carbon Nanotubes. Imperial College Press, London, 1998.
2. V. D. Sobolev, N. V. Churaev, M. G. Velarde, and Z. M. Zorin, Surface Tension and Dynamic Contact Angle of Water in Thin Quartz Capillaries, Journal of Colloid and Interface Science, vol. 222, pp. 51-54, 2000.
3. V. N. Bogomolov, Capillary Effects in Ultrathin Channels, Soviet Physics-Technical. Physics, vol. 37, pp. 79-82, 1992.
4. P.M. Ajayan and S. Iijima, Capillarity-induced Filling of Carbon Nanotubes, Nature, vol. 361, pp. 333-334, 1993.
5. D. Ugarte, T. Stockli, J.M. Bonard, A. Chatelain, and W.A. de Heer, Filling Carbon Nanotubes, Applied Physics A, vol. 67, pp. 101-105, 1998.
6. J. Sloan, D.M. Wright, H.G. Woo, S. Bailey, G. Brown, A.P.E. York, K.S. Coleman, J.L. Hutchison, and M.L.H. Green, Capillarity and Silver Nanowire Formation Observed in Single Walled Carbon Nanotubes, Chemical Communications., vol. 8, pp. 699-700, 1999.
7. J. Mittal, M. Monthioux, H. Allouche, and O. Stephan, Room Temperature Filling of Single-wall Carbon Nanotubes with Chromium Oxide in Open Air, Chemical Physics Letters, vol. 399, pp. 311-318, 2001.
8. Y.L. Hsin, K.C. Hwang, F.R. Chen, and J.J. Kai, Production and In-situ Metal Filling of Carbon Nanotubes in Water, Advanced Materials, vol. 13, pp. 830-833, 2001.
9. R. Allen, S. Melchionna, and J.P. Hansen, Intermittent Permeation of Cylindrical Nanopores by Water, Physical Review Letters, vol. 89, pp. 175502-1-4, 2002.
10. G. Hummer, J.C. Rasaiah, and J.P. Noworyta, Water Conduction through the Hydrophobic Channel of a Carbon Nanotube, Nature, vol. 414, pp. 188-190, 2001.
11. W.H. Noon, K.D. Ausman, R.E. Smalley, and J. Ma, Helical Ice-sheets inside Carbon Nanotubes in the Physiological Condition, Chemical Physics Letters, vol. 355, pp. 445-448, 2002.
12. J. Libera and Y. Gogotsi, Hydrothermal Synthesis of Graphite Tubules using Ni Catalyst, Carbon, vol.39, pp. 1307-1318, 2001.
13. Y. Gogotsi, J. Libera, A.G. Yazicioglu, and C.M. Megaridis, In situ Multiphase Fluid Experiments in Hydrothermal Carbon Nanotubes, Applied Physics Letters, vol. 79, pp. 1021-1023, 2001.
14. C.M. Megaridis, A.G. Yazicioglu, J.A. Libera, and Y. Gogotsi, Attoliter Fluid Experiments in Individual Closed-end Carbon Nanotubes: Liquid Film and Fluid Interface Dynamics, Physics of Fluids, vol. 14, pp. L5-L8, 2001.
15. Y. Gogotsi, T. Kraft, K.G. Nickel, and M.E. Zvanut, Hydrothermal Behavior of Diamond, Diamond and Related Materials, vol. 7, pp. 1459-1465, 1998.
16. Y. Gogotsi, N. Naguib, and J.A. Libera, In situ Chemical Experiments in Carbon Nanotubes, Chemical Physics Letters, vol. 365, pp. 354-360, 2002.
17. E. Dujardin, T. W. Ebbesen, H. Hiura, and K. Tanigaki, Capillarity and Wetting of Carbon Nanotubes, Science, vol. 265, pp. 1850-1852, 1994.
18. T. W. Ebbesen, Annual Review of Materials Research, vol. 24, p. 235, 1994.
19. H. Li, X. Wang, Y. Song, Y. Liu, Q. Li, L. Jiang, and D. Zhu, Super-"amphiphobic" Aligned Carbon Nanotube Films, Angewandte Chemie International Edition English, vol. 40, pp. 1743-1746, 2001.

20. M. Luna, J. Colchero, and A. M. Baro, Study of Water Droplets and Films on Graphite by Noncontact Scanning Force Microscopy, Journal of Physical Chemistry B, vol. 103, pp. 9576-9581, 1999.
21. Y. Gogotsi, N. Naguib, H. Ye, A.G. Yazicioglu, and C.M. Megaridis, Analysis of Fluid in Hydrothermal Fluids, Proc. Electrochemical Society, in Fullerenes vol. 12: The Exciting World of Nanocages and Nanotubes, pp. 451-459, 2002.

NANOFABRICATION OF CARBON NANOTUBE (CNT) BASED FLUIDIC DEVICE

M. Riegelman[1], H. Liu[1], S. Evoy[2], and H. H. Bau[1,*]
[1]*Department of Mechanical Engineering and Applied Mechanics, University of Pennsylvania, Philadelphia, PA 19104*
[2]*Department of Electrical and Systems Engineering, University of Pennsylvania, Philadelphia, PA 19104*

Abstract: A method for the fabrication of nanotube and nanopore based fluidic devices is described. With the aid of dielectrophoresis, a tube is assembled in the gap between two electrodes that were patterned on a silicon or glass wafer. Subsequently, standard microfabrication techniques were applied to fashion fluidic conduits and wells to facilitate fluid transport through the tube. The device will be used for fundamental studies of the transport of fluids and macromolecules under nanometer scale confinement and as a high sensitivity biosensor. We describe experimental observations of the trapping of a single tube, the fractal patterns formed when the process is allowed to proceed for an extended period of time, and the construction of a nanotube based fluidic device.

1. INTRODUCTION

Recent advances in the synthesis of nanotubes and nanofibers offer numerous opportunities for fundamental studies of transport and interactions under extreme confinement, and for the development of novel, highly sensitive sensors and measurement techniques. As one example of a nanotube-based sensor, we describe the fabrication of a Coulter counter-like device [1]. The device is described schematically in Figure 1.

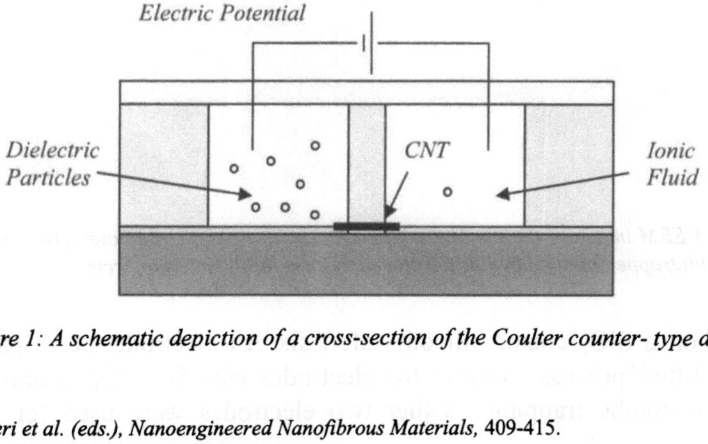

Figure 1: A schematic depiction of a cross-section of the Coulter counter- type device

S. Guceri et al. (eds.), Nanoengineered Nanofibrous Materials, 409-415.

The device consists of a nanotube positioned between two wells. Both the wells and the tube are filled with ionic solution. Functionalized particles or macromolecules are introduced in the left well. An electric field is applied across the tube, and the ionic electric current is continuously measured. The electric field causes the particles/macromolecules to (electrophoretically) migrate through the tube. As a particle enters the tube, it interferes with the current flow, causing a depression in the current's magnitude. By monitoring and analyzing the current's intensity as a function of time, one can obtain information on the particle's size and characteristics.

The Coulter Counter can be used as a chemical or biological sensor. As analytes bind to functionalized nanoparticles, the apparent size of the particles changes and can be detected through the monitoring of the alternation in the magnitude of the electric current as the particles pass through the pore. The smaller the pore diameter is, the greater the device's sensitivity.

In this paper, we briefly describe a novel technique for fabricating the nanotube-based device. Since the nanotubes are too small to manipulate mechanically, we utilize dielectrophoresis to assemble the tubes at predetermined locations. Subsequently, standard photolithographic techniques are used to fabricate the wells and the microfluidic conduits.

2. DEVICE FABRICATION

Our device, depicted in Figure 2, was fabricated through a combination of standard microfabrication processes and dielectrophoretic trapping of individual carbon nanotubes.

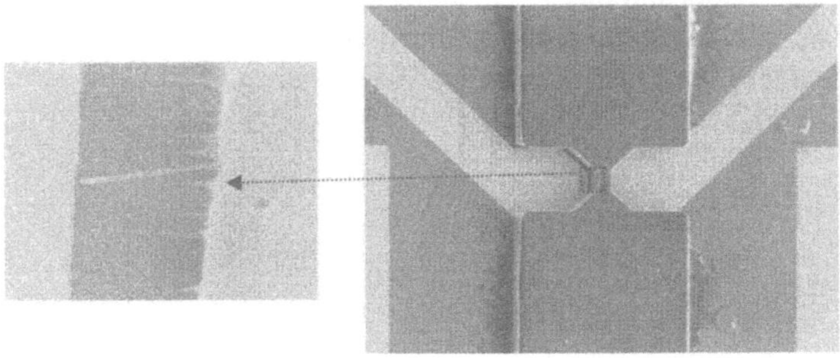

Figure 2: A SEM image of the nanotube-based fluidic device. SU-8 photoresist spun over a substrate with trapped nanotubes was developed to form fluidic interconnects.

First, on a SiO_2 coated silicon wafer, six Au electrodes were patterned using the liftoff process. Two of the electrodes with 8μm gap spacing, were used for nanotube trapping. Other two electrodes were used for driving particles through the carbon nanotube (CNT) pore, and the last two electrodes were used for sensing through a current or resistance spike, when the particles traveled through the CNT. The electrode layout is depicted in Figure 3.

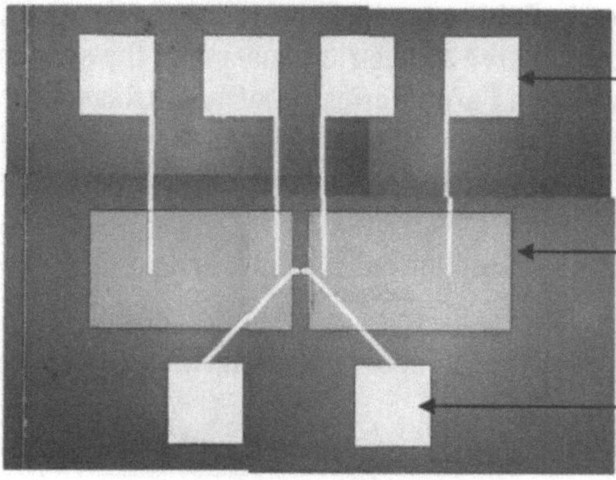

Figure 3: An image of the electrodes' layout. The trapping, driving, and sensing electrodes are photographed. Also shown are the locations of the liquid reservoirs.

Next the carbon nanotubes [2], synthesized using an alumina template, were suspended in an isopropanol solution to a concentration of approximately 0.1μg/L. These particular tubes were selected because they are open-ended, straight, and possess uniform pore geometry with outer diameters of approximately 250nm, wall thickness of 15nm, and lengths of up to 50μm.

The CNTs were then assembled between the two trapping electrodes using dielectrophoresis. This process is discussed in detail in the following section. Next, 200nm of SiO_2 was wet etched in hydrofluoric acid to liftoff the isopropanol residue. Finally, a 50μm thick layer of SU-8 (MicroChem Inc.) negative photoresist was spun onto the chip and patterned to define the reservoirs on both ends of the CNT.

3. DIELECTROPHORETIC TRAPPING -EXPERIMENTS

Dielectrophoresis (DEP) is the motion of uncharged, polarizable particles in a non-uniform electric field [3]. The dielectrophoretic forces result from the interaction of the electric field with a dipole induced in the particles. Depending on the particles' and the medium's dielectric properties, the motion is directed either toward (positive DEP) or away from (negative DEP) regions of high electric field intensity. While DEP occurs in both DC and AC electric fields, AC fields are often preferred in order to suppress undesired electrochemical interactions at the electrodes' surfaces and migration due to the particles' electric charge.

The dielectrophoretic force is usually approximated through its first dipole moment contribution [4]:

$$\mathbf{F} \sim c\varepsilon_0\varepsilon_m V_p \,\mathrm{Re}(f_{CM})\nabla\left|\mathbf{E}^2_{rms}\right|, \qquad (1)$$

where **E** is the electric field; ε_0 and ε_m are, respectively, the dielectric constant of free space and the relative dielectric constant of the suspending liquid; V_p is the particle's volume; c is a numerical coefficient on the order of unity; and $Re(f_{CM})$ is the real part of the relative particle polarization (i.e., the Clausius-Mossotti factor). Both c and f_{CM} depend on the particle's geometry. For example, in the case of a spherical particle, $f_{CM} = \dfrac{\varepsilon_p^* - \varepsilon_m^*}{\varepsilon_p^* + 2\varepsilon_m^*}$ and $c = 3/2$, where ε_p^* and ε_m^* are the complex permittivities of the particle and of the medium, respectively; $\varepsilon^* = \varepsilon - j\dfrac{\sigma}{\omega}$; and σ is the electric conductivity.

Equation (1) is valid only when the particle's dimensions are smaller than 10% of the length scale associated with the electric field ($|\mathbf{E}|/|\nabla\mathbf{E}|$) [5,6]. This is not the case in our experiments. More accurate calculations can be carried out utilizing either the principle of virtual work or the Maxwell stress tensor [7]. Nevertheless, we will use here equation (1) to qualitatively explain our experimental observations. Witness that the dielectrophoretic force's magnitude is proportional to the particle's volume, the difference between the particle's and medium's dielectric properties, and the gradient of the square of the electric field. This last term makes the method attractive for nanoassembly. By patterning small gaps between electrodes, one can obtain very high intensity electric fields with moderate potential differences.

CNTs with high permittivity are suspended in isopropanol solution with low permittivity thereby maximizing the permittivity difference. Large electric field gradients are generated through microfabricated electrodes with small electrode gap spacing. Figure 4 illustrates how microfabricated electrodes with small gap spacing are capable of producing large electric field gradients. The lines correspond to the normalized electric field as a function of x-at various horizontal planes above the microfabricated electrodes: y=1, 3,5, 10 μm. y=1μm corresponds to the plane of the electrodes' surface. The gap between the electrodes is 10 μm. Witness that at the electrodes' plane, the electric field has two maxima located at the electrodes' edges. As we move away from the electrodes, the two maxima merge into a single maximum located right above the gap's center.

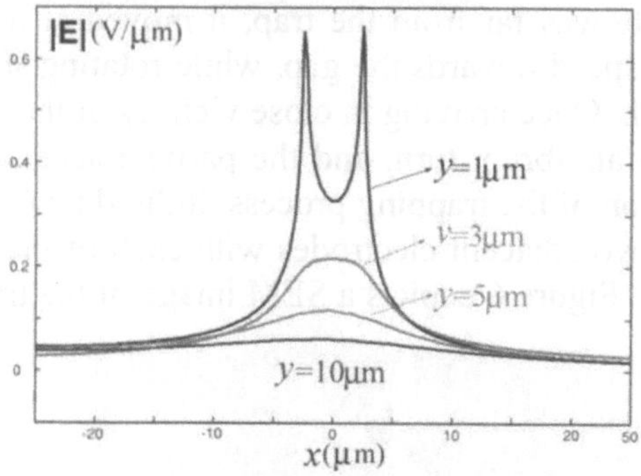

Figure 4: Through planar microfabricated electrodes with small electrode gap spacing, large electric field gradients are generated in the y direction which are represented by the normalized electric field lines. These electric field gradients generate a strong dielectrophoretic force on the CNTs and cause the CNTs to migrate towards the region of the highest electric field gradient and align themselves with the electric field.

In the experiments, we applied 12 V_{pp} potential at 2MHz between a pair of electrodes spaced 8μm apart. The CNT solution was exposed to the electric field for about 30 seconds. Photographs were taken with a Nikon CoolPix 995 camera mounted on an Olympus BX2 microscope at a magnification of 400x. The set-up allowed us to obtain only the top view of the particle's trajectory. Figures 4a, b, and c depict sequentially the position of the CNT at times 0, 0.2, and 0.28 seconds during the trapping process. For better visibility, we encircled the particle that we were tracking with an ellipse.

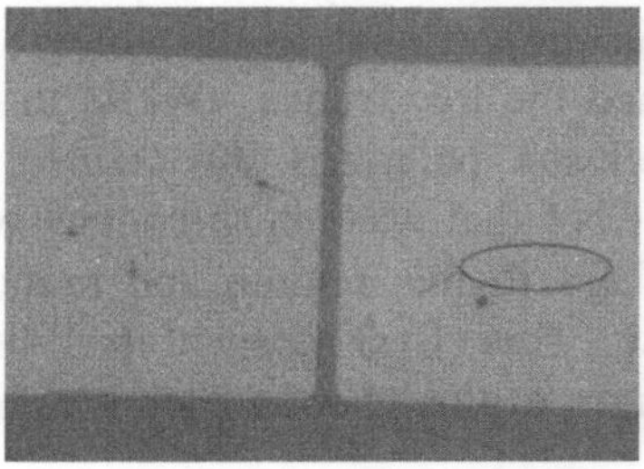

a) Time = 0s

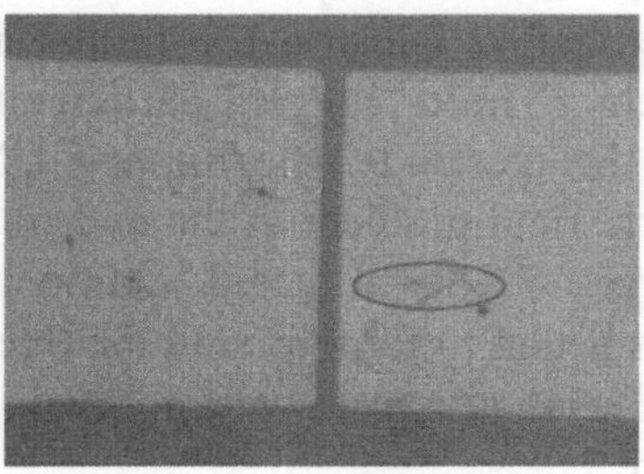

b) Time = .2s

c) Time = .28s

Figure 5: Images of the nanotube during various stages of the trapping process: 0s (a), 0.2s (b), and 0.28s (c). The nanotube is encircled with a black ellipse for better visibility.

When the particle was far from the trap, it moved at nearly level height with nearly uniform speed towards the gap, while rotating so as to align itself with the electric field. Once arriving in close vicinity of the gap, the particle's trajectory underwent an abrupt turn, and the particle accelerated towards the gap. At the conclusion of the trapping process, individual CNTs were trapped in the gap between two adjacent electrodes with each of their ends resting on a different electrode. Figure 6 depicts a SEM image of the trapped tube.

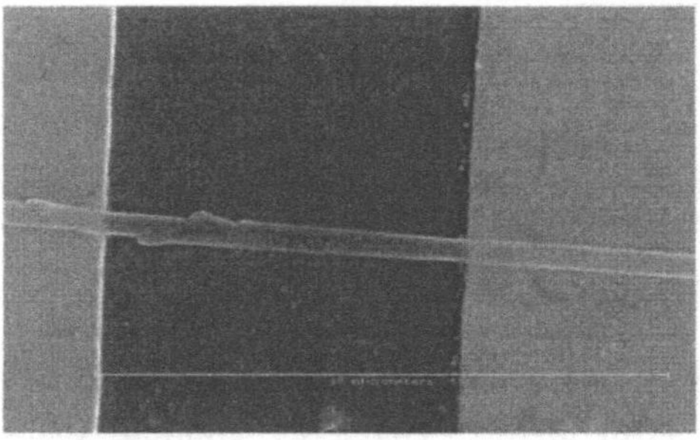

Figure 6: SEM Image of a trapped tube in the gap between a pair of electrodes

Another phenomenon which occurred when the trapping process was allowed to continue for a relatively long period of time was chain and tree formation. In the case of the CNT tree structures, positive DEP pulled CNTs suspended in isopropanol solution to the region of the highest electric field gradient located on the edge of a gold microfabricated electrode. Clustered tubes, which were present as a result of the tube synthesis process, were attracted first because the DEP force is proportional to the volume of the particle. These "bigger" particles formed the tree trunks. While in contact with the electrode, the trapped CNT cluster polarized and created a high electric field gradient at the end of the CNT that was not in contact with the electrode. This attracted additional CNTs to this region, and branches of CNTs began to form around the tree base. This effect repeated itself until full tree fractal structures were formed. Images of this phenomenon are shown in Figure 7.

a) *b)* *c)*

Figure 7: Images of CNT tree fractal structures; a) SEM of a chain formed by individual uniform size CNTs; b) SEM of CNT tree structure; c) CNT tree structure at 400X magnification in bright field.

4. CONCLUSION

The paper describes a practical means of constructing nanotube-based fluidic devices. Dielectrophoresis provides a convenient and efficient technique for the trapping of CNTs at pre-determined locations. We studied the motion of the particles during the trapping process experimentally and described CNT tree formations that developed during the process. Once the nanotube has been trapped at a known position, one can bring to bear the tools of photolithography to fabricate fluidic interfaces that will allow one to transmit various fluids, macromolecules, and nanoparticles through the tube. Such devices can be used for fundamental studies of the behavior of simple and complex fluids under extreme confinement. Moreover, these devices can be used as highly sensitive biological and chemical sensors. The fabrication technique can be readily scaled down to allow one to trap single-walled nanotubes.

Acknowledgements: M. Riegelman, H. Liu, and H. H. Bau are supported by the DARPA SIMBIOSYS program N66001-01-C-8056 and the NSF NIRT program CTS 0210579. M. Riegelman also acknowledges support from a GAANN fellowship. S. Evoy is supported by NSF #ECS-0225496.

References

[1] H. E. Kubitschek, Nature (London) 182, 234 (158)

[2] J.-C Bradley, S. Babu, S. Ndungu, P. Nikitin, Y. Gogotsi, 2003, Nanotube Synthesis using Alumina Template, , SMIRP Bradley Research Lab Knowledge Product 10975_0004

[3] H. A. Pohl, *Dielectrophoresis: The Behavior of Neutral Matter in Non-uniform Electric Fields,* Cambridge University Press, New York, 1978.

[4] T.B. Jones, *Electromechanics of Particles,* Cambridge University Press, New York, 1995.

[5] S. Evoy *et al.*, "Dielectrophoretic Assembly and Integration of Nanowire Devices with Functional CMOS Operating Circuitry", in press (2003).

[6] H. Liu and H. H. Bau, 2003, The Dielectrophoresis of Cylindrical and Spherical Particles Submerged in Shells and in Semi-infinite Media, in review.

[7] J. A. Stratton, *Electromagnetic Theory,* McGraw-Hill, Inc., New York, 1941.

OPENING MULTIWALL CARBON NANOTUBES WITH ELECTRON BEAM

H. YE, N. NAGUIB and Y. GOGOTSI
Department of Materials Science & Engineering and
A.J. Drexel Nanotechnology Institute, Drexel University
3141 Chestnut Street, Philadelphia, PA 19104, USA

Abstract: A new method of nanomachining and opening multiwall carbon nanotubes (MWNTs) has been demonstrated. The puncture is accomplished by moving a converged electron beam in transmission electron microscope (TEM) onto the water inclusion in the multiwall carbon nanotube. The high resolution TEM images show that the reaction between the water inclusion and the walls of the carbon nanotube is the main reason for the appearance of the puncture. The reaction is stimulated by the converged electron beam irradiation. The puncture in different electron microscopy modes occurs with the same mechanism and approximately the same size, independent on the diameter of the electron beam.

1. INTRODUCTION

Opening of MWNTs is of great value in many applications, such as electrochemical energy or gas storage systems, biological nanosensors, and nanofludic devices [1]. It is not easy to open the originally closed MWNTs owing to their high mechanical strength and chemical stability. Various methods have been developed to open MWNTs, including chemical treatment with nitric acid or oxidants [2], thermal treatment by heating the nanotubes in oxidizing atmosphere [3] and abrasive treatment by rubbing the nanotubes on diamond lapping film disks [4]. However, a common feature of these methods is that the opening only happens at the tips of nanotubes. There is no control over the location of the opening on the tube surface. Laser machining (Fig. 1) [5] or focused ion beam (FIB) which may be used for opening large tubes [6] does not have sufficient spatial resolution to machine small (<100 nm in diameter) tubes with high precision. Here, we report a new method which can open the sidewalls of large MWNTs (nanofibers) by using converged electron beams in transmission electron microscope. In addition, with this method we can specifically select the puncture position on the nanotube.

S. Guceri et al. (eds.), Nanoengineered Nanofibrous Materials, 417-424.

The interaction between electron beams and carbon nanotubes (CNTs) has been thoroughly investigated [7-10]. Strong electron irradiation was found to cause either disordering of the graphite tube walls or the shrinkage and collapse of the single-wall nanotubes. None of them, however, has led to the puncture of CNTs. The basic difference in our experiment from the previous investigations is that we conducted electron irradiation on liquid-containing CNTs, instead of hollow CNTs. The liquid-containing CNTs were produced hydrothermally at high temperature and high pressure from a supercritical C-H-O system using Ni catalyst [11, 12]. A unique and important feature of these CNTs is that they contain liquid inclusions which are composed of mainly water with dissolved gases. However, as our recent work shows, MWNTs of various diameters can be filled with water by treatment at elevated temperature and pressure. Thus, the proposed technique is applicable to various kinds of closed MWNTs. Movement and shrinkage/ expansion phenomena of the liquid inclusions under electron beam in TEM have been previously demonstrated [13, 14]. In this paper, opening the sidewalls of the liquid-filled CNTs is shown to be possible under an extremely converged electron beam with a high intensity.

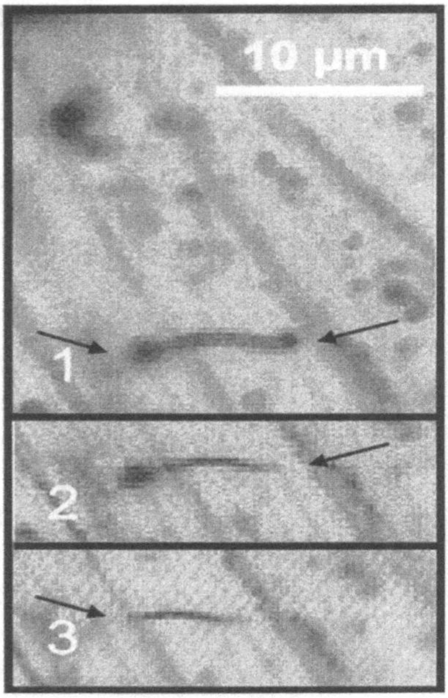

Figure 1. Light micrographs showing laser cutting of multiwall carbon nanotubes (nanofibers) in air using a focused laser probe of the Raman spectrometer. Air oxidation temperature established by Raman shift at onset of burning was ~700°C. 1 - Both ends closed; 2 - Right end cut off; 3 - Left end cut off.

2. EXPERIMENTAL

The hydrothermal CNTs were synthesized as described in Ref. [11, 12]. For synthesis, mixtures of ethylene glycol, water and Ni catalyst were sealed into gold capsules and reacted at temperature 700-800°C and pressure up to 100 MPa. The produced hydrothermal multiwall CNTs (MWNTs), were dispersed in ethanol and then placed on the carbon coated copper grid for TEM observation. All irradiation and imaging work was done in a field emission JEOL 2010F TEM with an accelerating voltage of 200 kV and a lattice resolution of 0.1 nm. The microscope could be operated in two modes: conventional TEM mode and scanning transmission electron microscopy (STEM) mode. When electron irradiation was performed in conventional TEM mode, the electron beam through the specimen could be either diverged with a low electron flux down to 2×10^{17} electrons/cm^{2}s (measured by a Faraday cup in TEM) or strongly converged, with the electron flux reaching 2×10^{21} electrons/cm^{2}s (beam size 0.1μm). When STEM mode was adopted, an extremely small electron probe (1 nm) could be formed for irradiation, with an electron flux up to 6×10^{23} electrons/cm^{2}s. After irradiation, the CNTs were imaged in conventional TEM mode.

3. RESULTS AND DISCUSSION

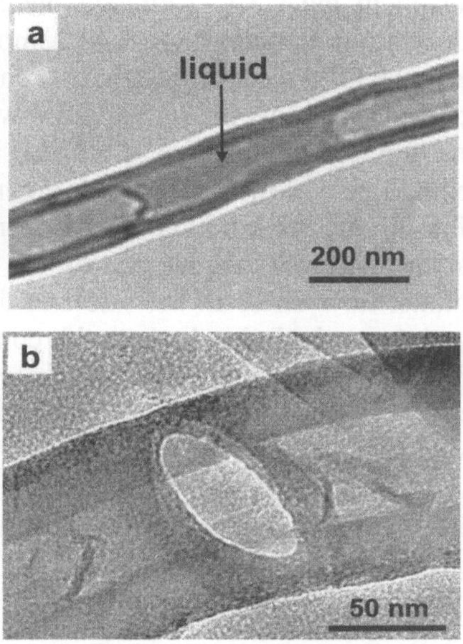

Figure 2. TEM image of a typical hydrothermal CNT with the liquid inclusion (a) before the converged electron beam irradiation and (b) after the irradiation. Puncture on the nanotube is visible after the strong beam irradiation.

The TEM image in Fig. 2a shows a typical CNT with liquid entrapped. The liquid is composed of water and a small amount of dissolved CO_2 [13]. When the electron beam in TEM is converged rapidly (within 1 second) and precisely on the region of the liquid inclusion, a puncture of the nanotube can be immediately observed, as shown in Fig. 2b. The hole formed after puncture has typically an elliptic shape crossing the nanotube and a size larger than the inner diameter of the CNT. Depending on the irradiation intensities, the holes can be produced either on the front wall (part of the nanotube wall directly facing electron beam irradiation, Fig. 2b) or on both front and back walls of the nanotube, as shown in Fig. 3a. Although it can be seen that the front hole has a bigger size than the back hole (Fig. 3a), this is not the case for all punctured CNTs and occasionally the back hole was observed to be larger. High resolution TEM image (Fig. 3b) of the left tube wall of the CNT in Fig. 3a shows that part of the inner wall was damaged as the hole was formed. The broken graphite layers bent toward the center of the CNT, as indicated by white lines in Fig. 3b.

Technically it is not easy to precisely converge an electron beam on such a tiny liquid inclusion (less then 50 nm in diameter) in the conventional TEM mode. Scanning transmission electron microscopy (STEM) mode, because of its ability to converge electron beam to 1 nm or less and control the movement of the probe conveniently, was thus adopted to study the puncture of CNTs. Fig. 4a shows the TEM image of a hole in a CNT which was formed nearly instantaneously when STEM probe was shifted onto the liquid inclusion in the CNT. Although the STEM probe is extremely small, the produced hole is quite large, and comparable in size to those formed by the converged TEM electron beam. This confirms that the tube opening is caused by chemical interaction of the liquid with tube walls rather than by electron beam damage of the wall. HRTEM image in Fig. 4b shows the broken graphite layers of the nanotube wall bent toward the inside of CNT, which is similar to the observation in the conventional TEM mode (Fig. 3b). It therefore indicates that the puncture in two modes has the same mechanism, no matter how different the size and the intensity of the electron beam are. Sometimes a single irradiation with converged TEM electron beam or STEM probe can cause the formation of two or more holes on the tube wall of CNTs, as shown in Fig. 5. In Fig. 5a, a converged electron beam in conventional TEM mode can cause the formation of two holes on the front wall of the CNT. Fig. 5b shows 7 holes formed simultaneously on the front wall of the CNT when the STEM electron probe was used. Three conditions must be satisfied to open the walls of CNTs in TEM.

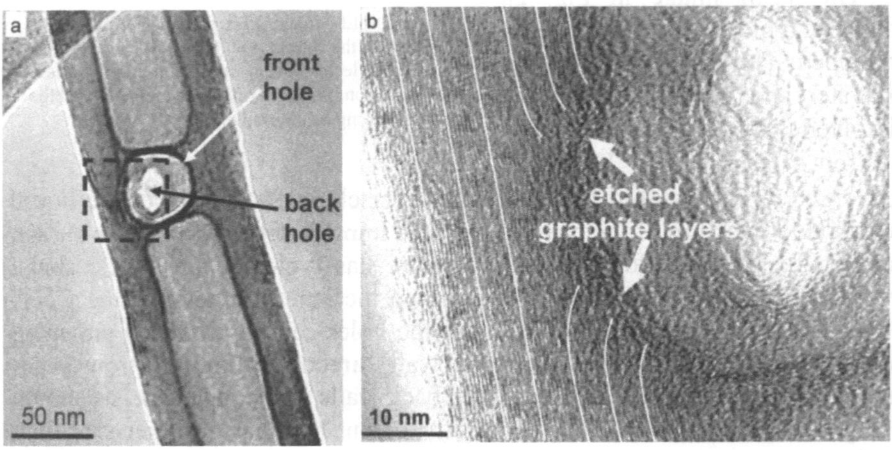

Figure 3. (a) TEM image of a punctured nanotube, showing that two holes can be produced at the same time, one on the front tube wall and the other on the back wall, by using a single beam irradiation. (b) A high resolution TEM image of the framed region in (a). The orientation of graphene layers in the tube wall is marked by white lines. The inner tube wall is found to be heavily etched and the broken graphite layers bend toward the center of the nanotube, as indicated by the white arrows in (b).

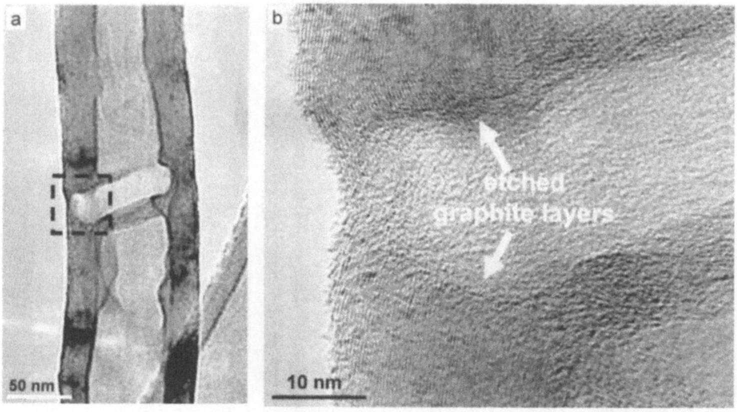

Figure 4. (a) A hydrothermal CNT punctured by an electron probe (1 nm) in STEM mode; (b) HRTEM image of the framed region in (a), showing the etched graphite layers which bend toward the center of the nanotube.

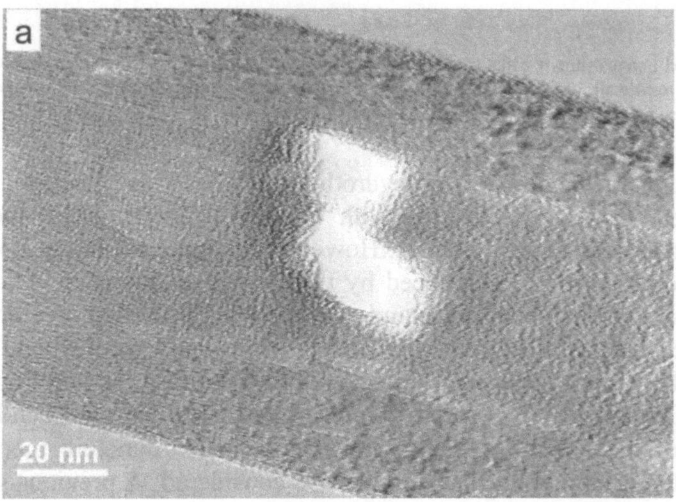

First, the CNTs should contain liquid inclusions which are the targets of the electron-beam irradiation. The irradiation effects on multiwall [7] and single-wall nanotubes [8, 9] in TEM have been investigated previously. The primary cause of irradiation damage to CNTs was revealed to be large-angle electron-nuclear collisions [10], while beam heating played a less important role to induce the structural failure. The rapid atomic diffusion and reconstruction following the irradiation damage, as exhibited in these previous reports, could cause disordering, shrinkage or collapse of CNTs. No observation of the puncture on the tube walls of CNTs was reported. The importance

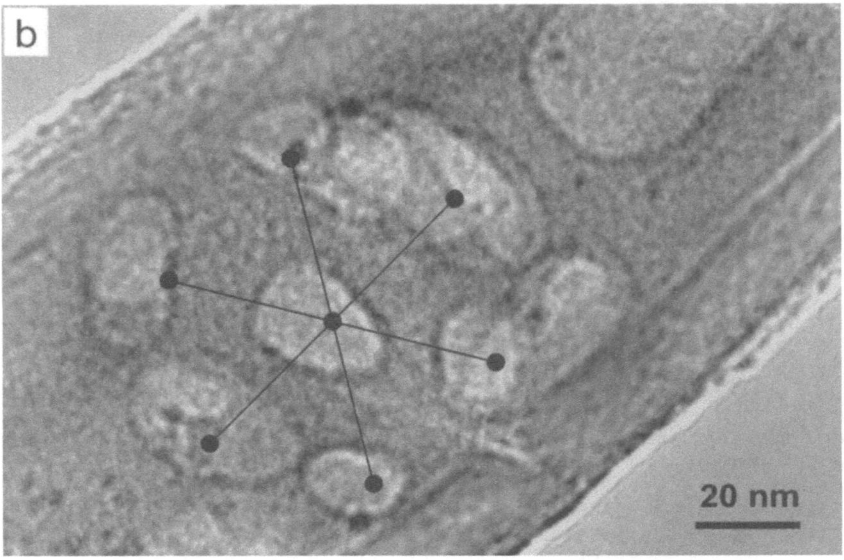

Figure 5. TEM images showing that several holes appear at the same time after a single exposure: (a) two holes on front wall, opened in conventional TEM mode. (b) 7 holes on front wall with different sies and shapes, opened in STEM mode.

of the liquid inclusion in the hydrothermal CNTs is therefore clearly demonstrated. Due to the heating with electron beam in TEM, the liquid inclusion undergoes an evaporation. However, the puncture might not be the result of the physical force induced by the inner gas explosion. This is not only because the CNTs can withstand extremely high internal pressures [15], but also because, in that case, the broken tube walls would bend toward outside as the vapor rushes out, which is not the case (Fig. 3b and Fig. 4b). In these HRTEM images, the broken graphite layers were found to bend toward the inside of the nanotube, disproving the explosion-puncture mechanism. Therefore, the chemical mechanism must be considered. A previous study on the hydrothermal CNTs [16] indicated that a chemical reaction between liquid inclusion and carbon tube walls could take place due to the electron beam irradiation in TEM:

$$C + 2H_2O \rightarrow CO_2 + 2H_2.$$

The reaction could result in dissolution of the graphite layers and etching of the inner nanotube walls. Since in the present work we performed an extremely strong electron beam irradiation, etching of the graphite layers could be accomplished through the tube walls of the CNTs until the puncture appeared and the residual liquid was lost to the microscope vacuum. This process was very fast because we were dealing with an extremely small liquid inclusion (in the order of 10^{-18} liters). The size and shape of the holes produced were determined by the area where the reaction took place, not by the size of the electron beam. This explains why similar size and shape of punctures have been observed when either converged electron beam (Fig. 3a) or STEM probe (Fig. 4a) were used.

The second condition required to open CNTs is that the beam intensity (in term of electron flux) through the specimen must be strong enough to stimulate the reaction. A parallel electron beam with large exposure area and low electron flux in conventional TEM mode was not found to be able to puncture the nanotube. A strongly converged beam in the same TEM mode, however, can violently stimulate the etching reaction and the puncture (Fig. 2 and 3). The electron probe in STEM mode, which is extremely converged electron beam and thus possesses high beam intensity, can puncture the nanotube instantly once being shifted upon the liquid inclusion (Fig. 4).

The third condition is that the operation of converging electron beam must be quick. This is especially crucial in conventional TEM mode. Only fast beam converging (within 1 second) can cause the puncture of CNTs. Slow beam converging would not cause sufficient etching reaction in a short time. It would instead lead to liquid transport away from the electron beam or/and its evaporation and condensation in a cold part of the nanotube [13]. It was also shown in our previous work [16] that a slower heating may lead to a local increase in internal tube diameter, thus producing a chamber inside the tube. This can be used for improving flow mixing inside the nanopipe.

Since the CNTs investigated are not always perfect, various shapes and distributions of punctures have been observed when the above three conditions are satisfied. The etching reaction may happen more easily at defects of the nanotube walls. If the defects are random point defects, the punctures caused by etching reaction are also randomly distributed, like in Fig. 5a. If the defects are caused by stacking faults of the nanotube walls during their formation, the distribution of the defects and thus the punctures may appear with some amount of periodicity. An example is shown in Fig. 5b where the punctures form a hexagonal structure, as indicated by lines and solid cycles.

4. SUMMARY AND CONCLUSIONS

It has been shown that the tube walls of the carbon nanotubes containing water can be opened by strongly converged electron beams. The puncture can be done in conventional TEM mode and STEM mode which produce similar results regardless of the large difference in beam size and intensity. The reaction between water and carbon tube walls, stimulated by electron beam irradiation, etches carbon nanotubes from inside until the puncture appears. The necessary conditions to initiate the puncture of the carbon nanotubes include the presence of water inclusion, the high beam intensity and the rapid beam-converging operation (in case of conventional TEM mode). This new method for nanomachining carbon nanotubes provides a way to produce carbon nanotubes with three or more entrances, instead of only two at the ends, which could be useful in many potential applications, for example, fabrication of nanofluidic devices.

Acknowledgements: We thank Prof. C.M. Megaridis and A.G. Yazicioglu, University of Illinois at Chicago, for helpful discussions. Figure 1 was provided by Joe Libera. The TEM and STEM work was done in the LRSM of the University of Pennsylvania. This work was supported by the US National Science Foundation under NIRT grant CTS-0210579.

References

1. Harris, P. (1999) *Carbon nanotubes and related structures*, Cambridge University Press, Cambridge.
2. Tsang, S.C., Chen, Y.K., Harris, P.J.F. and Green, M.L.H. (1994) A simple chemical method of opening carbon nanotubes, *Nature* **372**, 159-162.
3. Tsang, S.C., Harris, P.J.F. and Green, M.L.H. (1993) Thinning and opening of carbon nanotubes by oxidation using carbon dioxide, *Nature* **362**, 520-522.
4. Maurin, G., Stepanek, I., Bernier, P., Colomer, J.-F., Nagy, J.B. and Henn, F. (2001) Segmented and opened multi-walled carbon nanotubes, *Carbon* **39**, 1273-1278.
5. Libera, J.A. and Gogotsi, Y. (1999) Probing carbon nanostructures using Raman spectroscopy, *MRS Annual Meeting, Boston, MA, poster presentation.*
6. Muradov, N. and Schwitter, A. (2002) Formation of conical carbon structures on vapor-grown carbon filaments, *Nano Lett.* **2**, 673-676.
7. Crespi, V.H., Chopra, N.G., Cohen, M.L., Zettl, A. and Louie, S.G. (1996) Anisotropic electron-beam damage and the collapse of carbon nanotubes, *Phys. Rev. B* **54**, 5927-5931.
8. Ajayan, P.M., Ravikumar, V. and Charlier, J.-C. (1998) Surface reconstructions and dimensional changes in single-walled carbon nanotubes, *Phys. Rev. Lett* **81**, 1437-1440.
9. Smith, B.W. and Luzzi, D.E. (2001) Electron irradiation effects in single wall carbon nanotubes, *J. Appl. Phys.* **90**, 3509-3515.
10. Banhart, F. (1999) Irradiation effects in carbon nanostructures, *Rep. Prog. Phys.* **62**, 1181-1221.
11. Gogotsi, Y., Libera, J.A. and Yoshimura, M. (2000) Hydrothermal synthesis of multiwall carbon nanotubes, *J. Mater. Res.* **15**, 2591-2594.
12. Libera, J. and Gogotsi, Y. (2001) Hydrothermal synthesis of graphite tubes using Ni catalyst, *Carbon* **39**, 1307-1318.
13. Gogotsi, Y., Libera, J., Guvenc-Yazicioglu, A. and Megaridis, C.M. (2001) In situ multiphase fluid experiments in hydrothermal carbon nanotubes, *Appl. Phys. Lett.* **79**, 1021-1023.
14. Megaridis, C.M., Guvenc-Yazicioglu, A., Libera, J.A. and Gogotsi, Y. (2002) Attoliter fluid experiments in individual closed-end carbon nanotubes: liquid film and fluid interface dynamics, *Phys. Fluids* **14**, L5-L8.
15. Galanov, B.A., Galanov, S.B. and Gogotsi, Y. (2002) Stress-strain state of multiwall carbon nanotube under internal pressure, *J. Nanoparticle Res.* **4**, 207-214.
16. Gogotsi, Y., Naguib, N. and Libera, J. (2002) In situ chemical experiments in carbon nanotubes, *Chem. Phys. Lett.* **365**, 354-360.

MOLECULAR CHEMICAL CONCEPTS FOR THE SYNTHESIS OF NANOCRYSTALLINE CERAMICS

S. MATHUR
Institute of New Materials
CVD Division, 66041 Saarbruecken, Germany

1. INTRODUCTION

Nanomaterials research is seeking methods to achieve a precise control over composition, particle size, size-distribution and morphology [1, 2]. The conventional syntheses involving solid-state or liquid phase reactions of two or more components fail to meet the challenge. The chemical syntheses based on the controlled reactions (interactions) of atoms or molecules are becoming promising alternative for a rational synthesis of nanostructured materials. In this context, different chemical approaches have been put forward to synthesize well-defined materials due to the low temperatures required for the synthesis and the possibility of building extended solid-state structures from molecular building blocks [3-7]. When compared to the solid-state reactions, the solution methods allow a controlled interaction of atoms or molecules to form uniform films or particles. Further, the flexibility to combine different ligands or metal combinations allows the precursor designing to meet the demands of the target material [8]. Recently, a one-step strategy for the synthesis of nanomaterials involving ensemble of the phase-forming elements in a single molecular source, has gained significant attention because it enhances the advantages of chemical processing and simultaneously reduces the process parameters [8-14].

2. THE PRECURSOR CONCEPT

Although heterometal alkoxides are invariably formed in a mixture of constitutent alkoxides, the simple mixing of two alkoxides $M(OR)_x$ and $M'(OR)_y$ is no guarantee for the formation of a single precursor species $MM'(OR)_{x+y}$ mainly due to the structure potential of alkoxide ligand to bind one, two or three metal centers. As a result, besides the desired precursor, several other compounds can be present in an uncharacterized precursor cocktail. For example, Ba and Zr(Ti) alkoxides can

S. Guceri et al. (eds.), Nanoengineered Nanofibrous Materials, 425-442.

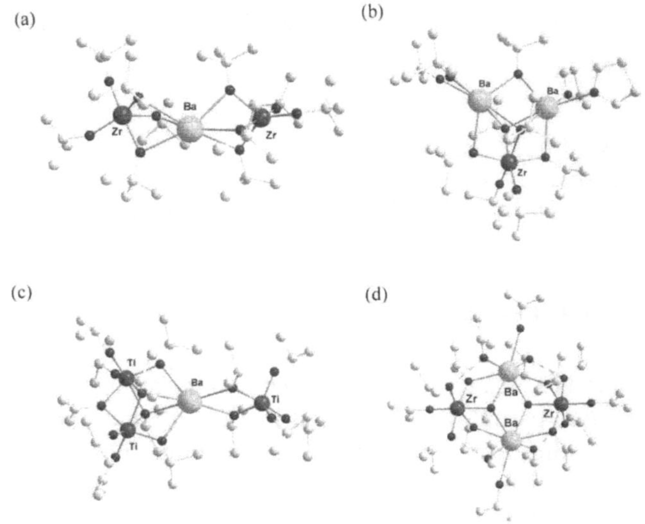

Figure 1. Molecular structure of different Ba-Zr(Ti) alkoxide clusters: (a) $[BaZr_2(OBu^t)_{10}]$ (Ba:Zr = 1:2), (b) $[Ba_2Zr(OBu^t)_8(HOBu^t)_2(THF)]$ (Ba:Zr = 2:1), (c) $[BaTi_3(OPr^i)_{14}]$ (Ba:Ti = 1:3) and (d) $[BaZr(OPr^i)_5(OH)(HOPr^i)_3]_2$ (Ba:Zr = 1:1).

combine in different ways to produce mixed-metal Ba-Zr(Ti) alkoxides with different structures and metal ratios [11]. Some examples of alkoxide clusters based on different Ba:Zr(Ti) ratios are shown in Figure 1a-d. Of course the formation of a particular framework is subject to the chemical nature (neutral or anionic) and size (steric profile) of the ligands.

In most of the chemical methods in practice, the starting materials are commonly available reagents, which are close to the desired composition but often need a heat-treatment step to attain the final composition. Since the material synthesis procedures (e.g., solid-state reactions, combustion reactions, etc.) rely on the intrinsic chemical behaviours (e.g., different hydrolysis rates of the reactants in solution phase reactions or different vapour pressure or thermal stability in the gas phase reactions) of the different components present in a reaction mixture, the stoichiometry of the target material is susceptible to inaccuracies due to the lack of chemical bonding. If the metal-ligand connectivity corresponding to the solid state structure is not given, the different compounds present in the reaction mixture randomly interact to form several intermediate species with metal ratios unfavorable for obtaining a single-phase material. As a consequence, phase separation and element segregation are present at the nanometer scale, although the global stoichiometry of the product may correspond to the desired composition (Scheme I). For instance, a simple mixture of two chemical precursors A and B (Multi Source Process, MSP), to produce the bimetallic oxide system $A_xB_yO_z$, may produce a visibly homogeneous mixture but the molecular level scenario can be different and despite a correct global stoichiometry the precursor 'cocktail' may have element segregation and non-ideal A:B ratios at the nanometer scale that may be carried forward to the end product (Scheme I).

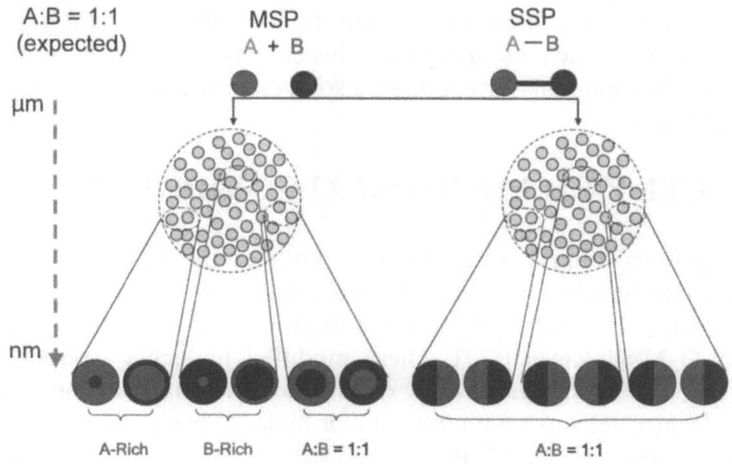

Scheme I

Our investigations on the application of heterometal alkoxides in the material synthesis have shown that **isolation and characterization of preceramic aggregates** is necessary for the phase selective synthesis of materials. For instance, high purity $BaZrO_3$ powders can be obtained by using a single–source Ba-Zr alkoxide, $[BaZr(OPr^i)_5(OH)(HOPr^i)_3]_2$, whereas perovskite obtained from a mixture

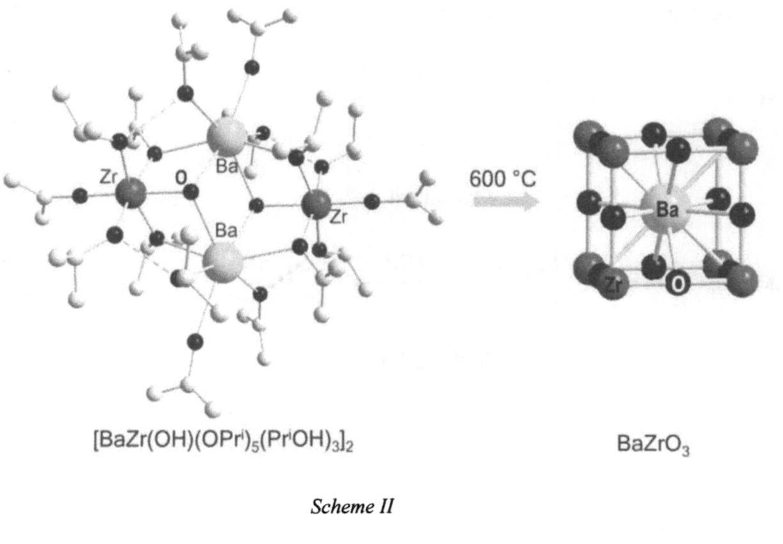

Scheme II

of Ba and Zr precursors was found to contain undesired phases. The main drawback of solution syntheses performed using mixtures of alkoxides is the differential hydrolysis behaviour (kinetics) of individual components, which often leads to the formation of secondary phases. This limitation can be overcome by employing well-characterised heterometal molecules ('single-source') with metallic elements present in the ratio compatible with the targeted ceramic. Such an inorganic cluster can be seen as a molecular template for initiating the formation of a solid-state structure at the nanometer scale (Scheme II).

3. SOL-GEL PROCESS USING ALKOXIDE PRECURSORS

The sol-gel technique using metal alkoxides is based on inorganic polymerisation reactions in which metal alkoxide derivatives such as $M(OR)_x$ or $MM'(OR)_{x+y}$ are carefully hydrolysed to produce species containing M–O–M or M–O–M' linkages [5-7]. These modified molecular entities arrange themselves *via* hydrolysis and cross-condensation reactions to form macromolecular networks with oxo- and/or hydroxo bridges thus favoring the gelation process. The partially hydrolysed alkoxides can be viewed as molecular building blocks for the development of sol (particulate) or gel (polymeric) structure, which can be dried and fired to obtain the desired oxide. Modifying the alkoxide precursors with less hydrolysable or polymerizable groups can chemically control the condensation reactions [5,6]. The metal species bearing a terminal hydroxy group react with other species in the solution to form oligomeric structures linked by oxo- (O^{2-}) or hydroxo (OH^-) ligands, which transforms into polymeric gels by aging. The polymerization reactions and the followed drying and thermal treatments result in the elimination of water (oxolation) and alcohol (alcoholation) to finally produce a solid material (Scheme III).

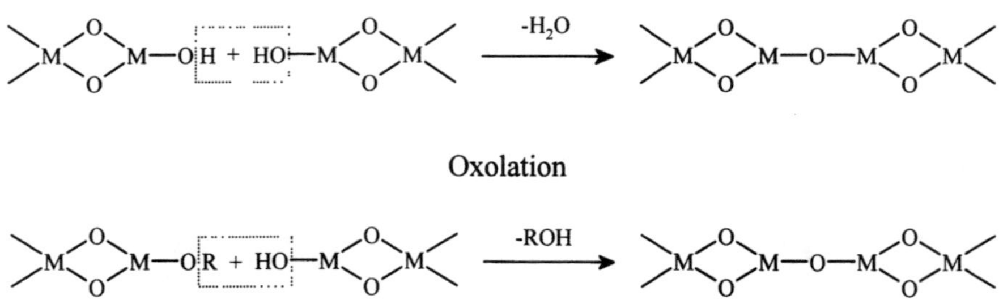

Oxolation

Alcoholation

Scheme III

4. NANOCERAMICS

This section contains the selected examples of the synthesis and characterization of nanoscaled powders of bimetallic oxides obtained by the sol-gel processing of single molecular precursors.

4.1. Perovskites ($BaMO_3$; M = Ti, Zr)

The dielectric ceramics $BaTiO_3$ and $BaZrO_3$ are finding extensive electronic applications due to their high dielectric constant and ferroelectricity [25, 26]. The powders obtained from conventional routes are agglomerated with large and unequal grain sizes and contain varying amounts of undesirable impurities due to incomplete reactions or undecomposed intermediates. The poor control over compositional homogeneity and phase-purity has an adverse effect on their electrical and mechanical properties [27]. Nanocrystalline $BaTiO_3$ and $BaZrO_3$ oxides [28] have been synthesised from single-source Ba-Ti(Zr) alkoxide precursors. The improved homogeneity, phase-purity, and lower crystallisation temperatures for gel $\rightarrow$ oxide conversion are attributed to the use of mixed-metal compounds containing preformed Ba–O–Ti and Ba–O–Zr bonds. The TEM images (Fig. 2) of barium-titanate powder obtained from single source route revealed homogeneous nanoparticles, whereas for powder from hydroxide route showed severe agglomeration and abnormal particle growth [28]. The strict control over the metal ratios through stoichiometric clusters allows the crystallisation of perovskite nanoparticles at low-temperature (ca. 600 °C).

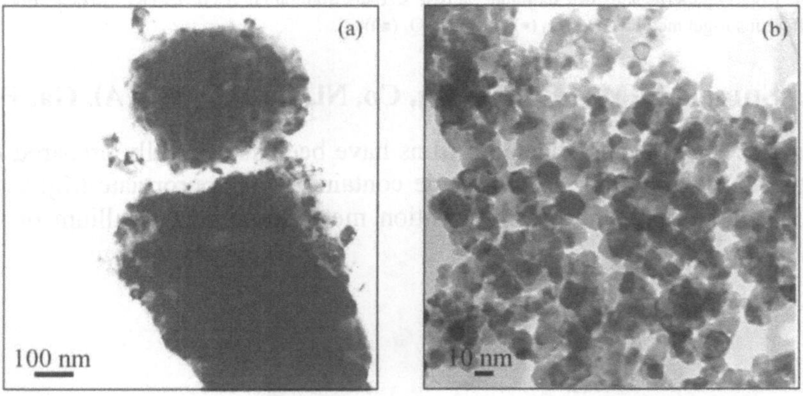

Figure 2. TEM images of $BaTiO_3$ ceramic obtained from (a) hydroxide and (b) single source routes.

The X-ray powder diffractograms (Fig. 3) reveal that the powder obtained from conventional method is contaminated even at 1200 °C by undesired side-products like $BaCO_3$ and Ba_2TiO_4 whereas a single-phase material is obtained using the mixed-metal alkoxide precursor. In addition, the volume- and number-weighted size dispersions are significantly narrow for the perovskite powders obtained from alkoxide precursors (Fig. 3), which is not the case for powders obtained by hydroxide route due to severe agglomeration and abnormal particle growth [28].

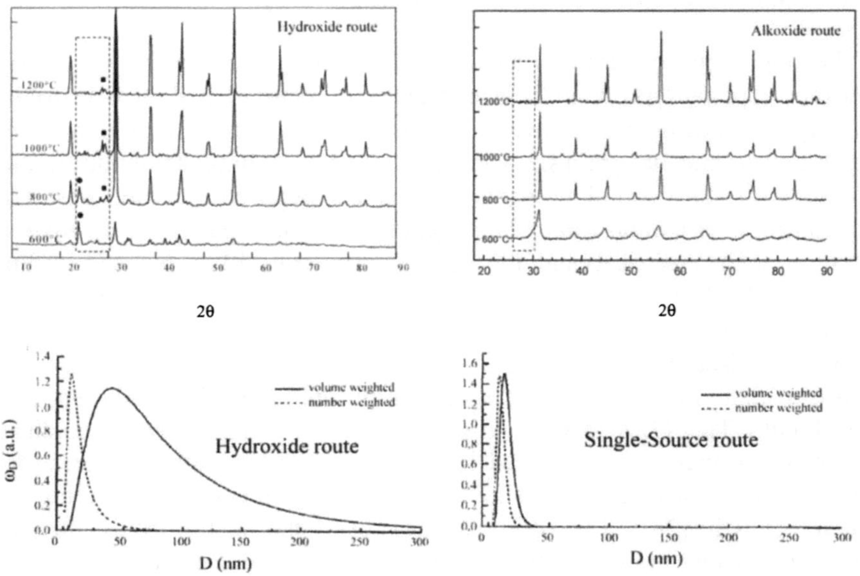

Figure 3. XRD patterns and corresponding particle size-distribution curves of $BaTiO_3$ particles obtained by different sol-gel methods ($BaCO_3$ (•) and Ba_2TiO_4 (■)).

4.2. Spinels (MM'_2O_4; M = Mg, Co, Ni, Cu, Zn; M′= Al, Ga, Fe)

Although $MgAl_2O_4$ powders and films have been successfully prepared from magnesium-aluminium iso-propoxide containing the appropriate Mg:Al ratio (1:2) [20], the preparation of transition metal-aluminium, -gallium or -iron spinels is

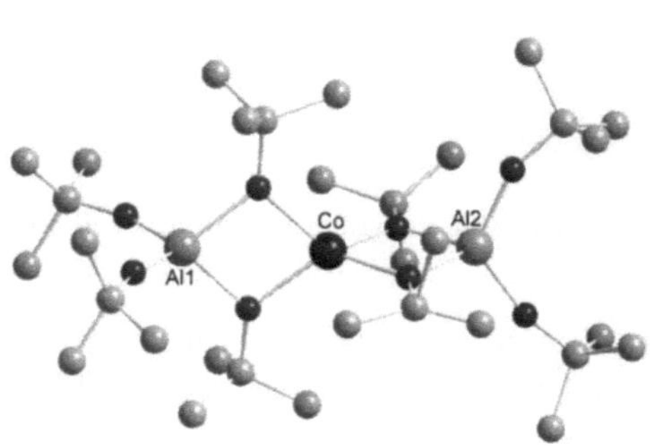

Figure 4. Molecular structure of $[CoAl_2(OBu^t)_8]$.

limited to the fusion of the two component oxides [29, 30] at high temperatures (1000-1600 °C) or by co-precipitation reactions [31, 32], in solution. We have used hetero-bimetallic precursors, $[MM'_2(OR)_8]$ (M(II) = Co, Zn; M' = Al, Ga, Fe), to obtain high purity spinels at low-temperature. The molecular structure (Fig. 4) of the Co-Al derivative reveals the Co:Al ratio suitable for $CoAl_2O_4$ [24]. For a comparative evaluation of the single- and multi-component approaches, $CoAl_2O_4$ spinel was synthesized from (i) $[CoAl_2(OBu^t)_8]$ and (ii) a stoichiometric mixture of $[Al(OPr^i)_3]_4$ and $Co(OR)_2$ ($R = (C_6H_5)_3C–$).

The XRD patterns (Fig. 5) of the two $CoAl_2O_4$ samples reveal that the single-source synthesis yields, under similar experimental conditions, higher phase

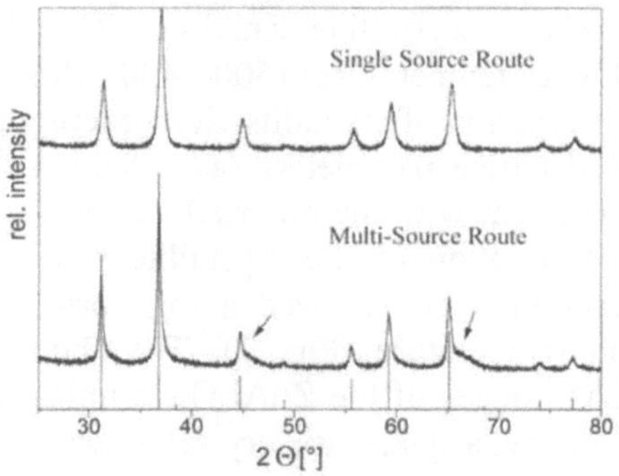

Figure 5. XRD patterns of $CoAl_2O_4$ obtained from (a) single- and (b) multi-source routes. Arrows in the case of multi-source synthesis indicate undesired phases.

purity and crystallinity when compared to the spinel obtained using two separate metal sources. It should be noted that in the later case the different hydrolysis kinetics of Co and Al alkoxides and formation of several species with different metal ratios can adversely affect the homogeneity (with respect to the cation distribution) of the precursor solution. Similarly, nanocrystalline $CuAl_2O_4$ and $NiAl_2O_4$ could be obtained by a single-step processing of $[CuAl_2(OBu^t)_8]$ and $[NiAl_2(OBu^t)_8]$, respectively [24].

Similarly zinc aluminate ($ZnAl_2O_4$), a well-known wide band gap semiconductor with a spinel structure, was prepared by the sol-gel processing [33]

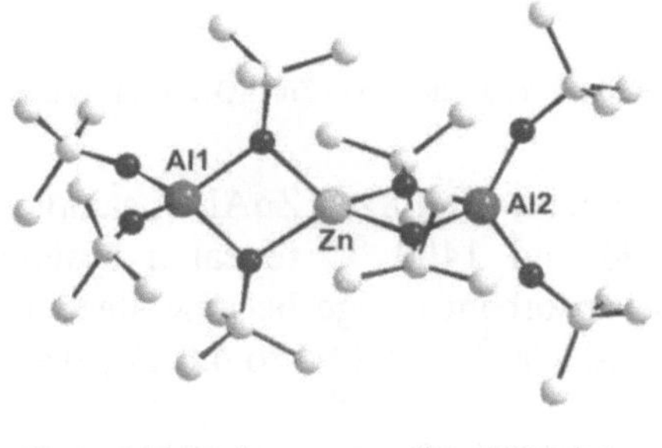

Figure 6. Molecular structure of $[ZnAl_2(OBu^t)_8]$.

of bimetallic precursor shown in Fig. 6. Recent investigations on ZnM_2O_4 (M = Al, Ga) compounds have shown these systems to be new transparent and conductive materials.

The optical band gap of polycrystalline $ZnAl_2O_4$ (3.8 eV) indicates the material to be transparent for the light possessing wavelength larger than 320 nm. Thus it can be utilized for UV-photoelectronic devices. Owing to the quantum confinement, the energy levels of a semiconductor nanoparticle (quantum dot) are 'blue shifted' with respect to those of the bulk. Moreover, $ZnAl_2O_4$ spinel is useful in many catalytic reactions such as cracking, dehydration, hydrogenation and dehydrogenation reactions [34]. The XRD pattern of the sample calcined at 400 °C, shows zinc aluminate (Gahnite) to be the only crystalline phase. This is the lowest reported temperature for the formation of single phase crystalline $ZnAl_2O_4$. The diffraction peaks in samples calcined at higher temperatures (600-1400 °C) show sharper profiles resulting from the continuation of crystallisation/ordering processes and grain growth. The calculated lattice parameters (a_0 = 8.0855 Å) for cubic spinel phase are in good agreement with the reported value (a_0 = 8.0848 Å, JCPDS File 5-669). The determination of the crystallite size from the line shape analysis of the diffraction peaks showed a log normal grain growth with increasing calcination temperature (Fig. 7). The above observations were confirmed by the TEM images of the $ZnAl_2O_4$ samples calcined at different temperatures, which showed a systematic crystal growth [33].

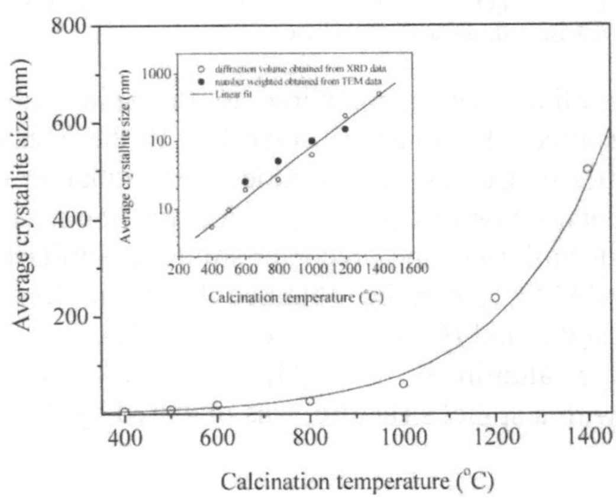

Figure 7. Exponential relationship of the crystallite size in $ZnAl_2O_4$ ceramic and calcination temperature.

The absorbance spectra (Fig. 8) of $ZnAl_2O_4$ samples heat-treated (6h) at 700, 800, 1000, 1200 and 1400 °C reveal a systematic shift in the absorption maxima. The absorbance edge became steeper on increasing the calcination temperature and a 'red shift' (4.5 to 4.0 eV) was observed (Fig. 8).

It is known that the band gap, E_g, of a nanocrystalline specimen scales with its diameter, d, as $E_g \sim 1/d$. It can be surmised that the observed variation in absorption is relevant to a strong change in the particle size reflecting the transition of spinel from nanocrystalline to the micron (bulk) state. The absorbance spectra of different $ZnAl_2O_4$ samples and ZnO, Al_2O_3 standards are shown in Fig. 8. The band gap

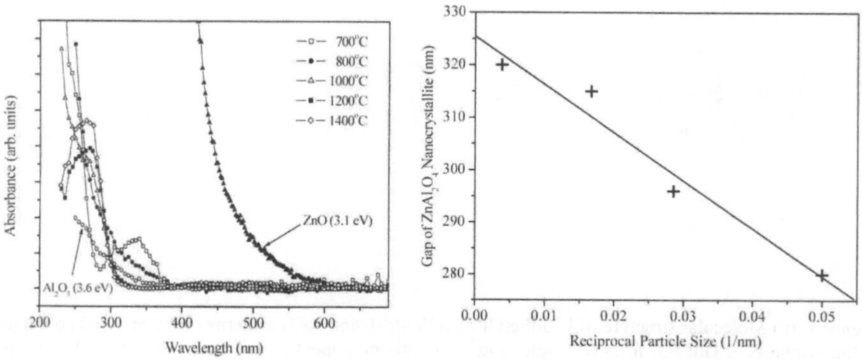

Figure 8. (a) Absorbance spectra of nanocrystalline $ZnAl_2O_4$ samples and (b) relationship between $ZnAl_2O_4$ bandgap and the particle size.

values (4.0–4.5 eV) were obtained from a linear extrapolation of the absorbance edge to cross the wavelength axis and are quite different from those of Al_2O_3 and ZnO in spite of the existence of the Al^{3+} and Zn^{2+} cations. This corroborates the phase purity of $ZnAl_2O_4$ samples and the absence of any micro-phase separation.

4.3 Orthoferrite ($GdFeO_3$)

A designed assembly of trivalent lanthanide and iron ions in a molecular framework (Fig. 9) provided, for the first time, access to metastable rare earth iron perovskites of general formula $LnFeO_3$ (Ln = any lanthanide ion). The unique magnetic properties such as high coercivity [35] and Faraday rotation [36], make orthoferrites interesting materials for magneto-optical data storage devices. The problem commonly encountered in the selective synthesis of $LnFeO_3$ compounds is the formation of thermodynamically favored garnet and secondary iron oxide phases caused by the use of a mixture of Ln^{3+} and Fe^{3+} constituents as the precursor. Since no chemical control (bonding) is offered, the lanthanide and iron compounds in the admixture randomly collide to form various intermediate species with metal ratios unfavorable for obtaining a single-phase material. The result is the co-existence of more stable garnet ($Ln_3Fe_5O_{12}$) and magnetite (Fe_3O_4) phases, which in view of their higher magnetic moments hinder specific investigations on the weak ferrimagnetic behaviour of the orthoferrite, $LnFeO_3$. In this context, the precursors with the required cation ratio and preformed Ln–O–Fe bonds are a major break-through in the selective synthesis of orthoferrite films and particles [37].

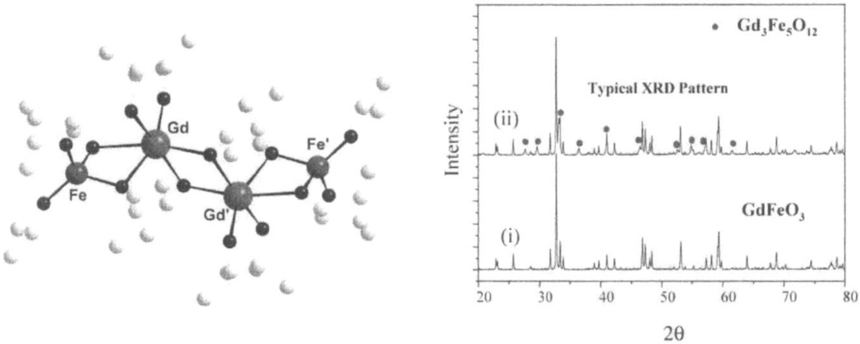

Figure 9. (a) Molecular structure of $[\{GdFe(OPr^i)_6\}(HOPr^i)]_2$ and XRD patterns of the materials obtained in the synthesis of $GdFeO_3$ from (i) single- and (ii) multi-component routes. Peaks due to $Gd_3Fe_5O_{12}$ are marked as •.

Nanoparticles of $GdFeO_3$ were obtained by a controlled hydrolysis of the heterometal alkoxide. The sol was freeze-dried to obtain the xerogel that was calcined at different temperatures (600, 800 °C) and characterized by powder X-ray diffratometry (Fig. 10a). The raw powder was amorphous till 600 °C. A well-developed orthoferrite phase is observed at 800 °C. The crystalline $GdFeO_3$ forms directly from an amorphous precursor without the crystallisation of any intermediate phases. The garnet composition, invariably formed in the synthesis of orthoferrites, is not observed when a single-source precursor is used. Moreover, no phase segregation or crystallisation of other stoichiometries was observed, which confirms the compositional purity of the sample and our contention that the chemical mixing

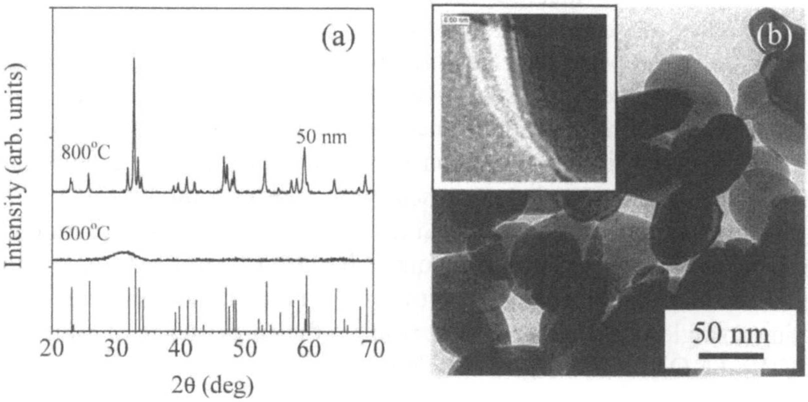

Figure 10. (a) XRD patterns and (b) normal and high resolution TEM images of $GdFeO_3$ ceramic.

of the ions is retained during the various stages of the processing. Further, the crystalline $GdFeO_3$ is formed at much lower temperature than those required for the solid-state reaction using Gd_2O_3 and Fe_2O_3 powders (~1800 °C). The TEM image (Fig. 10b) of the $GdFeO_3$ ceramic calcined at 1000 °C shows faceted crystallites of nearly uniform size (~ 60 nm). The high-resolution image of the $GdFeO_3$ particles revealed clear lattice fringes with interplanar spacings corresponding to the perovskite lattice [37].

Orthoferrites display antiferromagnetic order for T < 620-730 K, depending on the atomic size of the rare earth involved. The weak ferromagnetism arises from the low symmetry of the magnetic unit cell, producing a spin-canted structure of the Fe sublattice [38]. Both the temperature and field dependence of the magnetization in $GdFeO_3$ can be explained from the weak ferromagnetism [39] of these systems, and the very different ordering temperatures of Gd (~1 K) and Fe (~600 K) magnetic sublattices [40]. The magnetic susceptibility χ_{dc} displays a paramagnetic-like behaviour down to the lowest temperature available (T ~ 2 K). The fit of the experimental data using the Curie-Weiss law $\chi(T) = C/(T - \Theta)$ was used to calculate the effective magnetic moment $\mu_{eff} = 8.1(1)\ \mu_B$ and Curie-Weiss temperature Θ= -7(1)K. The value of effective moment corresponds to the paramagnetic contribution of Gd^{3+} ions (i.e., 8.0 μ_B for an $^8S_{7/2}$ state). On the other hand, the small (negative) value of Θsupports the fact that the Gd-Gd (antiferromagnetic) exchange

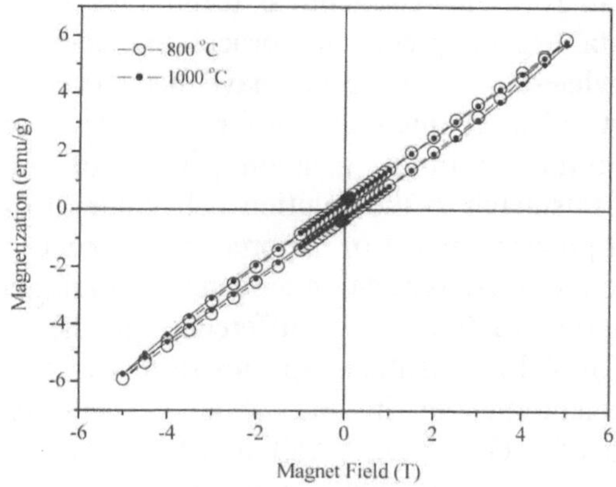

Figure 11. (a) Magnetization and inverse magnetic susceptibility (inset, a) curves and (b) M-H hysteresis (300 K) for the $GdFeO_3$ ceramic calcined at 800 and 1000 °C.

are remarkably weak. The M(H) curve (Fig. 11) shows a linear increase with field, without signs of saturation up to 5 T, resembling a paramagnet. However, the large coercivity (H_C = 2.6(1) kOe) observed indicates that the sample is not superparamagnetic in spite of the nanosized grains.

4.4 Pure and Lanthanide-Doped Garnet Powders ($Y_3Al_5O_{12}$ and Ln:$Y_3Al_5O_{12}$)

YAG is a promising high-temperature engineering material for insulating or refractory coatings because of its low creep rate, high oxidation resistance, and low electrical conductivity. Lanthanide-doped yttrium aluminum garnets (Ln:YAG; Ln:$Y_3Al_5O_{12}$, Ln = any lanthanide element) have been widely studied due to their application in solid-state lasers [41-45]. YAG is a robust host for luminescent ions because it does not damage easily under high irradiance with an electron beam. Therefore, rare-earth-doped YAG materials are promising phosphor candidates in cathode-ray tubes (CRTs), field-emission display (FED), scintillators, vacuum fluorescent displays (VFDs) and electroluminescent systems.

The solid-state synthesis of YAG ceramic from Al_2O_3 and Y_2O_3 powders usually requires extensive mechanical mixing and lengthy heat treatments [45, 46]. The yttria-alumina system has several stable phases, namely YAP ($YAlO_3$), YAM ($Y_4Al_2O_9$) and YAG ($Y_3Al_5O_{12}$). Commonly YAP and YAM phases are formed as intermediate byproducts even if YAG is synthesized with stoichiometric mixture of Y_2O_3 and Al_2O_3. Therefore extensive heat treatments at high temperatures (> 1600 °C) [45, 46] and repeated mechanical milling are required to decompose the intermediates so that a pure garnet phase can be formed. These processing conditions do not allow facile control over microstructure, grain size and grain size distribution in the resulting powders or shapes. Several wet-chemical techniques such as polymerized complex route, metal-organic preceramic processing, coprecipitation methods or yttrium carboxylate-alumoxane route have been used to produce YAG phases [47]. Most of these methods suffer from the complex and time consuming (long refluxing times, gelation periods of several days, etc.,) procedures and/or mismatch in the solution behaviour of different chemical species. Further, a precise control of cationic stoichiometry is not achieved, for instance, due to the phase separation tendencies of unimetal phases. In the case of precipitation reactions, the difference in the isoelectric points (different pH level) of the constituent species results in element segregation and non-ideal stoichiometries in the obtained ceramic. For example, significant amounts of Y_2O_3, Al_2O_3, $YAlO_3$ and $Y_2Al_4O_9$ phases are present in the synthesis of $Y_3Al_5O_{12}$ from aqueous yttria and alumina sols [48]. The problems of non-uniform mixing, residual phases, unreacted starting materials pose a challenge on the conventional methods used for the synthesis of doped materials where a few atomic weight percentage of a constitutent is to be homogenously dispersed in a host matrix. Since sol-gel methods are based on molecular precursors, they allow **chemical** cross-linking of different metal centers in the initial mixture of precursor species leading to a homogeneous distribution of dopant ions in a uniform matrix.

For comparative evaluation of the conventional solution approach and the one based on molecular alkoxide precursors, we have prepared garnet powders by employing common inorganic salts (chloride, nitrate, acetate, etc.) and alkoxide precursors [47]. Both pure and Nd-doped (1-5 wt.%) yttrium aluminum garnet powders were obtained by the complexation of Y^{3+}, Al^{3+} and Nd^{3+} ions with 1,2-ethanediol **(Polymeric Precursor Route)** in an aqueous media as well as by the controlled hydrolysis of Y, Al and Nd alkoxides **(Alkoxide Route)**.

The TG/DTA curves for the Y-Al-O gel synthesized by polymeric precursor and alkoxide routes are shown in Fig. 12. Whereas the crystallisation of the garnet phase occurs around 800 °C for the alkoxide-derived powder the onset of crystallisation is observed at higher temperature (> 900 °C) in the polymeric processing. Despite the lower crystallisation temperature, the residual carbon content in alkoxide-derived powders was significantly lower than that found in polymeric precursor case. The higher percentage of carbon impurities in the glycolate route was found to originate from the traces of undecomposed intermediates (e.g., yttrium carbonate or oxycarbonate) or trapped complexing agents. This observation was confirmed by recording the infrared spectra of heat-treated samples that showed the presence of residual organics and carbonate species. The X-ray diffraction profiles of the powders reveal the formation of monophasic garnet powders in both the cases, however the microstructural features of the two samples were found to be significantly different.

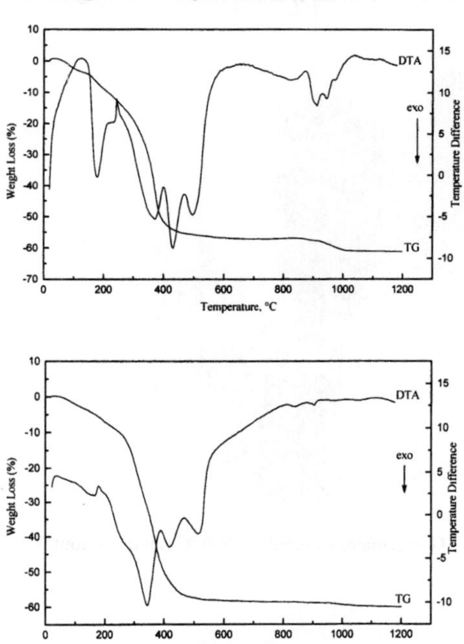

Figure 12. TG/DTA profiles of Nd:YAG ceramics obtained from (a) glycolate and (b) alkoxide routes.

The TEM images of the powders heated at 1000 °C showed high agglomeration effects in the case of polymeric precursor synthesis while well-dispersed particles of regular size were obtained in the case of alkoxide processing (Fig. 13). The high C and H contents in YAG powders obtained by using a complexing agent together with the observed morphology suggest that the residual organic materials act as a 'binder' for the nanoparticles. A high-resolution TEM image shows that the sintering in alkoxide-derived YAG powder begins at relatively low temperature as revealed by the fusion of YAG crystallites to form sinter-necks (Fig. 14).

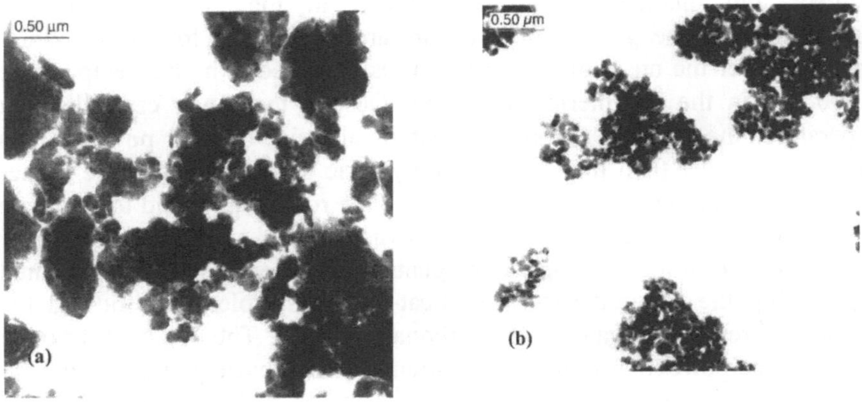

Figure 13. Transmission electron micrographs of YAG ceramic, sintered at 1000 °C: (a) glycolate and (b) alkoxide route.

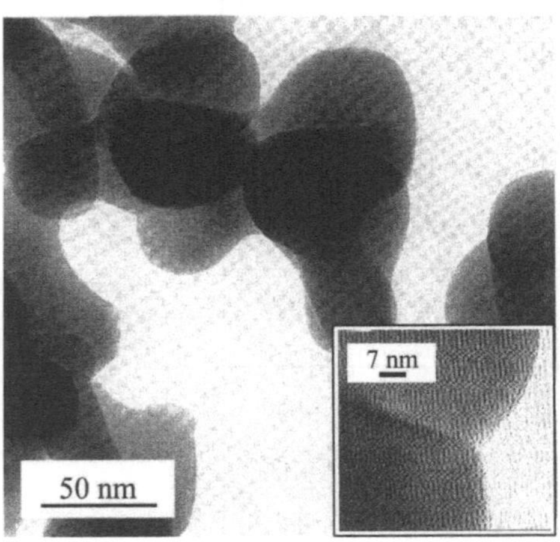

Figure 14. HRTEM image of a YAG specimen, sintered at 1000 °C (alkoxide route).

In the case of Nd-doped (1 wt.%) YAG powders, the XRD results show that although a pure garnet matrix is formed in the case of polymeric precursor route, a homogeneous distribution of Nd ions is not achieved. This is evident in the observance of small peaks in the diffractogram, which correspond to residual neodymium oxide (Fig. 15). A higher homogeneity is achieved in the case of alkoxide synthesis because all the metal precursors bear similar ligands and can form a homogeneous network during the hydrolysis reactions. Consequently, the dopant Nd ions are chemically 'fixed' in a Nd-O-Y-O-Al network which leads to the formation of monophasic Nd:YAG ceramic (Fig. 15).

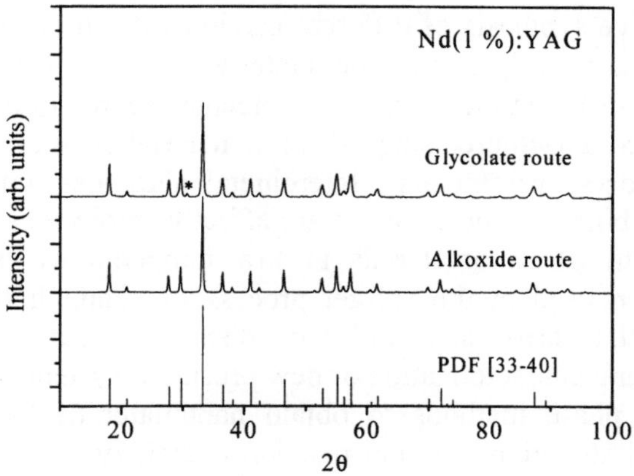

Figure 15. Room temperature XRD traces of Nd:YAG ceramics obtained from glycolate and alkoxide routes (Nd_2O_3 phase (*)).

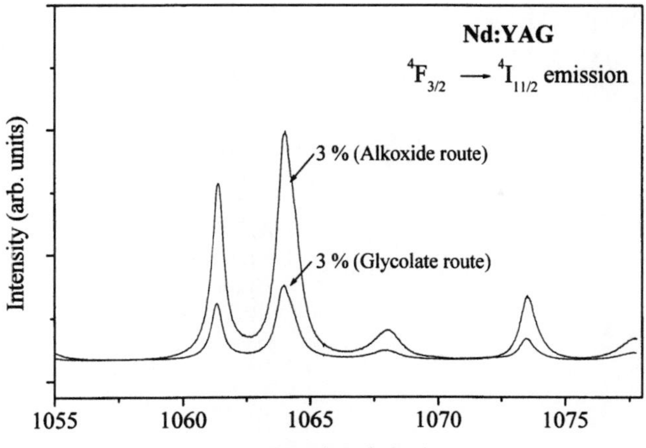

Figure 16. Room temperature emission spectra ($^4F_{3/2}$ to $^4I_{11/2}$ transition) of Nd(3%):YAG ceramics obtained from glycolate and alkoxide routes.

The better dispersion of Nd ions in the case of alkoxide processing is also observed in the emission spectra of the compacts obtained from the two different powders. A comparison of the $^4F_{3/2} \rightarrow {}^4I_{11/2}$ emission of Nd(3 wt.%):YAG samples (Fig. 16) shows a significantly higher emission efficiency for alkoxide-derived powder, which can be attributed to an ultra-homogeneous incorporation of Nd ions in the garnet matrix when using alkoxide route.

5. CONCLUSIONS AND PERSPECTIVES

The hydrolytic transformation of molecular alkoxide derivatives used for the phase selective synthesis of different ceramics illustrates the potential of alkoxide precursors to 'preform' the material on a molecular level. The incorporation of different elements in a molecular source to obtain the target material from a Single-Source simplifies the material synthesis by drastically reducing the process parameters. Heterometal alkoxides with metal ratios compatible with binary oxide ceramics are effective molecular templates for a controlled growth of nanomaterials at low temperatures. These building blocks have been used in the sol-gel process to obtain high purity oxide nanopowders with a narrow size- and shape-dispersion. Our continuing efforts in the synthesis and characterization of new precursor systems and their use in solution or gas phase methods to obtain nanomaterials have shown that predefined metal stoichiometry and reaction chemistry of the precursor can enforce a molecular level homogeneity in the obtained materials [49-55].

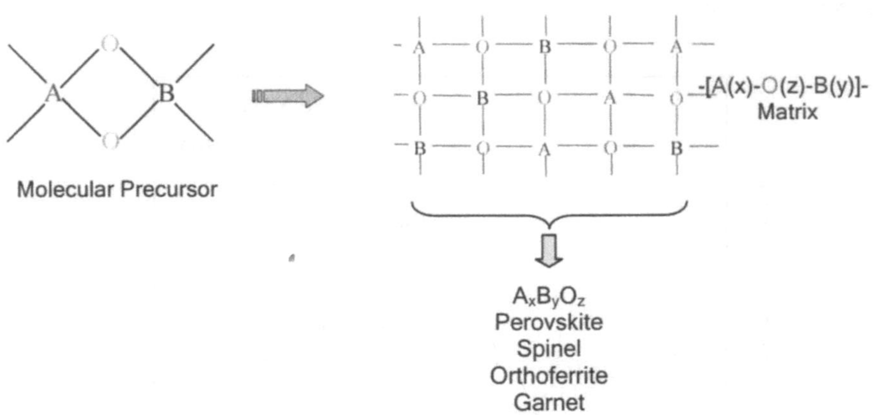

Scheme IV

In the conventional material synthesis procedures (e.g., solid-state reactions, coprecipitations, polymeric precursor methods, etc.), the intrinsic behaviours of the different components make a precise control over the phase formation rather difficult. In this context, the single-source precursors based on predetermined metal-ligand interactions (e.g., metal-carbon for carbides, metal-nitrogen for nitrides, metal-oxygen for oxides, etc.) are an attractive choice to achieve homogeneous cation distribution at the atomic level and a well-defined stoichiometry.

Acknowledgments: Author kindly acknowledge Prof. H. Schmidt and Prof. M. Veith, Institute of New Materials, Saarbruecken for their continuing support and for providing the necessary infrastructural facilities.

References

1. Edelstein, A. S. and Cammarata, R. C. Eds. (1996) *Nanomaterials: Synthesis, properties and applications,* Institute of Physics Pub., Bristol and Philadelphia.
2. Fendler, J. H. Ed. (1996) *Nanoparticles and Nanostructured Films,* WILEY-VCH, Germany.
3. Segal, D. L. (1989) *Chemical Synthesis of Advanced Materials Cambridge* University Press, Cambridge.
4. Cheetham, A. K. Brinker, C. J. McCartney, M. L. and Sanchez, C. (1994) Better Ceramic Through Chemistry Mat. Res. Soc. Symp. Proc. 360.
5. Klein, L. C. Ed. (1988) *Sol-Gel Technology for Thin Films, Fibers, Preforms, Electronics and Speciality Forms* Noyes Pub., Park Ridge, NJ.
6. Brinker, C. J. Scherer, G. W. (1990) *Sol-Gel Science: The Physics and Chemistry of Sol-Gel Processing* Academic Press, New York.
7. Hench, L. L. West, J. K. (1990) Chem. Rev. **90**, 33.
8. Mathur, S. (2002) in NATO ASI Series: *Chemical Physics of Thin Film Deposition for Micro- and Nano-Technologies*, Y. Pauleau (ed.), Kluwer Academic Publications, 91.
9. Veith, M. (2002) J. Chem. Soc. Dalton Trans. 2405.
10. Mathur, S. Veith, M. Shen, H. and Hüfner, S. (2002) Mat. Sci. Forum **386-388**, 341.
11. Veith, M. Mathur, S. Huch V. and Decker, T. (1998) Eur. J. Inorg. Chem. 1327.
12. Narayanan, R. Laine, R. M. (1997) Appl. Organomet. Chem. **11**, 919.
13. Jones A. C. (1997) Chem. Soc. Rev. 101.
14. Coles, M. P. Lugmair, C. G. Terry, K. W. Tilley, T. D. (2000) Chem. Mater. **12**, 122.
15. Bradley, D. C. (1989) Chem. Rev. **89**, 1317.
16. Mehrotra, R. C. (1990) Chemtracts **2**, 389.
17. Veith, M. Mathur, S. Mathur, C. (1998) Polyhedron **17**, 1005.
18. Mehrotra, R. C. Singh, A. (1997) Prog. Inorg. Chem. **46**, 239.
19. Xu, R. (1997) J. Mater. Res. **49**, 1.
20. Jones, K. Davies, T. J. Emblem, H.G. Parkes, P. (1986) Mat. Res. Soc. Symp. Proc. **73**, 111.
21. Hirano, S. Hayashi, T. Noaski, K. Kato, K. (1989) J. Am. Ceram. Soc. **72**, 707.
22. Bradley, D. C. Mehrotra, R. C. Gaur, D. P. (1978) *Metal Alkoxides*, Academic Press, London.
23. Meyer, F. Dierstein, A. Beck, C. Hempelmann, R. Mathur, S. Veith, M. (1999) Nanostruct. Mater. **12**, 71.
24. Meyer, F. Hempelmann, R. Mathur, S. Veith, M. (1999) J. Mater. Chem. **9**, 1755.
25. Riman, R. E. (1993) *High-performance Ceramics* Eds. Pugh, R. and Bergstroem, L. Marcel-Dekker, New York.
26. Goodman, G. (1986) in *Ceramic Materials for Electronics* ed. Buchanan, R. C. Marcel-Dekker, New York.

27. Deb, K. K. Hill, M. D. and Kelly, J. F. (1992) J. Mater. Res. **7**, 3296.
28. Veith, M. Mathur, S. Lecerf, N. Huch, V. Decker, T. Beck, H.P. Eiser, W. Haberkorn, R. (2000) J. Sol-Gel Sci.Tech. **17**, 145.
29. Mellor, J. W. (1937) Trans. Cer. Soc. **36**, 1.
30. Hedvall, J. A. and Heuberger, J. (1921) Z. Anorg. Allg. Chem. **116**, 137.
31. Baker, E. and Burch, R. Yugin, N. (1991) Appl. Cat. **73**, 135.
32. Ohgushi, T. and Umeno, S. (1987) Bull. Chem. Soc. Jpn. **60**, 4457.
33. Mathur, S. Veith, M. Haas, M. Shen, H. Hüfner, S. Haberkorn, R. Beck, H. P. Jilavi, M. (2001) J. Amer. Ceram. Soc. **84**, 1921.
34. Valenzuela, M. A. Bosch, P. Aguilar-Rios, G. Montoya, A. and Schifter, I. (1997) J. Sol-Gel Sci. Tech. **8**, 107.
35. Schmool, D. S. Keller, N. Guyot, M. Krishnan, R. Tessier, M. (1999) J. Magn. Magn. Mater. **195(2)**, 291.
36. Schmool, D. S. Keller, N. Guyot, M. Krishnan, R. Tessier, M. (1999) J. Appl. Phys. **86(10)**, 5712.
37. Mathur, S. Shen, H. Lecerf, N. Fjellvag, H. Goya, G. F. (2002) Adv. Mater. **14**, 1405.
38. Dzyaloshinsky, I. E. (1958) J. Phys. Chem. Solids **4**, 241.
39. Bozorth, R. M. (1958) Phys. Rev. Lett. **1**, 362.
40. Treves, D. (1965) J. Appl. Phys. **36**, 1033.
41. Geusic, J. E. Marcos, H. M. van Uitert, L. G. (1964) Appl. Phys. Lett. **4**, 182.
42. Machan, J. Kurtz, R. Bess, M. Birnbaum, M. (1987) J. Opt. Soc. Am. **20**, 134.
43. Shkadarevich, A. P. (1989) J. Opt. Soc. Am. **22**, 60.
44. Izumitani, T. (1991) Bull. Ceram. Soc. Jpn. **26**, 108.
45. Ikesue, A. Kamata, K. Yoshida, K. (1995) J. Am. Ceram. Soc. **78**, 2545.
46. Yang, J.-M. Jeng, S. M. Chang, S. (1996) J. Am. Ceram. Soc. **79**, 1218.
47. Veith, M. Mathur, S. Kareiva, A. Jilavi, M. Zimmer, M. and Huch, V. (1999) J. Mater. Chem. **9**, 3069.
48. King, B. H. Halloran, J. W. (1995) J. Am. Ceram. Soc. **78**, 2141.
49. Mathur, S. Veith, M. Sivakov, V. Shen, H. Gao, H. B. (2001) J. Phys. IV **11**, 487.
50. Mathur, S. Lecerf, N. Shen H. Veith, M. and Huefner S. (2001) Scripta Mater. **44**, 2157.
51. Veith, M. Mathur, S. Lecerf, N. Bartz, K. Heinz, M. Huch, V. (2000) Chem. Mater. **12**, 271.
52. Mathur, S. Veith, M. Shen, H. Lecerf, N. Hüfner S. and Jilavi, M. (2001) Chem. Mater. **13**, 4041.
53. Mathur, S. Veith, M. Shen, H. Hüfner, S. and Jilavi, M. (2002) Chem. Mater. **14**, 568.
54. Shen, H. and Mathur, S. (2002) J. Phys. IV **12**, PR4-1.
55. Mathur, S. and H. Shen. (2002) J. Sol-Gel Sci. Tech. **25**, 143.

NANOSIZED FINE PARTICLES AND FIBERS AS REINFORCING MATERIALS SYNTHESISED FROM SEPIOLITE

A.O. KURT
Sakarya University
Engineering Faculty, Esentepe Kampüsü,
54187 - ADAPAZARI - TURKEY

Abstract: Sepiolite has very small (several hundred nanometre thick and a few micron long) fibers with high internal and external surface areas. It was found to be a good Si source for the thermal reduction- nitridation reaction (with a reducing agent) to produce silicon nitride. Sepiolite of Turkish origin was successfully purified by a hydrometallurgical process using controlled additions of concentrated HCl acid. The crystal structure, morphology and mineral composition of the sepiolite was preserved for later treatment by intercalation and nitridation. Various types of silicon nitride having different morphologies were obtained, ranging from long whiskers to equiaxed fine particles using sepiolite and various reducing agents. Silicon nitride, in the form of nanosized fine particles and fibrous form, has been produced by pyrolysis of purified sepiolite and polyacrylonitrile intercalation complex.

1. INTRODUCTION

Sepiolite is a hydrated magnesium-silicate type clay mineral. Its chemical formula, for a half unit cell, is given as $[(Si_{12})(Mg_8)O_{30}\,(OH)_4\,(OH_2)_4\,8H_2O]$ by Brauner and Preisinger [1]. The unit cell parameters are given as a=13.4Å, b=26.8Å, c= 5.28Å [1]. Typically, sepiolite has a fibrous morphology with micro-channels along the fiber axis. The cross section of such a channel is 13.4Å by 6.7Å [1]. Each macro-fiber consists of tens of hundreds of sub-nanometer fibers [2]. The parallel `piped` fibers are bundled together along the *c*-axis to form microscopically fibrous structure, which is porous in its crystalline form (Fig. 1-b). The generic molecular structure of sepiolite is given in Figure 1-a.

The crystal structure of sepiolite consists of two tetrahedral sheets that are continuous along the *b*- and *c*-axes and an octahedral layer that lies in one dimension is sandwiched between the two tetrahedral sheets (Fig. 1-a). Micro-channels exist along the *c*-axis and contain free water (zeolitic water) molecules attached to the edges of the octahedral layer. This gives sepiolite

S. Guceri et al. (eds.), Nanoengineered Nanofibrous Materials, 443-449.

molecular-sieving properties since only certain size molecules can either pass through or be trapped in these channels. In industry, sepiolite and its derivatives have been widely used for adsorption and molecular sieving applications in chemical industries [3-5].

It is a well-known application to exchange some Mg^{2+} for other cations and with that of heavy metals in order to increase sepiolite`s sorptive and catalitic properties [6,7]. Mostly the exchange takes place at easily accessible sites at the broken bonds on the fiber surface and in the octahedral sites at the channel edges. It is also possible to disintegrate the bundles in the individual fibers and align them in one direction after replacing some of the Mg^{2+} ions with Fe^{2+} in an ion-exchanged process using aqueous solution while applying

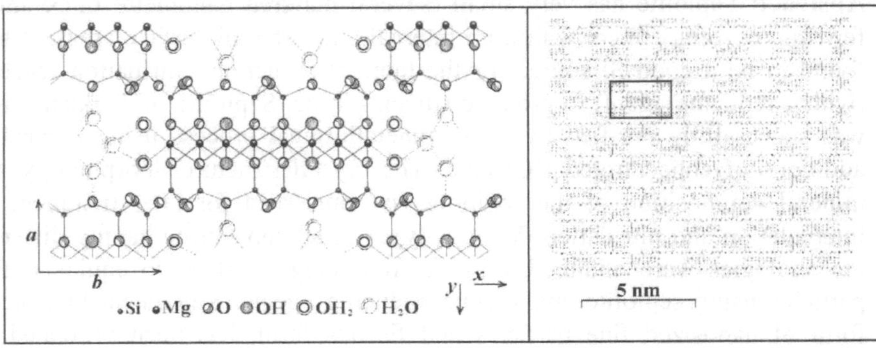

(a) (b)

Figure 1. Simplified ionic structure of sepiolite (a), after Brauner and Preisinger [2]. The repeat of the molecular unit in *x* and *y*-axis forms the pattern in (b). To give a clear view in b, zeolitic water molecules (H_2O) are not shown in the system.

a continuous magnetic field [8].

Positively several numbers of successful research work could be found in literature where the clay minerals and some other unusual sources of precursors were used to produce ceramic powders. For instance, Si_3N_4 of α-phase was synthesised using rice husk, digested with NHO_3 and pyrolysed before heat treating at 1450 °C in flowing nitrogen [9]. Short needle-like and wool-like whiskers of Si_3N_4 were obtained using a quartzite, diatomaceous earth, terra abla and moulding sand, at 1370 °C in an NH_3-N_2 gas-stream [10]. Without paying much attention on the purity level and the rheological properties, the uses of sepiolite (Japanese and Turkish origin clay minerals) as a Si source for the synthesis of Si_3N_4 were previously explored [11-13]. In this paper, it was aimed to provide an overview of the results obtained from the study of the synthesing Si_3N_4 ceramics from sepiolite clay mineral and the potential utility of the findings.

2. MATERIALS

As-received sepiolite of brown colour was obtained from Eskişehir-Turkey. Sepiolite samples were received in the form of coarse particles. These were crushed and sieved to different size fractions. Charcoal, carbon black and petroleum coke were primarily used as reducing agents (Table 1). Carbon black was used as-received without any further processing. Charcoal, however, was supplied in the form of big granular shaped particles; this was ball-milled to a fine powder with a high surface area. Acrylonitrile monomer supplied by *Hopkin & Willams ltd.* was used as an alternative carbon source in a polyacrylonitrile (PAN)-sepiolite intercalation compound. Azo-iso-butyronitrile (*GPR*) was used as an initiator (0.5wt.%) in acrylonitrile monomer. To stimulate the formation of α-phase, fine α-Si_3N_4 powders (average particle size, 0.5μm) (NU10, *TENMAT Ltd.*) was added in some of the reactants before the carbothermal reaction.

TABLE 1. Reducing agents used in processing.

Materials	Supplied by	Purity (%)	Pour Density (kg/m^3)	Tap Density (kg/m^3)	Mean Particle size (μm)	Surface Area (m^2/g)
Carbon Black (Vulcan-10H)	Cabot	99.7	345#	390	5	110
Charcoal, (C/4120/53)*	Fisher Scientific	92.8	333	488	30	777
Petroleum Coke D44 GAO 43B	James & Durans ltd.	99.0	761	904	70	60

* Charcoal received in the form of granular was ball milled (4h). # Values underlined are supplier data.

3. EXPERIMENTAL METHODS

Mineral and chemical compositions of sepiolite are given in Table 2 and Table 3, respectively. Sepiolite clay mineral comes with high amount of impurity content (i.e., dolomite). Sepiolite in the form of course particles was sieved to different size fractions. Powders of -150μm+75μm size fraction were purified using 0,21 M HCl acid solutions. Techniques to use and the details of the purification process have been previously reported [14]. Purified samples were used as a precursor for later use in carbothermal reduction-nitridation process and in sepiolite-polyacrylonitrile compound. Analysing of the starting materials, precursors and the reaction products using standard characterisation techniques have been thoroughly described [14,15].

TABLE 2. Mineral composition and features of the clay sample used in this study.

Mineral	Sepiolite (wt.%)	Dolomite (wt.%)	Total Weight Loss at 1000 °C (%)	Surface Area at 25 °C (m^2 g^{-1})
As-received	81.52 ±1.00	18.48 ±1.00	25.25	87
After purification	100	0	-	161

TABLE 3. Chemical composition (wt.%) of sepiolite samples, as-received material -150 +75μm size fraction after sieving.

SiO_2	MgO	CaO	Al_2O_3	Fe_2O_3	K_2O	TiO_2	Na_2O	P_2O_5
47.11	20.48	5.62	0.94	0.44	0.08	0.05	0.02	0.01

4. RESULTS AND DISCUSSION

4.1 Sepiolite

Dolomite impurity, constitutes in approx. 20 % of the as-received mineral, was fully eliminated by treating with 0,21M HCl for 2h at 60 °C (solid to liquid ratio 1/15 $g.ml^{-1}$) without any obvious effect on the molecular framework of fibrous sepiolite structure. This was confirmed by XRD, IR and SEM data (Fig. 2). The specific surface area of sepiolite is doubled after the elimination of the impurity, from 87 $m^2.g^{-1}$ for sepiolite as-received material to 161 $m^2.g^{-1}$ for the sample after leaching as expected since dolomite occurs as a cement in-between the parallel pipe-shaped sepiolite fibers.

An initial trial was made to produce silicon nitride from the HCl acid treated sepiolite and a polyacrylonitrile intercalation compound. The purified sepiolite was soaked in either a fixed amount or excess of an acrylonitrile monomer after the dehydration of zeolitic water molecules (at 140 °C, 2h). This precursor was subsequently polymerised and cyclized to produce stabilised PAN (to be a source of carbon in the later stage of pyrolysis) so that the resulting carbon could be used as a reducing agent in the carbothermal reduction-nitridation (CRN) process. The main purpose of using sepiolite-PAN was to obtain an intimate mixture of the reducing agent with the oxide phase(s) so that sepiolite and carbonised organic molecules (i.e., PAN) would contact at a molecular level during the carbothermic-reduction process. In

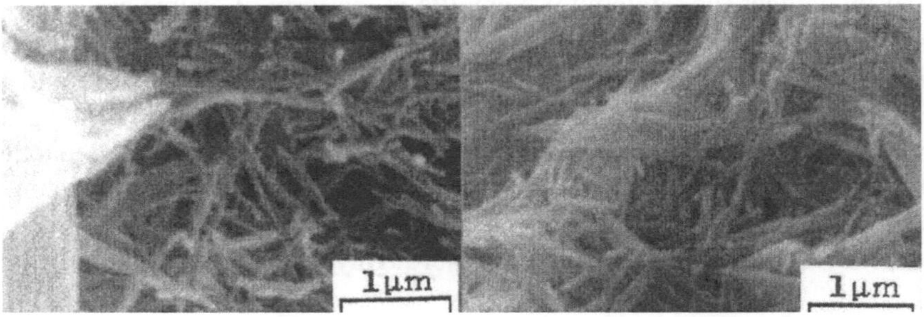

(a) (b)

Figure 2. SEM micrographs of the sepiolite fibers as-purified (a) and after polymerisation and cyclization (b) using the monomer at a weight to volume ratio of ½ (g/ml).

addition, the use of a high purity monomer virtually free from inorganic matter would eliminate any ash content. Figure 2-b shows sepiolite fibers after polymerization and cyclization stages. The specific surface area of the cyclized sepiolite-PAN sample was measured as 6.30 $m^2.g^{-1}$, which is well below the value of the total external surface area (161 $m^2.g^{-1}$) of an equal mass of pure sepiolite fibers. This suggests that although there is no optically visible external polymer coating on the fibers, i.e., the sepiolite`s fibrous morphology was preserved, a thin coating of polymer must exist over the bundle of fibers. Also, the channels and pores of the fibers, which were previously open, must have been filled by the monomer and the subsequent polymerisation would then block and cover the sepiolite fibers with a thin coating of polymer.

Acrylonitrile monomer having a free radical initiator is successfully impregnated and polymerizased into the channels of sepiolite fibers. An expanded account of the polymerisation of sepiolite-AN monomer is given somewhere [15].

Sepiolite as purified form or after polymerization or cyclization stages could be of some use, such as, in polymeric systems where fine fibers as reinforcing or filler elements are needed. Due to very cheap and relatively unsophisticated production processes purified sepiolite having individual fibers (or bundles of fibers) aligned in one direction could be used in, for example micro chemical sensors and membranes for molecular level filtration. Sepiolite fibers were used commercially as components for a carrier in chemical industry, such as for pharmaceuticals, cosmetics, and pesticides. However, application of sepiolite in chemical industry, such as in the pharmaceutical products, could be well explored after fully searching its health related biological behavior. There is a limited research on this therefore it has yet to be demonstrated whether the inhalation of sepiolite fibers is harmful to humans or not [16].

4.2 Silicon Nitride

Fine sized (uniaxial or long whiskers) and surface area up to 11.5 $m^2.g^{-1}$ silicon nitride powders were synthesised from sepiolite-PAN and sepiolite-solid carbon precursors; preparations of the precursors and details on the carbothermic reactions have been previously explained [14]. It was found that the amount and the morphology of the products after nitridation were dependent upon several factors in the synthesis and on the constituents, as well as the method of preparation of reagents. The state of the starting materials was found to have a major effect, also impurities, on the formation of Si_3N_4 and the α/β phase ratio. Under the identical testing conditions (i.e. the same purified sepiolite sample used and all samples nitrided identically, for example, at 1400 °C for 4h with 300 °C/h heating rate and 1000 ml/min N_2-flow rate); the best results were obtained with sepiolite-PAN and sepiolite-charcoal mixtures. Sepiolite-petroleum coke, on the other hand, yielded only small amounts of β-Si_3N_4 in CRN process. This effect was somewhat attributed to the specific surface area of the reducing agents (Table 1). It was observed that samples ran at relatively low temperatures (1300 - 1323 °C) with long nitridation time (8 - 16h.) resulted in the formation of Si_3N_4 whiskers (Fig. 3-a), which were identified as α-form constituted 40wt.% of the product. Either increasing the temperature or lowering the reaction time results in the formation of uniaxed fine grains that mainly of β-phase (approx. 74-81wt.%). Changes in powder morphology and particle size with a change in using different reducing agent was obvious since sepiolite-PAN intercalation compound yielded very fine particles mainly of β-Si_3N_4 (Fig. 3-b). Due to the nature of using clay mineral as a main source for Si, small amount of the retained magnesium silicate phases remained in the final product after nitridation. However, they were found beneficial since Si_3N_4 powders require such additives, which forms liquid phase on sintering.

Microscopic examinations using SEM and TEM showed that both precursor and end products after nitridation were submicron (100-200nm) in size (or cross section) that may be utilised as reinforcing material in metallic and polymeric systems. Sepiolite alone or combined with PAN could be of some use in polymeric systems as a cheap filler material. The further work is projected on concentrating the benefits of using such materials as strength inducing mechanisms. Silicon nitride synthesised from sepiolite contains inevitably some retained oxides phases that are not tolerable to make them use in high temperature applications. Therefore, the synthesised powders could be of an alternative source for low temperature applications such as, thermocouple protection sheaths, riser and delivery tubes in low-pressure diecasting operations and cutting edges for scissors. Some powders having high amount of silicon nitride phase were pressureless sintered at 1700 °C for 1h without using any additives. Powders, retained their small average grain size after sintering produced up to 85% theoretical density [17].

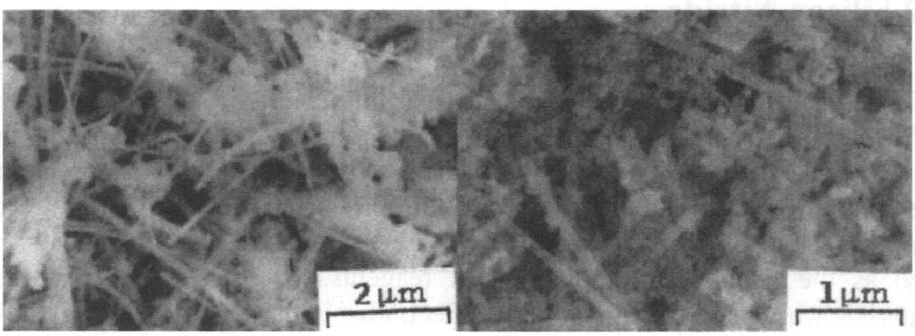

(a) (b)

Figure 3. The particle morphology of Si_3N_4 whiskers after 16h reaction at 1300 °C using sepiolite-charcoal mixture (a). Mainly fine β-phase particles with some lath-like shaped α-phase short fibers produced from sepiolite-PAN (b). (Carbothermal reduction-nitridation process took place at 1350 °C for 4h.)

5. CONCLUSION

Sepiolite clay mineral retained its fibrous microporous morphology after controlled hydrometallurgical processes. It could be of some use as its pure mineralogical form or after modifying it with hydrometallurgical processes. Sepiolite possesses microfibers, which have parallelepiped tunnels along the fiber axis that were filled with a suitable monomer for the preparation of fibrous silicate - polymer nanocomposite materials. Research and progress on the behaviour of an individual and the group of the fibers with or without intercalation will open new perspective of using sepiolite.

Silicon nitride powders of fine uniaxed grains or long whiskers were produced using sepiolite fibers with a reducing agent under reduction-nitridation atmosphere. The grain size and morphology of silicon nitride were found to be dependent on the reaction conditions and starting reagents.

References

1. Brauner, K. and Preisinger, A. (1956) Structure of Sepiolite, *Miner. Petrogr. Mitt.*, **6**, 120-140.
2. Rautureau, M. and Tchoubar, C. (1976) Structural analysis of sepiolite by selected area electron diffrac-tion relations with physico-chemical properties, *Clays and Clay Minerals*, **24**, 43-49.
3. Daza, L. Mendioroz, S. and Pajares, J.A. (1989) Influence of texture and chemical composition on sulphur deposition onto sepiolites, *Applied Clay Science*, **4**, 389-402.
4. Fujiwara, I. and Sato, M. (1992) Adsorption of water vapor on sepiolite for chemical heat pumps, *Journal of Chemical Engineering of Japan*, **25**, (5), 609-611.
5. Grillet, Y. Llewellyn, P.L. Tosi-Pellenq, N. and Rouquerol, J. (1993) Adsorption of argon, methane, nitrogen, carbon monoxide and water vapour on sepiolite and ALPO4-5 as studiesd by isothermal microcalorimetry. *Fundamentals of Adsorption, Proc. IVth Int. Conf.* on Fundamentals of Adsorption, p.235-242, International Adsorption Society, Kyoto, May 17-22, 1992.
6. Brigatti, M.F., Lugli, C. and Poppi, L. (2000) Kinetics of heavy-metal removal and recovery in sepiolite, *Applied Clay Science*, **16**, 45–57.
7. Galan, E. (1996) *Clay Minerals*, **31**, 443-453.
8. Esmer, K. and Yeniyol, M. (1999) Current-voltage characteristics and aging of sepiolite oriented by mag-netic field, *Materials Letters*, **38**, 445-449.
9. Rahman, I.A. (1994) *Ceramic International*, **20**, 195-199.
10. Mizuhara, Y. et al. (1994) *J. Ceram. Soc. Japan, Int. Edi.*, **102**, 489-494.
11. Sugahara, Y. *et al.* (1985) *J. Mater. Sci. Lett.*, **4**, 928-931.
12. Arık, H. *et al.* (1995) "Sepiyolitten Karbo-thermal Indirgeme Yöntemiyle Silisyum Nitrür (Si_3N_4) Dönü-şümünün Araştırılması", VII Ulusal Kil Sempozyumu, (VII National Clay Symozium), 27-30 September, Ankara, Turkey.
13. Arık, H. (1996) Ph.D. Thesis, Gazi Üniversitesi, Ankara, Turkey.
14. Kurt, A.O. and Davies, T.J. (2001) Synthesis of Si_3N_4 using sepiolite and various sources of carbon, *Journal of Materials Science*, **36**, 5895-5901.
15. Kurt, A.O. and Davies, T.J. (2001) Sepiolite-PAN intercalation used as Si_3N_4 forming precursor, *Journal of Materials Science*, **36**, 957-962.
16. Bellmann, B. Muhle, H. and Ernst, H. (1997) Investigations on health-related properties of two sepiolite samples, *Enviromental Health Perspectives*, **105**, (S5), 1049-1052.
17. Kurt, A.O. and Davies, T.J. (2001) Pressureless sintering of Si_3N_4 powders obtained from carbother-mally reduced sepiolite, *Journal of Materials Science Letters*, **20**, 1067-1070.

THERMAL CONDUCTIVITY OF PARTICLE REINFORCED POLYMER COMPOSITES

I.H.Tavman
Mechanical Engineering Department
Dokuz Eylül University, 35100 Bornova /Izmir – Turkey.

Abstract: In this paper a survey of literature on thermal conductivity of nanometer sized particle and fiber filled polymer composites are given as well as thermal conductivity of nanofluids (nanoparticles in suspension in fluids). Also models predicting the effective thermal conductivity of composites filled with particles are stated and are compared with experimental data with micro-sized Al_2O_3 particle filled HDPE composite.

1. INTRODUCTION

In order to obtain materials of desired thermal, mechanical and electrical properties, matrix materials, such as polymers, are processed with different kind of fillers. Fillers may be in the form of fibers or in the form of particles uniformly distributed in the polymer matrix material. Many of the properties of fibrous composite materials are strongly dependent on the fiber properties as well as on microstructural parameters such as fiber diameter, fiber length, fiber length distribution, volume fraction of fibers and the alignment and packing arrangement of fibers. It is evident that thermophysical properties of fiber filled composites are anisotropic, except for the very short, randomly distributed fibers. Whereas, thermophysical properties of particle filled polymers are isotropic. Many theoretical, numerical and experimental studies exist about thermal conductivity of micrometer sized particles and fiber filled polymer composites.

Recently, techniques have been developed to process matrix materials with nanometer sized grains, particles and fibers. These filler may be used as an effective strategy for either increasing or reducing heat transfer rates of composite materials. There are very few publications about thermal conductivity of nano particle and fiber filled polymer composites, a recent research by Putman et al. [1] used the 3ω method to study the thermal conductivity of composites of nanoscale alumina particles in polymethylmethacrylate (PMMA) matrices in the temperature range 40<T<280 K. For 10% of 60 nm of alumina particle filler by weight (3.5% by volume) thermal conductivity of the composite is slightly decreased at low

S. Guceri et al. (eds.), Nanoengineered Nanofibrous Materials, 451-457.

temperatures (below 100 K), the maximum decrease taking place at 40K which is about 2.5% less than the pure PMMA. Whereas, above 100 K, thermal conductivity of the nanocomposite is slightly increased, maximum increase of 4% at room temperature. They also did measurements with micrometer sized alumina particles, thermal conductivity of the composite is increased about 2 to 3% in the whole temperature range.

Another publication by Kruger and Alam [2] is about the thermal conductivity of aligned, vapor grown carbon nanoscale fiber reinforced polypropylene composite. They measured thermal conductivity by Holometrix μ Flash Thermal Properties instrument in the longitudinal and transverse directions for 9%,17% and 23% fiber reinforcements by volume. The values of thermal conductivity are 2.09, 2.75, 5.38 W/m.K for the longitudinal directions and 2.42, 2.47, 2.49 W/m.K for the transverse direction respectively, while the thermal conductivity of unfilled PP is 0.24 W/m.K. There is anisotropy in thermal conduction, there is increase in both direction.

A recent area of research is about the effective thermal conductivity of liquid with nanoparticle inclusions. The effective thermal conductivity of nanoparticle suspension can be much higher than the normally used industrial heat transfer fluid, such a fluid has terminologized as "nanofluid", and considered to be a novel enhanced heat transfer fluid. Very recently, Keblinski et al. [3] reported their idea on the possible mechanisms of en-hancing thermal conductivity, and suggested that the size effect, the clustering of nanoparticles and the surface adsorption could be the major reason of enhancement, while the Brownian motion of nanoparticles contributes much less than other factors. In particular, it was demonstrated that solid nanoparticle colloids (i.e., colloids in which the grains have dimensions of ≈10 - 40 nm) are extremely stable and exhibit no significant settling under static conditions, even after weeks or months [4]. Furthermore, the enhancement of thermal-transport properties of such "nanofluids" was even greater than that of sus-pensions of coarse-grained materials [5].For example, the use of Al_2O_3 particles ≈13nm in diameter at 4.3% volume fraction increased the thermal conductivity of water under stationary conditions by 30% [6].Use of somewhat larger particles (≈40 nm in diameter) only led to an increase of less than ≈10% at the same particle volume fraction [5]; more in accord with theoretical predictions[7]. An even greater enhancement was recently reported for Cu nanofluids, where just a 0.3% volume fraction of 10 nm Cu nanopar-ticles led to an increase of up to 40% in thermal conductivity [8], a result that is more than an order of magnitude above the increase predicted by macroscopic theory. Currently, the origin of such remarkable increases in the thermal conductivity of nanofluids eludes theoretical understanding.

2.THERMAL CONDUCTIVITY MODELS FOR TWO PHASE SYSTEMS

Many theoretical and empirical models have been proposed to predict the effective thermal conductivity of two phase mixtures. The upper or lower bounds of effective thermal conductivity are given when materials are arranged in either parallel or series with respect to heat flow. For the parallel conduction model:

(1) $$k_c = \phi.k_f + (1-\phi).k_m$$

and for series conduction model:

(2) $$\frac{1}{k_c} = \frac{\phi}{k_f} + \frac{1-\phi}{k_m}$$

where, k_c , k_m and k_f are thermal conductivities of composite, matrix and filler, respectively and ϕ is the volume fraction of filler.

In the case of geometric mean model, the effective thermal conductivity is given by:

(3) $$k_c = k_f^{\phi}.k_m^{(1-\phi)}$$

Using potential theory, Maxwell [9] obtained a simple relationship for the conductivity of randomly distributed and non-interacting homogeneous spheres in a homogeneous medium:

(4) $$k_c = k_m \frac{k_f + 2.k_m + 2.\phi.(k_f - k_m)}{k_f + 2.k_m - \phi.(k_f - k_m)}$$

Based on Tsao's probabilistic model Cheng and Vachon [10], assumed a parabolic distribution of the discontinuous phase. The constants of the parabolic distribution were evaluated as a function of the discontinuous phase volume fraction. The effective thermal conductivity is given for the case $k_f > k_m$:

$$\frac{1}{k_c} = \frac{1}{\sqrt{C.(k_f - k_m)(k_m + B.(k_f - k_m))}} \ln \frac{\sqrt{k_m + B(k_f - k_m)} + B/2\sqrt{C.(k_f - k_m)}}{\sqrt{k_m + B.(k_f - k_m)} - B/2\sqrt{C.(k_f - k_m)}} + \frac{1-B}{k_m}$$

(5)

where, $B = \sqrt{\frac{3.\phi}{2}}$, $C = -4.\sqrt{\frac{2}{3.\phi}}$

Lewis and Nielsen [11] derived a semi-theoretical model by a modification of the Halpin-Tsai equation to include the effect of the shape of the particles and the orientation or type of packing for a two-phase system:

$$k_c = k_m \frac{1 + A.\beta.\phi}{1 - \beta.\phi.\psi}$$

(6) $$\beta = \frac{k_f/k_m - 1}{k_f/k_m + A} \quad \text{and} \quad \psi = 1 + \frac{1-\phi_m}{\phi_m^2}\phi$$

The constant A depends upon the shape and orientation of the dispersed particles. ϕ_m is the maximum packing fraction of the dispersed particles. For randomly packed spherical particles A=1.5 and ϕ_m=0.637, whereas for randomly packed aggregates of spheres or for randomly packed, irregularly shaped particles A=3 and ϕ_m=0.637.

Based on the generalization of models for parallel and series conduction in composites, Agari and Uno [12] propose a new model for filled polymers:

(7) $$\log k_c = \phi.C_2.\log k_f + (1-\phi).\log(C_1.k_m)$$

where, C_1, C_2 are experimentally determined constants of order unity. C_1 is a measure of the effect of the particles on the secondary structure of the polymer, like crystallinity. C_2 measures the ease of the particles to form conductive chains, the more easily particles are gathered to form conductive chains, , C_2 becomes closer to 1. However, experimental data is needed for each type of composite in order to determine the necessary constants.

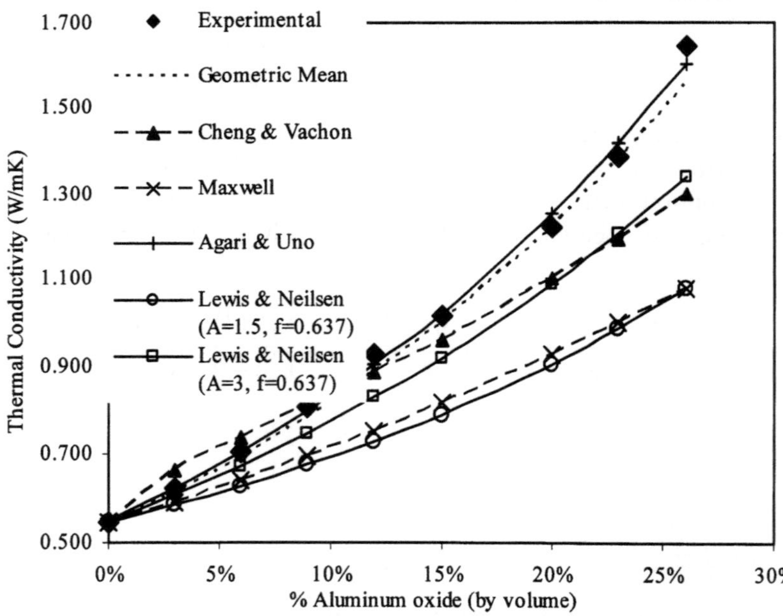

Fig. 1. Comparison of the experimental and predicted thermal conductivity values for high density polyethylene filled with Al_2O_3 micro particles at 15°C.

3. EXPERIMENTAL

Samples are prepared by mold compression process. The matrix material is commercial high density polyethylene in powder form, density 0.968 g/cm^3, melt index 5.8 g/10min and measured thermal conductivity 0.545 W/m.K. The filler is Al_2O_3 powder with particle size 10-20 microns, the solid density of Al_2O_3 is 3.99 g/cm^3 and its thermal conductivity 31 W/m.K. Thermal conductivity measurements are performed using the Shotherm QTM thermal conductivity meter working with a modified hot wire method. Thermal conductivity measurements are performed on HDPE filled with Al_2O_3 particles up to a volumetric fraction of 26%, at 15°C. The results are plotted in Fig.2 and are compared to calculated values from Maxwell, Cheng & Vachon, Lewis & Nielsen, Agari & Uno, geometric mean models. Geometric mean and Agari & Uno models predict quite well the thermal conductivities of the composite in the whole range. In Agari & Uno model the values of the coefficients are calculated as C_1=1.013 and C_2=1.021.

4. THERMAL CONDUCTIVITY OF POLYMER NANOCOMPOSITES

The existing understanding of the effective thermal conductivity of composites and mixtures is derived from continuum-level phenomenological formulations that typically incorporate only the particle shape and volume fraction as variables and assume diffu-sive heat transport in both phases; no effects of interfaces or particle mobility are taken into account. This approach, while providing a good description of systems with micro-meter or larger-size particles, fails to describe thermal transport in nanocomposites. Interfaces between materials become increasingly important on small length scales. The thermal conductance (G) of many solid-solid interfaces have been studied experiment-tally but the range of observed interface properties is much smaller than predicted by simple theory. Classical molecular dynamics simulations are emerging as a powerful tool for calculations of thermal conductance and phonon scattering, and may provide for a lively interplay of experiment and theory in the near future.

For sufficiently small particles, the properties of the polymer/nanoparticle interface also control thermal transport in the composite. Interface effects in thermal transport can be captured by effective medium models if the introduction of particle fillers does not significantly alter the thermal conductivity of the matrix material. Since local vibrations of the atoms and molecules dominate heat transport in an amorphous polymer, this assumption should be well satisfied in a PMMA composite. In the limit of a low volume fraction ϕ of high thermal conductivity spherical particles, the thermal conductivity of the composite k_c is [1]:

(8) $$k_c - k_m = 3\Phi k_m \frac{\gamma - 1}{\gamma - 2}$$

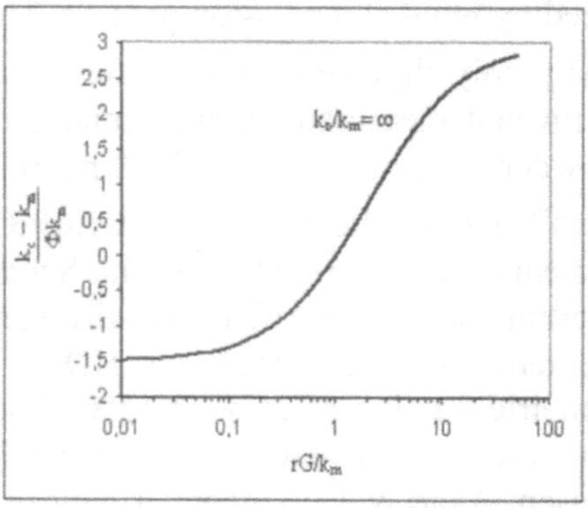

Fig.2. Normalized changes in thermal conductivity of the composite as a function of normalized thermal conductance of the nanoparticle interface

where, k_m is the thermal conductivity of the matrix and $\gamma = r.G/ k_m$; r is the radius of the particle and *G* is the thermal conductance per unit area of the particle/marix interface. This equation is plotted in Fig. 2. A critical particle radius is defined as $r_c = k_m/G$; the thermal conductivity of the composite is increased by particles of size $r > r_c$ and decreased for $r < r_c$. The changes in conductivity are a relatively weak function of (rG), but the thermal conductivity is sensitive to *G* over a wide range of particle sizes $0.3 < r/r_c < 8$. Fig.3. shows experimental resuts by Putman et al. [1] of normalized difference in the thermal conductivity $(k_c-k_m)/k_m$ of nano and miro-scales Al_2O_3 particles in PMMA matrices. In this figure the small differences in conductivity with respect to pure PMMA are highlighted.The conductivity is slightly decreased by the particles fillers at $T < 100$ K and modestly increased at $T > 150$ K. The anomalous increases in conductivity that have been reported recently for nanoparticles in fluids[9] is not observed in this case. As for the μm-sized Al_2O_3 particles filled composite, the constant slight increase in normalized thermal conductivity is roughly in agreement wit Eq.8 for the limit of large r or large thermal conductance G

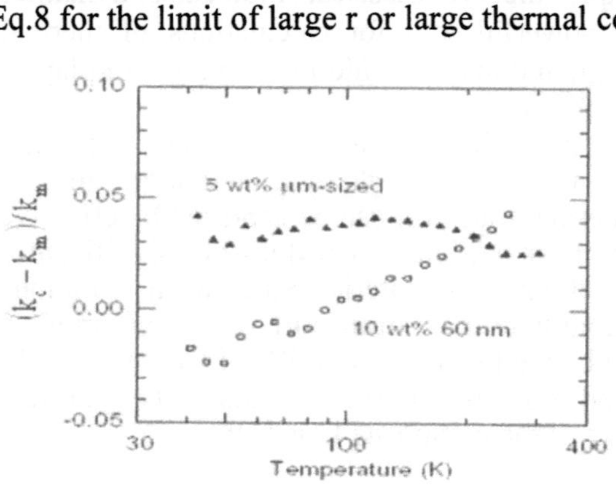

Fig.3. Normalized changes of the measured thermal conductivity of Al_2O_3 particles filled PMMA composites.

References

[1] Putnam, S.A., David G. Cahili,D.G., Ash, B.J. and Schadler, L.S. (2003), High-precision thermal conductivity measurements as a probe of polymer/nanoparticle interfaces
[2] Kuriber R.J., Alam M.K. (2002) Thermal conductivity of thermoplastic composites with submicrometer carbon fibers, Exp. Heat Transfer 15, 19-30
[3] Keblinski, P; Phillpot, SR; Choi, SUS; Eastman, JA, Mechanisms of heat flow in suspensions of nano-sized particles, 2002, Int. J. Heat and Mass Transfer, 45(4):855-863
[4] J.A. Eastman, U.S. Choi, S. Li, G. Soyez, L.J. Thompson, R.J. Di Mem, Novel thermal properties of nanostructured materials, Mater. Sci. Forum 312-314 (1999) 629-634.
[5] S. Lee, S.U.S. Choi, S. Li, J.A. Eastman, Measuring thermal conductivity of fluids containing oxide nanoparticles, ASME J. Heat Transfer 121 (1999) 280-289.
[6] H. Masuda, A. Ebata, K. Teramae, N. Hishinuma, Alteration of thermal conductivity and viscosity of liquid by dispersing ultra-fine particles (dispersions of $Y\text{-}A1_2O_3$, SiOi, and TiO_2 ultra-fine particles), Netsu Bussei (Japan) 4 (1993) 227-233.
[7] R.L. Hamilton, O.K. Crosser, Thermal conductivity of heterogeneous two-component systems, I & EC Fundamentals 1 (1962) 187-191.
[8] Eastman, JA; Choi, SUS; Li, S; Yu, W; Thompson, LJ, 2001, Anomalously increased effective thermal conductivities of ethylene glycol-based nanofluids containing copper nanoparticles, Appl. Phys Lett. 78(6):718-720
[9] Maxwell JC. A treatise on electricity and magnetism. Dover (3rd Ed.), New York, Ch.9. 1954.
[10] Cheng SC, Vachon RI. The prediction of the thermal conductivity of two and three phase solid heterogeneous mixtures. Int. J. Heat Mass Transfer 1969; 12: 249.
[11] Lewis T, Nielsen L. Dynamic mechanical properties of particulate-filled polymers. J. Appl. Polym. Sci. 1970; 14: 1449.
[12] Agari Y, Uno T. Estimation on thermal conductivities of filled polymers. J. of App. Poly. Comp. Sci. 1986; 32: 5705-5712.

FLUORESCENCE QUENCHING BEHAVIOUR OF HYPERBRANCHED POLYMER TO THE NITRO-COMPOUNDS

HONGXIA WANG[1], TONG LIN[1*], FENG LIAN BAI[2] AND AKIF KAYNAK[1],

School of Engineering and Technology, Deakin University, Geelong, Victoria 3217 Australia[1]

The Centre of Molecule Science, Chinese Academic of Sciences. Beijing, 100080, PR China[2]

Abstract: Fluorescence quenching behaviour of hyperbranched polymer film to various aromatic nitro-compounds is presented. The quenching efficiency of 1% and 1 ppm polymer solution in ethanol is 80% and 40%, respectively, for the nitro aromatic compounds. It has been established that aromatic structure and numbers of nitro groups have almost no effect on the quenching properties. The quenching is concentration dependent. The quenching for the vapour of nitro compounds is fast and related to the vapour pressure. The quenching mechanism was attributed to the strong electron-withdrawing property of the nitro groups. Presence of an electron-donating group showed fluorescence enhancement to hyperbranched polymer. Molecular simulation of the polymer structure has shown that the hyperbranched conjugated polymer has a well-developed network structure, with cavities being large enough for the nitro aromatic molecule to penetrate and interact with the polymer linkage.

Keywords: hyperbranched conjugated polymer, fluorescent quenching, nitro-aromatic compounds, chemical sensor.

1. INTRODUCTION

Recently, accurate and rapid detection of the trace amounts of nitro-compounds in the environment has attracted considerable attention because of its potential applications in exploring the 2,4,6, trinitro toluene (TNT) or picric acid, often occurring in explosives. The sensor device is based on the fact that the nitro-compounds quench the florescence of luminescent materials due to the strong electron-withdrawing nature of nitro-groups. The sensor

* Corresponding Author: Tong Lin, Deakin University, School of Engineering and Technology, Geelong, Victoria, 3217, Australia. Email: tongl@deakin.edu.au, Fax: (03) 5227 2539

S. Guceri et al. (eds.), Nanoengineered Nanofibrous Materials, 459-468.

material includes highly luminescent conjugated polymers and polymer nanowires [1,2].

Porosity of the structure in the solid state and strong interaction with the analysed molecules are very important for developing the highly efficient sensor materials. To improve the absorption ability, large steric substituents were introduced to the side-chain of linear polymers material [3,4]. In addition, the substituents that have special affinity to the nitro compounds were introduced into the polymer to "amplify" the sensor signal [5,6].

A hyperbranched polymer shows a considerably different structure from a linear polymer. The numbers of branches increases in every repeated unit and exhibits a three dimensional network structure, irrespective of existence of large steric substituents in the side chain. In our former research, we have reported the synthesis, photophysical properties and electroluminescent applications of the hyperbranched conjugated polymers. Those materials showed a high fluorescence quantum yield and a network structure, which stimulated us for the further investigation of their potential applications as chemical sensors.

In this report, the fluorescent quench properties of the hyperbranched poly(p-phenyl vinylene) (PPV) in various nitro-compounds are presented.

2. EXPERIMENTAL

All the compounds were reagent grade and used as received. The organic solvent was dried according to the standard dehydrating methods before use. The hyperbranched polymer was synthesized in the Centre for Molecule Science, The Chinese Academic of Sciences, China, according to our former method [7] and the polymer structure was characterized by NMR and GPC.

The polymer films were fabricated on glass slides by spin-coating method. The concentration of the hyperbranched polymer was about 1mg/ml in chloroform. UV-VIS absorption and fluorescence spectra were measured on a HITACHI U-3010 spectrophotometer and a fluorescence spectrophotometer, respectively.

The fluorescent quenching properties of the film device in solution was determined by immersing the device into the ethanol containing the analysed compound for a period of time followed by drying in nitrogen atmosphere for 1 minute. The film fluorescence spectra before and after immersing into the ethanol were recorded and the quenching efficiency η was calculated by the following formula:

$$\eta = \frac{I_0 - I}{I_0} \times 100\% \qquad (1)$$

Where, I_0 and I are the emission intensities of the device at the same emission wavelength, before and after the quenching treatment, respectively.

The fluorescence quenching by the nitro compound vapour were obtained by inserting the device into a sealed glass bottle (200 ml size) that contained solid nitro compound. A cotton gauze was used to prevent the direct contact of the polymer film with the nitro compound and maintain a constant vapour pressure. After a period of exposure, the film was taken out and the fluorescence spectrum was recorded immediately. The quenching efficiency was calculated using equation (1).

3. RESULTS AND DISCUSSION

3.1. Photo physical properties of hyperbranched polymer

The structure of hyperbranched PPV (HP1) is shown in Figure 1. The polymer contains a phenyl vinylene unit, which is similar as in the linear poly(p-phenyl vinylene)(PPV). In contrast to the linear PPV, the HP1 had a 1,3,5-trivinly phenyl moiety in every repeated unit that led the polymer increase its linkage in the hyperbranched way and produce many terminal groups. In this report, HP1 was used with 4-tert-Butyl phenyl vinyl as terminal groups.

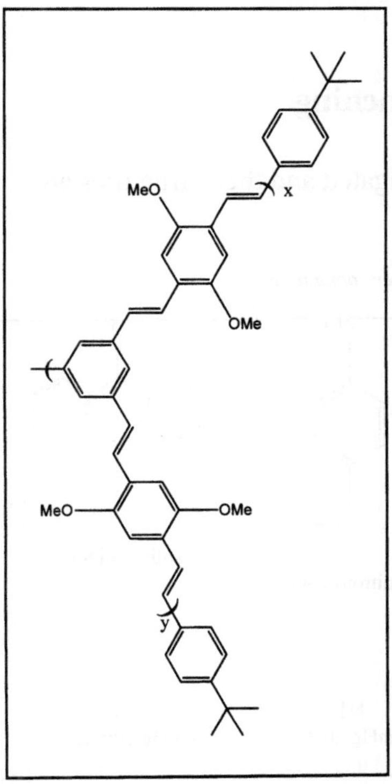

Figure 1.

The UV-visible absorption of HP_1 in CH_2Cl_2 shows maxima at 323 and 405nm, with a molar absorption coefficient in the order of 10^4. Upon UV excitation, polymer solution exhibited intense fluorescence with the maxima at 457 and 480nm. The large Stokes shift (57nm) suggests that a structural relaxation occurs in the singlet excited states.

Upon exciting at 340nm, the fluorescence quantum yield was measured as 95% by the relative method using the 1,6-diphenyl-1,3,5-hexatriene as standard (φ=80%) [6].

By spin coating, an optically transparent film was obtained. No crystallization appeared even when the film was exposed to air for a long time. The film shows absorption maxima at 313 and 392 nm and an emission maximum at 502nm, about 20 nm red-shifted with respect to the solution state. The small difference in emission spectra between the solution and solid film states indicated a weak inter-chain interaction in the solid film.

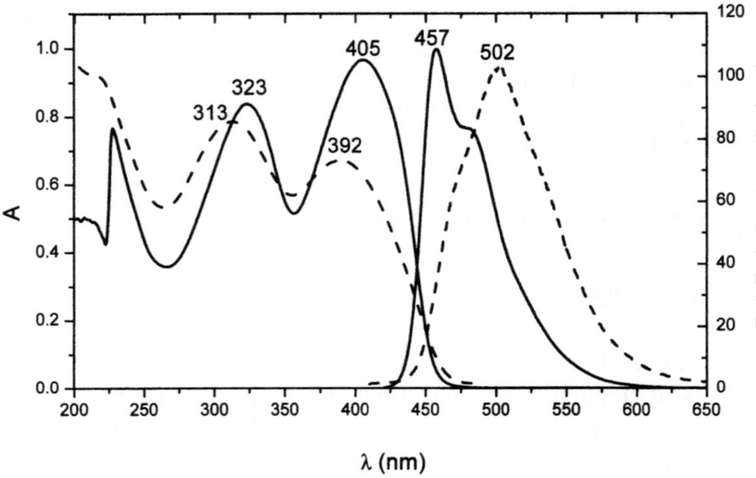

Figure 2. *UV-VIS and fluorescence spectra of HP in dichloromethane (solid line) and solid film state (dotted line).*

3.2. Solution State Fluorescence Quenching

Ten different aromatic compounds were investigated and their structures are listed in Table 1.

Table 1. *Molecular structure and vapour pressure of the quencher compounds*

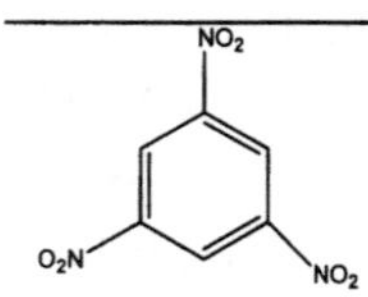	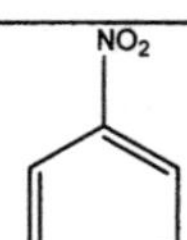	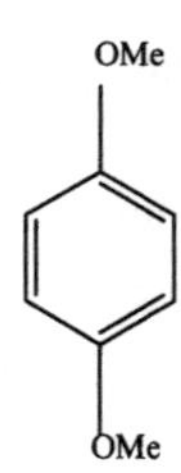	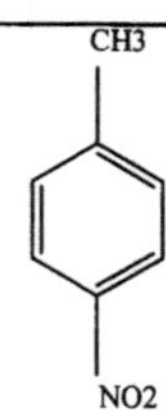	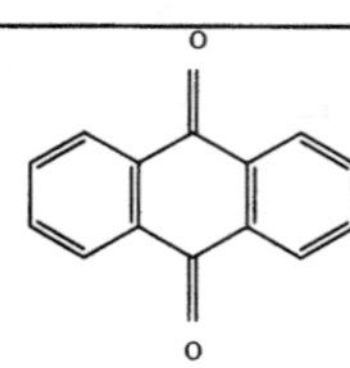
1,3,5-trinitrobenzene	nitrobenzene	p-dimethoxy benzene	4-nitrotoluene	anthraquinone
TNB	NB	DMB	NT	AQ
3.2 10-6 mmHg 25°C	0.27mmHg 25 °C	0.08mmHg 25 °C	0.1mmHg 20 °C	1.8mmHg 200 °C

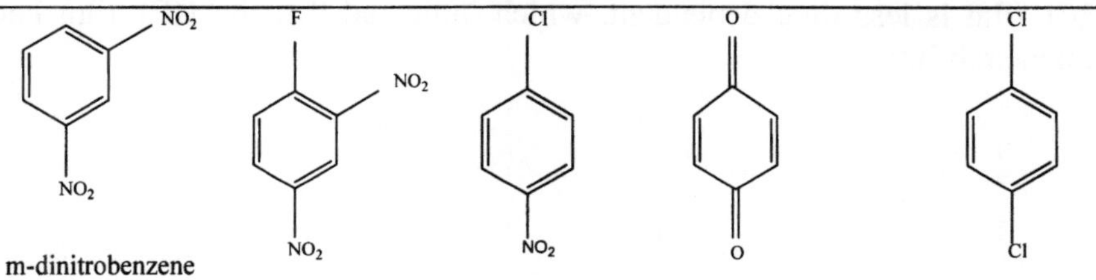

m-dinitrobenzene	2,4dinitrofluoro benzene	p-chloronitro benzene	1,4-benzoquinone	p-dichlorobenzene
mDNB	DNFB	CNB	BQ	DClB
Less than 1mmHg 20 °C	0.75 mmHg 20 °C	0.09mmHg 25 °C	0.8mmHg 25 °C	10mmHg 54.8 °C

Because HP1 is not soluble in ethanol, the ethanol solutions of the quenchers were used in the quenching test. The change in quenching efficiency with the concentration of NB can be seen in Figure 3. Concentration dependence was observed with the efficiency decreasing from 75% to 40% as the concentration changed from 1% to 1ppm. As the fluorescence quenching involves an electron transfer between the NB and the polymer, NB molecules should penetrate into the HP1 film and reach luminescent centres. The diffusing speed of the quencher was determined by its concentration and affinity to the polymer.

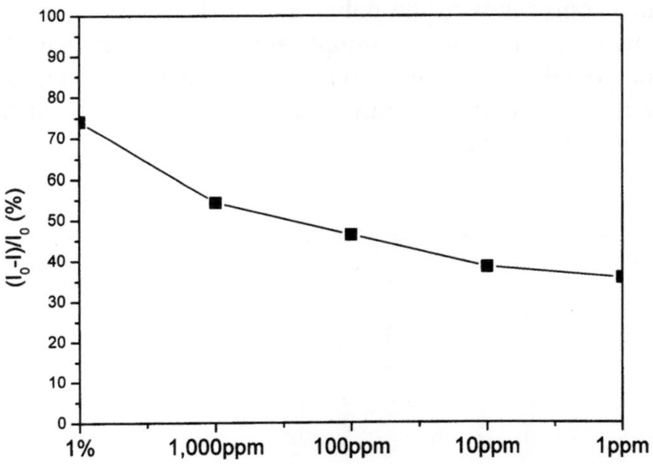

Figure 3. *Quenching efficiency versus NB concentration*

When immersed into the NB solution for 5 minutes, the quenching efficiency of the HP1 film reached to 35% (Figure 4). Longer period of immersion, even as long as 1 hour, did not cause a large increase on the quenching efficiency. So the diffusion of NB molecule from ethanol solution to HP1 film is less time dependent, which indicated that the HP1 film had high penetrability.

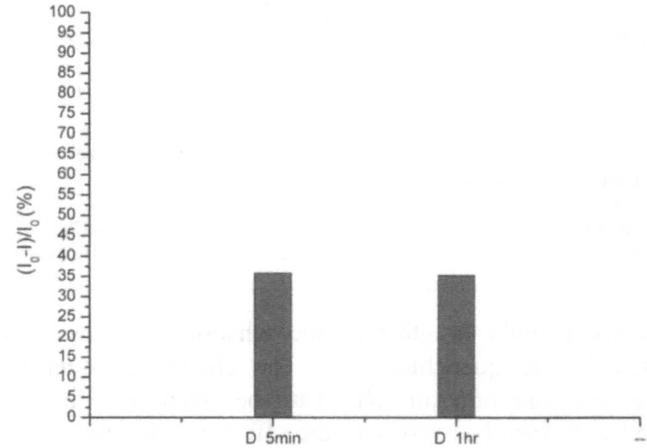

Figure 4. Quenching *efficiency versus quenching time*

Different quencher molecules were used to interact with the HP1 film. The quenching efficiencies of the 1% of nitro compound solutions were around 80% (Figure 5). The aromatic structure had less effect on the quenching efficiency, which suggests that aromatic structure had a minor effect on the diffusion of the nitro compounds to the polymer film. In addition, aromatic compounds with two or three nitro groups showed similar quenching efficiency to the compounds with a single nitro group in the molecule, as the space structure might allow only one nitro group of the molecule to have quenching interaction with HP1 polymer.

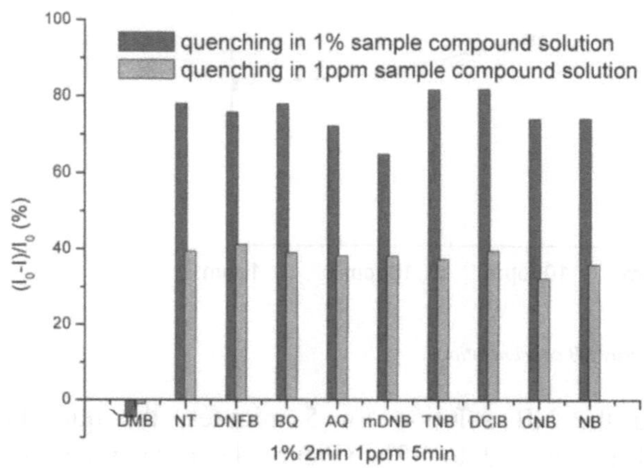

Figure 5. *Fluorescence quenching yield of different nitro compounds*

Quinone was found to quench the fluorescence of HP1 film and revealed a similar quenching efficiency to the nitro compounds. As both the quinone and nitro group are strong electron withdrawing groups, the similarity in fluorescence quenching behaviour suggested strong electronic interaction between the electron-withdrawing groups and the optically excited HP1 molecule. A photo induced electron transport occurred between the excited polymer and the grounded quencher molecule [5,6]. In order to confirm the electronic interaction, an electron-donating compound, p-dimethoxyl benzene was used in the quenching test. The fluorescence determination revealed that the emission intensity increased after the same quenching treatment. A similar result was reported for the linear conjugated polymers, which was attributed to electron-donating interaction of the DMB molecule [8]. The solvent effect was excluded by using a pure ethanol to interact with the HP1 film.

The quenching efficiencies in 1ppm solution showed lower quencher efficiency, around 40%, when compared with the 1% solutions (Figure 5).

3.3 Gas fluorescence quenching

Figure 4 shows the quenching efficiency of HP1 film in mDNB vapour. The quenching efficiency increased as the exposing time of HP1 film in the vapour. The quenching efficiency after a one minute exposure to vapour was 40%, which is very close to the result in the 1ppm solution. After five minutes, the quenching efficiency reached 90%, which corresponded to the HP1 film in 1% solution. The similar quenching efficiency values suggested that similar quantities of quencher molecules interacted with the HP1. So the HP1 film was capable to detect ppm concentration of mDNB vapour in a very short time.

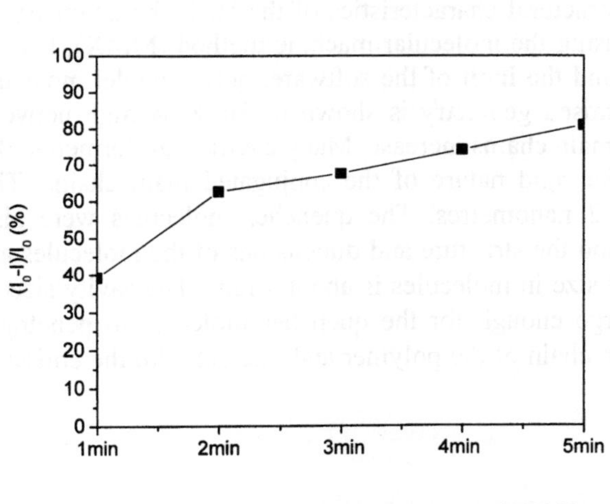

Figure 6. *Quenching efficiency versus the exposure time of HP1 film in the mDNB vapour.*

The fluorescence quenching efficiencies in other quencher vapours were shown Figure 7. Due to the different vapour pressures, these compounds had different concentrations in air, which led to the different diffusion speeds. In the same exposure time, 5 minutes, the quenching efficiencies of DNFB, mDNB, NB, CNB, BQ reached to 90%, which were higher than results obtained in 1% solution. The quenching efficiency was 70% for NT and 60% for TNB and DClB. In contrast, the AQ showed low quenching efficiency of 20%. The quench behaviour in the vapours seems relate to the vapour pressure of the quencher compounds. Similar to the result in the solution state, HP1 film showed a fluorescence enhancement in the DMB vapour.

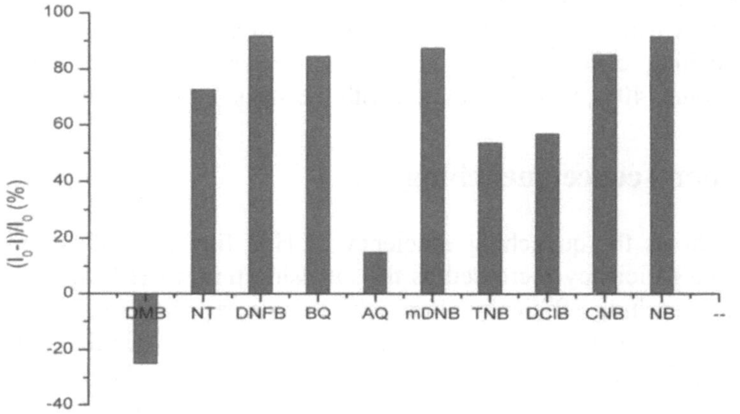

Figure 7. *Fluorescence quenching efficiency of the HP1 film in various compounds vapours for 5 minutes.*

In order to understand the structural characteristics of the HP1, the geometry of the HP1 was simulated by using the molecular machine method (MMX). Due to the macromolecular nature and the limit of the software, only a model molecule was simulated and the optimised geometry is shown in Fig 8. A high network structure was formed as the main chains increase. Many cavities are formed in the polymer molecule due to the rigid nature of the conjugated main chains. The diameter of the cavity is ~2 nanometres. The quencher molecules were also simulated in the same way and the structure and dimensions of the molecules are shown in Fig. 9. The largest size in molecules is about 1 nm. The cavity size in the polymer molecule is large enough for the quencher molecule to penetrate, rotate and approach the main chain of the polymer and interact with the emission centres.

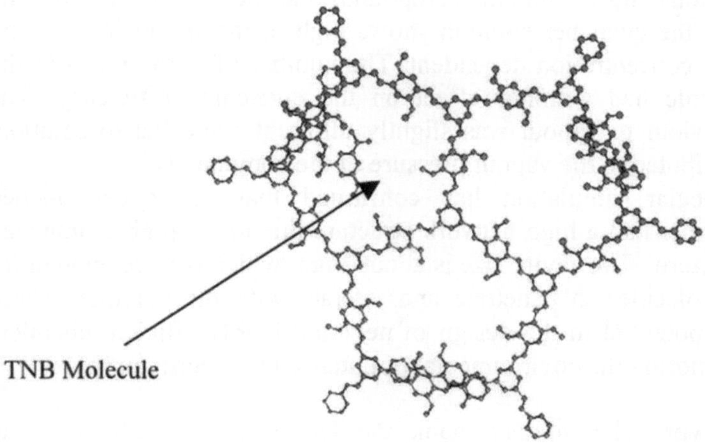

Figure 8. *Network structure of the optimised geometry of HP1 section and trinitro benzene(TNB) molecule*

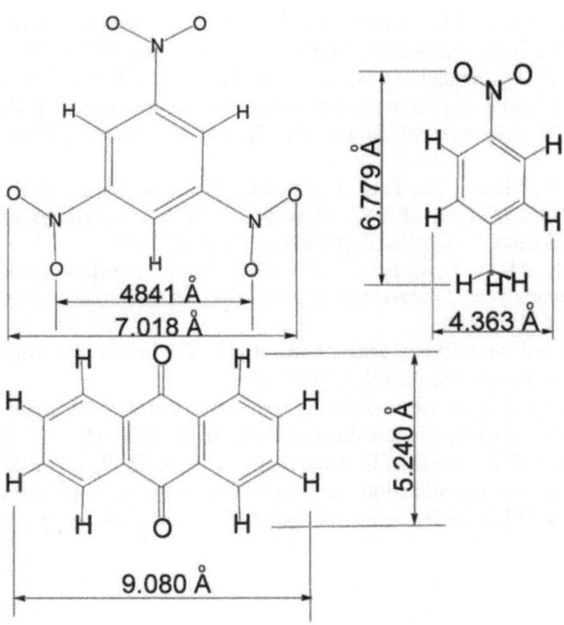

Figure 9. *Simulated dimensions of quencher molecules*

4. CONCLUSION

The fluorescence quenching behaviour of the hyperbranched conjugated polymer to various nitro aromatic compounds has been investigated. The polymer film in the quencher solution shows high quenching efficiency and the efficiency is concentration dependent. The number of nitro groups in the quencher molecule had a minor effect on the quenching efficiency. The quenching behaviour to vapour was slightly different from that to solution. This may be attributed to the vapour pressure of the compounds.

MMX molecular simulation has confirmed that the hyperbranched conjugated polymer had a high network structure due to the highly branched main chain structure. The cavity size is about 2 nm, which is large enough for the quencher molecules to penetrate and interact with the polymer. These results indicate potential in the design of new and highly efficient chemical sensors for monitoring the environmental pollutants and security purposes.

Acknowledgement: The authors thank the Department of Chemistry at Deakin University for their support in spectroscopy.

References

1. Yang, J.-S., Swager, (1998) T. M. Porous Shape Persistent Fluorescent Polymer Films: An Approach to TNT Sensory Materials, *Journal of the American Chemical Society*, *120*, 5321-5322.
2. Wang, J., Bhada, R. K., Lu, J., MacDonald, D (1998) Remote electrochemical sensor for monitoring TNT in natural waters, *Analytica Chimica Acta*, *361*, 85-91.
3. Williams, V. E.; Yang, J. S., Lugmair, C. G., Miao, Y. J., Swager, T. M. (1999) Design of novel iptycene-containing fluorescent polymers for the detection of TNT, *Proceedings of SPIE-The International Society for Optical Engineering*, *3710*, 402-408.
4. Rose, A., Lugmair, C. G., Miao, Y.-J., Kim, J., Levitsky, I. A., Williams, V. E., Swager, T. M. (2000) Optimization of TNT sensory polymers, *Proceedings of SPIE-The International Society for Optical Engineering*, *4038*, 512-518.
5. Hancock, L. F., Deans, R., Moon, J., Swager, T. M. (2002) Amplifying fluorescent polymer detection of bioanalytes, *Proceedings of SPIE-The International Society for Optical Engineering*,*4575*, 78-82.
6. Swager, T. M. (2000) Self-amplifying sensory materials: The molecular wire concept, *Abstr. Pap. - Am. Chem. Soc.*, *220th*, ORGN-305.
7. He, Q., Lin, T., Bai, F. (2001) Synthesis and photophysical properties of linear and hyperbranched conjugated polymer, *Chinese Science Bulletin*, *46*, 636-641.
8. Chambers, W. B., Rodacy, P. J., Jones, E. E., Gomez, B. J., Woodfin, R. L (1998) Chemical sensing system for classification of mine-like objects by explosives detection, *Proceedings of SPIE-The International Society for Optical Engineering*, *3392*, 453-461.

NANOTECHNOLOGY FOR PHOTONICS: RECENT TRENDS ON NEW LIGHT SOURCES

Roberto Cingolani
National Nanotechnology Laboratory
University of Lecce – 73100 Lecce, Italy

Abstract: We briefly review the recent advances of nanotechnology in the field of photonics. Alternative materials are being developed, such as semiconductor quantum dots integrable on GaAs, and optically-active organic molecules, which should allow easier and cheaper fabrication of more efficient devices on plastic substrates. The basic physical principles and the main technological issues of these devices are discussed in view of their application to telecom and optical networks and illumination technology.

1. INTRODUCTION

The increasing demand for faster and cheaper telecommunication devices can hardly be satisfied by the conventional integration methods and by the standard optical materials. III-V InP based semiconductors have clearly shown their main limitations due to the expensive processing and fabrication steps when fast and sophisticated devices have to be built. One the other hand, sub 100 nm fabrication technologies become very critical and expensive, regardless of the specific technique adopted for the patterning. Therefore highly integrated, fast and low consumption optical devices need new concepts and new materials to fullfil the technology requirements.

In Fig.1 we exemplify the exponential rise of the datacom bit-rate required by the massive use of internet in the last years, as opposed to the almost constant rate of the conventional voice traffic rate (telephone). In order to guarantee bit rates in the THz range, and to keep the power dissipation as low as possible meanwhile keeping the efficiency and the cost effectivness at the best level, further miniaturization and integration of optical components is necessary. Fig.2 shows qualitatively that all the known lithographic methodologies are expected to reach the resolution limit of about 60 nm in a few years (primarily for the physical-chemical limitations of the

S. Guceri et al. (eds.), Nanoengineered Nanofibrous Materials, 469-480.

photoresists), while keeping an ever increasing cost which makes their cost-effectiveness less and less convenient.

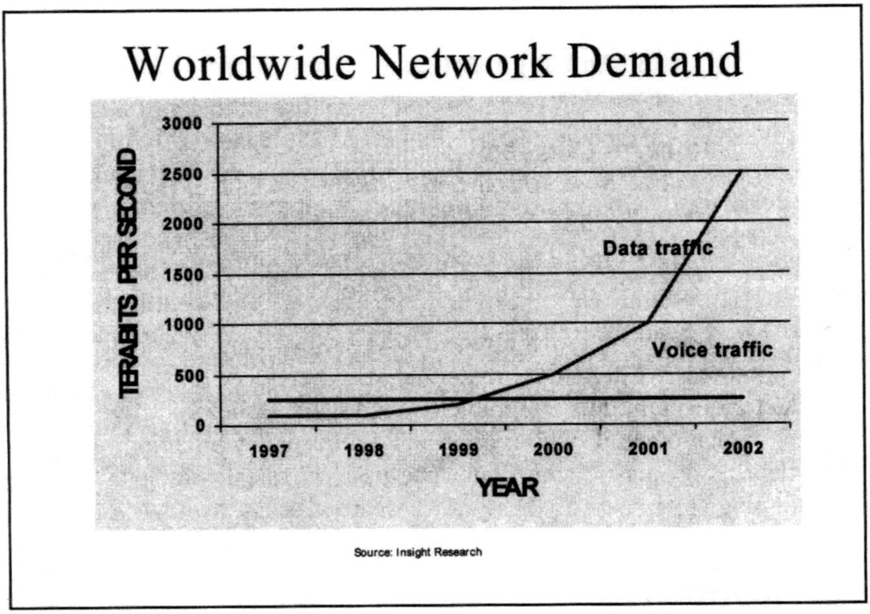

Fig.1 Worldwide network demand for datacom and voice traffic over the years.

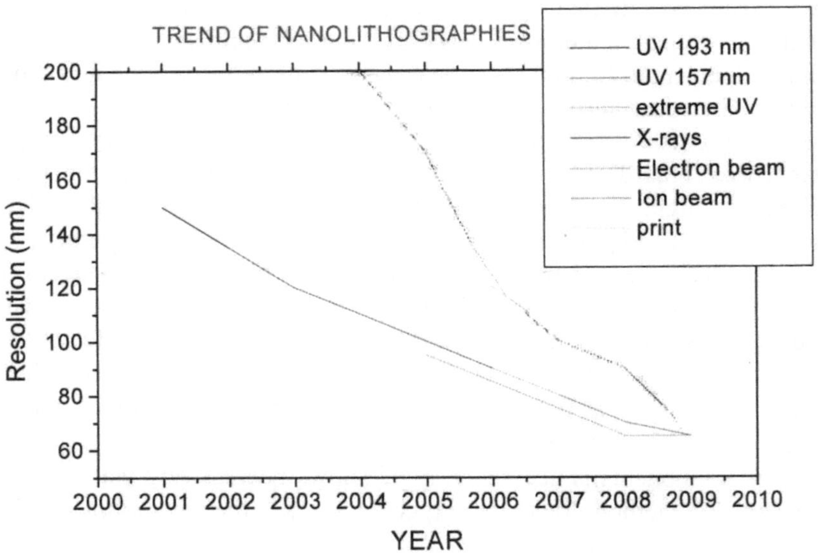

Fig.2 Trend of the lithographies expected for the next years. "UV" indicates the ultraviolet optical lithographies (the number is the optical wavelength), "X-rays" refers to proximity X-ray lithography, "electron" and "ion" beam indicate the usual electron and ion beam induced patterning, whereas "print" includes all the presently available soft lithographies, contact lithographies and hot embossing methods.

In view of these problems, it is therefore necessary to envisage new solutions and new materials which may outperform the existing photonic technologies concomitantly reducing the fabrication costs. The basic structure of a photonic network consists of a laser source, a modulator and a receiver. Normally the laser and the detector share a similar structure, namely a p-i-n structure, which is used to generate light under direct bias, and to detect light under reverse bias. The modulator is normally a non linear system, which can either be integrated or external to the laser source.

In this work we focus our attention on the light sources. In particular we will discuss new III-V lasers integrable on GaAs substrates, new molecular light emitters and a few architectures capable of efficiently confine and drive the light in the device.

2. THE QUANTUM DOT LASER

The main requirement of future telecom technologies is to have lasers with reduced threshold, high critical temperature T_0 (threshold weakly dependent on the device temperature), wide tunability, large gain and no Peltier cooler. This means to move towards devices which are integrated on GaAs instead of InP. The InP substrates are more expensive and difficult to process (as compared to GaAs) and can hardly be made larger than 4 inches. InGaAs quantum dots (QD) can be fabricated directly on GaAs substrates by means of Stransky-Krastanov epitaxy (either MOCVD or MBE) , exploiting the strain-induced self assembling of the nanocrystal. The sharp, zero-dimensional-like , density of states of QDs results in a low lasing threshold , I_T, and a high T_0, and therefore permits to avoid the cooling stage in the device.

Fig 3 (bottom) shows the high resolution TEM cross section of a quantum dot grown by MOCVD. A clear lateral faceting can be seen, resulting in the characteristic truncated pyramid shape [1]. The AFM

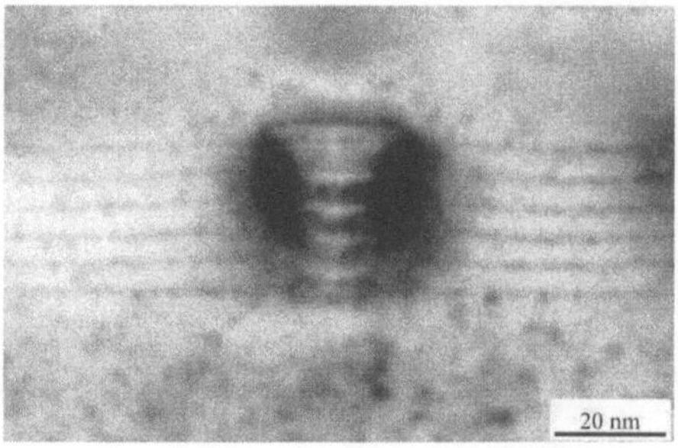

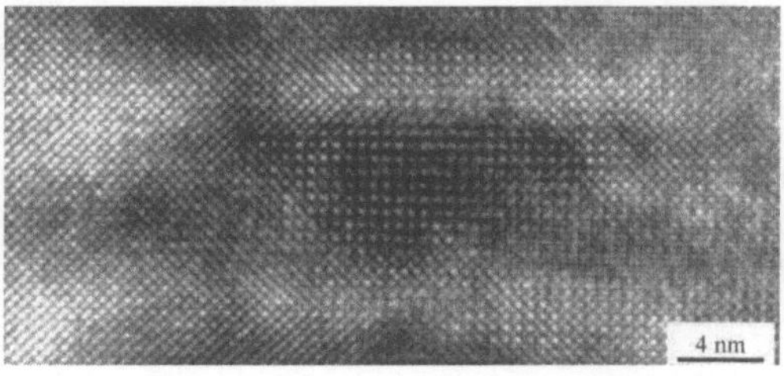

Fig.3 (bottom) High resolution TEM cross section of a InGaAs single dot in a GaAs matrix. The bright spots represent the diffraction sites of the individual atoms constituting the dot. (top) Vertically stacked quantum dots used as active layers of a QD laser

thopography of a single dot, and of a grand ensamble of QD are shown in Fig4.

In order to have large optical gain the density of QD has to be large, possibly in the 10^{11} cm^{-2} range. However, vertical stacking of several layers is often used to increase the volume density of dots (Fig.3, top). The number of layers to be stacked, the quality of the dots and their reproducible shape and size depend critically on the growth and on the overall strain in the active layer.

The electronic states in the dots behave just like the electonic states in the atom. The three-dimensional confinement induced by the surrounding host matrix (typically GaAs) on the carriers in the QD (typically InGaAs) results in a set of confined states from which radiative recombination occurs. The ground level of the dot is ultimately set by the size, composition and shape of the nanocrystal, and can be tuned in the range 900 – 1500 nm, of interest for

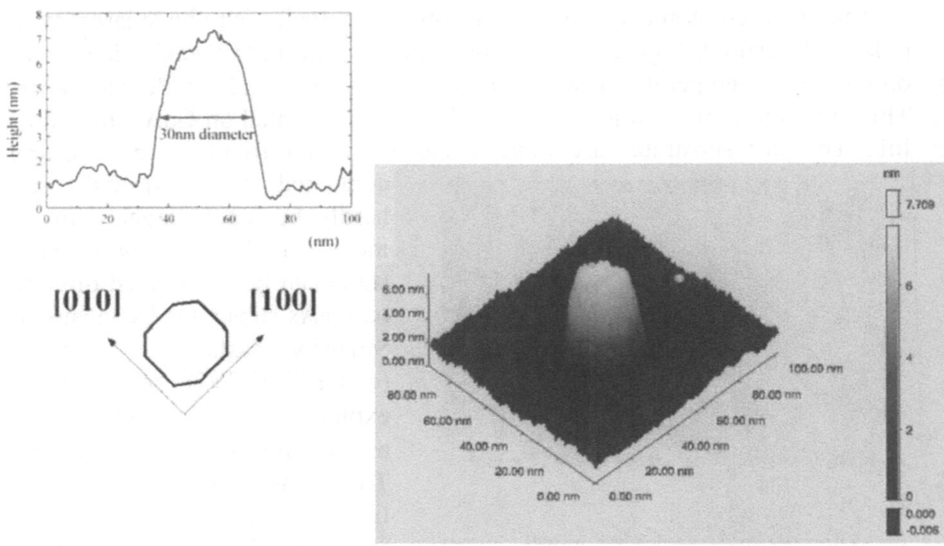

Fig.4 (top) AFM topography of a single quantum dot and its cross section (bottom) AFM topography of a grand ensamble of quantum dots (the density is of the order of 10^{10} cm$^{-2)}$. The small circles indicate the occurrence of small dots, large dots and laterally coupled dots, resulting in a polydispersed size distribution over a length scale of half a micron (large circle).

photonic applications.

The typical structure of a QD laser grown by MBE [2] is shown in Fig.5, together with the room temperature lasing spectra [2,6]. In this specific example the laser operates around 1.3 micron at room temperature, which is particularly appealing for telecom applications and to replace the conventional InP-based technologies. A threshold as low as 45 Acm^{-2} per layer of Dots is achieved in this device. Fig.6.shows the trend of the lasing threshold over the last decades, clearly showing how the reduced dimensionality, i.e. the reduction of the density of state dispersion in the energy space, has led to a remarkable decrease of the threshold. Needless to say, these principle studies, though promising, do not necessarily imply that QD technologies may be realistically exploitable, as many other issues such as cost effectivness, packaging, reliability , reproducibility etc. have to be taken into account before a technology transfer towards mass production is done [3-6]

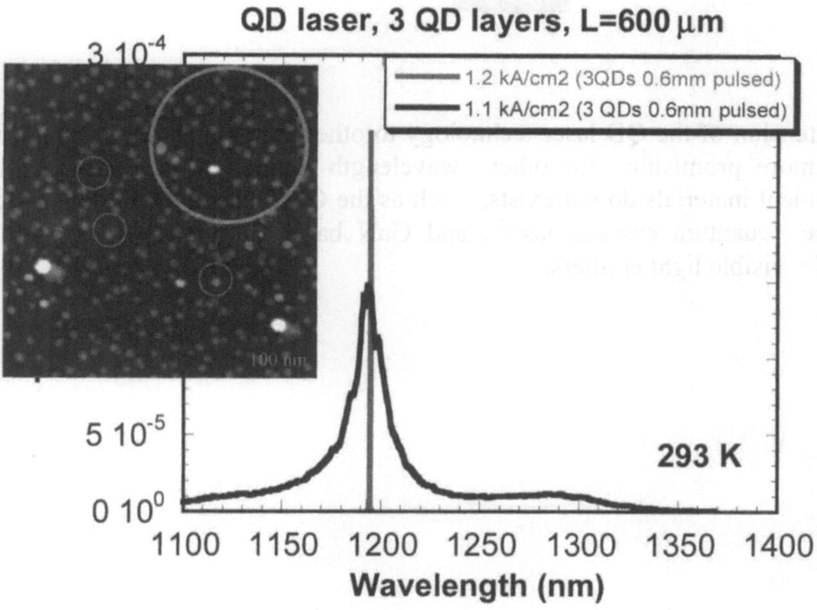

Fig.5 Lasing spectrum of a QD laser device below and above threshold. The scheme shows the layer sequence of the device, fabricated at EPFL-Lausanne (CH). See ref.[2].

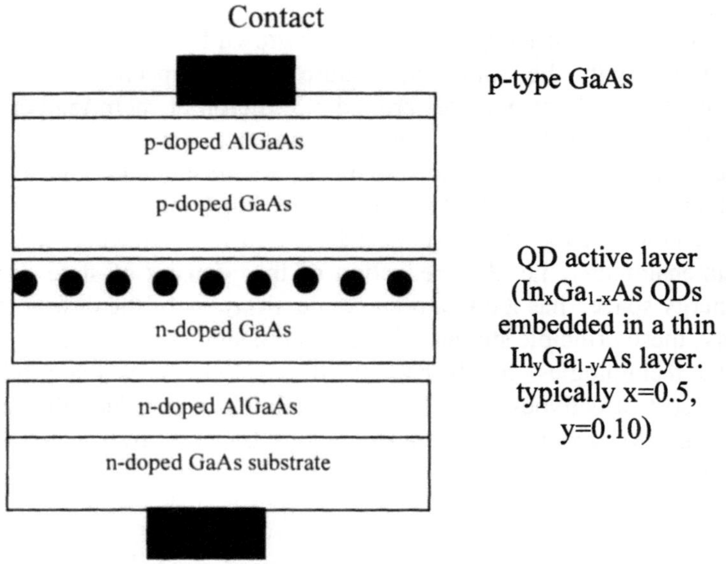

Extension of the QD laser technology to other material systems could be even more promising for other wavelength regions where sufficiently convenient materials do not exists, such as the GaSb based QD systems to replace quantum cascade lasers, and GaN based QD systems for fully tunable visible light emitters.

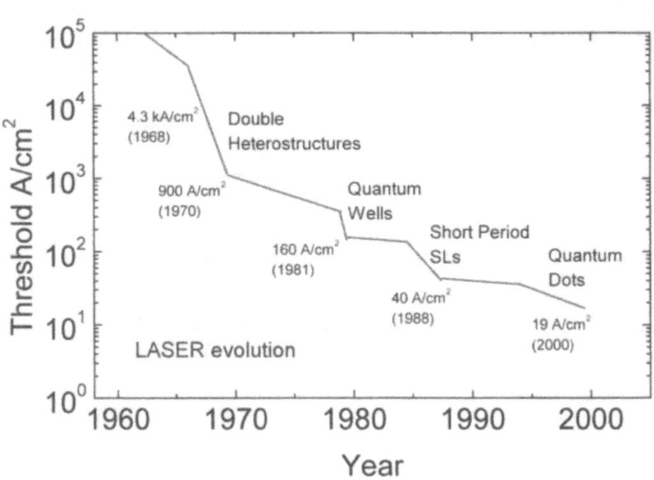

Fig.6 Trend of lasers threshold over the last decades.

3. ORGANIC MOLECULAR LEDS

Another important field of nanotechnology applied to photonics regards the use of organic molecules to fabricate lasers and LEDs. The basic target is to get fully plastic, flexible devices, exploiting optically-active molecules deposited on glass or plastic substrates by means of very cheap techniques such as spin coating or evaporation. The molecules behave qualitatively like a semiconductor. The optical transitions occur between the highest occupied molecular orbital (HOMO, equivalent to the valence band) and the lowest unoccupied molecular orbital (LUMO, equivalent to the conduction band). The variation of the HOMO-LUMO gap of the molecules is obtained by chemical functionalisation, such as grasping different functional groups to the molecule, or by changing the length of the molecule, i.e.by varying the orbital delocalisation along the molecule backbone. This results in a wide tunability , covering the entire range from the blue to the near infrared [7].In addition, the solubility of most molecules and the possibility to blend them allows one to create composite colors thus covering the entire chromaticity diagram, including white light generation [8]. This opens up important applications even in the field of the illumination technology. The LED fabrication is quite simple, as it requires a glass or a plastic substrate onto which a hole injector contact is deposited (normally ITO), followed by the molecular layer, deposited by evaporation or spin coating, and by the electron injector contact (nromally a metal whose electron affinity is suitably chosen to inject electrons in the molecular LUMO). The device is then encapsulated by an epoxy resin, to prevent oxidation and degradation of the molecular layer (see left image in Fig.7). Various technologies are being explored worldwide for multicolor LEDs fabrication, with different advantages and drawbacks, including Foerster transfer in blends [9], blends of different molecules [8,10] and multilayered microcavities [11]. Fig.7 shows the fine tuning of the organic LEDs achieved by blending different molecules in the active layer of the device [8]. The position of the color coordinates on the International Chromaticity Diagram (CIE) clarifies the potential impact of these technologies.

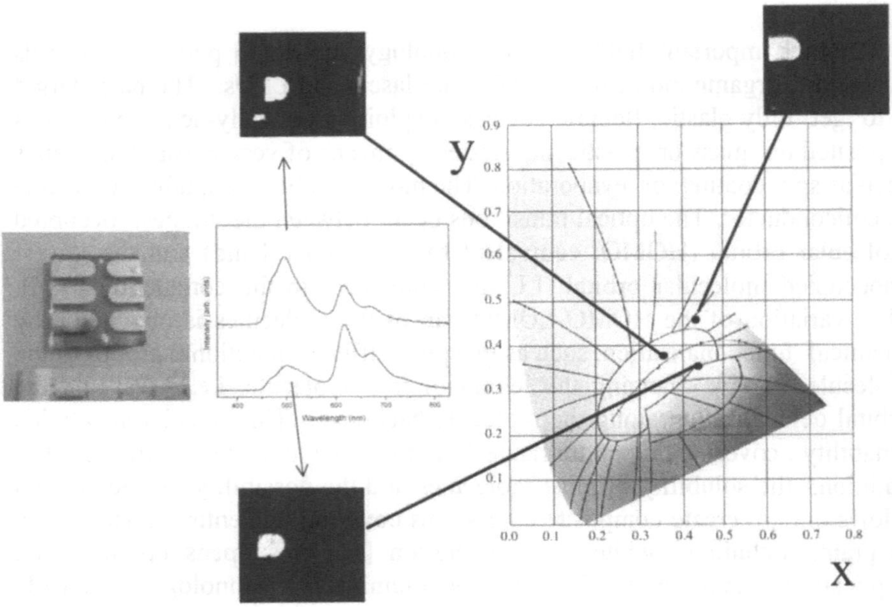

Fig.7 Photographs of three different LEDs emitting pink, white and yellowish light (in clockwise sequence), fabricated by blending different organic compounds. The optical spectra show the spectral components of the molecules constituting the blends. The white emission originates from a combination of green (around 500 nm) and red (600-700 nm) emission bands with dominant green component. The pink spectrum has similar components but of inverse relative intensity. The three colours correspond to the chromaticity coordinates indicated on the CIE diagram, and can be easily tuned by varying the relative concentration of the molecules constituting the blend. The device is shown in the left picture (See ref.8).

The final emission wavelength of the blend is found to depend on the energy separation between the HOMO-LUMO gap of the constituting oligomers: namely, it is influenced primarily by the Foerster transfer occurring between the high energy and the low energy components of the blend when the energy separation is below 0.6 eV, whereas it follows the efficiency ratio of the constituents for energy differences exceeding 0.6 eV. The blend of properly functionalised molecules [8] thus provides a wide tunability across the entire visible range, high quantum efficiency, up to 65 % , due to the reduction of the non-radiative processes in the functionalized molecules induced by the supramolecular organization in the solid state, and a marked increase of the electron affinity.

4. ORGANIC MOLECULAR LASERS

Concerning molecular lasers, since the demonstration of optically pumped lasing from a PPV layer incorporated in an optical microcavity [12], great attention has been paid to the research of good active materials for solid state organic lasers. This has led to the demonstration of optical gain in different conjugated materials [13-20] and of optically pumped lasers built using many different cavity geometry [21-28]. Most of the effort worldwide is devoted to the identification and to the synthesis of new molecular materials combining high emission efficiency, easy processability (eg. solubility and good film forming properties), good stability (low oxidation and long lifetime) and strong optical gain. The latter is a primary requisite to achieve light amplification and lasing in such materials. Besides some polymers, recently a new class of modified short oligomers has been demonstrated to match the previous requirements, namely the soluble thiophene-S,S-dioxide compounds [7]. In Fig.8 we show a quinque-thiophene-S,S-dioxide (T5oA, constituted by a sequence of 5 aromatic rings) having high PL efficiency (>50%) and good chemical stability. The study of this compound was performed in films of thickness around 375 nm, prepared by spin coating from a chloroform solution on a Corning glass substrate [29]. The optical gain capability of the

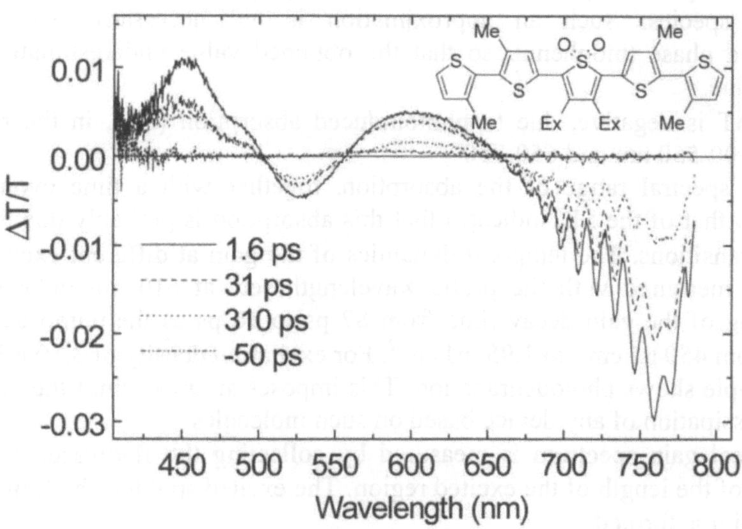

Fig.8 Time resolved pump and probe transmission spectra of a T5oA film at room temperature. The inset shows the thiophene molecule.

film is probed by pump and probe experiments, in which the transmission of the sample is probed at different delays from the intense excitation pulses provided by the pump beam. Fig.8 displays the differential transmission

$\Delta T/T$, obtained by subtracting the pump-on and the pump-off transmission spectra, at different time delays. The pump pulse energy density was 920 μJcm^{-2}. Two main features can be distinguished in the spectra:

1. The differential transmission spectrum exhibits two positive features at different wavelengths.

The high energy one, in the range 410-500 nm, falls in the spectral region of the singlet $S_0 \rightarrow S_1$ absorption resonance, and it is attributed to absorption bleaching. The low energy one, in the range 550-650 nm, is spectrally overlapped with the T5oA emission, and is then attributed to Stimulated Emission (gain). A spectrum taken at negative time delay between the pump and the probe was collected to measure the positive signal due to PL, which was then subtracted from all the spectra.

The gain cross section σ_g is estimated by the relation:

$$\sigma_g \approx \left(\frac{\Delta T}{T}\right)_M \frac{1}{Nd} \qquad (1)$$

Where $(\Delta T/T)_M$ is the maximum value of the differential transmission in the gain region, N is the density of photoexcited states involved in the transition, and d is the sample thickness. The obtained value was σ_g=4*10^{-18} cm^{-2} at λ=600 nm. In the calculations it is assumed that N coincides with the number of absorbed photons per unit volume, i.e. unity quantum efficiency for the emitting species. Such an approximation is not necessarily valid for condensed phase thiophenes, so that the obtained value underestimates the actual value.

2. ΔT is negative, due to photoinduced absorption (PA), in the range 500-550 nm and 650-750

nm. The spectral range of the absorption, together with a time evolution similar to that of the SE, indicates that this absorption is probably due to the $S_1 \rightarrow S_n$ transitions. The temporal dynamics of the gain at different excitation densities measured with the probe wavelength set at 610 nm indicates a shortening of the gain decay time from 87 ps to 16 ps as the pump density varies from 450 $\mu J\ cm^{-2}$ to 1.95 $mJ\ cm^{-2}$. For excitation density of 3.10 $mJ\ cm^{-2}$ the sample shows photodegradation. This imposes an upper limit the actual power dissipation of any device based on such molecules.

The optical gain spectrum is measured by collecting the fluorescence as a function of the length of the excited region. The excited spot has the form of a stripe and it is formed.

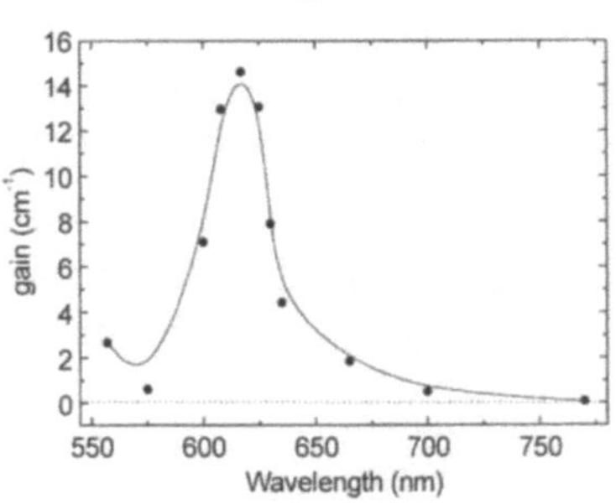

Fig.9.Optical gain spectrum of the T5oX molecular film measured at room temperature

by a cylindrical lens focusing the laser onto the sample surface spatially filtered by a variable slit to change the stripe length. The typical gain spectrum measured at room temperature under excitation density of about 1.53 mJ cm^{-2} is shown in Fig.9. The maximum value of the gain is 14.5 cm^{-1} at λ=616 nm, while the linewidth of the gain resonance is about 20 nm. The measured gain is the largest ever reported for short oligomers, and among the largest for any molecular compounds, clearly indicating the great potentiality of these organic molecules [29].

Due to the excellent plastic properties of the molecular film, the active materials can be patterned by inprint lithography to realize a quarter/wavelength grating directly onto the laser active material (See. left scheme in Fig.10). This was used to realize the first plastic Distributed Feedback Laser (DFB) shown in Fig.10 [30]. The optical spectrum under optical excitation shows a sharp and narrow single-mode lasing line, emerging in the low energy wing of the spontaneous emission band of the molecular layer, at a wavelength set by the grating periodicity.

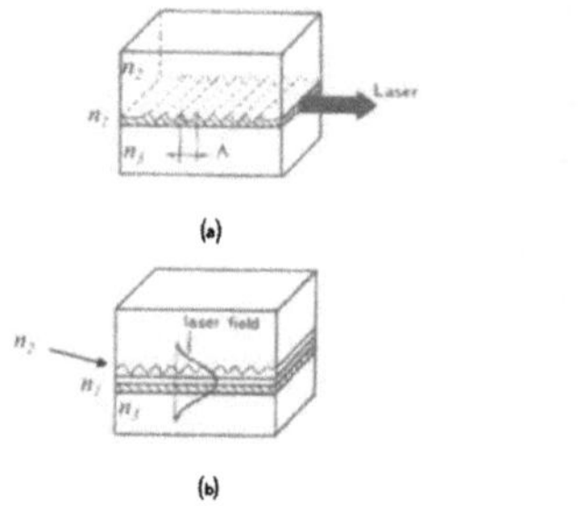

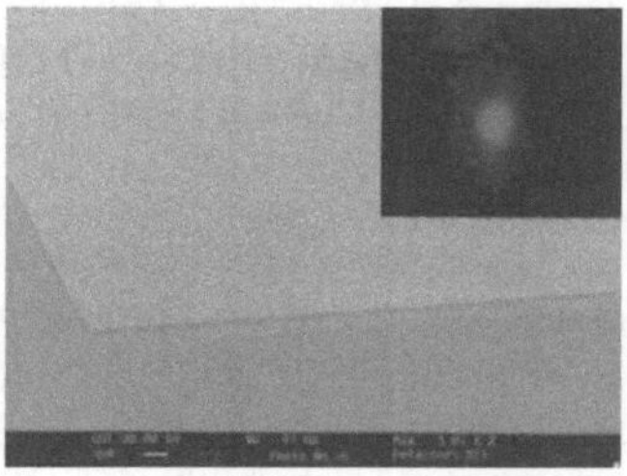

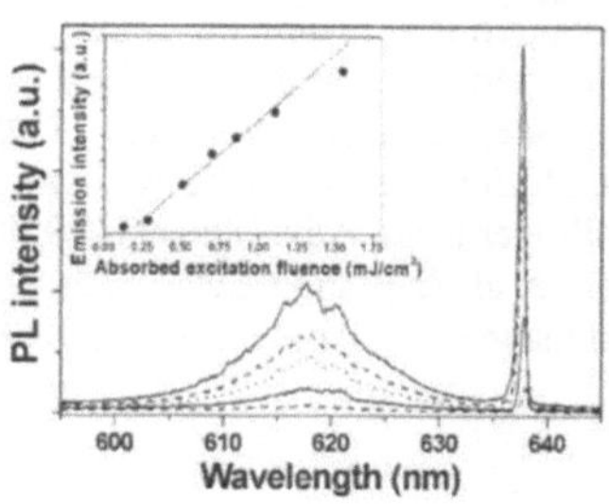

Fig.10 SEM image of the patterned molecular layer and single mode spectrum of the DFB laser. The inset shows the far field spot of the laser. The scheme of the DFB laser is displayed on the left.

5. CONCLUSIONS

Nanotechnology has recently enabled the fabrication of novel photonic devices of great potentiality. The control of the self-organisation of semiconductors at nanoscale, as well as the molecular engineering in combination with alternative patterning techniques, such as soft lithography on plastic materials, are very promising for the next generation of cheap and efficient devices to be used in optical networks. This work was partially supported by INFM and MIUR. The author acknowledges the collaborative work of the NNL staff at Lecce University. Special thanks to A.Passaseo (NNL), M.DeVittorio (NNL), G.Gigli (NNL), M.Anni (NNL), D.Pisignano (NNL) and to A.Fiore (EPFL-CH) for providing for useful discussions about their data.

References

1. A.Passaseo et al. Appl. Phys.Lett. 78, 1382 (2001)
2. Fiore et al. Appl.Phys.Lett. 81, 1756 (2002)
3. A.Markus et al. Appl. Phys.Lett. 82, 1819 (2003)
4. H.Y.Liu et al. Appl.Phys.Lett. 79, 2863 (2001)
5. A. Passaseo et al., Appl.Phys.Lett. 78, 1382 (2001)
6. A. Passaseo et al., Appl.Phys.Lett. 82, 3632 (2003)
7. G.Gigli et al. Appl. Phys. Lett. 75, 439 (1999)
8. M.Anni et al. Appl.Phys.Lett. 77, 2458 (2001)
9. C-I Chao and S-A. Chen, Appl. Phys. Lett. 73, 426 (1998)
10. R.S.Deshpande et al., Appl. Phys. Let. 75, 888 (1999)
11. P.E.Burrows, V.Khalfin, G.Gu and S.Forrest, Appl. Phys. Lett. 73, 435 (1998)
12. N. Tessler et al. Nature 382, 695 (1996).
13. W. Graupner et al. Phys. Rev. Lett. 76, 847 (1996).
14. S. V. Frolov et al., Phys. Rev. Lett. 78, 729 (1997).
15. G. J. Denton et al., Adv. Mat. 9, 547 (1997).
16. G. H. Gelink et al., Chem. Phys. Lett. 265, 320 (1997).
17. S. Stagira et al., Chem. Phys. Lett. 289, 205 (1998).
18. B. Schweitzer et al., Appl. Phys. Lett. 72, 2933 (1998).
19. H. J. Brouwer et al., Appl. Phys. Lett. 73, 708 (1998).
20. C. Spiegelberg et al. Appl. Phys. Lett. 75, 748 (1999).
21. S. V. Frolov et al., Phys. Rev. B 56, 4363 (1997).
22. V. G. Kozlov et al., Nature 389, 362 (1997).
23. M. Berggren et al., Nature 389, 466 (1997).
24. M. Berggren et al., Appl. Phys. Lett. 71, 2230 (1997).
25. M. Berggren et al., Appl. Phys. Lett. 72, 410 (1998).
26. S. Stagira et al., Appl. Phys. Lett. 73, 2860 (1998).
27. M. D. McGehee et al Appl. Phys. Lett. 72, 1536 (1998).
28. C. Kallinger et al., Adv. Mater. 10, 920 (1998)
29. M.Anni et al.,Appl.Phys.Lett. 78, 2679 (2001)
30. D.Pisignano et al., Appl.Phys.Lett. 83,in press (2003)

MAGNETIC NANOSCALE PARTICLES AS SORBENTS FOR REMOVAL OF HEAVY METAL IONS

M. Vaclavikova, S. Jakabsky and S. Hredzak
Institute of Geotechnics, Slovak Academy of Sciences
Watsonova 45, 043 53, Kosice, Slovakia

Abstract: The contribution presents the method of preparation and the detailed characterization of the structural and magnetic properties of nano-scale magnetite particles playing a major role in determining the observed magnetic properties. Aqueous suspensions of magnetic particles were prepared by co-precipitation of Fe(III) and Fe(II) in the presence of NH_4OH. Heavy metal ions adsorption from single metal aqueous solutions was investigated in batch adsorption – equilibrium experiments. Magnetic particles were used in the adsorption of selected bivalent heavy metal ions, i.e. Cu(II), Cd(II) and Pb(II) from aqueous media containing different amounts of these ions (20-400 $mg.L^{-1}$) and at different pH values 2.0-8.0 at sorbent concentration 2 $g.L^{-1}$. The model solutions of $Cu(NO_3)_2$, $Cd(NO_3)_2$, $Pb(NO_3)_2$ were used to realize the adsorption tests.

1. INTRODUCTION

Heavy metals have become an ecotoxicological hazard of prime interest owing to their tendency to accumulate in living organisms. The necessity to reduce the amount of heavy metal ions in wastewater streams from hydrometallurgical and other industries, and subsequent possible re-use of these metal ions, has led to an increasing interest in selective sorbents. The mobility and bio-availability of aqueous metal cations in soils and sediments is often controlled by sorption onto iron oxides. A molecular understanding how to sorb the metals onto mineral surfaces is needed before we can reliably model equilibria from sorption isotherm measurements. Sorption of ions onto mineral surfaces may occur via outer-sphere complexes, inner-sphere complexes, surface precipitates or via ion exchange [1].

Recently, there has been increased interest in the use of magnetic carriers in the removal of heavy metals. Magnetic carriers can be produced using inorganic materials or a number of synthetic and natural polymers [2]. Iron oxides are common in the environment, occurring either naturally or as results of human activities. The most common Fe(III)- hydroxides, -oxides and oxyhydroxides include ferrihydride ($Fe_5HO_8.4H_2O$, but often written as $Fe(OH)_3$) which transforms to hematite (alpha- Fe_2O_3) and/or goethite (alpha-FeOOH), depending on solution composition, temperature and pH. So called "Green rust" are Fe(II, III)-hydroxy salts, which contain

S. Guceri et al. (eds.), Nanoengineered Nanofibrous Materials, 481-486.

anions such as CO_3^{2-}, SO_4^{2-} or Cl^- in the interlayers. Depending on the composition of the solid and solution, oxidation can transform the green rust to lepidocrocite (gamma-FeOOH) or magnetite ($Fe(II)Fe(III)_2O_4$) [3]. Magnetite crystal habits are typically octahedrons but rarely rhombododecahedrons and other isometric forms, most commonly found massive or granular. Twinning of octahedrons into spinel law twins is observed occasionally.

In the references [4-7], numbers of experimental as well as modelling studies have been reported on the adsorption of cations, particularly heavy metal ions on to pure soil component such as oxides and clay minerals at different values of pH, metal ion, S/L ratio and temperature. These studies in general show that the adsorption of cations favored at higher pH and the maximum enhancement in adsorption occurs over a narrow region of 1-2 pH units. Langmuir as well as Freundlich adsorption isotherms have been used to describe the equilibrium.

In this work, magnetic nano-scale particles have been used for batch adsorption studies of contaminants from aqueous model solutions.

2. EXPERIMENTAL

2.1. Materials and Methods

All chemicals were used of analytical grade. Deionized water was used throughout the experiments.

The pure iron oxide was directly prepared under identical reaction conditions by co-precipitation of Fe(III) and Fe(II) in the presence of NH_4OH. The following method of preparation is similar to those described in Refs. [3, 8]. A solution of mixture of Fe(III) and Fe(II) ions in equal molar ratio was prepared from $FeCl_3.6H_2O$ and $FeSO_4.7H_2O$. An equal volume of 1M aqueous ammonia solution was then added dropwise to the iron mixture to maintain the pH at 11.5. The suspension was finally decantated using deionized water to approximately pH 6.5 and centrifuged to remove the solid material. The clear sample was then collected for further analysis.

Scanning electron micrographs were obtained using a Tesla BS 340 scanning electron microscope (SEM) equipped with a Link ISIS 300 EDS. The specific surface area was determined by the low temperature nitrogen adsorption method using a GEMINI 2360 apparatus.

Zeta potential measurements were conducted on a microelectrophoretic zeta potential analyzer (Apparatus Mark II, Rank Brothers). 25 ml of metal solutions containing 100 $mg.L^{-1}$ of selected cations with 2 $g.L^{-1}$ of magnetite were shaken for 24 h, over a range of pH values (2-9) and then the zeta potential of the particles was measured.

Salts used in the preparation of model contaminant aqueous solutions are $Pb(NO_3)_2$, $Cd(NO_3)_2.4H_2O$, $Cu(NO_3)_2$ .3H $_2$O, obtained from the Merck.

All the adsorption experiments were conducted at ambient temperature in a rotary shaker (about 20°C). Initial total metal ion concentration range was

20-400 mg.L^{-1}. The sorbent concentration was 2 g.L^{-1}. The quantity of heavy metal ions in solutions were determined both before the introduction of sorbent and after the reaction with the metals in solution by flame atomic absorption spectroscopy Varian Spectr AA-30 spectrometer. All experiments were carried out in constant ionic strength (0.01M $NaNO_3$). The solutions pH was adjusted before starting the adsorption experiments with suitable concentrations of NaOH and HNO_3.

2.2. Results and Discussions

Figure 1 presents SEM image of magnetic nano-scale preparation

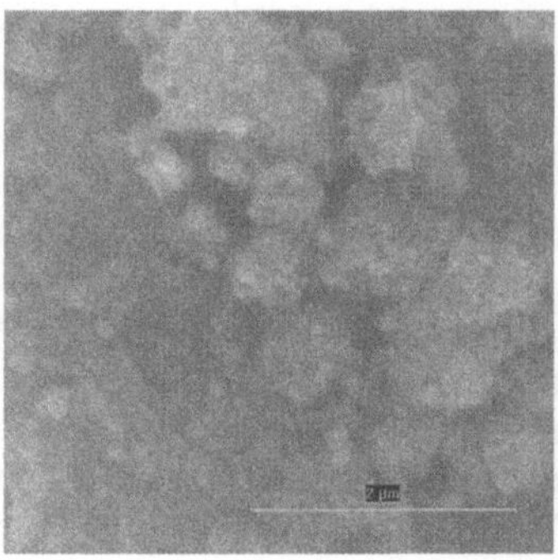

Figure 1: SEM image of magnetite particles

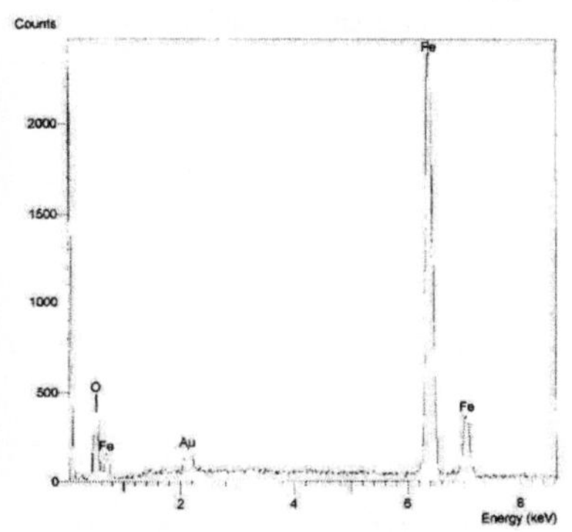

Figure 2: The energy dispersive X-ray spectroscopy (EDS) spectrum of synthetic magnetite

It shows evidence of cluster aggregates. The original magnetite particles attain of 10-40 nm. Due to agglomeration during the air-drying real grain size ranges from 20-900 nm. BET surface area was below 3.8 m^2.g^{-1}.
It is well known that heavy metal ions hydrolyse in water solutions and precipitate as insoluble hydroxides at appropriate pH values. It is very difficult to determine what amount is removed by sorption and what is abstracted from the solution due to precipitation. In order to choose a suitable pH value for studying the metal ions adsorption on synthetic magnetite the thermodynamic equilibrium diagram of each cation for a total concentration of 100 mg.L^{-1} and a constant ionic strength (0.01 M $NaNO_3$) has been constructed with the computer program Mineql Plus, see *Figure 3-5*.

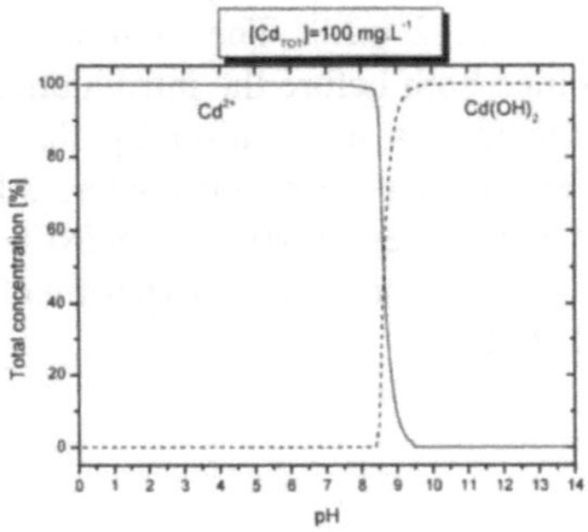

Figure 3: Thermodynamic equilibrium diagram of cadmium

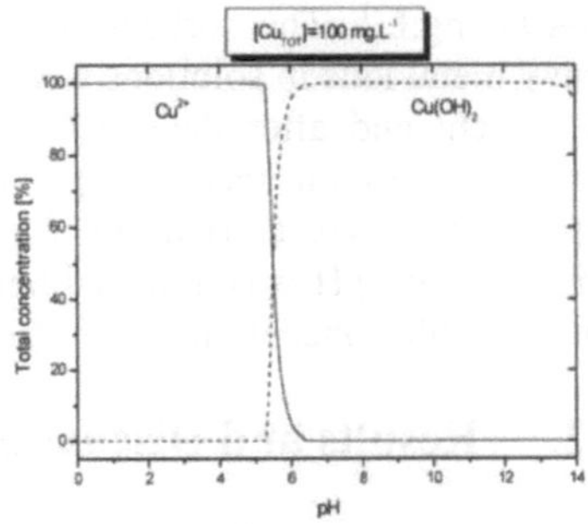

Figure 4: Thermodynamic equilibrium diagram of copper

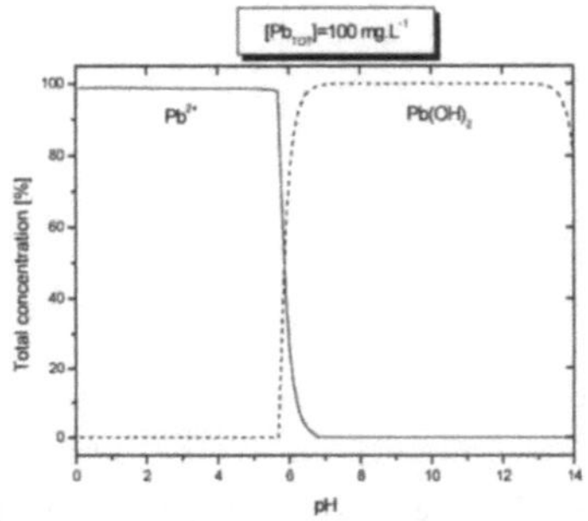

Figure 5: Thermodynamic equilibrium diagram of lead

Figure 3 shows that for these conditions cadmium remains soluble up to pH value of about 8.4 where it starts to precipitate as insoluble $Cd(OH)_2$. At pH 9.1 more then 90% of cadmium has been precipitated. Similarly, the *Figure 4* shows the images of copper forms in all pH range. It is observed that copper remains soluble up to pH value of about 5.4 where it is starts to precipitate as insoluble $Cu(OH)_2$.

Figure 5 shows that lead remains soluble up to pH value of about 5.7 where it starts to precipitate as insoluble $Pb(OH)_2$. At pH 6 more then 60% of lead has been precipitated. In this study a nano-scale magnetite prepared synthetically was used to remove heavy metal ions from aqueous model solutions using batch adsorption procedure. Sorption isotherms constructed and the experimental results were modelled by Langmuir-type equations (*Figure 6*).

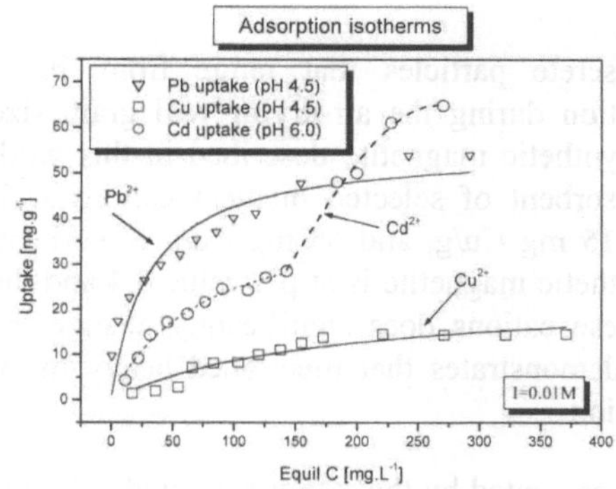

Figure 6: Pb, Cu and Cd uptake by synthetic magnetite modelled by Langmuir-type equatation.

Figure 6 presents adsorption isotherms (equilibrium concentration versus uptake). The synthetic magnetite (SM) was found to be a good lead sorbent with sorption capacity of ca. 54 mg Pb/g in solution of constant ionic strength (0.01M $NaNO_3$) at pH 4.5. Otherwise, synthetic magnetite showed lower sorption capacity for copper. There was only about 15 mg Cu/g of sorbent.

In the case of cadmium removal, two-stage metal uptake was observed. The first one at lower cadmium initial concentration with a capacity of ca. 29 mg Cd/g of sorbent and the second one at higher cadmium initial concentration with a capacity of 65 $mg.g^{-1}$ of sorbent. It follows, that the sorbent surface is heterogeneous with two different types of active sites. Finally, the zeta potential of synthetic magnetite in the absence and presence of mentioned cations has been measured. The effect of pH on zeta potential of synthetic nano-scale magnetite is shown in *Figure 7*.

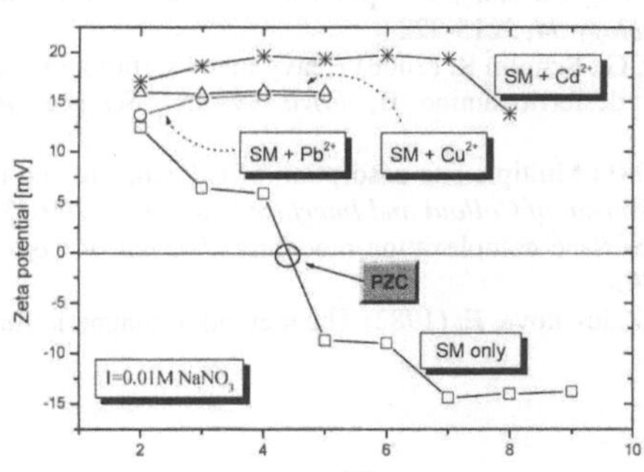

Figure 7: Effect of pH on the zeta potential of magnetite in the absence and presence (100 $mg.L^{-1}$) of selected cations.

From the zeta potential measurement results it is observed, that the point of zero charge (PZC) of SM is at pH value 4.4. At lower pH the SM is positively charged and negatively charged at higher pH values. When it is treated by mentioned cations (100 $mg.L^{-1}$), the zeta potential curves changed the signs (from negative to positive) and this indicates the presence of specific sorption.

3. CONCLUSION

Iron oxides precipitate as discrete particles that range from 10-40 nanometers. Due to agglomeration during the air-drying real grain size ranges from 20-900 nm. The synthetic magnetite described in this work was found to be an effective sorbent of selected bivalent cations with sorption capacity 54 mg Pb/g, 15 mg Cu/g, and 65 mg Cd/g of sorbent. The point of zero charge of synthetic magnetite is at pH value 4.4 and the presence of 100 mg. L^{-1} of these cations does significantly change the surface charge, which clearly demonstrates that mentioned heavy metal ions are sorbed by specific sorption.

Acknowledgements: The work presented by this paper was funded by the Slovak Grant Agency VEGA under Project No 2/2105/22 and by the Greek–Slovak Project, "Development, Testing and Application of Magnetic Sorbents for Removal of Heavy Metals from Wastewater and Soil".

References

1. Peacock, C.L., Sherman, D.M, Todd, E., Heasman, D.M. (2000) Mechanism of Cu sorption onto iron oxides: Results from Sorption Isotherms and Spectroscopy, *Journal of Conference Abstract* ***5***(2) Cambridge Publications, 774.
2. Moffat, G., Williams, R.A., Webb, C.and Stirling, R. (1994) Selective separation in environmental and industrial processes using magnetic carrier technology, *Minerals Engineering 7*, 1039-1056.
3. Swertmann, U., Cornell R.M. (1991) Iron Oxides in the Laboratory: Preparation and Characterization, WCH Publisher, Weinheim.
4. Trivedi, P., Axe, L. (2000) Modeling Cd and Zn sorption to hydrous metal oxides, *Environmental Science and Technology* ***34***, 2215-2223.
5. Neubauer, U., Nowack, B., Furrer, G., Schulin R. (2000) Heavy metal sorption on clay minerals affected by sidephore desferrioxamine B, *Environmental Science and Technology* ***34***, 2749-2755.
6. Benjamin, M.M., Leckie, J.O. (1981) Multiple-site adsorption of Cd, Cu, Zn, and Pb on amorfous iron oxyhydroxide, *Journal of Colloid and Interface Science* ***79***, 209-221.
7. Katz, L.E., Hayes, K.F. (1995) Surface complexation modeling, *Journal of Colloid and Interface Science* ***170***, 477-490.
8. Jakabsky, S., Zatko, S, Bakos, J., Zalesakova, E. (1982) The method of magnetic fluid production, *SR Patent No. 22369.*

SIMULATION OF CRYSTALLIZATION AND GLASS FORMATION PROCESSES FOR BINARY Pd-Ag METAL ALLOYS

H. H. Kart[1,a], M. Uludoğan[2], T. Çağın[3], and M. Tomak[1]

[1]Department of Physics, Middle East Technical University 06531 Ankara, Turkey
[2]School of Materials Science and Engineering, Georgia Institute of Technology 771 Ferst Dr. Atlanta, GA 30332-0245, U.S.A
[3]Material and Process Simulation Center, California Institute of Technology Pasadena, CA 91125, U.S.A

Abstract: Glass formation and crystallization process of Pd-Ag metallic alloys are investigated by means of molecular dynamics simulation. This simulation uses the quantum Sutton-Chen (Q-SC) potential to study structural and transport properties of Pd-Ag alloys. Cooling rates and concentration effects on the glass formation and crystallization of binary alloys considered in this work are investigated. Pd-Ag alloys show the glass structure at fast cooling rates while it crystallizes at slow cooling rates. Increment of concentration of Ag in Pd-Ag alloys gives rise to glass structure at cooling rate 0.5 K/ps.

Keywords: glass formation, molecular dynamics, quantum Sutton-Chen potential.

1. INTRODUCTION

The understanding of the structure and thermal stability of metallic glasses is of great importance from both the fundamental and practical viewpoints [1,2]. Since their discovery in 1960 by Duwez and co-workers [3], much research has been devoted to the study of bulk metallic glasses [4,5,6,7,8,9,10,11].

The molecular dynamics (MD) method is currently one of the most powerful tools for obtaining the macroscopic and microscopic features of the materials, and has been used to investigate the glass formation and crystallization processes of liquids by using different types of interatomic potentials. MD can provide important insight by allowing one to determine

---

[a] Corresponding author: H. H. Kart, E-mail: kart@metu.edu.tr

S. Guceri et al. (eds.), Nanoengineered Nanofibrous Materials, 487-493.

quantities which are difficult to access in real experiments and hard to obtain with reasonable precision [12,13,14,15].

MD is also extensively used in simulating nano-clusters [16,17]. In this work, we are using a finite system of 864 atoms. Although this is still in the nano-particle range, it shows all the properties of bulk materials. But our simulation method can be used for clusters containing small number of atoms.

There is considerable progress in development of empirical and semi-empirical many-body potentials for accurately reproducing thermodynamics and structural properties of most transition metals [18,19,20,21]. Sutton-Chen potential (SC) is used to study thermal and mechanical properties of some transition metals and alloys via molecular dynamics simulations [22,23] and it's new version, called quantum Sutton-Chen potential which is used to investigate crystallization and glass formation properties of Cu-Ag and Cu-Ni alloys [24]. This quantum Sutton-Chen potential is also used to predict the viscosity as a function of share rates for Au-Cu alloys by means of non-equilibrium molecular dynamics [25].

In this study, our main aim is to investigate the melting, crystallization, and glass formation properties of Pd-Ag metallic alloys. We aim to see if the Q-SC potential can correctly and efficiently describe the melting, crystallization and glass formation during the heating and cooling processes of Pd-Ag alloys. Cooling rate effects on the physical properties of the Pd-Ag alloy modeled by quantum Sutton-Chen potential are also studied.

2. METHOD AND MODEL

Total potential energy of the metals and metal alloys in the SC formalism is given as follows [21];

$$U_{tot} = \sum_i U_i = \sum_i \left[\sum_{j \neq i} \frac{1}{2} D_{ij} \left(\frac{\alpha_{ij}}{r_{ij}}\right)^{n_{ij}} - c_i D_{ii} \left(\sum_{i \neq j} \left(\frac{\alpha_{ij}}{r_{ij}}\right)^{m_{ij}} \right)^{\frac{1}{2}} \right], \qquad (1)$$

where the first term is a pair potential accounting for the repulsion between the i and j atomic cores, and the second term is a local density accounting for the cohesion associated with atom i.

In Eq.1, r_{ij} is the distance between atoms i and j , a_{ij} is a length parameter scaling all spacings, c_i is a dimensionless parameter scaling the attractive terms, D_{ij} sets the overall energy scale.

The combinations rules in defining the interaction between Pd-Ag atoms are given in Refs. [24,26].

Çağın and co-workers [27] reparametrized the SC potential by fitting to such 0 K experimental properties as density, cohesive energy, moduli, and phonon frequencies at the X point (at room temperature) while including the zero-point energy (thus it is called quantum Sutton-Chen potential (Q-SC)) and employed this potential in series of important problem in metal physics [24,25,28,29]. Q-SC potential parameter sets for Pd and Ag are given in the Table 1 [27].

The simulation in the present paper is based on constant temperature and constant thermodynamics tension (TtN) molecular dynamics simulation [30,31,32,33]. TtN simulation started from a cubic box subject to periodic boundary conditions for a system with 864 particles. Computational details of this study can be found in Ref. [24].

Table I. Q-SC potential parameters

metal	n_i	m_i	D_{ij}(eV)	c_i	a_i(A^0)
Pd	12	6	3.2864E-3	148.205	3.8813
Ag	11	6	3.9450E-3	96.524	4.0691

3. RESULTS AND DISCUSSIONS

Fig. 1 a) presents the variation of volume of $Pd_{0.8}Ag_{0.2}$ as a function of temperature and different cooling rates. As shown in Fig. 1 a), there is a sharp jump during the heating and cooling processes. This sudden increase and decrease is due to the first order transition such as melting and crystallization. The melting temperature of $Pd_{0.8}Ag_{0.2}$ is 1695±10 K which is in reasonable agreement with the experimental melting temperature, 1704 K. Among the reasons for the melting temperature to be a little bit different than the experimental value, are that the system is homogeneous without any free surface and that we started the simulation with a perfect crystal. The system might not have had time to rearrange an equilibrium distribution of defects. These might cause the melting point to be different from the experimental value. This alloy changes into crystalline form at 0.5 K/ps while it transforms into a glass formation at faster cooling rates, that is, there is no sudden change in the volume at these cooling rates. This is a sign of glass formation. Fig. 1 b) shows the pair distribution function during the heating and cooling processes at different cooling rates and different temperatures to investigate the cooling rates effects on the $Pd_{0.8}Ag_{0.2}$. As seen in Fig. 1 b), pair distribution functions have sharp peaks during the heating and cooling rate of 0.5 K/ps at 300 K. After the cooling of the model system with the cooling rate of 0.5 K/ps, alloy system recrystallizes and shows the same structure it had during heating processes. However, other cooling rates give rise to glass formation of chosen system. Moreover, faster cooling rates lead to glass structure, i.e., there is a splitting in the pair distribution function at 300 K, this is a characteristic of a metallic glass.

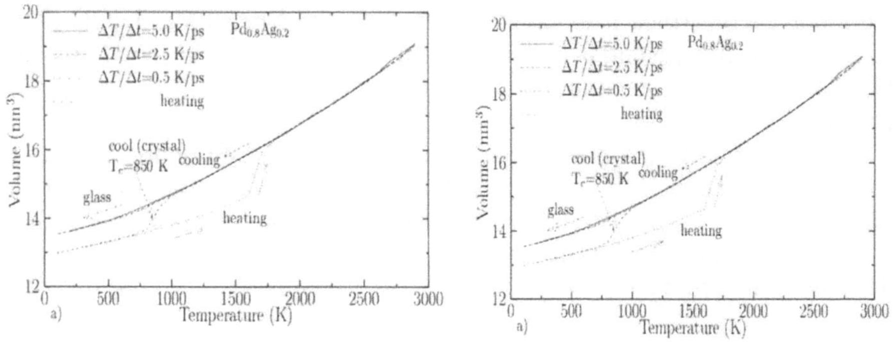

Figure 1. a) Volume of $Pd_{0.8}Ag_{0.2}$ as a function of temperatures during the cooling and heating processes at different cooling rates. b) Pair distribution function at 300 K for structure obtained from heating and cooling cycles.

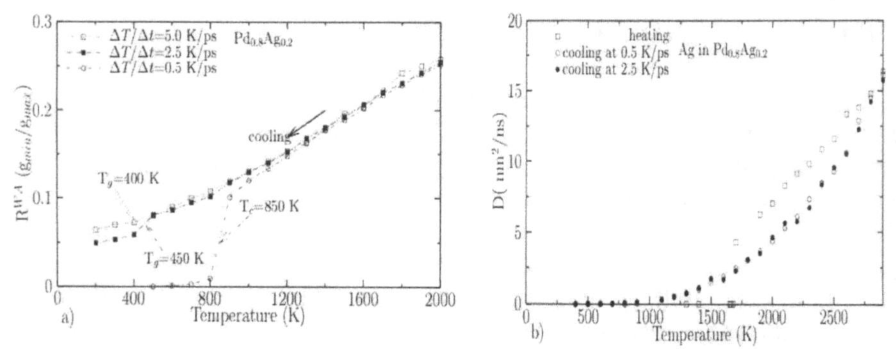

Figure 2. a) Wendt-Abraham parameter R versus temperature. 0.5 K/ps lead to crystallization of the material at 850 K. The calculated glass temperature is 400 K and 450 K at other cooling rates. b) Diffusion coefficients versus temperature during heating processes in EVN dynamics and cooling processes for Ag in the $Pd_{0.8}Ag_{0.2}$ alloy at 2.5 K/ps and 0.5 K/ps cooling rates in the TVN dynamics.

Glass and crystallization temperatures can also be predicted at different cooling rates by means of the Wendt-Abraham parameter [34], defined by $R^{WA}(g_{min}/g_{max})=g_{min}/g_{max}$. The Wendt-Abraham parameter describes the local properties of pair distribution function which allows one to estimate the glass and crystallization temperatures. The Wendt-Abraham glass and crystallization temperatures are displayed in the Fig. 2 a) at different cooling rates. The model system shows a glass formation at temperatures of 400 K and 450 K while the slower cooling rate of 0.5 K/ps leads to crystallization at 850 K. At fast cooling rates the material does not have enough time to relax. This leads to an amorphous structure, but as the simulation time increases the material crystallizes.

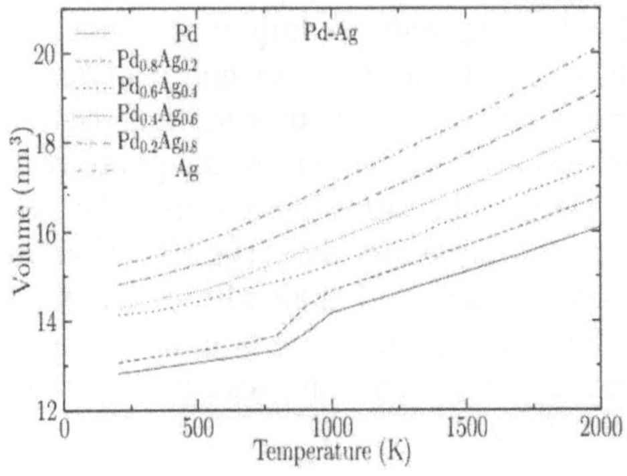

Figure 3. Concentration effect on the crystallization and glass formation of Pd-Ag alloys at 0.5 K/ps.

The Einstein relation is given as [35];

$$<|r(t)-r(0)|^2>=6Dt+C, \quad (2)$$

where $<|r(t)-r(0)|^2>$ is the mean square distance. D and C are constants. The constant D is the diffusion coefficient. As shown in Fig. 2 b) diffusion coefficient increases linearly with heating while it decreases exponentially during the cooling as a function of temperature. The diffusion coefficients during the cooling show the same behavior at two different cooling rates. At the crystallization point diffusion coefficient diminishes sharply. This is an expected result since the system freezes and stays that way. The atoms lose their configurational freedom and vibrate around their local positions in the crystalline phases. Fig. 3 shows the concentration effects on the crystallization and glass formation of Pd-Ag alloys. The increment of the concentration of Ag in the alloy systems cause to be amorphous phase. This is due to the difference between the atomic sizes.

4. CONCLUSION

We have studied the melting, cooling, and crystallization of binary Pd_xAg_{1-x} alloys using the quantum Sutton-Chen potential. As shown in Fig. 1 a) glass transition is not a first-order transition. Indeed there is not a unique glass state since its properties depend on how it is observed. The agreement between the estimated melting points and experimental values is quite good. As the simulation time of the system increases the corresponding crystallization temperature also increases. A long simulation time is necessary to crystallize the system with increasing concentration of Ag in Pd-Ag alloy. This is due to the increment of atomic size of Ag. We have also calculated diffusion constants during the heating and cooling processes. It is seen that Pd-Ag forms a metallic glass at fast cooling rates and it crystallizes at slow cooling rates. The cooling rates determine whether the material changes into glass or not. The glass transition and crystallization temperatures depend strongly on the chosen cooling rates. This study shows that the quantum Sutton-Chen potential is practical for studying the crystallization and glass formation properties of transition metals and their alloys. In the absence of specific experimental data for Pd-Ag alloys, it would be interesting to test the predictions of the present work with the results of first-principles simulation study or experimental study.

References

1. Guntherodt, H. I. and Beck H. Beck *Glassy Metals I*, (Springer, New York, 1981).
2. Shelby, J. E. *Introduction to Glass Science and Technology*, (The Royal Society of Chemistry, 1997).
3. Klement, W., R. Wilens and P. Duwez *Thermophysical properties of bulk metallic glass-forming liquids* Nature **187** 869, 1960.
4. Faupel, F., W. Frank, M. P. Macht, H. Mehrer, V. Naundorf, K. Ratzke, H. R. Schober, S. K. Sharma and H. Teichler *Diffusion in metallic glasses and Supercooled melts* Rev. Mod. Phys. **74** 237, 2003.
5. Schneider, S. *Bulk metallic glass* J. Phys.:Condens. Matter **13** 7723, 2001.
6. Löffler, J. F. *Bulk metallic glass* Intermetallics *11* 529, 2003.
7. Egami, T. *Atomistic mechanism of bulk metallic glass formation* J. Non-Cryst, Solids **317** 30, 2003.
8. Kob, W. *Computer simulations of supercooled liquid and glass* J. Phys.:Condens. Matter 11 R85, 1999.
9. Hausleitner, C., M. Tegze and J. Hafner *Structural and electronic properties of y-based metallic glasses* J. Phys.:Condens. Matter **4** 9557, 1992.
10. Hafner, J. *Theory of the formation of metallic glasses* Phys. Rev. B **21** 406, 1980.
11. Ediger, M. D., C. A. Angell and S. R. Nagel *Supercooled liquid and Glasses1* J. Phys. Chem. **100** 13200, 1996.
12. Wang, L., X. Bian and J. Zhang *Structural simulation of amorphization and crystallization in liquid metals* J. Phys. B: At Mol. Opt. Phys. **35** 3575, 2002.
13. Liu, C. S., Z. G. Zhu, J. Xia and D. Y. Sun *Cooling rate dependence of structural properties of aluminum during rapid solidification* J. Phys.:Condens. Matter **13** 1 1873, 2001.
14. Qi, Y., T. Çağın , Y. Kimura and W. Goddard III *Molecular-dynamics simulations of glass formation and crystallization in binary liquid metals: Cu-Ag and Cu-Ni* Phys. Rev. B **59** 3527, 1999.
15. Wang, L., H. Liu, K. Chen and Z. Hu *The local orientational orders and structures of liquid and amorphous metals Au and Ni during rapid solidification* Phys. B **239** 267, 1997.

16. Dong, K. J., R. S. Liu, A. B. Yu, R. P. Zou, and J. Y. Li *Simulation study of the evolution mechanism of clusters in a large-scale liquid Al system during rapid cooling processes* J. Phys.: Condens. Matter **15** 743, 2003.
17. Liu S. R., K. J. Dong, A. B. Yu, R. P. Zou, and J. Y. Li *Formation characteristics of nano-clusters during rapid solidification process of liquid Al* J. Mat. Scien. lett. **22** 779, 2003.
18. Daw, M. S. and M. I. Baskes *Semiempirical, quantum mechanical calculation of hydrogen Embrittlement in metals* Phys. Rev. Lett. **50** 1285, 1983.
19. Daw M S and Baskes M *Embedded-atom method: Derivation and application to impurities, surfaces, and other defects in metals* Phys. Rev. B **29** 6443, 1984.
20. Finnis, M. W. and J. E. Sinclair *A simple empirical n-body potential for transition metals* Phil. Mag. A **50** 45, 1984.
21. Sutton, A. P. and J. Chen *Long-range Finnis-Sinclair potentials* Phil. Mag. Lett. **61** No:3 139, 1990.
22. Çağın, T., G. Dereli, M. Uludoğan, and M. Tomak, *Thermal and mechanical properties of some fcc transition metals*, Phys. Rev. B **59** 3468-3472, 1999.
23. Dereli, G., T. Çağın, M. Uludoğan, and M. Tomak *Thermal and mechanical properties of Pt-Rh alloys* Phil. Mag. Lett. **75** 209, 1997.
24. Qi, Y., T. Çağın , Y. Kimura and W. Goddard III *Molecular-dynamics simulations of glass formation and crystallization in binary liquid metals: Cu-Ag and Cu-Ni* Phys. Rev. B **59** 3527, 1999.
25. Qi Y., T. Çağın, Y. Kimura and W. A. Goddard III *Viscosities of liquid metal alloys from nonequilibrium molecular dynamic* J. Comp. Aid. Mat. Des. **8** 233-243, 2002.
26. Rafii-Tabar, H. and A. S. Sutton *Long-range Finnis-Sinclair potentials for fcc metallic Alloys* Phil. Mag. Lett. **63** 217, 1991.
27. Çağın T, Y. Qi, H. Li, Y. Kimura, H. Ikeda, W. L. Jonhson, and W. A. Goddard III MRS Symp. Ser. **554** 43-48, 1999.
28. Qi Y., T. Çağın, W. L. Johnson and W. A. Goddard III Melting and Crystallization in Ni nanoclustes: The mesoscale regime J. Chem. Phys. **115** 385, 2001.
29. Ikeda H.,Y. Qi, T. Çağın, K. Samwer, W. L. Jonhson, and W. A. Goddard III *Strain Rate Induced Amorphization in Metallic Nanowires* Phys. Rev. Lett. **82** 2900, 1999.
30. Ray J R and Rahman A *Statistical-ensembles and molecular-dynamics studies of anisotropic solids 2* J. Chem. Phys. **82** 4243, 1985.
31. Nose, S. *A unified formulation of the constant temperature molecular dynamics methods* Mol. Phys. **52** 255, 1984; S. Nose J. Chem. Phys. **81** 511, 1984.
32. Parinello, M. and A. Rahman *Crystal structure and pair potentials: A molecular dynamics study* Phys. Rev. Lett. **45** 1196, 1980.
33. Parinello, M. and A. Rahman *Polymorphic transition in single crystals: A new molecular dynamics method* J. Apply. Phys. **52** 7182, 1981.
34. Wendt, H. R. and F. F. Abraham *Empirical criterion for the glass transition region based on Monte Carlo simulations* Phys. Rev. Lett. **41** 1244, 1978.
35. Allen, M. P. and D. J. Tildesley, *Computer Simulation of Liquids* (Clarendon Press, 1987).

HYDROTHERMALLY PREPARED NANOCRYSTALLINE Mn-Zn FERRITES: SYNTHESIS AND CHARACTERIZATION

D.N. BAKOYANNAKIS, E.A. DELIYANNI, K.A. MATIS*
Division of Chemical Technology, School of Chemistry, Aristotle University (Box 116), GR-54124 Thessaloniki, Greece

L. NALBANDIAN and V.T. ZASPALIS
Lab. Inorg. Materials, Inst. Chem. Proc. Eng., Center Res. Tech., P.O. Box 361, GR-57001 Thessaloniki, Greece

1. INTRODUCTION

In this study of nanosized manganese-zinc ferrites, particularly, two aspects were investigated in-depth for the main product being $Mn_{0.65}Zn_{0.35}Fe_2O_4$: i) The influence of the method followed during the material preparation, and ii) The effect of the varying stoichiometry of the powder compared with its properties. Earlier work in the area of the group involves, among other, the preparation of nanocrystalline hydroxy-ferric oxides and their application as adsorbents in environmental technology – see for instance [1-3}.

2. EXPERIMENTAL

Mn-Zn ferrite powders were prepared from metal chloride precursors through precipitation using aqueous ammonia (separately of iron, at first) and then a hydrothermal processing route in an autoclave at 180 °C for 2 hr; the products were freeze dried. After precipitation, the chloride ions were removed and their concentration controlled.

ICP-AES was used to find the stoichiometry of the material produced in each case. Characterisation of the ferrites was conducted by X-ray diffraction of samples (XRD), Transmission Electron Microscopy (TEM), Scanning Electron Microscopy with microanalysis (SEM-EDS), and Infrared spectroscopy with Fourier transform (FTIR).

Pressed compacts made from the synthesized in the lab powders have been sintered under controlled partial pressure conditions. Noting that the required sintering temperature is inversely proportional to the specific surface of the material. Therefore, the latter was taken as criterion of the work - the surface was measured with the BET method.

S. Guceri et al. (eds.), Nanoengineered Nanofibrous Materials, 495-500.

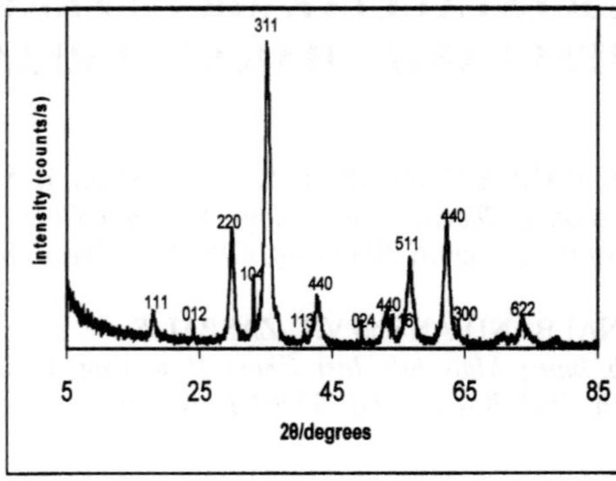

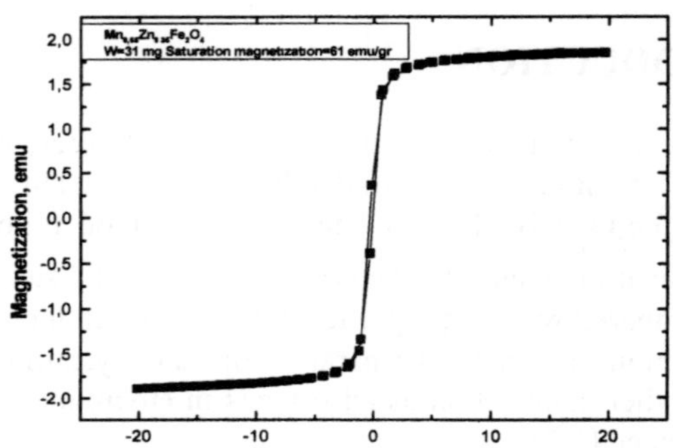

Figures 1 & 2. XRD pattern (top) and magnetization vs. applied magnetic field (bottom) for the main product.

3. RESULTS AND DISCUSSION

3.1. Notes on Chemistry

The introduction of a non-magnetic ion, like Zn in $MnFe_2O_4$ results in lowering the total anisotropy. In this way, with the changes in the composition, the coercivity is influenced and hence, the saturation magnetization [4]. A difference in cation distribution (apart from the particle size), according to the literature, also affects the magnetic properties [4-6].

The achievement of very fine grained microstructures with simultaneously high densities became possible. Ring shaped fired specimens of these powders were characterized for their electromagnetic performance: they exhibited low electromagnetic power losses and resonance frequencies in the MHz range.

The use of ammonia was preferred because it is known that Na (if sodium hydroxide is applied), as an impurity, decreases the initial permeability of the ferrite materials. Significant role for the achievement of the desired stoichiometry in the nanoparticles play the concentration of chloride ions in the solution and the pH of precipitation. During the precipitation of Fe, Mn and Zn from their mixed chloride solutions by aqueous ammonia at different pHs and reaction times the result is only a partial precipitation of Mn and Zn, as their respective hydroxides. These could then redissolve in solution (the reactions are shown in the next page) forming complexes with ammonium salts; hence, an existing competition between the formation and redissolution of the precipitates.

$$Mn/Zn(OH)_2+4NH_4Cl \leftrightarrow [Mn/Zn(NH_3)_4Cl_2]+2HCl+2H_2O$$

At a lower pH these ammoniacal complexes can be hydrolysed: $[Mn/Zn(NH_3)_4Cl_2] + 4H_2O \leftrightarrow Mn/ZnCl_2 + 4NH_4OH$

At higher pH (>7) the following reaction takes place: $Mn/Zn(OH)_2+2NH_4Cl+2NH_3 \leftrightarrow [Mn/Zn(NH_3)_4Cl_2]+2H_2O$

The complexes of zinc with ammonia are stable at higher pHs, while those of Mn at values > 9 are instable. So, to inhibit the redissolution of Mn and Zn and assist the stoichiometric precipitation, it is necessary to reduce the concentration of NH_4Cl in the system, i.e. that formed mainly during the iron hydroxide precipitation.

3.2 Product Characterization

The XRD spectra (see Fig. 1) showed that there is a quantity of hematite; which is reduced with lowering the Mn quantity in the compound stoichiometry. The IR spectra for the samples are almost identical. In all of them appear the typical peaks characteristic of the existence of bonds of compounds, from which the ferrite is composed. Also, some peaks were observed showing the existence of certain hydroxide or chloride ions.

The existing size of crystallites is not large (wide peaks), which is apparent and from the TEM pictures, too (Fig. 3); as it is also shown by the fact that the pattern of the rings was not intense. The SEM microphotographs (Fig. 4) showed all the grains to have approximately the same spherical shape, while their fine size was certified.

The magnetic properties of the Mn-Zn ferrite nanoparticles powder were measured by VSM, as shown in Fig. 2. From VSM measurement estimated the magnetic parameters such as saturation magnetization (Ms = 60 emu g-1) and coersivity (He = 125 Oe). Though the particles are of nanosize, they are not superparamagnetic as confirmed from the high Hc values. The Hc values of the order of >10 Oe as seen in these particles are in fact much higher compared to that of bulk ferrites (0.1 – 1 Oe). In summary, nanoparticles obtained by this process present higher saturation magnetization and coersivity compared with same materials from literature data.

The results of surface area, pore volume and pore size distribution obtained from nitrogen adsorption-desorption isotherms are presented in Tables 1 and 2.

TABLE 1. Physical characteristics of the powders (BET results): effect of the applied synthesis method

Sample	Specific surface area (m^2/g)	Porosity (cm^2/g)	Range of pore diameters (Å)

A (160)	66	0.26	10-1060
B (120)	36	0.19	10-1040
C (110)	65	0.12	10-1060
D (120)	63	0.17	10-1050

TABLE 2. Cumulative results changing the ferrite stoichiometry

Sample*	Particle dimension (nm)	Cell constant (Å)	Specific surface area (m^2/g)	Porosity (cm^2/g)	Predominent pore diameter (Å
A	119	8.46	66	0.26	160
A1	99	8.46	94	0.43	160
A2	104	8.43	87	0.67	160
A3	109	8.44	77	0.75	160

* A: $Zn_{0.35}Mn_{0.65}Fe_2O_4$, A1: $Zn_{0.14}Mn_{0.86}Fe_2O_4$, A2: $Zn_{0.66}Mn_{0.34}Fe_2O_4$, A3: $Zn_{0.73}Mn_{0.27}Fe_2O_4$

4. SUMMARY AND CONCLUSIONS

Nanosized manganese-zinc ferrite powders with cubic spinel crystal structure and a significantly high surface area have been synthesized from metal chloride precursors through a hydrothermal precipitation route using aqueous ammonia followed by freeze drying of the products. Pressed compacts made from those powders have been subsequently sintered under controlled oxygen partial pressure conditions.

The ferrite formation was found to be quite sensitive to the procedures adopted for preparing the hydroxide slurry prior to hydrothermal treatment. The chlorine ion concentration in the solution and the pH of the precipitation is shown to play a crucial role in retaining the initial stoichiometry of the solution in the nanoparticles. The produced powders were examined by X-ray diffraction for identification of the crystalline phases present, by Scanning and Transmission Electron Microscopy for identification of their morphological structure and properties, nitrogen sorption for determination of the BET surface area and by thermogravimetric and Differential Scanning Calorimetry for identification of the oxidation/reduction behavior upon firing.

All powder characterization results point out to the formation of nanosized manganese-zinc ferrite crystals ranging from 5 to 30 nm with a relatively high surface area (~75 m^2/g) when compared to this of conventionally prepared powders by the mixed oxide methods (~5-7 m^2/g). Compacts made from those powders have been sintered to various temperatures using manganese-zinc ferrite equilibrium oxygen partial pressures atmospheres. It appeared that for the same final densities the sintering temperatures are significantly lower when compared to those of conventionally prepared ferrite powders.

Moreover, the achievement of very fine grained microstructures and simultaneous high densities becomes possible. Ring shaped fired specimens of the previously mentioned nanocrystalline powders have been characterized for their electromagnetic performance. They exhibit low electromagnetic power losses and resonance frequencies in the MHz range. As can be concluded from the processing and characterization results, in terms of industrial applications, hydrothermally prepared nanocrystalline manganese-zinc ferrites may offer the following advantages:

a) lower manufacturing cost of inductive electromagnetic components since the energy consuming prefiring step can be eliminated and the firing temperatures can be significantly reduced.

b) Manufacturing of high frequency inductive electromagnetic components because of their fine grained microstructure, extending therefore the frequency application range of manganese-zinc ferrites and at the same time meeting the market trends for high frequency operation and miniaturization of inductive components.

Acknowledgements: This paper is gratefully coming from our chemistry student's, Ms. Lamprini Tzelepi, dissertation. Many thanks are also due to the Directors of this NATO ASI and the Turkish hospitality.

References

1. Deliyanni, E.A., Bakoyannakis, D.N., Zouboulis, A.I., Matis K.A. and Nalbandian L. (2001) Akaganéite-type β-FeO(OH) nanocrystals: preparation and characterzation, *Microporous and Mesoporous Materials* **42**, 49-57.
2. Bakoyannakis, D.N., Deliyanni, E.A., Zouboulis, A.I., Matis, K.A., Nalbandian, L. and Kehagias, Th. (2003) Akaganéite and goethite-type nanocrystals: synthesis and characterization, *Microporous and Mesoporous Materials* **59**, 35-42.
3. Zouboulis, A.I. and Matis, K.A. (1997) Removal of metal ions from dilute solutions by sorptive flotation, *Crit. Rev. Envir. Sci. Tech.* **27**(3), 195-235.
4. Rath, C., Sahu, K.K., Anand, S., Date, S.K., Mishra, N.C. and Das, R.P. (1999) Preparation and characterization of nanosize Mn-Zn ferrite, *J. Magnetism & Magn. Mater.* **202**, 77-84.
5. Jeyadevan, B., Tohji, K., Nakatsuka, K. and Narayanasamy, A. (2000) Irregular distribution of metal ions in ferrites prepared by co-precipitation technique structure analysis of Mn-Zn ferrite using extended X-ray absorption fine structure, *J. Magnetism & Magn. Mater.* **217**, 99-105.
6. Gillot, B. and Guendouzi, M.El. (1993) Effect of the preparation method on the oxidation-reduction mechanism and the cation distribution of Mn-Zn ferrites, *J. Solid State Chem.* **106**, 443-450.

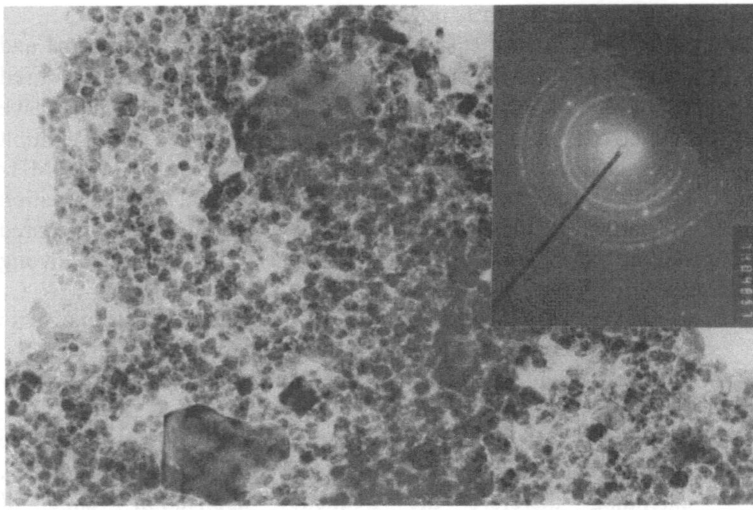

Figures 3 & 4. TE micrograph and O-ring pattern (top) and SEM image (bottom).

DEPOSITION OF SUB-MICRON Ni DROPLETS ON GLASS SUBSTRATES BY A COMBINATION OF PLASMA ASSISTED CVD AND PVD

H. AKBULUT, C. BINDAL
Sakarya University, Engineering Faculty, Dept of Metallurgy and Materials, Esentepe Campus, 54187, SAKARYA - TURKEY

O.T. INAL
New Mexico Institute of Mining and Technology Socorro, New Mexico 87801, USA

Abstract: In the present study, Ni coatings have been deposited on glass microscope slides at substrate temperatures of 300, 350, 400 and 450 °C. The depositions were performed by evaporating the metallic Ni source in a low-voltage (600-1000V) plasma of argon:hydrogen (2:1) gas. Morphology of the Ni coatings was altered by changing the substrate temperature whereas the coating thickness was kept nearly constant. X-ray and XPS analyses showed that the increase in the substrate temperature resulted in the formation of NiO and Ni_2O_3 on the deposited layer. Resistivity of the deposited layers and the adhesion between the substrate and coatings were increased at high temperatures.

Key words: Metallization, Ni coating, plasma deposition

1. INTRODUCTION

Metallization of microelectromechanical systems (MEMS) devices and multilevel metallization of very large scale integrated (VLSI) circuits entail design and processing issues beyond those of concern in IC technology [1, 2]. Metallizations generally require good conductivity, electromigration resistance, controllable contact performance, corrosion resistance, adherence, thermal stability, bondability, ability to be patterned into desirable geometry and must be economically feasible to implement. Metallization of the ceramics and glasses provided to obtain neutron optical elements, such as neutron guides, are nowadays indispensable at a neutron source especially for cold and thermal neutrons. The neutron mirrors, usually used in the construction of such elements, are often made

S. Guceri et al. (eds.), Nanoengineered Nanofibrous Materials, 501-507.

of highly polished nickel plated glass which are relatively close to unity [3].

Traditionally, copper and nickel films have been deposited by electroless coating or electron beam evaporation. However, these deposition techniques suffer from inadequate step coverage; this often results in the formation of microcracks and discontinuities at holes and deep trenches [4]. Most metal deposition processes can be divided into PVD and CVD type processing. The most common techniques that are being used for metallization are the CVD, MOCVD and R.F. [5]. Laser chemical vapor deposition (LCVD) is another metallization technique applied especially for Ni [6]. The major objective of our work was to investigate the effect of substrate temperature on the structure, resistivity, and adhesion between substrate and the deposited layer.

2. EXPERIMENTAL METHOD

The substrates used in the present work were glass microscope slides. A d.c. plasma system was used for the coating of glass substrate surfaces with pure nickel. The details of the plasma apparatus as well as the uniquenesses of plasma assisted surface processing are given elsewhere [7]. Prior the coatings, the system was evacuated at a base pressure of 10^{-3} Torr and then filled with argon and hydrogen gases for sputtering. Each sample was subjected to ion bombardment, sputter cleaning, in an argon and hydrogen (1:1) atmosphere at a gas pressure of 300 mTorr and a cathodic voltage of 600 - 1000 V for 20 min. The sputtering of the substrates were made at the substrate temperatures of 300, 350, 400 and, 450 °C. A tungsten boat, coated with pure alumina, and amenable to resistive heating, was used as the source for Ni evaporations inside the vacuum system. The boat temperature was kept at 1800 °C for evaporation of pure nickel during the total deposition period of 6 minutes. Following placement in the vacuum chamber, the glass substrates were then coated with nickel. An $Ar:H_2$ (2:1) gas mixture was used to provide a clean nickel deposition on the glass substrates. All coatings were investigated for layer thickness, as measured with metallography following grinding and polishing of the samples by conventional methods. SEM analysis was carried out with a Hitachi HHS-2R system employing a 20 keV electron beam. The room temperature resistivity of the deposited Ni layer was measured by a low-fequency a.c. method using a four-probe technique. X-ray diffraction and XPS analysis of the specimens were carried out. X-ray photoelectron spectroscopy (XPS, ESCA750) was employed to analyze the surface components of the nickel coatings. Surface profilometry of uncoated glass substrates and the coatings were carried out for determining the roughness of the coatings. Scratch testing was performed using a Romulus III, Universal Scratch Tester (dead load) fitted with a diamond stylus of a radius of 533 μm. At least four scratches were performed on a given sample. Normal loads, effective friction, transverse loads and acoustic emission outputs were recorded by a computer, attached to the scratch tester. A standard loading speed of 0.15 N/min. was used; the scratches were about 2.5-4.0 mm long. The critical load (L_c) that gives coating failure was estimated by using the information obtained from microscopic observations and from the data on acoustic emission and effective friction.

3. RESULTS

Grain size of the Ni films showed a gradual increase with increasing substrate temperature. The measured grain sizes are as follows: 0.52 μm ± 0.085 for 300 °C; 0.58 μm ± 0.05 for 350 °C; 0.63 μm ± 0.04 for 400 °C and 0.68 μm ± 0.04 for 450 °C. SEM micrographs of Ni coatings at 300 and 450 °C are presented in Fig. 1, showing a continuous, smooth polycrystalline structure and a uniform, dense surface with a grain size of 0.52-0.68 μm. As seen in Fig. 1, the Ni deposits that coated at 300 °C exhibit different morphology when compared with Ni deposits coated at 450 °C substrate temperature. Fig. 1a shows that at 300 °C the deposition of Ni at 300 °C substrate temperature gives very fine equiaxed crystallites of Ni, whereas Ni particles (in Fig. 1b) showed triangular, prismatic shape on the glass substrate at 450 °C. When the substrate temperatures were taken into account, the coating thickness values revealed an approximately 8 ± 0.8 μm range; no significant differences in the thickness have been observed by changing substrate temperature. The coating thickness values were measured as: 7.2 μm at 300 °C, 7.9 μm at 350 °C, 7.6 μm at 400 °C and 8.8 μm at 450 °C. X-ray analysis of the Ni layers at different substrate temperatures exhibited different

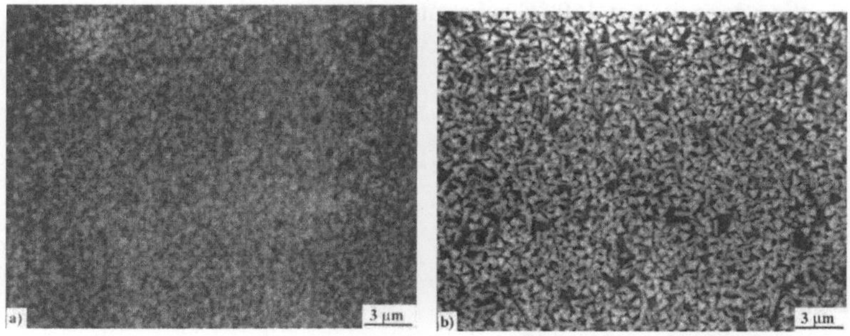

Figure. 1. SEM micrographs of surface morphologies of Ni coatings, deposited on glass substrate at 300 mTorr total gas pressure; a) 300 °C and b) 450 °C substrate temperatures.

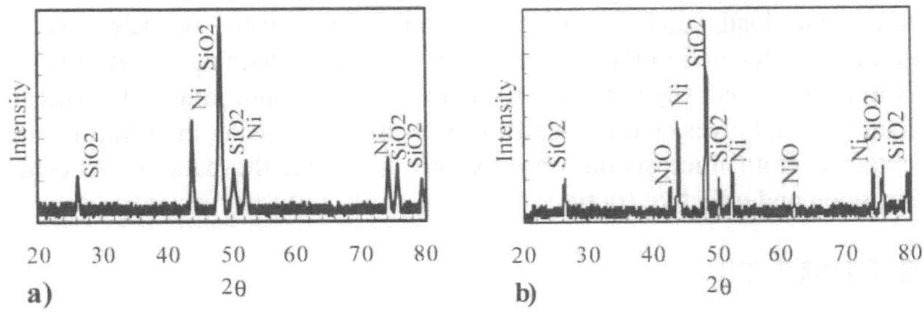

Figure 2. XRD patterns of Ni deposits: a) 300 °C and b) 450 °C substrate temperatures.

phases. In general, increasing substrate temperature from 300 °C to 450 °C resulted in the formation and a gradual increase the NiO phase. The XRD patterns of Ni films deposited on glass substrates are given in Fig. 2. For brevity, only the x-ray patterns of the films deposited at 300 °C and 450 °C substrate temperature are presented. The film deposited at 300 °C exhibited only the diffraction peaks corresponding to Ni cubic phase. In particular, diffraction peaks indicating NiO phase were never detected in the XRD data (Fig. 2a). An XRD pattern at 300 °C, showed a clean Ni phase with major diffraction peaks appearing 2θ = 43.955° (111), 50.299° (200) and 74.100° (220). Ni deposition at 450 °C resulted in relatively high intensity NiO peaks on glass substrate coated with Ni (Fig. 2b).

The observation of the peaks that were obtained from the XPS analyses of the coatings showed that increasing substrate temperature resulted in an increasing amount of the oxide peaks at the deposited Ni layers. Fig. 3 represents XPS spectra and the relation between resistivity and substrate temperatures of Ni films. Three peaks are identified as Ni, NiO and Ni_2O_3. As seen in Fig. 3a, the Ni coating deposited at 300 °C revealed small NiO and Ni_2O_3 peaks (Fig. 3a). The intensities of the oxide peaks are larger in the coating deposited at 450 °C substrate temperature.

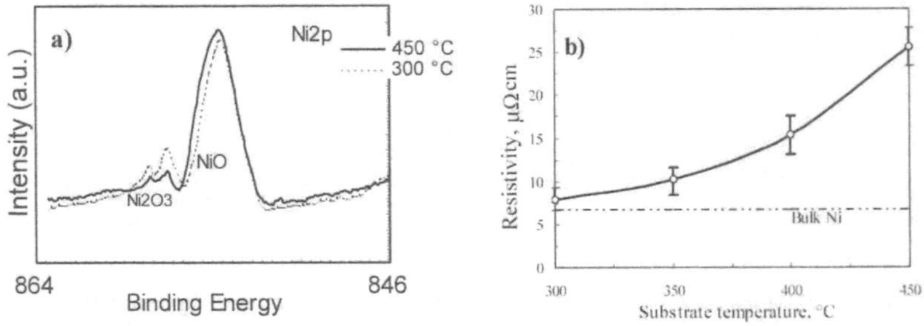

Figure 3. a) XPS analysis of the Ni deposits on the glass substrate, b) Four-point probe measurements resistivity of Ni films as a function of glass substrate temperature.

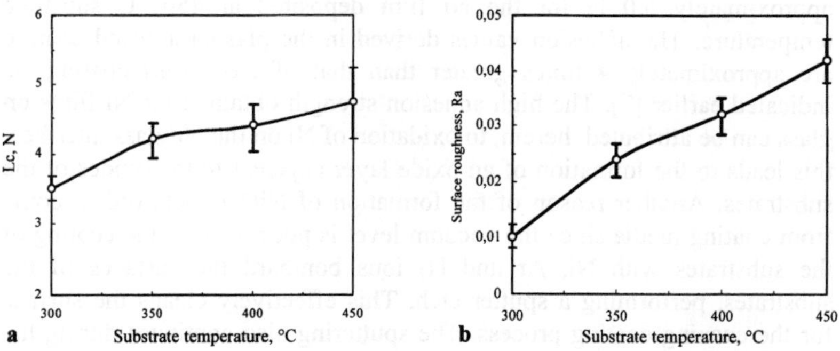

Figure 4. a) Critical Load (L_c) values versus substrate temperature, b) The relationship between the substrate temperature versus surface roughness values.

As seen in Fig. 3b, the resistivity is apparently higher for high temperatures. The bulk resistivity of Ni is also shown as a reference value in Fig. 3b. The resistivity of Ni film is 7.90 μΩ.cm at 300 °C substrate temperature and gradually increases until reaches to 25.6 μΩ.cm at 450 °C. Experimental observations on the scratch test curves showed that increasing the substrate temperature caused to higher critical load (L_c) values. The L_c values that were obtained from the scratch test curves are illustrated in Fig. 4a as a function of substrate temperature. A gradual increase in the roughness of the Ni deposited surfaces with increasing substrate temperature was observed (Fig. 4b).

4. DISCUSSION

Ni coating on glass microscope slides, in the present study, showed uniform coverage of the substrate. However, increasing substrate temperature resulted in an increase in grain size of the Ni films. The grain morphology has been observed to be changed by substrate temperature. It is believed that the change of the morphology in the Ni films has been caused the growth mechanism of the film. At low substrate temperatures relatively pure Ni film grows as nucleation of small equiaxed grains. The Ni grains show formation of columnar and triangular transition as increasing substrate temperature. As was also stated by Lu et al [8], Ni based films deposited at low substrate temperature have smaller grains. At higher substrate temperatures, the grains are larger with strong columnar structure. Wong and co workers [9] obtained rectangular and triangular grain, and observed increased surface roughness with increasing deposition rate in the Ni films that were deposited with electroforming.

Coating of pure metallic conductors with high coating rates requires high vacuum systems. In this study pure coatings were achieved with a relatively poor vacuum level. No separation of the film from the substrate was possible. Measurement using scratch test has given a L_c of approximately 5.0 N for the Ni film deposited at 450 °C substrate temperature. The adhesion values derived in the plasma assisted coating are approximately 4 times greater than that of electroless coating, as indicated earlier [7]. The high adhesion strength obtained for Ni films on glass can be attributed, herein, to oxidation of Ni on the Ni/glass interface, this leads to the formation of an oxide layer adjacent to the oxides of the substrates. Another reason of the formation of NiO is believed to come from coating media since the vacuum level is poor. During the coating of the substrates with Ni, Ar and H_2 ions bombard the surfaces of the substrates, performing a sputter etch. This effectively cleans the surface for the ensuing coating process. The sputtering also continues during the deposition. It was reported that, in the absence of H_2, the metallic films usually contain significant amounts of impurities [10]. To decrease the impurities and increase deposition rate, an Ar:H_2 (2:1) gas mixture was used to provide a clean coating on the glass substrates. In the present work, since the vacuum level of the plasma chamber was low 10^{-3} Torr a remarkable amount of oxygen remained in the system. Hydrogen is believed to combine the oxygen according to the following reaction:

$$2Ni_{(Vapor)} + O_2 + H_2 \longrightarrow Ni_{(Deposit)} + Ni(OH)_2 \qquad (1)$$

Based on AES measurements, it has been concluded that H_2 must be present in the carrier gas to obtain pure films when organo-metallic Cu precursors are used [11]. As in the case of PACVD, it is thought that using H_2 as carrier gas, together with Ar, is necessary to obtain pure coating layers. Since vacuum level of the plasma system used in present work was relatively low (10^{-3} torr), a 1:1 (Ar:H_2) plasma gas ratio was used in sputtering and a 2:1 (Ar:H_2) ratio was used during Ni coating.

5. CONCLUSIONS

1. Thin film Ni depositions were performed on glass substrates by using a combination of PVD and PACVD processes in a d.c. plasma source with the use of a filament as an evaporator at a relatively low vacuum level.
2. All coatings showed continuous, smooth, polycrystalline structures with a dense surface at all the substrate temperatures. Increasing substrate temperatures resulted in triangular Ni grains and increased surface roughness of the films.
3. A short deposition period of 6 min. is sufficient to obtain a thickness of approximately 8 μm coatings on the glass substrates.
4. Oxygen free deposition of Ni is achieved even with relatively poor vacuum by the judicious use of hydrogen in the sputtering and deposition gas mixtures at 300 °C substrate temperature.

5. At high substrate temperatures the resistivity and the adhesion of the Ni films increased due to oxidation of deposited Ni by the oxygen diffusion from glass substrate to the film.

References

1. Carraro, C., Magagnin, L and Maboudian, R., (2002), Selective metallization of silicon micromechanical devices, Electrochimica Acta **47**, 2583-2588.
2. Baselt, D.R., Fruhberger, B., Klaassen, E., Cemalovic, S., Britton, C.L., Patel, S.V. Mlsna, T.E., McCorkle, D., and Warmack, B. (2003), Design and performance of a microcantilever-based hydrogen sensor, Sensors and Actuators **B88**, 120-131.
3. Maayouf, R.M.A, (2000), On the neutron reflectivity of Ni-coated mirrors, Physica **B291**, 54-58.
4. Coupeau, C., Naud, J.F., Cleymand, F., Goudeau, P. and Grilhe, J. (1999) Atomic force microscopy of in situ deformed nickel thin films, Thin Solid Films **353**, 194-200.
5. Thompson, A. G., (1997), MOCVD technology for semiconductors, Mater. Lett. **30**, 255-263.
6. Mikami, Y., Yamada, K. Ohnari, A., Degawa, T., Migita, T., Tanaka, T., Kawabata, U.K. and Kajioka, H., (2000), Effect of DC bias voltage on the deposition rate for Ni thin films by RF–DC coupled unbalanced-magnetron sputtering, Surf. Coat. Technol. **133/134**, 295-300.
7. Akbulut, H., and Inal, O.T., (1998), Plasma Assisted Deposition of Metal and Metal Oxide Coatings J. Mater. Sci. **33**, 1189-1199.
8. Lu, Y.M., Hwang, W.S. Yang, J.S. and Chuang, H.C., (2002) Properties of nickel oxide thin films deposited by RF reactive magnetron sputtering Thin Solid Films, **420/421**, 54-61.
9. Wong, K.P., Chan, K.C. and Yue, T.M., (1999), Study of surface finishing in pulse current electroforming of nickel by utilizing different shaped waveforms, Surf. Coat. Technol. **115**,132-139.
10. Cohen, S.L., Liehr, M. and Kasi, S. (1992), Mechanisms of copper chemical vapor depositionAppl Phys. Lett. **60**, 50-52.
11. Murarka, S. P. and Hymes, S.W., (1995), Copper metallization for ULSI and beyond, Critic. Rev. Solid State Mater. Sci. **20**, 87-124.

FORMATION OF FIBROUS-STRUCTURED MULTI-LAYER OXIDE COATING ON SI_3N_4-TIN AND SI_3N_4-TIB_2 CERAMICS BY ELECTROCHEMICAL POLARIZATION

V.A.Lavrenko[(1)], M.Desmaison-Brut[(2)], V.A.Shvets[(1)], J.Desmaison[(2)]

1) Institute for Problems of Materials Science, 3 Krzhyzhanovsky str., Kiev, 03680 Ukraine

2) University of Limoges, Science des Procédés Céramiques et de Traitements de Surface, SPCTS, UMR CNRS 6638,123 Avenue Albert Thomas, F-87060 Limoges cedex, France

Abstract: Fibrous-structured multi-layer oxide coatings of nano-dimension thickness (5-8 nm) have been formed on the surfaces of Si_3N_4-TiN and Si_3N_4-TiB_2 ceramic samples under their electrochemical polarization in 3 % NaCl solution. On the surface of Si_3N_4-TiN ceramics a very thin (~ 5 nm of total thickness) oxide coating was formed. This coating consisted of three nano-layers containing $TiN_{0.7}O_{0.3}$ titanium oxynitride and Si_3N_4 in its inner part, TiO phase, mixed with Si_3N_4, in the intermediate part, and TiO_2 (rutile) of fibrous structure in the outer one. In the case of a Si_3N_4-TiB_2 ceramics a more friable oxide coating of 7-8 nm thickness was formed. The first surface layer of coating (nanofibrous structure of ~ 3 nm thickness) contains, mainly, the TiO_2 rutile fibers mixed with silicon oxynitride Si_2N_2O particles; the intermediate one was the mixture of TiO_2 with an oxygen solid solution (5 – 10 mass.% O) in Si_3N_4 and TiB_2 traces; and the inner one (at the coating-substrate boundary) consisted of TiO_2 in the mixture with Si_3N_4 and TiB_2. The nanofibrous coating on Si_3N_4-TiN ceramics obtained was characterized by a higher density and the level of adhesion to the substrate, as well as more stable structure compared with the coating on Si_3N_4-TiB_2 ceramics.

1. INTRODUCTION

There are literature data [1-6] concerning the formation of surface oxide layers under high-temperature oxidation of multiphase non-oxide ceramic materials, such as Si_3N_4-TiN, AlN-TiB_2, AlN-TiN, AlN-SiC, AlN-SiC-TiB_2, AlN-TiB_2-$TiSi_2$. Conversely, there is only rare information [7-8] about the anode formation of oxide coatings on these electroconductive ceramic composites under their electrochemical polarization in a sodium chloride solution.

S. Guceri et al. (eds.), Nanoengineered Nanofibrous Materials, 509-518.

2. MATERIALS AND METHODS

The initial ceramic samples were prepared using the HIP method. The powders, sintering parameters and relative density of samples prepared are given in Table 1.

Table 1. The powders, sintering parameters and relative densities.

	Powders	Degassing	HIP cycle	$d_{relative}$%
SN-TiN	Si_3N_4 65 vol%, TiN 35 vol% Y_2O_3 0,26 wt%, Al_2O_3 0,13 wt%	600°C 10h	1725°C 1h 180 MPa	100
SN-TiB$_2$	Si_3N_4 65 vol%, TiB_2 35 vol% Y_2O_3 0,5 wt%, Al_2O_3 0,25 wt%	600°C 10h	1710°C 1h 170 MPa	98

The following powders were used to obtain the composite materials: TiN Starck Grade C, TiB_2 Starck Grade F, Si_3N_4 UBE SN10. The experimental processing procedure was the following: the two powders and sintering aids (Al_2O_3, Y_2O_3) were dispersed by ultrasonic assistance in ethanol. The mixture was then dried at 100°C, sieved at 32 μm and cold isostatically pressed. The crude cylinder was introduced inside a silica container and sealed. The final relative densities were respectively 100 and 98%. The density was determined using the Archimede principle.

The microstructures of initial composites are compared on Fig. 1, the good dispersion of the secondary phase being noticed (Fig. 1, a-b). The phase and microstructure analyses of surfaces were carried out using XRD (Philips-CGR), SEM and EDS (Philips XL30) methods. The scanning electron microscopy of oxidized sample cross section was performed using a DSM 982 device.

The Si_3N_4-TiN (**SN-TiN**) and Si_3N_4-TiB_2 (**SN-TiB$_2$**) samples were subjected to electrochemical polarization in 3% NaCl solution at 20°C. A ceramic anode and a Pt cathode constitute the electrolytic cell. The anode polarization curves of electrolytic oxidation (Fig. 2) were obtained by a potential-dynamic method at a potential sweep rate of 5 mV/s using a P 5848 potential-static device and an Ag/AgCl/KCl reference electrode. In this paper all the potentials are measured versus this silver-silver chloride electrode.

The phase composition of thin oxide films formed as a result of electrochemical oxidation of both composite samples was determined on the basis of the elements' molar ratio, by Auger electron spectroscopy, using a LAS-2000 device (Riber, France). Hereby the Auger analysis was carried out after gradual etching of film layers at the rate of 5 nm/min under Ar^+- ions bombardment. The Auger spectra of initial samples were used as a reference. The computer program using the spectra of standard Ti- and N- containing samples was applied for the correct separation of Ti and N peaks which are close both in position and energy.

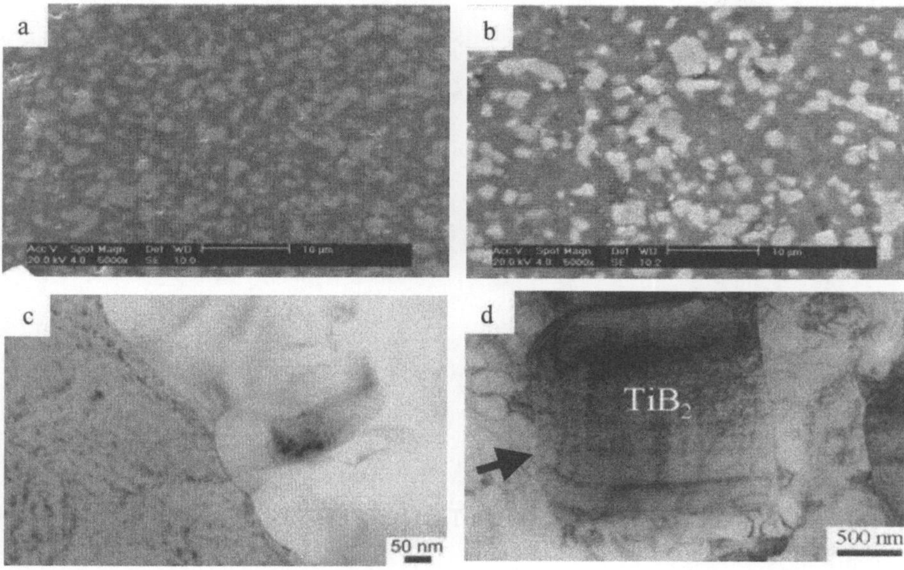

Figure 1. SEM (a,b) and TEM (c,d) images of SN-TiN (a,c) and SN-TiB_2 (b,d) composite microstructure.

3. RESULTS AND DISCUSSION

The measured stationary potentials, i. e.: + 0.25 V for (SN-TiN) and – 0.10 V for (SN-TiB_2), showed that, without polarization, the SN-TiN ceramics had a lower oxidation rate in 3 % NaCl solution than SN-TiB_2 ones (Fig. 3).

In the case of SN-TiB_2, the total negative charge of the double layer on the electrode surface is determined by Cl^- ion both in the compact layer and in the diffuse layer (Fig. 3, a). The measured Nernst's potential E is defined by the sum $E = \varphi + \xi$ of the difference of potential through these two layers (Fig. 3, a and b). As one can see, the SN-TiN system differs from the SN-TiB_2 one (Fig. 3, b). Indeed, resulting from a greater value of positive potential ξ, in comparison with the negative φ one, the measured stationary potential E proved to be positive.

It must be noted that by EDAX analysis the presence of chlorine adsorbed particles on the surfaces of both ceramic electrodes sunk in the electrolyte has been confirmed.

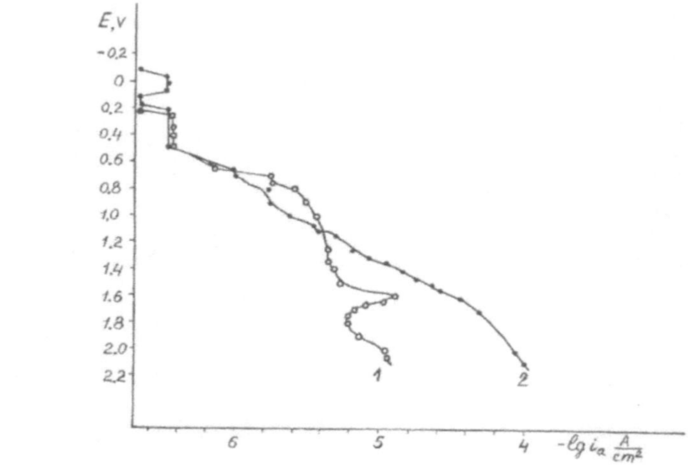

Figure 2. Anodic polarization curves of SN-TiN (1) and SN-TiB_2 (2) ceramics oxidation in 3% NaCl solution

The increase of anode potential on Fig. 2 shows that the oxidation process of both ceramics proceeds in three stages. At every stage, chemical analyses were used [7, 8] for the determination of Ti (IV) ions, i.e. TiO^{2+} titanyl-ions, and Ti (III), i.e. Ti^{3+}-ions, which come into solution from the TiN or TiB_2 electrodes due to the anode dissolution. The results were in accordance with the following equations (1-4):

$$2\,TiN + 2\,H_2O = 2\,TiO^{2+} + N_2 + 4\,H^+ + 8\,e \qquad (1)$$

$$2\,TiN = 2\,Ti^{3+} + N_2 + 6\,e \qquad (2)$$

$$TiB_2 + 7\,H_2O = TiO^{2+} + 14\,H^+ + 2\,BO_3^{3-} + 10\,e \qquad (3)$$

$$TiB_2 + 6\,H_2O = Ti^{3+} + 12\,H^+ + 2\,BO_3^{3-} + 9\,e \qquad (4)$$

Auger spectra of samples subjected to oxidation during 2 h in 3 % NaCl solution under the potential of + 1.8 V for the SN-TiN composite and + 2.0 V for the SN-TiB_2 one are given on Fig. 4, a, and 4, b, respectively, as well as the Auger spectra of initial samples.

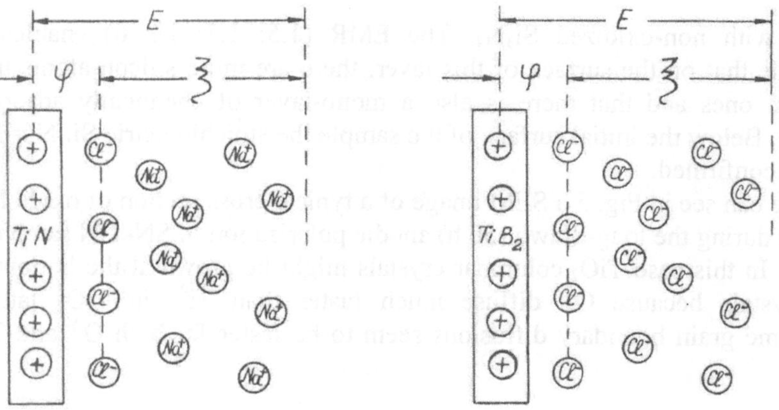

Figure 3. Scheme of the double electric layer on the electrode surface in 3 % NaCl solution: SN-TiN; (b) SN-TiB_2.

<u>**SN-TiN.**</u> At the first oxidation stage, from + 0.25 to + 0,5 V potentials, a thin (~ 2.5 nm) $TiN_{0.7}O_{0.3}$ oxynitride layer was formed on the sample surface. The Ti: N: O: Si element molar ratio (EMR) is, mainly, equal to 3: 3.7: 1: 1. Si_3N_4 in all parts of the surface film is in the non-oxidized state. However, in the vicinity of the boundary of this layer with the substrate, the EMR approaches to 10.5: 14.7: 1: 5 suggesting the formation of an oxygen solid solution (0.1 mol.% O) in TiN. The phase composition of thin oxide films formed as a result of electrochemical oxidation of both composite samples was determined on the basis of elements molar ratio, by Auger electron spectroscopy analysis. It was carried out after gradual etching of film layers at a rate of 5 nm/min under Ar^+- ions bombardment.

The potential range from + 0.5 to + 0.7 V corresponds to the TiN dissolution in the electrolyte according to (1). However, with a further increase of the anode potential up to + 1.5 V, titanium dissolves into the solution already in the trivalent state, according to (2), and on the surface, a very thin non-stable TiO film (EMR: 2,2: 1.5: 2.0: 1) is formed according to equation (5). The passivity section is only sharply observed at potentials > 1.6 V (Fig. 1). The sample dissolution is entirely stopped, and the oxide outer layer (~ 2 nm thickness) is composed of TiO_2 rutile (see equation 6):

$$2\ TiN + 2\ H_2O = 2\ TiO + N_2 + 4\ H^+ + 4\ e \qquad (5)$$

$$2\,TiN + 4\,H_2O = 2\,TiO_2 + N_2 + 8\,H^+ + 8e \quad (6)$$

mixed with non-oxidized Si_3N_4. The EMR (1.5: 1.5: 12: 6) enables to conclude that on the surface of this layer, there are more silicon atoms than nitrogen ones and that there is also a mono-layer of chemically adsorbed oxygen. Below the initial surface of the sample the stoichiometric Si: N = 3: 4 ratio is confirmed.

One can see in Fig. 5 a SEM image of a typical cross section of oxide film formed during the long-drawn (50 h) anodic polarization of SN-TiN sample at 1.8 V. In this case TiO_2 columnar crystals might be grown at the bottom of the crystals because O^{2-} diffuse much faster than Ti^{4+} in TiO_2 lattice. Meantime grain boundary diffusions seem to be faster for both O^{2-} and Ti^{4+} ions.

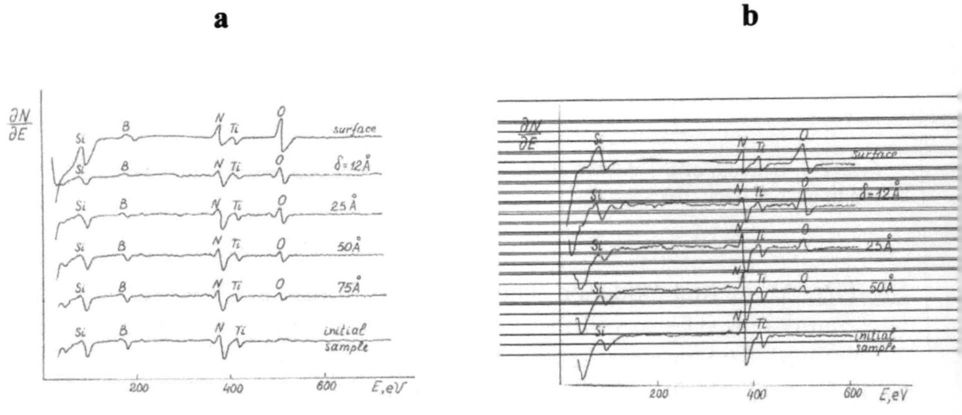

Figure 4. Auger spectra of elements distribution along cross section of oxide film (δ - thickness of oxide layer): a – SN-TiN; b – SN-TiB_2.

On the base of Fig. 5 one can suppose that in the case of SN-TiN oxidation, the oxide layers were formed in the accordance with a scheme of Fig. 6. Hereby the processes of ion diffusions in the solids (oxides, oxynitrides) play the exceptionally important role, not less important than the anodic electrochemical reactions themselves.

The microstructures of initial and oxidized SN-TiN surfaces are presented in Fig. 7, a and b. After oxidation a great amount of finely-dispersed white fibers of TiO_2 (rutile) is noticed on the electrode surface (Fig. 7, b).

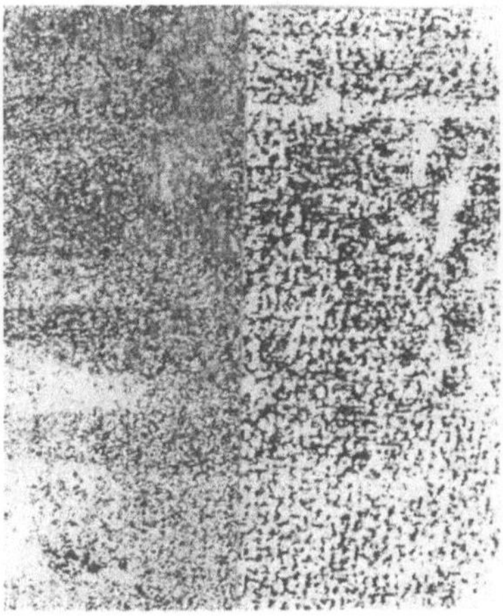

Figure 5. SEM image of a cross section of oxide film on SN-TiN sample (on the left, TiO_2 column-like nanocrystals grown out in the outer oxide layer; on the right, TiO_xN_y + TiO intermediate phases layer; x 160000).

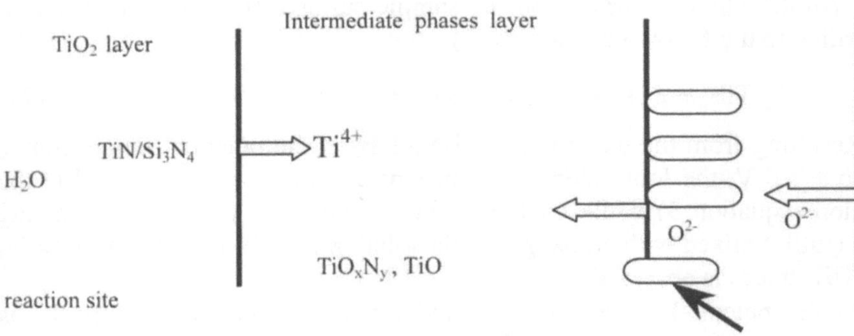

Figure 6. Scheme of diffusion processes in the multi-layer oxide film formed under SN-TiN anodic polarization.

TiO_2 rutile is a semiconductor of n-type, in which the oxygen vacancies ensure the primary, compared with Ti^{4+} ions, diffusion of oxygen ions through the oxide crystal lattice.

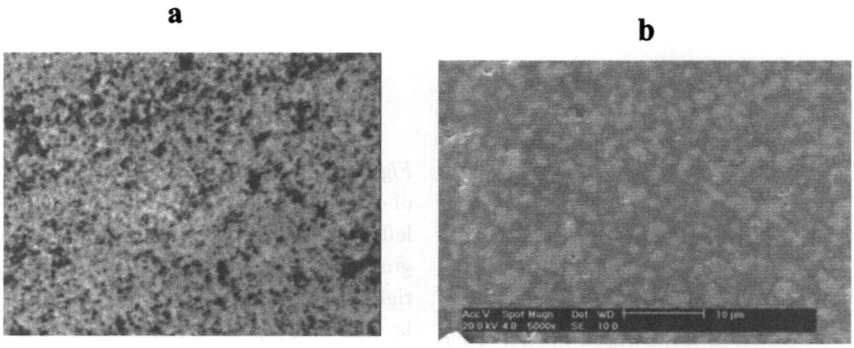

Figure 7. SN-TiN: a – a SEM image of a polished surface before oxidation; b – optical microscope image of an oxidized surface (x 400).

SN-TiB_2. The oxidation mechanism (Fig. 2) is significantly different than the previous one. In this case, in the potential range from – 0.04 to + 0.5 V, a TiO_2 (rutile) phase is formed on the sample surface (thickness up to 7 nm) according to the following equation (7):

$$TiB_2 + 8\,H_2O = TiO_2 + 16\,H^+ + 2\,BO_3^{3-} + 10\,e \qquad (7)$$

Resulting from the oxidation of Si_3N_4-TiB_2 at the potential range from + 0.5 to + 1.6 V, the four-valent titanium ions and boric acid are found in the solution (equation 3) while on the surface a thin (~ 3 nm) film containing TiO_2 (rutile) mixed with an oxygen solid solution (5 – 10 mass.% O) in Si_3N_4 and TiB_2 traces is observed.

At the potentials > + 1.6 V the oxidation of Si_3N_4-TiB_2 composite is slowed down (Fig. 2), and the polarization curve is evolving from an almost linear shape to a reciprocal parabolic dependence of the potential E versus the logarithm of the anode current density. Along with the formation of the friable enough TiO_2 film, the TiB_2 dissolution takes place. The trivalent ions are passing into the electrolyte (according to equation 4). The obtained Auger spectrum of the outer surface film (~ 2 – 3 nm) is characterized by a Ti: B: N: O: Si = 1.0: 0: 10.4: 10.0: 7.5 mole ratio of elements, suggesting that its composition is a mixture of TiO_2 and Si_2N_2O, the latter having the stoichiometric formula.

In the case of SN-TiB_2, the TiO_xB_y oxynitride cannot be formed. The formation of Si_2N_2O in the outer film layer may be explained according to the following reaction:

$$2\,Si_3N_4 + 3\,O = 3\,Si_2N_2O + N_2 \qquad (8)$$

taking place due to an appearance of pores resulting from too speedy dissolution of TiB_2 with a transition of boric acid into electrolyte. The rate of TiB_2 dissolution, for example, at the potential of 1.8 V, was by 10 times more than that of TiN (Fig. 2). Herewith the atomic oxygen moving to the sample substrate through the pores in oxide film and reacting with Si_3N_4 was formed at E > 1.8 V under electrochemical reaction

$$TiB_2 + 9\ H_2O = TiO_2 + 2\ BO_3^{3-} + 18\ H^+ + O\ (\text{in situ}) + 12\ e \quad (9)$$

Our observations (an evolution of gas bubbles on the anode) may testify that in the potential range pointed out both reactions – (7) and (9) – were proceeding almost simultaneously.

On the sample surface there is also a mono-layer of chemically-bonded oxygen. The microstructures of both initial and partially oxidized surfaces of Si_3N_4-TiB_2 composite are given in Fig. 8, a and b, respectively.

One can see in Fig. 7, b and 8, b that the oxidized ceramics surfaces are characterized by a nano- fibrous structure.

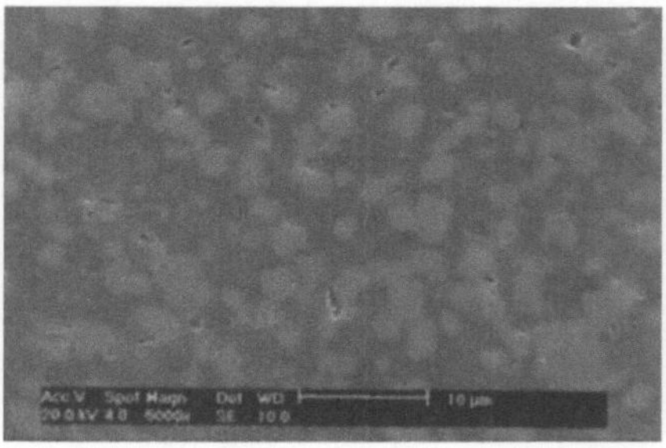

a

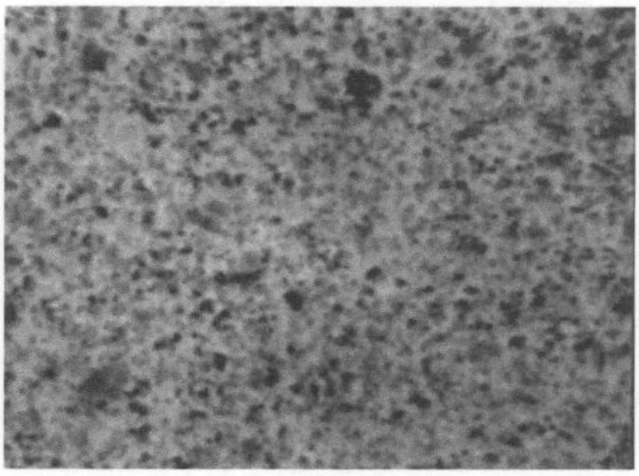

b

Figure 8. SN-TiB_2: a – SEM image of a polished surface; b – optical microscope image of an oxidized surface (x 400).

4. CONCLUSION

The behavior of two silicon nitride-based composite materials under their electrochemical oxidation in the 3% NaCl solution has been studied. It has been established that in both cases the fibrous-structured multi-layer oxide coatings are formed. The anode polarization curves and the nature of the double electric layer on and near the electrode surface have been analyzed. The results obtained have shown that the Si_3N_4-TiN electroconductive ceramic material is more stable in the studied medium than the parent Si_3N_4-TiB_2 composite even if a thicker nano-crystal oxide film is formed on its surface.

References

1. Yu.G. Gogotsi, V.A.Lavrenko. Corrosion of High-Performance Ceramics, Berlin, Springer, 1992.
2. S.V. Schneider, M. Desmaison-Brut and Yu.G. Gogotsi, J. Desmaison, Key Eng.Mater., v. 113 (1996), 49-58.

3. V.A Lavrenko, T.V. Mosina, O.N. Grigoriev et al. Proc. Academy Sci. Ukraine, N 12 (1997), 139-143.
4. V.A Lavrenko, M. Desmaison-Brut, A.D. Panasyuk, J..Desmaison, J. Europ. Ceram. Soc., v.18 (1998), 2339-2343.
5. V.A. Lavrenko, J. Desmaison et al. British Ceramic Proc., v.1 (1999), 445-446.
6. V.A. Lavrenko, J. Desmaison, A.D. Panasyuk, M. Desmaison-Brut, J. Europ. Ceram. Soc., v. 23 (2003), 357-369.
7. V.A Lavrenko, S.B. Prima, V.A Shvets, V.N. Talash, British Ceramic Proc., v.1, N 60 (1999), 233-234.
8. V.A Lavrenko, N.V. Boshitskaya, V.A Shvets, Key Eng.Mater., v. 206-213 (2002), 1571-1574.

FORMATION OF HIGH-TEMPERATURE NANOFIBROUS-LIKE COATING STRUCTURE UNDER MAGNETRON SPUTTERING OF ALN-(TI,CR)B_2 TARGET

A.D.Panasyuk, I.A.Podchernyaeva, V.A.Lavrenko
Frantsevich Institute for Problems of Materials Science,
3 Krzhyzhanovsky str., 03680 Kiev, Ukraine

Abstract: The structure, composition and properties of AlN-(Ti,Cr)B_2 coatings on Si, Al_2O_3 and GaAs mono-crystals were studied. These coatings were deposited using a radio-frequency magnetron sputtering of AlN-(Ti,Cr)B_2 target manufactured by powder metallurgy methods. It was established that the coating phase composition was differed from the target one due to the partial aluminum nitride oxidation. The coatings above were characterized by ultra-dispersion structure and great resistance to high-temperature oxidation owing to the partial formation of Al_2O_3-TiO_2, Al_2O_3-Cr_2O_3 and Al_2O_3-TiO_2- Cr_2O_3 solid solutions. Hereby the mass gain of target material proved to be only 1.4 mg/cm^2 at 1300°C. After oxidation, until 1500°C the nanofibrous structure was reinforced with β-Al_2TiO_5 and Al_2O_3 fibers was formed. It was shown that the AlN-(Ti,Cr)B_2/ Al_2O_3 (110) and AlN-(Ti,Cr)B_2/ GaAs (100) coatings were thermal-stable up to 900°C; they had high magnitudes of both microhardness (30 GPa) and fracture toughness (3.3 – 4.7 MN/m$^{3/2}$). The mechanical properties of coatings changed for the worse only after annealing at the temperature higher than 900°C. The composition and structure of coatings were investigated using an XRD, EPMA, Auger spectroscopy, and SEM methods. TG- and DTA- oxidation curves were obtained using the Setaram device with the rate of temperature change of 15 °/min. The AlN-(Ti,Cr)B_2 target material can be recommended for deposition of wear- and corrosion- resistant coatings on parts working under the extreme conditions.

1. INTRODUCTION

The magnetron sputtering is widely used for the deposition of ultra-dispersion films with a high level of properties, including a corrosion resistance and a thermal conductivity. It is known [1-3] that the TiAlN coating as (Ti_{1-x} Al_x)N solid solution being formed at the sputtering of Al and Ti blocks of multi-section target has a good application. In electronics, the magnetron coatings are used as the barrier layers for the averting of diffusion processes at the "substrate-coating" boundary when

S. Guceri et al. (eds.), Nanoengineered Nanofibrous Materials, 519-527.

the silicon, aluminum oxide (sapphire), and GaAs gallium arsenide monocrystals are traditionally used as a substrate [4,5]. In this case a thermal stability of a barrier layer is extremely important.

One must pay attention that the magnetron coatings with high tribological and mechanical characteristics have a high resistance to high-temperature oxidation find an application in aeronautics, medicine, device- and machine- building as well.

As a rule, only composite ceramics containing three or more components may satisfy the demands concerning these material applications in electronics. The use of a multi-section target cannot ensure a variation of coating properties in a wide range; it doesn't enable to obtain the coating composition corresponding entirely to the target composition. What is why it is more expedient to use one-section targets of composite ceramics with a low porosity prepared by powder metallurgy methods.

The high corrosion resistance and thermal conductivity of aluminum nitride in a combination with a high hardness and electric conductivity of borides of transition metals allow us to consider the "AlN-refractory boride" ceramics as a perspective kind of target materials, mainly, for deposition of barrier layers from a gas phase.

The goal of this paper is the study of structure, composition, mechanical properties and a resistance to high-temperature oxidation of films being obtained using the magnetron sputtering of AlN-$(Ti,Cr)B_2$ composite ceramics. The double titanium-chromium boride is a solid solution of CrB_2 in TiB_2, corresponding to TiB_2 + 20% CrB_2 composition, the latter being characterized by a higher, compared with TiB_2, wear resistance and hardness. In a course of high-temperature oxidation of the material pointed out, the Al_2O_3-TiO_2, Al_2O_3-Cr_2O_3 and Al_2O_3-B_2O_3 solid solutions and then the Al_2TiO_5 (β-tialite), $(Al,Cr)_2O_3$, and aluminum borates (high-temperature compounds) are formed. Hereby, the β-tialite may play a role of solid lubricant, similar to $FeAl_2O_4$ and $FeAlO_3$ compounds [4], under condition of a dry friction.

2. METHODS

The AlN-$(Ti,Cr)B_2$ targets of equimolar composition (a diameter of 60 mm and highness of 2 mm) were fabricated by hot pressing in graphite moulds with a boron nitride plaster at 1860-1880°C and a pressure of 150 MPa. The residual porosity of targets did not exceed 2-3%. The powder of demanded composition as the mechanical mixture of components was obtained as a result of grinding in the planetary mill for 8 h, for its homogenization and crumbling up.

The thin coatings were deposited on Si, Al_2O_3 (110) and GaAs (100) monocrystals using the method of radio-frequency magnetron sputtering in the medium of purified Ar. The installation for coating deposition was a metallic chamber, in which the magnetrons were placed. The substrates were placed on the table that was put on the revolving disk. The sputtering was carried out at the argon pressure of 5.10^{-5} mm Hg and the discharge current of 0.4 A. The rate of coating deposition was ~ 120 A^0/min, the variation of thickness from a center to the edges being ~ 1 %.

The mechanical properties of coatings (microhardness H_μ and fracture toughness K_{1c}) were determined using a PMT-3 device at the load to an indentor from 0.1 till 2 N. The coating thickness was ~ 0.6 µm.

The fracture toughness was determined using a formula [5]:

$$\beta K_{1c} = P C^{3/2} \quad (1)$$

Here C – a length of radial cracks, $\beta = \pi^{3/2}$ tg θ = 13.8 – a Vickers pyramid coefficient.

H_μ and K_{1c} magnitudes were measured for initial samples and samples after annealing in the temperature range from 500 to 1000°C.

The resistance to high-temperature oxidation of target materials and coatings was studied under non-isothermal conditions by TG- and DTA-methods at the rate of temperature change of 15°/ min up to 1500°C using a Q-1500 device.

The structure and composition of coatings were investigated by XRD, SEM, EPMA and AES methods using a DRON-3, Camebax SX-50 and LAS-2000 (Riber) device, respectively.

The effect of $AlN-(Ti,Cr)B_2$ high-temperature oxidation on composition and structure of films. The reactions of $AlN-(Ti,Cr)B_2$ film components with a medium coincide with those of target material, viz.:

$$4\,AlN + 3\,O_2 = 2\,Al_2O_3 + 2\,N_2 \quad (2)$$

$$2\,TiB_2 + 5\,O_2 = 2\,TiO_2 + 2\,B_2O_3 \quad (3)$$

$$4\,CrB_2 + 9\,O_2 = 2\,Cr_2O_3 + 4\,B_2O_3 \quad (4)$$

An availability of solid solutions in the Al_2O_3-TiO_2, Cr_2O_3 –TiO_2, Al_2O_3-Cr_2O_3 and Al_2O_3-B_2O_3 systems at high-temperature oxidation of a film leads to the formation in a target/ film scale of following high-temperature compounds: β-Al_2TiO_5, Cr_2TiO_5, $(Al,Cr)_2O_3$ and, maybe, aluminum borates.

The oxidation curves of $AlN-(Ti,Cr)B_2$ material (Fig. 1) in a combination with the data of EPMA and SEM analysis of a surface of material oxidized at different temperatures [6] enabled us reach the following conclusions.

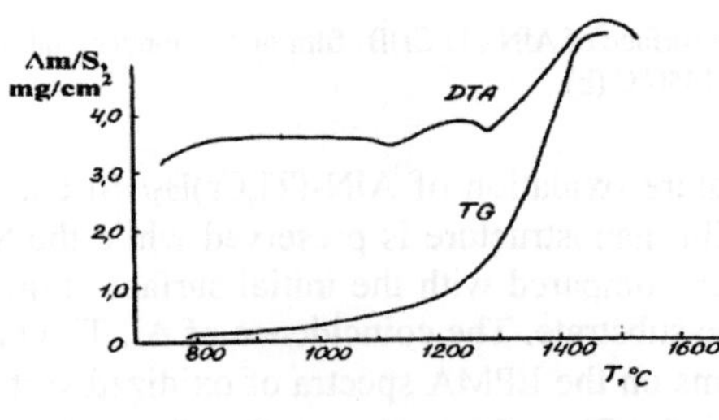

Fig. 1. TG- and DTA- curves of high-temperature oxidation of $AlN-(Ti,Cr)B_2$ target.

Oxidation proceeds in the three stages. In the first stage (1000-1120°C), the oxidation of a double titanium-chromium boride with a formation of TiO_2 (rutile), Cr_2O_3 and B_2O_3 oxides takes place. The heterogeneous structure of oxide film corresponds to a formation of prolonged TiO_2 grains alloyed with Cr_2O_3, the latter disposing at the boundary with AlN phase. In the second stage (1120-1280°C), the aluminum nitride is oxidized to α-Al_2O_3. In the third stage of oxidation (1280-1520°C), β-Al_2TiO_5 and Cr_2TiO_5 compounds are formed resulting from a solid-phase interaction in the Al_2O_3-TiO_2 and Cr_2O_3–TiO_2 systems.

The substrate material influences the film composition in the film-substrate system for both initial and its oxidized state. So, according to EPMA data (Fig. 2, a), in the film (a thickness of 3-5 μm) deposited at the AlN-(Ti,Cr)B_2 target, sputtering on Si monocrystal in the medium of a technical Ar there are Ti, Al, Cr, O, and also Si (a small amount). The uniformity of the element line intensity and the small wideness of intensity peaks along the electron probe scanning, especially for Al and testifies to the formation of a film nanostructure (the grain size < 100 nm).

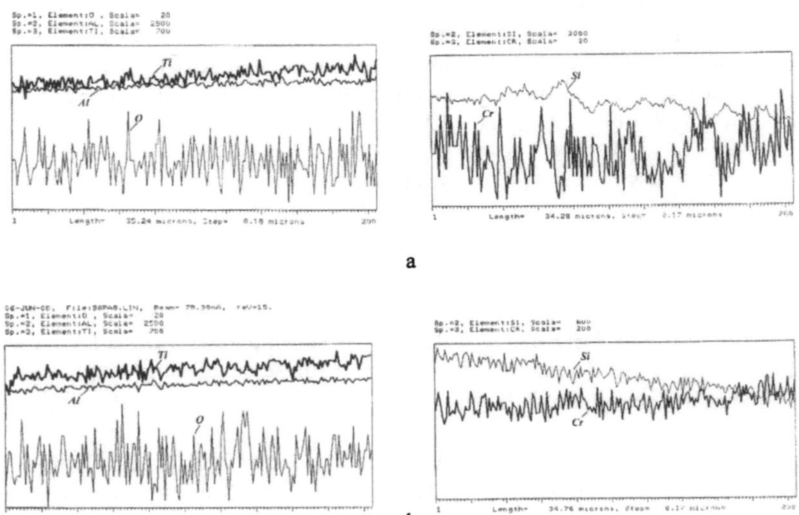

Fig. 2. EPMA spectra of the surface of AlN-(Ti,Cr)B_2 film on Si monocrystal before (a) and after oxidation in air at 1450°C (b).

After high-temperature oxidation of AlN-(Ti,Cr)B_2/ Si composite up to 1450-1500°C, the film nanostructure is preserved while the Si amount grows three times when compared with the initial surface (Fig. 2) owing to its diffusion from the substrate. The coincidence of Al, Ti, O and Al, O concentration maximums on the EPMA spectra of oxidized surface points towards the presence in the film of two phases: β-tialite and α-aluminum oxide. As a result of oxidation, the film acquires the character of a structure reinforced with Al_2O_3 nanofibers (Fig. 3), the length-diameter ratio being 1: 10, 1: 20 and more. The average grain length was 180-350 nm.

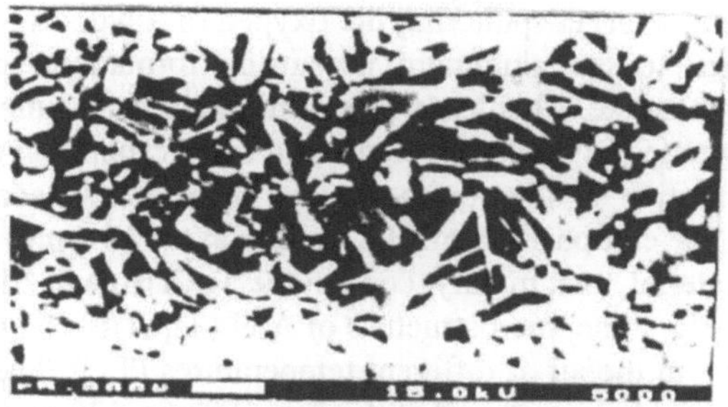

Fig. 3. The microstructure of AlN-(Ti,Cr)B_2/ Si film after high-temperature oxidation at 1450°C.

The formation of Al_2O_3-TiO_2 and Al_2O_3-SiO_2 solid solutions at high-temperature oxidation of a film-substrate composite (Fig. 2, b) ensures the high adhesion strength of a "silicon-oxide film" bond. At temperatures > 1350°C, the oxide film that prevents silicon diffusion into a scale is formed.

The nanofibrous film is also formed on a Si monocrystal at the sputtering of AlN target in the medium of a technical (non-purified) Ar [7]. After high-temperature oxidation of AlN/ Si composite until 1450°C, the distribution of Al and O on the surface EPMA spectra corresponds to aluminum oxide. The silicon was not found in the film, and, consequently, the Si diffusion through the film did not take place. Hereby, the mass gain and heat evolution were barely observed. This led to the formation of Al_2O_3 film of high density, which prevented oxygen diffusion.

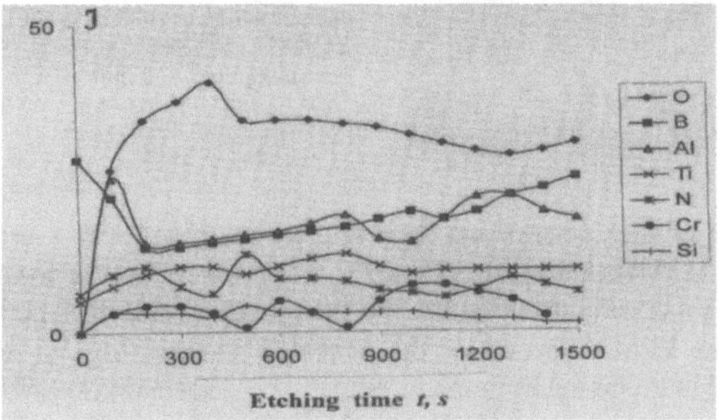

Fig. 4. Auger spectra of AlN-(Ti,Cr)B_2/ Al_2O_3 (110) film.

The results of layer-by-layer AES analysis obtained for the different time (t) of Ar^+-ion etching of AlN-(Ti,Cr)B_2/ Al_2O_3 (110) composite surface testifies to the partial formation of an oxide phase in the film formed at the sputtering of the target even in the medium of high-purity argon (Fig. 4) The Al is present in the film in a bound state, mainly, as oxynitride. Hereby, the (Ti,Cr)B_2 and aluminum oxynitride coexist throughout the film thickness. However, it must be noted that the amount of oxygen in the thin surface layer corresponding to t < 50 s is sharply decreased because at a target sputtering the oxygen of a working medium is intensively spent to the aluminum oxynitride formation, the latter depositing on the substrate. This leads to a diminution (in a time course) of O_2 partial pressure in the working volume and a corresponding decrease of oxygen concentration in the surface layer. Therefore, in accordance with AES spectra, the composition of film surface at t < 10 s proved to be close to a target material, and, mainly, contain the Ti,Cr)B_2 and AlN.

One can see in Fig. 5 the microstructure of AlN-(Ti,Cr)B_2/ Al_2O_3 (110) film after annealing in the air at different temperatures ($T_{an.}$). The film in initial state is an amorphous one, and at the increase of x 13000 the traces of crystalline components are not observed on its surface. At $T_{an.}$ = 600°C the separate centers of crystallization of oxide phase are appeared on the sample surface. Their size is equal to 1-2 μm (Fig. 5, a).

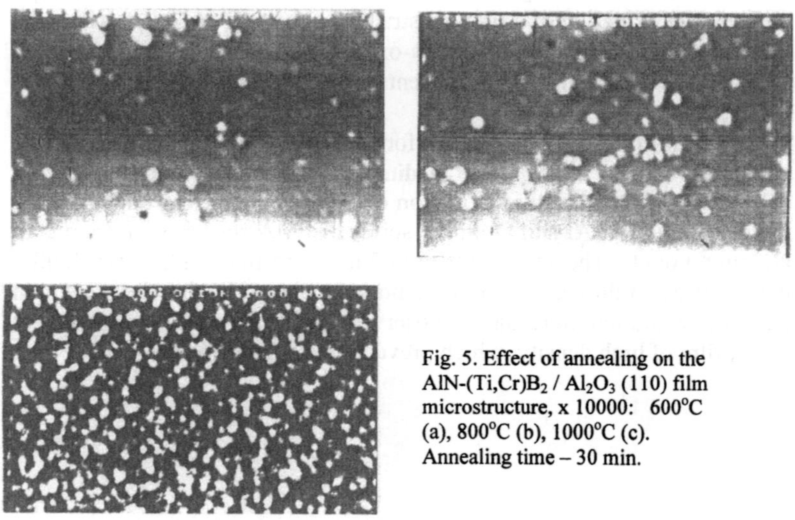

Fig. 5. Effect of annealing on the AlN-(Ti,Cr)B_2 / Al_2O_3 (110) film microstructure, x 10000: 600°C (a), 800°C (b), 1000°C (c). Annealing time – 30 min.

The amount of such centers and their size are increased with a growth of $T_{an.}$(Fig. 5, b). After annealing at 1000°C a thin oxide layer is formed, the latter having a crystalline structure, the average grain size does not exceed 2.5 μm. The EPMA spectra of the annealed surface (Fig. 6) show an existence of aluminum and titanium oxides, the dispersion of Al_2O_3 being higher. The coincidence of Al, Ti and O concentration maximums for the surface of composite annealed at 1000°C (Fig. 6, b) testifies to the appearance of β-Al_2TiO_5 in the film at the lower temperatures, in comparison to the temperature of its formation under stationary conditions (~1350°C).

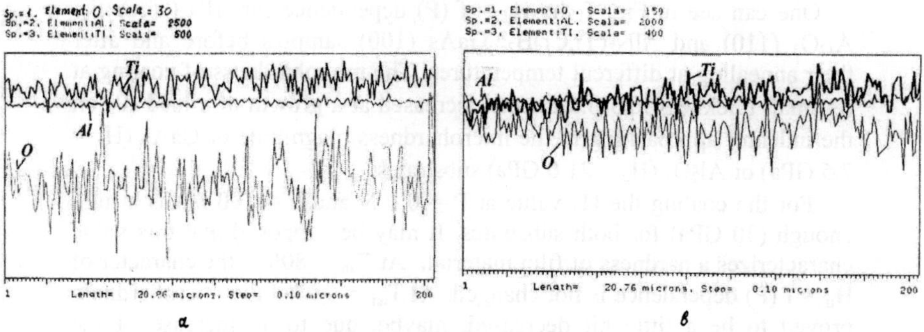

Fig. 6. EPMA of AlN-$(Ti,Cr)B_2$ /Al_2O_3(110) surface after annealing at 600°C (a) and 1000°C(b).

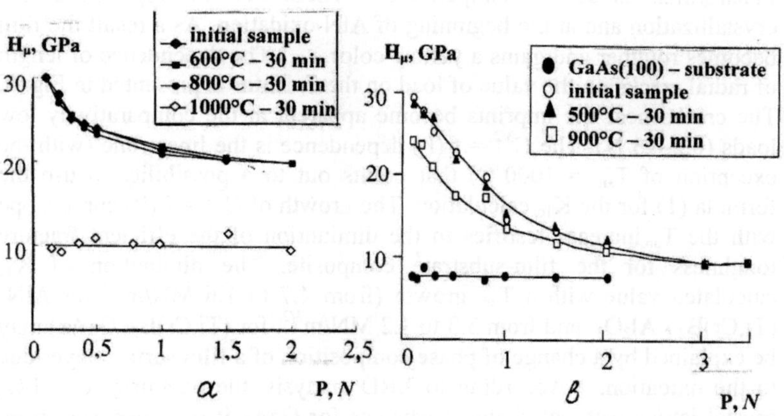

Fig. 7. H_μ (P) dependence for AlN-$(Ti,Cr)B_2$ / Al_2O_3 (110) (a) and AlN-$(Ti,Cr)B_2$/ GaAs (100) (b) composites before and after their annealing.

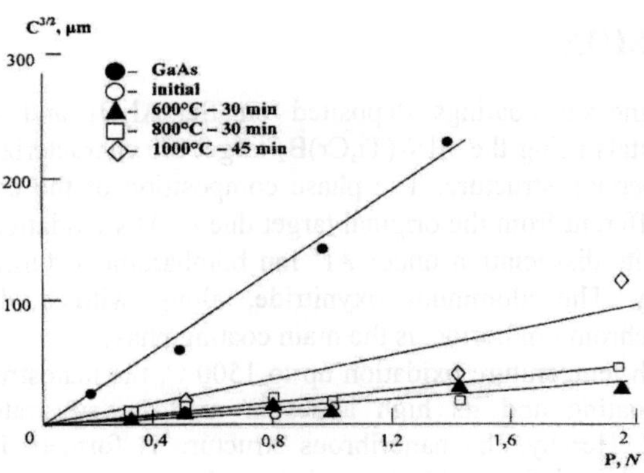

Fig. 8. $C^{3/2}$(P) dependence for AlN-$(Ti,Cr)B_2$ / Al_2O_3 (110) film before and after annealing. The data for initial GaAs (100) substrate are presented for a comparison.

One can see in Fig. 7 the H_μ = f (P) dependence for AlN-(Ti,Cr)B_2/ Al_2O_3 (110) and AlN-(Ti,Cr)B_2/ GaAs (100) samples before and after their annealing at different temperatures. The microhardness of coating at its small thickness (0.5-0.6 μm) is decreased at a growth of a load (P) on the indentor approaching to the microhardness magnitude of GaAs (H_μ = 7.5 GPa) or Al_2O_3 (H_μ = 21.0 GPa) substrates.

For the coating the H_μ value at P = 0.1 N and h = 100 nm is a high enough (30 GPa) for both substrates. It may be supposed that this value characterizes a hardness of film material. At $T_{an.}$ = 800°C the character of H_μ = f (P) dependence is not changed. At $T_{an.}$ = 900°C the microhardness proved to be a little bit decreased, maybe, due to an increase of the amount of crystallization centers of oxide phases in the coating (Fig. 5, a, b). At $T_{an.}$ = 1000°C the microhardness is sharply decreased and does not depend on a load value. This testifies to the phase and structural transformations at this temperature. These become apparent at the crystallization and at the beginning of AlN oxidation. As a result the film becomes rougher and gains a yellow color. The dependence of length of radial cracks on the value of load on the indentor is presented in Fig. 8. The cracks near the imprints become apparent at the comparatively low loads (0.2-0.3 N). The $C^{3/2}$ = f (P) dependence is the linear one (with the exception of $T_{an.}$ = 1000°C) that points out to a possibility to use the formula (1) for the K_{1c} calculation. The growth of $C^{3/2}$ = f (P) curve slope with the $T_{an.}$increase testifies to the diminution of the efficient fracture toughness for the film-substrate composite. The diminution of K_{1c} calculated value with a $T_{an.}$ growth (from 4.7 to 1.6 $MN/m^{3/2}$ for AlN-(Ti,Cr)B_2 / Al_2O_3 and from 3.3 to 1.2 $MN/m^{3/2}$ for (Ti,Cr)B_2/ GaAs) may be explained by a change of phase composition of a film surface layer due to the oxidation. According to XRD analysis, the parameter of Al_2O_3 crystal lattice was not changed whereas for GaAs it was increased from 5.646 A to 5.957 A on the account of dissolution of coating material components in the gallium arsenide as well as an appearance of thermal strains of tension in the substrate.

3. CONCLUSION

1. The magnetron coatings deposited on Si, Al_2O_3 and GaAs monocrystals using the AlN-(Ti,Cr)B_2 target are characterized by ultra-dispersion structure. The phase composition of the coating can be different from the original target due to AlN oxidation as a result of its dissociation under Ar^+ ion bombardment during the sputtering. The aluminum oxynitride, along with a double titanium-chromium boride, is the main coating phase.
2. After high-temperature oxidation up to 1500°C, the nanostructure of the coating and its high adhesion as to a substrate are preserved. Hereby, the nanofibrous structure is formed. In the structure, the reinforcing Al_2O_3 grains are drawn out as the fibers closely interwoven with each other. In a course of high-temperature oxidation the Al_2O_3-TiO_2-Cr_2O_3, Al_2O_3-B_2O_3 solid

solution fibers as well as β-Al_2TiO_5 (tialite) crystals are formed, the latter ensuring a high resistance of material to high-temperature corrosion.

3. The AlN-$(Ti,Cr)B_2$/ Al_2O_3 (110) and AlN-$(Ti,Cr)B_2$/ GaAs (100) coatings are thermal and stable until 1000°C. They have the high enough magnitudes of microhardness (H_{μ} = 30 GPa) and fracture toughness (K_{1c} = 3.3 – 4.7 $MN/m^{3/2}$). At the annealing temperature $T_{an.}$ = 1000°C the structural and phase transformations accompanied by the mechanical properties, worsening take place in the coatings.

Andrievski R. A., Lavrenko V. A., Desmaison J., Desmaison-Brut M, Kalinnikov G.V., Panasyuk A.D. High-temperature oxidation of aluminum nitride-based films / Proc. Ac. Sci. Russia, 2000, v. 373, N 1, p.p. 60-62.

References

1. Properties of $(Ti_{1-x}Al_x)N$ coatings for cutting tools prepared by the cathodic arc ion plating method: 38th Nat. Symp. Amer. Vac. Soc., Wash., 1991 / Tanaka Y., Gur T. M., Kelly M., Hagstrom S. B., Ikeda T., Wakihira K., Saton H. / J. Vac. Sci. and Technol., A., 1992, v. 10, N4, p.p. 1749-1756.
2 . Harte Schichten Zahlen Sich Aus / Techn. Rept.,1992, v. 19, N5, p.44.
3 Characterization of $(Ti_{1-x}Al_x)N$ films prepared by radio-frequency reactive magnetron sputtering / J. Y. Rauch, Ch. Rousselot, N. Martin, Ch. Jacquot, J. Takadoum / J. European Ceram. Soc., v. 20, 2000, p.p. 795-799.
4. Ravikiran A., Effect of interface layers formed during dry sliding of zirconia toughened aluminia against steel, Wear, 1994, v. 171, p.p. 129-134.
5. Lawn B. R., Fracture of brittle solids, Second Edition, Cambridge University Press, Cambridge, 1993, 378 p.
6. Lavrenko V. A. , Panasyuk A. D., Desmaison-Brut M, Desmaison J. Development of AlN-$(Ti,Cr)B_2$ composite ceramics and study of their corrosion resistance in air up to 1550°C, Key Eng. Mater., 2002, v. 206-213, p.p. 1069-1072.

PATTERNING SUB-100NM FEATURES FOR SUB-MICRON DEVICES

H. Kavak*, J. G. Goodberlet
*Research Laboratory of Electronics, MIT, Cambridge, MA 02139, *Physics Department, Çukurova University, Adana, Turkey 01330*

Abstract: Contact photolithography with an embedded amplitude conformable photomask has been used to pattern sub-100 nm features. A conformable quartz, embedded-amplitude mask is brought into intimate contact with a resist-coated substrate and exposed to 220 nm of radiation from an Xe arc lamp. Numerical simulations are used to design the photomask, which takes advantage of optical interference in the very near field. This technique offers high resolution sub-micron sized features at a low cost. A variety of useful structures can be patterned, including lines, holes and annuli. These features can be used to fabricate nanometer-scale vacuum electronic devices. Some patterned features will be presented in this paper.

1. INTRODUCTION

Numerous techniques have been proposed recently to perform nanolithography at a lower cost than electron-beam, ion beam or x-ray techniques [1]. Techniques such as nano-imprint lithography, microcontact printing, near-field phase shift lithography offer low-cost and high resolution patterns when compared with electron-beam or scanning techniques. In each of these low cost methods, a patterned mask is brought into contact with a substrate, and the pattern is replicated. These techniques can also be used to fabricate nanometer scale structures for sub-micron devices.

Conformable contact photolithography has been used to pattern features at the micron level as early as 1969 [2]. In this initial work 0.25 mm thin glass with an absorber was used as an optical mask and was brought into intimate contact with a resist-coated substrate. In later work, the resolution of contact photolithography was extended to the sub-micron level [3]. Some innovative variations of the conformable mask have also been developed; the light coupling masks, the membrane masks, and the embedded amplitude mask used in this work was for near field optical lithography [4-8].

The primary research has just started to apply conformable contact photolithography to fabricate advanced submicron devices. This device would enable on-chip, ultrasensitive and ultrafast

S. Guceri et al. (eds.), Nanoengineered Nanofibrous Materials, 529-534.

photodetection for potential applications in optical, biochemical sensing, communications or low light level imaging technologies.

2. EXPERIMENTAL DETAILS

The conformable Embedded Amplitude Mask (EAM) consist of quartz or deep UV transparent fused silica substrate, 150 μm thick, with a Cr absorber embedded into etched silica regions via a trilayer resist process. The embedded pattern was obtained using an electron-beam lithography, reactive ion etching steps and lift-off steps [5].

In the near field, the distance between the mask and substrate is shorter than the wavelength of the incident light. By placing an aperture a small distance from a surface, it is possible to create lithographic structures that are significantly smaller than the wavelength. Interference effects in the near field of a photomask are utilized to obtain feature sizes below the diffraction limit of the exposing radiation. The photomask designs are optimized through numerical simulations to take advantage of the optical interference, and to generate the desired pattern in resist. Full vector electromagnetic simulation tools, the finite-difference time-domain field solver TEMPEST, available over the internet were used extensively to optimize particular mask design [9]. These simulations were also used to achieve a high resolution variety of pattern.

For all patterning experiments with the EAM, trilayer resist stack was used on the silicon substrate. 100 nm thick antireflection coating ARC (XHRi-11) spun on the substrate and baked. ARC is used to minimize reflected power into the resist. 35-40 nm thick SiO_2 interlayer evaporated on ARC using e-beam evaporation technique. 30-60 nm thick poly(methyl methacrylate) (PMMA) resist spun on top SiO_2 and baked. This stack optimized to get a better resolution, considering the refractive index and thickness of the layers. To obtain intimate contact, a vacuum fixture is used to pull the mask onto the resist coated substrate. The contact pressure is about 0.7 atm. After accomplishing intimate contact, patterns were exposed in PMMA with a 220 nm mercury-xenon arc lamp. The doses used to expose the PMMA were varied about median value of 350 mJ/cm^2. And after the exposure the pattern immediately developed for 60-90 s in a 2:1 mixture of isopropyl alcohol:methyl-isobutyl ketone. Patterns were transferred to the substrate using reactive-ion etching and a chrome lift-off step. The oxide interlayer was etched with CHF_3 plasma, and the ARC layer was etched with an O_2 plasma. All processes were carried out in a class 10 clean room. Some are illustrated in Figure 1.

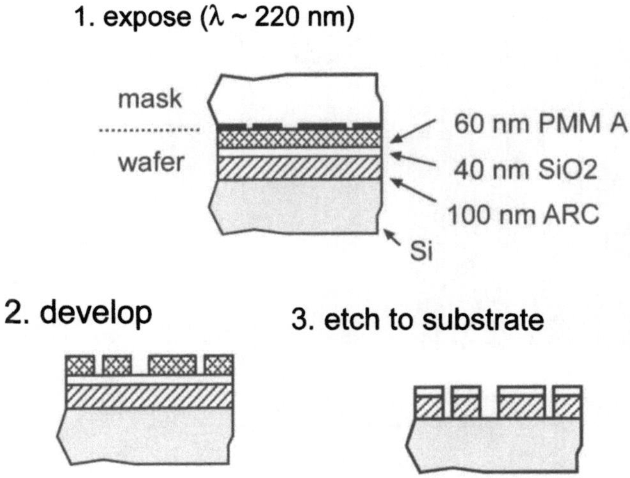

Figure 1. Patterning with a trilayer resist

3. RESULTS AND DISCUSSION

The application of near-field microscopy concepts to photolithography offers the possibility of developing high resolution, low-cost photolithography processes by utilizing the optical near field, rather than driving the exposure wavelength deeper into the ultraviolet.

Commercially available resists like PMMA, Shipley UV5 and S1813 were studied to obtain high resolution images. PMMA is a high resolution resist that has additional benefit of negligible outgassing during exposures. Substantial outgassing can cause the separation of mask and substrate during exposure. Numerical simulation technique (TEMPEST) used to optimize the three layer stack. The simulation and experimental results for a circular aperture in the photomask are shown in Figure 2. The hole in the embedded attenuator which was determined with the finite-difference time domain software, shown in Fig. (2a). The simulated instantaneous optical field for one polarization, shown in Fig (2b), was recorded 30 nm beyond the aperture. The simulated instantaneous optical field for one polarization used to calculate the time-averaged intensity, and this intensity is plotted in Fig. (2c). Scanning electron microscope (SEM) images in Fig (3a) shows the annular aperture was patterned on a photomask and the resulting processed pattern is shown in Fig. (3b). A sub 50 nm diameter hole produced using these apertures, which were fabricated using near field photolithography [10].

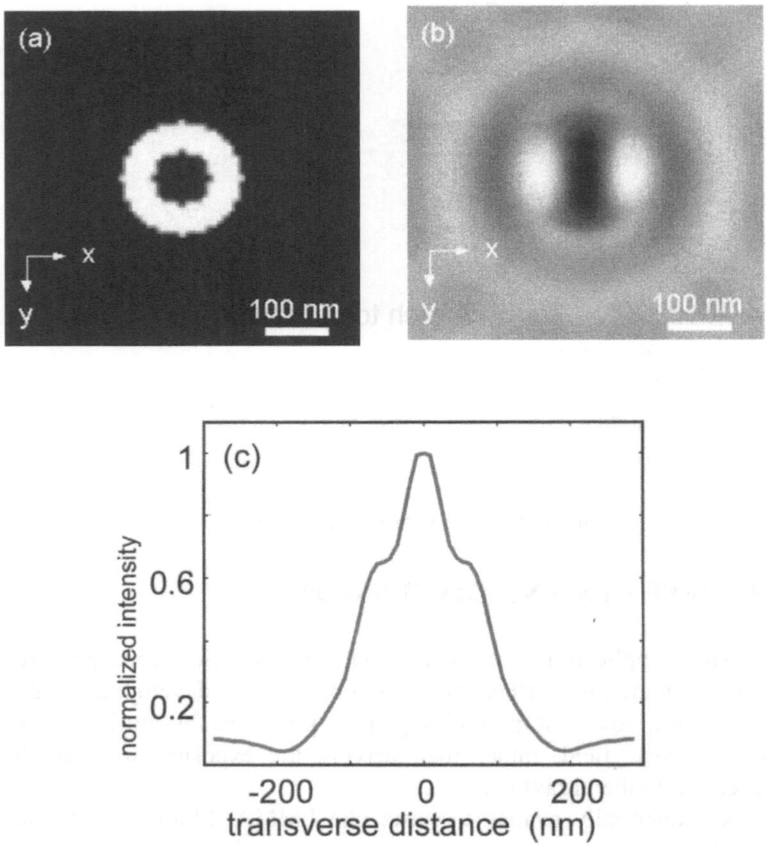

Figure 2. The simulation results for an annular aperture. a) the hole in the embedded attenuator b) instantaneous field 30 nm beyond the aperture of the mask. c) time averaged intensity.

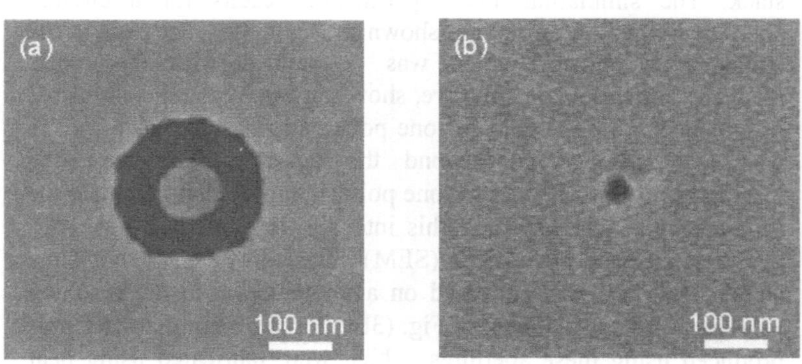

Figure 3. Experimental results for a) an annular aperture was patterned on a photomask b) sub-50 nm hole patterned in oxide and polymer layers.

EAM conformable contact photolithography can be used to pattern high resolution grating structures, Fig. (4a) and a dense triangular lattice of holes, Fig (4b).

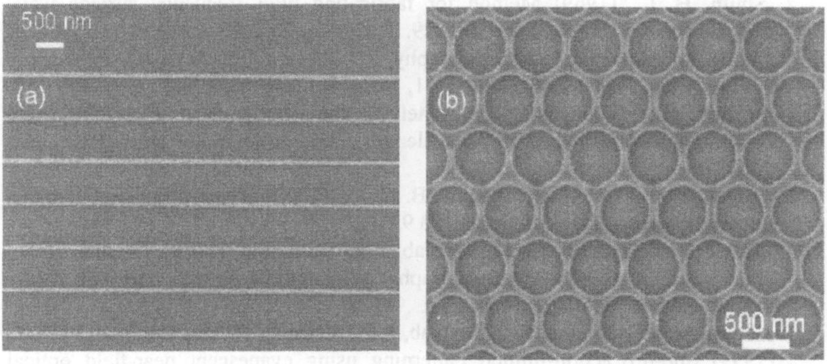

Figure 4. An SEM images of the resulting printed pattern a) grating b) dense triangular lattice of holes.

A successfully patterned variety of structures show the technique's potential for the fabrication of electron field emitting devices, integrated optical devices and etc. Multi-level patterning has been successfully achieved to fabricate a multichannel nanometric photomultiplier. For the first fabrication attempts the microphotomultiplier's channel width was chosen to be about 125 nm, and several device lengths, ranging from 3 μm to 50 μm were trailed. There were typically 200 or more channels in a single device. The channels were defined in the first level of lithography and subsequently etched into an oxide substrate. In the second level of lithography, a resistive strip was aligned to the channels with better than 200 nm overlay accuracy. In the third level of lithography and subsequent processing, the photocathode, anode and signal collector were defined in gold. The research is on progress.

4. CONCLUSION

The experimental and numerical results show that high resolution features (Sub-100 nm) can be created with embedded-amplitude mask using optical interference in the near field. This technique utilizes available simulation tools for designing photomask pattern, which take advantage of near-field interference. The technique offer some flexibility in mask design, and improvement are expected when mask material varied. Fine features were produced with a low cost, unpolarized arc-lamp source and further improvements can be expected with a more coherent and well polarized source. This low cost, high resolution patterning technique may find application in the rapidly growing area of nanotechnology such as nano-electronics devices.

References

1. Smith, H. I. (2001) Low cost nanolithography with nanoaccuracy, *Physica E* 11, 104-109.
2. Smith, H. I. (1969) Method for fabricating high frequency surface wave transducers, *Rev. Sci. Instrum.* **40**, 729.
3. Lin, B. J. (1980) Fine Line Lithography, R. Newman, Ed., *Materials Processing Theory and Practices Series*, Vol. 1, North-Holland, New York, , pp. 105-150.
4. Schmid, H., Biebuyck, H., Michel B., and Martin, O. J. F. (1998) Light coupling masks for lenses, sub-wavelength optical lithography, *Appl. Phys. Lett.* **72**, 2379.
5. Goodberlet J. G., and Dunn, B. L. (2000) Deep-ultraviolet contact photolithography, *Microelec. Eng.* **53** 95.
6. Blaikie, R. J., Alkaisi, M. M., McNab, S. J., Cumming, D.R.S., Cheung, R. D., and Hasko, G. (1999) Nanolithography using optical contact exposure in the evanescent near field *Microelec. Eng.* , **46**, 85.
7. Alkaisi, M. M., Blaikie, R. J., McNab, S. J., Cheung, R., and Cumming, D.R.S. (1999) Sub-diffraction-limited patterning using evanescent near-field optical lithography, *Appl. Phys. Lett.* **75**, 3560.
8. Goodberlet, J. G. (2000) Patterning 100 nm features using deep-ultraviolet contact photolithography, *Appl. Phys. Lett.* **76**, 667.
9. TEMPEST can be run at http://cuervo.eecs.berkeley.edu/Volcano/
10. Goodberlet, J. G., and Kavak, H. (2002) Patterning sub-50 nm features with near-field embedded-amplitude mask, *Appl. Phys. Lett.* **81**, 1315.

SOLID AND LIQUID PROPERTIES OF Pd-Ni METAL ALLOYS USING QUANTUM SUTTON-CHEN MANY-BODY POTENTIAL

S. ÖZDEMIR KART [1,a], M. ULUDOĞAN[2], T. ÇAĞIN[3], AND M. TOMAK[1]
[1]*Department of Physics, Middle East Technical University 06531 Ankara, TURKEY*
[2]*School of Materials Science and Engineering, Georgia Institute of Technology 771 Ferst Dr. Atlanta, GA 30332-0245, U.S.A*
[3]*Material and Process Simulation Center, California Institute of Technology Pasadena, CA 91125, U.S.A*

Abstract: The performance of the quantum Sutton-Chen many body potential on the solid and liquid properties of Pd-Ni fcc transition metal alloys is investigated by molecular dynamics simulations. Elastic constants, phonon dispersion relations and diffusion coefficients are studied. The melting temperatures of Pd-Ni alloys are predicted. The transferability of the potential is also tested by simulating the liquid state. Comparison of the molecular dynamics results with available experimental data shows that the quantum Sutton-Chen potential gives a reasonable description of the solid and liquid properties of alloys considered, in spite of having been parameterized on the experimental data of pure metals at solid phase.

Keywords: quantum Sutton-Chen potential, elastic constant, phonon dispersion, diffusion coefficient.

1. INTRODUCTION

Recent advances in numerical methods and computer technology have reinforced attempts to study the physical properties of fcc transition metal alloys. A detailed research on the development of materials is of great interest for their industrial applications. Palladium and its alloys have industrial importance because of their hardness and resistance to corrosion.

The most important problem in a computer simulation is to choose the interatomic potential model which provides an accurate description of atomic interactions. In metals and their alloys, the many body density term plays a significant role in the interactions. Therefore the model potential selected for transition metals should include many-body interactions as well as pairwise

[a] Corresponding Author: ozsev@newton.physics.metu.edu.tr (S. Özdemir Kart)

S. Guceri et al. (eds.), Nanoengineered Nanofibrous Materials, 535-540.

interactions. There are several empirical and practical approaches that can afford to investigate many systems and trends in physical properties. The empirical potentials provide a chance of studying systems with large number of atoms in a short time and with satisfactory accuracy. Recently, a many-body potential has been introduced [1] within the context of the tight binding approach and reparametrized by including quantum correction [2]. Because of computational efficiency and fairly long-range properties, Sutton-Chen potential (SC) has been used in many studies and successfully applied to a range of problems [4,5,6,7] and quantum Sutton-Chen potential (Q-SC) has been used for predicting the viscosity of liquid metal alloys [8] and alloy melting and solidification [9].

In this study, we perform molecular dynamics (MD) simulations by using SC and Q-SC potentials. The goal of this study is to examine the solid and liquid properties of Pd-Ni binary fcc metal alloys. The transferability of the Q-SC many body potential is an important question to be answered. Q-SC parameters are tested by computing the elastic properties and phonon dispersion curves at definite temperatures as solid properties and by calculating the diffusion coefficients as liquid properties. The melting temperatures for Pd-Ni alloys are investigated.

2. METHOD

The total potential energy of the system in SC type can be described by a sum of pair-wise repulsion term and many-body density dependent cohesion term. It has the following form;

$$U_{tot} = \sum_i U_i = \sum_i \left[\sum_{j \neq i} \frac{1}{2} \varepsilon_{ij} \left(\frac{a_{ij}}{r_{ij}}\right)^{n_{ij}} - c_i \varepsilon_{ii} \left(\sum_{i \neq j} \left(\frac{a_{ij}}{r_{ij}}\right)^{m_{ij}} \right)^{\frac{1}{2}} \right]. \tag{1}$$

These SC parameters ε, c, m and n are optimized to fit to the 0 K properties based on the experimental lattice parameter (at 300 K), cohesive energy, and bulk modulus. Q-SC parameters are obtained by fitting additional experimental properties, such as phonon frequencies at the X point (at 300K), vacancy formation energy, and surface energies. In Table I, the values of the Q-SC and SC parameters for Pd and Ni are listed. We use the random binary fcc metal alloy method [10], in which cites are occupied by two types of atoms completely randomly.

Table I. Quantum Sutton-Chen (Q-SC) and Sutton-Chen (SC) potential parameters.

Metal		N	m	ε(eV)	c	a(A^0)
Pd	Q-SC	12	6	3.2864E-3	148.205	3.8813
	SC	12	7	4.1260E-3	108.526	3.8900
Ni	Q-SC	10	5	7.3767E-3	84.745	3.5157
	SC	9	6	1.5714E-2	39.756	3.5200

MD simulations start from a cubic box with 864 atoms subject to periodic boundary conditions. The details of the computational procedure followed in this study can be found in Refs. [6,7].

Table II. Comparison of calculated and experimental (Exp) values for elastic constant (C_{ij}) and bulk modulus (B) at 0 K in the units of GPa and transverse (T) and longitudinal (L) phonon frequencies at the Brillouin zone boundaries X and L at 300 K in the units of THz.

Metal	Model	C_{11}	C_{12}	C_{44}	B	υ_{XT}	υ_{XL}	υ_{LT}	υ_{LL}
	Exp	234.10	176.10	71.20	195.00	4.56	6.70	3.21	6.86
Pd	Q-SC	216.00	150.25	91.66	172.17	4.25	6.19	2.83	6.17
	SC	248.20	175.90	93.29	200.00	4.30	6.20	2.88	6.16
	Exp	261.20	150.80	131.70	188.00	6.17	8.55	4.24	8.88
Ni	Q-SC	219.59	165.34	99.57	183.62	5.68	8.48	3.69	8.50
	SC	226.61	178.77	79.24	194.73	5.04	7.48	3.29	7.48

3. RESULTS and DISCUSSION

The elastic properties of bulk fcc Pd-Ni alloys are calculated using both SC and Q-SC parameters over 50000 steps of microcanonical (EVN) ensemble. The results for Pd and Ni at 0 K are given in Table II, along with comparisons with experimental data [11]. While SC describes the elastic constants of C_{12} for Pd well, C_{12} for Ni obtained by Q-SC is more consistent with experiment. Our results are close to the values calculated by Kimura *et. al.* [2]. In Fig. 1, the elastic constants of Pd-Ni alloy at six different concentration values are shown. The figures are plotted at 300 K and 1500 K to see the effect of temperature on the elastic properties. As temperature increases, thermal softening of alloy occurs.

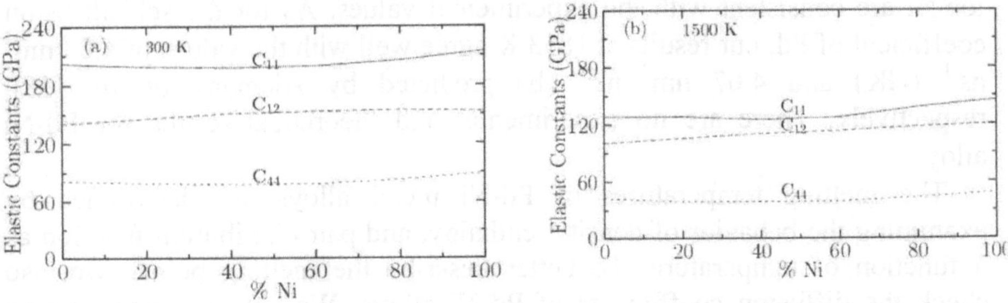

Figure 1. Elastic constants as a function of concentration of Ni in Pd a) at 300 K and b) at 1500 K.

We have also studied the phonon dispersion relation. The phonon dispersion curves for Ni at room temperature are presented in Fig. 2, together with the experimental data [12]. The overall structure of dispersion curves is well reproduced. Transverse (T) and longitudinal (L) phonon frequencies for Pd and Ni at the X and L points at the Brillouin zone are listed in Table II, together with experimental data [13]. Our results are more consistent with the experimental values than the other many-body calculations found in the literature [14,15,16]. The best agreement with the experiment is the longitudinal frequency at X point, as expected, due to fitting of Q-SC parameters to experimental phonon frequencies.

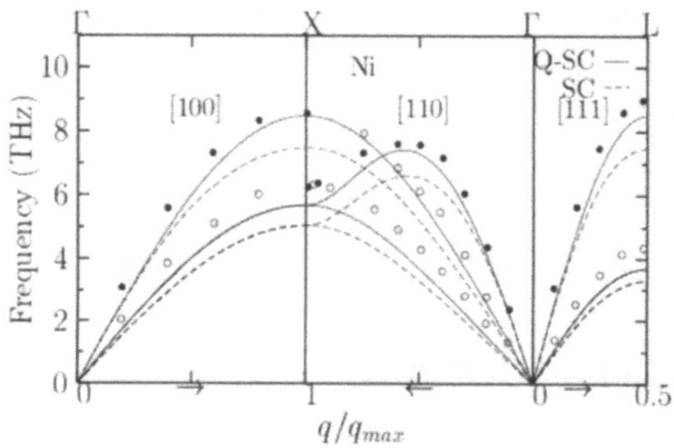

Figure 2: Phonon dispersion curves of Ni along symmetry directions. The solid curves represent the present calculations from Q-SC parameters. The dashed curves show the SC calculations. The points are experimental data [12] at room temperature.

The diffusion coefficients (*D*) of Pd-Ni alloy are computed from Green-Kubo (GK) and Einstein (E) relations to study liquid properties. Table III lists the computed values of *D* for Pd, Ni and Pd-Ni alloys, along with the available experimental data [17]. The only metal alloy studied here for which the experimental values of *D* are available is Ni [17]. Our simulation results for Ni are consistent with the experimental values. As for the self-diffusion coefficient of Pd, our results at 1853 K agree well with the values of 4.03 nm^2 ns^{-1} (GK) and 4.07 nm^2 ns^{-1} (E) predicted by Alemany *et. al.* [18], respectively. There are no experimental and theoretical results for Pd-Ni alloy.

The melting temperatures of Pd-Ni metal alloys are determined by examining the behavior of density, enthalpy, and pair distribution function as a function of temperature. To better describe the melting points, we also check the diffusion coefficients of Pd-Ni alloys. We get the same melting temperatures from all these physical properties. We find the melting temperatures for Pd, $Pd_{0.8}Ni_{0.2}$, $Pd_{0.6}Ni_{0.4}$, $Pd_{0.4}Ni_{0.6}$, $Pd_{0.2}Ni_{0.8}$, Ni to be 1820 ±10 K, 1680 ±10 K, 1600 ±10 K, 1570 ±10 K, 1610 ±10 K, 1710 ±10 K, respectively. These values are consistent with experimental values which are 1825 K, 1633 K, 1519 K, 1507 K, 1552 K, and 1726 K, respectively [19]. We find eutectic region between concentrations around $Pd_{0.5}Ni_{0.5}$ and $Pd_{0.3}Ni_{0.7}$, supporting the critical experimental concentration of $Pd_{0.45}Ni_{0.55}$.

Table III. Diffusion coefficients D in $nm^2\ ns^{-1}$ as evaluated by using the Green-Kubo (GK) and Einstein (E) relations at the shown temperatures for Pd and Ni and Pd-Ni alloys. Here calculated diffusion coefficients are obtained from Arrhenius equation.

			Self-Diffusion (nm^2 /ns)			
			Green-Kubo		Einstein	
Metals	Temp. (K)	Exp.	D for Pd	D for Ni	D for Pd	D for Ni
Pd	1853	-	4.98	-	4.96	-
$Pd_{0.6}Ni_{0.4}$	1853	-	4.84	5.78	4.85	5.88
Ni	1873	5.96		6.39	-	6.38
	1773	4.61		5.48	-	4.96

4. CONCLUSION

The transferability of the potential is an important conclusion which can be made from this work. Although the potential parameters were fitted to 0 K experimental properties of the solid pure system, the Q-SC model describes the properties of liquid Pd, Ni and Pd-Ni well.

One of the achievements of this work is to predict the melting temperatures of Pd-Ni metal alloys with the same trend as the existing experimental curve. The experimental data on the diffusion coefficients D for Pd-Ni alloys except for Ni are not available for comparison, but the values of D computed from Einstein and Green-Kubo relations are nearly equal to each other. These data may encourage the experimentalists to verify our results.

References

1. A. P. Sutton and J. Chen. Long-Range Finnis Sinclair Potentials. *Philos. Mag. Lett.* 61:139-146, 1990.
2. Y. Kimura, T. Çağın, Y. Qi, and W. A. Goddard III. The Quantum Sutton-Chen Many Body Potential for Properties of fcc Metals, unpublished.
3. K. D. Hammonds and R. M. Lynden-Bell. A Computational Study of Metal Stepped Surfaces. *Surf. Sci.*, 278:437-456, 1992.
4. B. D. Todd and R. M. Lynden-Bell. Surface and Bulk Properties of Metals Modeled with Sutton-Chen Potentials. *Surf. Sci.*, 281:191-206, 1993.
5. K. D. Shiang. Theoretical Studies of Adatoms Diffusion on Metal Surfaces. *Phys. Lett. A,* 180:444-452, 1993.
6. Dereli, G., T. Çağın, M. Uludoğan, and M. Tomak. Thermal and Mechanical Properties of Pt-Rh Alloys, *Philos. Mag. Lett.*, 75:209-217, 1997.
7. T. Çağın, G. Dereli, M. Uludoğan, and M. Tomak. Thermal and Mechanical Properties of

Some fcc Transition Metals. *Phys. Rev. B* , 59:3468-3473, 1999.

8. Y. Qi, T. Çağın, Y. Kimura, and W. A. Goddard III. Viscosities of Liquid Metal Alloys from Nonequilibrium Molecular Dynamics, *J. Comp . Aid. Mat. Des*., 8:233--243, 2002.
9. Y. Qi, T. Çağın, Y. Kimura, and W. A. Goddard III. Molecular-Dynamics Simulations of Glass Formation and Crystallization in Binary Liquid Metals: Cu-Ag and Cu-Ni. *Phys. Rev. B,* 59:3527-3533, 1999.
10. H. Rafii-Tabar and A. P. Sutton. Long-Range Finnis-Sinclair Potentials for fcc Metallic Alloys. *Philos. Mag. Lett.*, 63:217-224, 1991.
11. G. Simmons and H. Wang. Single Crystal Elastic Constant and Calculated Aggregate Properties. MIT Press, Cambridge, 1971.
12. R. J. Birgeneau , J. Cords, G. Dolling, and A. D. W. Woods. *Phys. Rev*., 136:A1359, 1964.
13. Landolt-Bornstein. Zahlenwerte und Funktionen aus Naturwisso und Technik, New Series III 13 a, Springer-Verlag, Berlin, 1981.
14. F. Cleri and V. Rosata. Tight Binding Potentials for Transition Metals and Alloys. *Phys. Rev. B,* 48:22-33, 1993.
15. M. S. Daw and M. I. Baskes. Embedded-Atom Method - Derivation and Application to Impurities, Surfaces, and Other Defects in Metals. *Phys. Rev. B*, 29:6443-6453, 1984.
16. M. S. Daw, S. M. Foiles, and M. I. Baskes. The Embedded-Atom Method - A Review of Teory and Applications. *Mater. Sci. Rep*., 9:251-310, 1993.
17. P. Protopapas, H. C. Andersen, and N. A. D. Parlee. Theory of Transition in Liquid Metals. 1. Calculation of Self-Diffusion Coefficients. *J. Chem. Phys*., 59:15-25, 1973.
18. M. M. G. Alemany, O. Diéguez, C. Rey, and L. J. Gallego. Molecular Dynamics Study of the Dynamic Properties of fcc Transition and Simple Metals in the Liquid Phase Using the Second-Moment Approximation to the Tight-Binding, *Phys. Rev. B*, 60 :9208-9211, 1999.
19. R. Hultgren, D. D. Desai, and D. T. Hawkins. Selected Values of Thermodynamic Properties of Binary Alloys. Metal Park, OH:ASM, 1973.

Subject Index

MIX
Papier aus verantwortungsvollen Quellen
Paper from responsible sources
FSC® C105338

If you have any concerns about our products,
you can contact us on
ProductSafety@springernature.com

In case Publisher is established outside the EU,
the EU authorized representative is:
Springer Nature Customer Service Center GmbH
Europaplatz 3, 69115 Heidelberg, Germany

Printed by Libri Plureos GmbH
in Hamburg, Germany